AF681221

Food, Energy, and Water Nexus 4.0

This book examines the complex interactions between food, energy, and water (FEW) resources within the framework of the Fourth Industrial Revolution (4IR). It traces the historical evolution of FEW 4.0, analyses its key components, and explores its applications in reshaping resource management practices by integrating emerging technologies like Internet of Things (IoT), artificial intelligence (AI), and blockchain technologies. By harnessing the intricate nexus of FEW resources and emerging technologies, it offers opportunities for sustainable development and addresses pressing challenges.

Features:

- Explores the historical development and integration of food, energy, and water systems within the context of the 4IR.
- Discusses the disruptive effects of Industry 4.0 on FEW systems, emphasising digitalisation, AI, IoT, and sustainable practices.
- Examines the use of emerging technologies such as precision agriculture, blockchain, and green energy integration in optimising FEW systems.
- Analyses FEW 4.0 governance frameworks, policy implications, and socio-economic and environmental impacts to offer actionable insights.
- Provides theoretical frameworks, real-world case studies, and forward-looking perspectives to understand FEW 4.0's transformative potential.

This book is aimed at undergraduate and postgraduate students of Computer Engineering, Electrical and Electronics Engineering, Environmental Engineering, and Computer Science as well as policymakers and industry practitioners focused on sustainability and resource management.

Food, Energy, and Water Nexus 4.0
Concepts and Applications

Omolola Ogbolumani and Nnamdi Nwulu

CRC Press
Taylor & Francis Group
Boca Raton London New York

CRC Press is an imprint of the
Taylor & Francis Group, an **informa** business

Designed cover image: © www.pixabay.com

First edition published 2025
by CRC Press
2385 NW Executive Center Drive, Suite 320, Boca Raton FL 33431

and by CRC Press
4 Park Square, Milton Park, Abingdon, Oxon, OX14 4RN

CRC Press is an imprint of Taylor & Francis Group, LLC

© 2025 Omolola Ogbolumani and Nnamdi Nwulu

Reasonable efforts have been made to publish reliable data and information, but the author and publisher cannot assume responsibility for the validity of all materials or the consequences of their use. The authors and publishers have attempted to trace the copyright holders of all material reproduced in this publication and apologize to copyright holders if permission to publish in this form has not been obtained. If any copyright material has not been acknowledged please write and let us know so we may rectify in any future reprint.

Except as permitted under U.S. Copyright Law, no part of this book may be reprinted, reproduced, transmitted, or utilized in any form by any electronic, mechanical, or other means, now known or hereafter invented, including photocopying, microfilming, and recording, or in any information storage or retrieval system, without written permission from the publishers.

For permission to photocopy or use material electronically from this work, access www.copyright.com or contact the Copyright Clearance Center, Inc. (CCC), 222 Rosewood Drive, Danvers, MA 01923, 978-750-8400. For works that are not available on CCC please contact mpkbookspermissions@tandf.co.uk

For Product Safety Concerns and Information please contact our EU representative GPSR@taylorandfrancis.com. Taylor & Francis Verlag GmbH, Kaufingerstraße 24, 80331 München, Germany.

Trademark notice: Product or corporate names may be trademarks or registered trademarks and are used only for identification and explanation without intent to infringe.

ISBN: 9781041044697 (hbk)
ISBN: 9781041061151 (pbk)
ISBN: 9781003633853 (ebk)

DOI: 10.1201/9781003633853

Typeset in Times
by codeMantra

Contents

Preface xiii
Author Biographies xv

Chapter 1 Food, Energy, and Water Nexus 1

1.1 Introduction to Food, Energy, and Water Nexus 1
1.2 Goals of Food, Energy, and Water Nexus 2
1.2.1 Sectoral Collaboration 2
1.2.2 Sustainable and Equitable Resource Management 3
1.2.3 Resource Security 3
1.3 Key Components of the Food, Energy, and Water Nexus 5
1.3.1 Food Production 5
1.3.2 Energy Production 5
1.3.3 Water Management 6
1.4 Summary 8
References 9

Chapter 2 Food, Energy, and Water Nexus 4.0 13

2.1 The Fourth Industrial Revolution and the Rise of FEW Nexus 4.0 13
2.2 The Evolution of FEW Nexus: Historical Development of the FEW Nexus Concepts, from FEW Nexus to FEW Nexus 4.0 18
2.2.1 FEW Nexus 4.0 Contribution to Economic Growth 22
2.2.2 FEW Nexus 4.0 Contribution to Productivity 22
2.2.3 FEW 4.0 Contribution to Cybersecurity Concerns 23
2.2.4 FEW 4.0 Contribution to Skilled Workforce 24
2.3 Disruptive Changes in the FEW Nexus, Driven by Industry 4.0 24
2.4 Use Cases of Industry 4.0-Based Production Companies 27
2.5 The Building Blocks of FEW Nexus 4.0 29
2.5.1 Food 4.0 30
2.6 Energy 4.0 37
2.6.1 Smart Grid System 39
2.6.2 Internet of Things in Energy 39
2.7 Water 4.0 42
2.7.1 Smart Water Management Systems 42
2.8 Key Innovations and Strategies of FEW Nexus 4.0 44
2.8.1 Digitalisation and Automation 45

2.8.2 Artificial Intelligence (AI) ... 47
2.8.3 Internet of Things (IoT) ... 47
2.8.4 Collaboration and Stakeholder Engagement ... 48
2.8.5 Integrated Resource Management ... 51
2.8.6 Decision-Making Tools in FEW Systems ... 53
2.8.7 Resource Efficiency ... 57
2.8.8 Circular Economy ... 59
2.8.9 Decentralisation ... 59
2.8.10 Sustainable Practices ... 60
2.8.11 Resilience and Adaptability ... 60
2.9 Goals of FEW Nexus 4.0 ... 62
2.9.1 Enhancing Resource Security ... 62
2.9.2 Sustainable Resource Production ... 63
2.9.3 Equity in Allocation of Resources ... 63
2.10 Summary ... 65
References ... 65

Chapter 3 Emerging Technologies and Innovations in FEW Nexus 4.0 ... 88

3.1 Emerging Technologies and Innovations in FEW 4.0 Applications ... 88
3.1.1 Definition of Emerging Technologies ... 88
3.1.2 Drivers of Emerging Technologies ... 90
3.1.3 Technologies Driving FEW Nexus 4.0 ... 91
3.2 Integrating IoT and AI in FEW Nexus 4.0 Systems ... 92
3.2.1 IoT Architecture ... 94
3.2.2 Integrating Artificial Intelligence in FEW 4.0 Systems ... 98
3.2.3 Deep Learning vs. Machine Learning ... 101
3.2.4 Differences between Machine Learning and Deep Learning ... 102
3.2.5 Deep Learning and Machine Learning in FEW 4.0 ... 102
3.3 Applications of Blockchain in Food, Energy, and Water Systems ... 105
3.3.1 Applications of Blockchain Technology in FEW Nexus 4.0 ... 106
3.4 Smart Cities and FEW Nexus 4.0: A Sustainable Future ... 109
3.4.1 Applications of Smart Cities ... 111
3.5 Precision Agriculture and FEW Nexus 4.0: Advancements and Challenges ... 114
3.6 Data Analytics and Machine Learning in FEW Nexus 4.0 ... 118
3.6.1 Characteristics of Big Data ... 119

3.6.2 Types of Big Data Analytics 121
3.6.3 Applications of Big Data Analytics in FEW 4.0 126
3.7 Applications Geographic Information System (GIS) in FEW Nexus 4.0 Systems 128
3.8 Green Energy and the Integration of Renewables in FEW Nexus 4.0 Systems 134
3.8.1 Deployment of Renewable Energy for Water Supply 136
3.8.2 Renewable Energy's Role in the Food-Energy Nexus 142
3.8.3 Dual Use of Land in FEW Nexus 146
3.8.4 Comparison of Renewables and Conventional Energy Generation in the FEW Nexus 147
3.9 Virtual Reality, Augmented Reality, and FEW Nexus 4.0 – Enhancing Human Interaction with Technology 147
3.9.1 Virtual Reality 148
3.9.2 Augmented Reality 150
3.9.3 Virtual Reality Display 152
3.10 Advancements in Sensors and Monitoring Technologies for FEW 4.0 Systems 156
3.10.1 Sensing, Smart, and Sustainable Technologies 157
3.10.2 Sensing Technologies 158
3.10.3 Smart Technologies 159
3.10.4 Sustainable Technologies 159
3.10.5 Integration of S3 technologies for the Agri-Food Industry 160
3.10.6 Case Studies of S3 Technologies 162
3.11 Summary 165
References 165

Chapter 4 Application of FEW Nexus 4.0 in Food 195

4.1 Food Security and Resilience 195
4.1.1 Food Security 196
4.2 Interconnection between Food Security and Sustainability 201
4.3 Achieving Food Security through a Food Systems Approach 201
4.3.1 Food Systems Interactions with Food Chains 202
4.3.2 Connections between Food Security and Food Insecurity 203
4.3.3 Resilience Thinking in Food Systems 204

4.4 Applications of Food Security and Resilience to the FEW 4.0 Nexus 204
4.5 Summary 207
References 207

Chapter 5 Application of FEW Nexus 4.0 in Energy 214

5.1 Energy Efficiency and Optimisation 214
5.1.1 Environmental and Economic Impacts of Energy Use 214
5.1.2 Energy Efficiency 215
5.1.3 Energy Efficiency Concepts 215
5.1.4 Benefits of Energy Efficiency 216
5.1.5 Energy Efficiency Strategies 218
5.2 Energy Optimisation Techniques 221
5.3 Application of FEW Nexus 4.0 225
5.4 Summary 226
References 226

Chapter 6 Application of FEW Nexus 4.0 in Water 233

6.1 Water Management and Treatment Strategies 233
6.1.1 Objectives of Water Management and Treatment 233
6.1.2 Water Demand and Usage 235
6.1.3 Water Management Strategies 237
6.1.4 Water Treatment Process 239
6.2 Application of FEW Nexus 4.0 in Water Management and Treatment 243
6.3 Summary 244
References 245

Chapter 7 Application of FEW Nexus 4.0 in Food, Energy, and Water Systems 252

7.1 Systems Thinking and Integration 252
7.2 Circular Economy and Achieving Zero Waste 257
7.3 Waste Management and FEW 4.0 Nexus 260
7.4 Summary 265
References 266

Chapter 8 Sustainability Metrics and Assessments For FEW Nexus 4.0 270

8.1 Sustainability in Food, Energy and Water Systems 270
8.2 LIFEWAY Indicators 274

8.2.1 Cases of Undernourishment ... 274
8.2.2 Cereal Production ... 274
8.2.3 Arable Land ... 274
8.2.4 Share of Local Breeds at Risk ... 274
8.2.5 Access to Reliable Electricity ... 274
8.2.6 Access to Eco-Friendly Fuels and Technologies for Cooking and Other Domestic Heating Purposes ... 275
8.2.7 Utilisation of Energy ... 275
8.2.8 Share of Energy Derived from Renewables ... 275
8.2.9 People Using at Least Basic Drinking Water Services ... 275
8.2.10 People Using at Least Basic Sanitation Services ... 275
8.2.11 Renewable Internal Freshwater Resources ... 276
8.2.12 Water Body Size ... 276
8.3 Summary ... 277
References ... 277

Chapter 9 Policy and Governance Frameworks For FEW Nexus 4.0 ... 281

9.1 Food, Energy, and Water Nexus 4.0 Governance Frameworks ... 281
9.1.1 Governance Structure of FEW Nexus 4.0 ... 283
9.1.2 Economic and Environmental Impacts of FEW Nexus 4.0 ... 284
9.1.3 Future Consideration of FEW Nexus 4.0 ... 284
9.1.4 Governance Challenges within the Food-Energy-Water Nexus ... 286
9.1.5 Tools for FEW Implementations ... 288
9.2 Summary ... 292
References ... 292

Chapter 10 Cultural and Societal Factors in Technology Adoption in FEW Nexus 4.0 ... 296

10.1 Cultural and Societal Influences on Technology Adoption ... 296
10.1.1 Cultural Beliefs and Energy 4.0 Adoption ... 297
10.2 Impact of Climate Change on Food, Energy, and Water System 4.0 Globally ... 310
10.2.1 Climate Change and Its Impacts on Global Food Systems ... 311

10.2.2 Climate Change and Its Impact on Global Energy Systems ... 312
10.2.3 Climate Change and Its Impact on Global Water Systems ... 312
10.3 Social, Economic, and Environmental Benefits of Food, Energy, and Water Nexus 4.0 ... 313
10.3.1 Social Benefits of the FEW Nexus 4.0 ... 314
10.3.2 Economic Benefits of the FEW Nexus 4.0 ... 314
10.3.3 Environmental Advantages of the FEW Nexus 4.0 ... 315
10.4 Food-Energy-Water Nexus Synergetic Interaction ... 316
10.5 Summary ... 316
References ... 317

Chapter 11 Case Studies and Best Practices in FEW Nexus 4.0 ... 322

11.1 Case Studies and Best Practices from Around the World ... 322
11.1.1 Best Practice Nexus Cases ... 322
11.2 Case Studies from Africa ... 323
11.3 Case Studies from Europe ... 330
11.4 Case Studies from North America ... 333
11.5 Case Studies from South America ... 336
11.6 Case Studies from the Middle East ... 339
11.7 Case Studies from Asia ... 342
11.7.1 Future Scenarios Possible for Central Asia ... 344
11.8 Case Studies from Australia ... 346
11.9 Summary ... 347
References ... 348

Chapter 12 Challenges, Opportunities, and Future Directions of FEW Nexus 4.0 ... 351

12.1 Challenges in the Food, Energy and Water Nexus 4.0 ... 351
12.1.1 Current Challenges in Water Management and Treatment ... 351
12.1.2 Current Challenges in Energy Efficiency and Optimisation ... 352
12.2 Opportunities in the FEW Nexus 4.0 ... 352
12.2.1 High Growth of Urbanisation ... 352
12.2.2 Urban Metabolism, Input-Output Models, and the Food-Energy-Water Nexus ... 353

12.2.3 Food, Land Use, and Energy 354
12.2.4 Urban Food Demand and Its Impact on Agricultural Land 354
12.2.5 Urbanisation, Changes in Dietary Patterns, and Energy Production 355
12.2.6 Water, Food, and Nutrient Flows 355
12.2.7 Dietary Shifts and Water 356
12.2.8 Water and Wastewater Management 356
12.2.9 Energy Generation and Water 357
12.2.10 Desalination 357
12.2.11 Refrigeration 357
12.2.12 Infrastructure 357
12.3 Future Directions 358
12.3.1 Future Directions of the FEW Nexus 4.0 358
12.3.2 Future Directions in Water Management and Treatment 359
12.3.3 Future Directions in Energy Efficiency and Optimisation 360
12.4 Governance, Impacts, and Future Directions of FEW Nexus 4.0 360
12.5 Summary 361
References 361

Index 369

Preface

Resources like food, energy, and water are essential for human survival. Although these three resources are getting more complex to manage, we still depend on them daily because they are important for maintaining the stability of our communities, boosting our economy, and promoting our health. Regretfully, these essential resources are beginning to run out. Our current infrastructure and methods of resource management are under more strain than ever before due to factors such as population growth, urbanisation, climate change, and the depletion of natural resources. These resources are closely related. For instance, the production of food and energy requires water. Food production uses a lot of energy and water, and energy is also required to treat, transport, and distribute food. For a sustainable future, it is necessary to comprehend these connections.

To address several issues, the Food-Energy-Water (FEW) Nexus is becoming more and more significant. It exhorts practitioners, researchers, and decision-makers to avoid working in silos. Rather, it advocates for integrated strategies that strike a balance between security, sustainability, and equity. The Fourth Industrial Revolution is what we are currently living through. This shows that technological innovations that use digital, physical, and biological systems such as the Internet of Things, artificial intelligence, big data, robotics, machine learning, blockchain, and virtual reality are developing at a rapid pace. Integrating the FEW Nexus into technological advancements leads to a new way of thinking: Food, Energy and Water Nexus 4.0. Building a more resilient and sustainable future, this new initiative enables us to reconsider and adjust to shifting resource consumption. This book investigates the relationship between the Fourth Industrial Revolution and FEW Nexus.

Food, Energy, and Water Nexus 4.0 provides a thorough and methodical analysis of this emerging field. With its introduction to fundamental ideas, examination of enabling technologies, and presentation of real-world case studies and useful examples, it is meant to be a comprehensive guide. The goal is to give professionals, researchers, policymakers, and students the knowledge they need to create and carry out integrated, forward-thinking solutions.

The structure of this book is intended to guide the reader from fundamental ideas to practical technologies and, ultimately, to futuristic viewpoints on sustainable development. Chapter 1 establishes the foundation for the journey by outlining the traditional FEW Nexus, its objectives, and its essential elements. The basic concepts that serve as the foundation for all ensuing discussions are established in this chapter. This book's central thesis, the emergence of FEW Nexus 4.0 due to the Fourth Industrial Revolution, is then presented in Chapter 2. It outlines the main strategies and innovations, from digitalisation and automation to the circular economy and decentralisation, and charts the development of the original Nexus concept to this new 4.0 paradigm, defining its essential building blocks: Food 4.0, Energy 4.0, and Water 4.0.

After establishing what and why of FEW 4.0, Chapter 3 of this book dives deep into the new innovations and technologies that power FEW 4.0. IoT and AI integration, the use of blockchain in FEW systems, the function of Geographic Information Systems, the significance of green energy, the possibilities of virtual and augmented reality, and the developments in sensor and monitoring technologies are all covered in this chapter.

We moved from discussing general technologies to their domestic use in the three primary sectors of food, energy, and water in this book's later chapters. We explore how the FEW 4.0 framework can transform water treatment and management in Chapter 4. Chapter 5 then discusses energy, specifically energy efficiency and usage optimisation. We explore the example of redefining food security and crisis management using technology and techniques in Chapter 6, which is entirely devoted to food systems. We present a modern perspective on food systems while also describing how technology is revolutionising the problems of the present and how it can help create sustainable practices. By examining how the concepts of FEW 4.0 can support the resilience of food systems, this book encourages a more connected and introspective way of participating. This idea is continued in Chapter 7, which highlights the importance of considering all sectors in a comprehensive manner while bringing the main ideas from the earlier chapters together. The circular economy is presented in this chapter as a key idea to support the fundamental idea of waste reduction and the efficiency of food, energy, and water in alignment with sustainability principles.

Chapter 8 describes how performance evaluation and sustainability metrics can be used to gauge success. The LIFEWAY Indicators, which offer quantifiable standards for access to dependable electricity, essential drinking water services, and agricultural output in the production of cereal crops, constitute a substantial component of this chapter. These indicators can be used to gather data and track the system's long-term performance. The governance and policy frameworks that are thought to be both the obstacles and the facilitators of the success of the FEW 4.0 solutions are examined in detail in Chapter 9. Additionally, the path of interagency coordination, supportive regulation, and cogent policy design is reflected. By highlighting the cultural values, societal behaviours, and regional norms that influence the potential for technology adoption, Chapter 10 provides additional analysis of the horizon. Whether or not the phenomenon affects the world's food, energy, and water systems, the issue of climate change is also a conscious one. It also demonstrates the significant social, environmental, and financial benefits that come from a properly managed transition.

In Chapter 11, a few global cases are used to implement the theory. In addition to being a future expectation, this book demonstrates that the FEW 4.0 framework already exists in some regions across the world, including Africa, Asia, Europe, the Americas, the Middle East, and Australia. Chapter 12 concludes with a discussion of examples and shifts to a future perspective. It finds new research opportunities and tackles enduring problems. This final chapter lays out a plan for developing the FEW Nexus 4.0 in a world that is constantly changing as cities expand and pressures increase.

Author Biographies

Nnamdi Nwulu is the Chair Holder, South Africa/Switzerland Bilateral Research Chair in Blockchain Technology (SARChI Chair), Professor of Sustainable Cyber-Physical Engineering Systems and the Director of the Centre for Cyber-Physical Food, Energy & Water Systems at the University of Johannesburg.

Prof Nwulu's research focuses on Food, Energy, and Water Systems (FEWS) and the inherent nexus between all three. Specifically, his research focuses on applying FEW-N thinking, emerging technologies, mathematical optimisation techniques, and machine learning algorithms in FEWS. This research focus has produced several books, inventions and over 200 publications, winning him the 2021 NRF Research Excellence Award for Early Career/Emerging Researchers in Engineering/Technology. He was also a 2020/2021 TW Kambule-NSTF-South32 Award finalist in the Emerging Researcher category and won the 2020 UJ Vice-Chancellors' Most Promising Young Researcher award. He is a member of the prestigious South African Young Academy of Science (SAYAS).

Prof Nwulu has received several research grants worth over 60 million rands from governmental and industrial sponsors. In 2022, Prof Nwulu led a consortium of academics from Greece, France, and South Africa to receive funding from the European Union Erasmus program. Prof Nwulu was announced in December 2022 as the South Africa/Switzerland (SA-Swiss) Bilateral Research Chair in Blockchain Technology, focusing on blockchain integration in South Africa and Switzerland's agri-food and medical supply chains.

He is a Professional Engineer registered with the Engineering Council of South Africa (ECSA), a Senior Member of the Institute of Electrical and Electronics Engineers (SMIEEE), and a Fellow of the South African Institute of Electrical Engineers (FSAIEE).

Prof. Nwulu is the Editor-in-Chief of the *Journal of Digital Food Energy and Water Systems* (JD-FEWS), Associate Editor of the *IET Renewable Power Generation* (IET-RPG) and the *African Journal of Science, Technology, Innovation and Development* (AJSTID).

Omolola Ogbolumani is a computer engineer licensed to operate in Nigeria by the Council for the Regulation of Engineering in Nigeria (COREN). She holds a PhD degree in Electrical and Electronic Engineering from the University of Johannesburg, South Africa, with a thesis titled *A Hybrid Decision-Making Framework for the Food-Energy-Water Nexus.*

Dr. Ogbolumani's research focuses on smart sustainability using FEW-Nexus thinking, mathematical optimisation techniques, and 4IR technologies, including the Internet of Things (IoT), machine learning (ML) algorithms, and blockchain technology applied in waste management systems for cleaner environment and food, energy, and water systems for increased productivity. Her research has produced several publications in peer-reviewed journals and inventions. She is affiliated with the Centre for Cyber-Physical Food, Energy & Water Systems (CCP-FEWS) at the University of Johannesburg South Africa and the TETFund Centre of Excellence for Biodiversity Conservation and Ecosystem Management (TCEBCEM) at the University of Lagos, Nigeria.

She is a member of the Nigerian Society of Engineers (MNSE), the Nigeria Computer Society (MNCS), and the Institute of Electrical and Electronics Engineers (MIEEE).

1 Food, Energy, and Water Nexus

1.1 INTRODUCTION TO FOOD, ENERGY, AND WATER NEXUS

In today's interconnected world, the relationship between food, energy, and water is more critical than ever. The increasing demand for these resources and the decreasing supply are a pressing concern. This interdependence underscores the necessity of understanding the food-energy-water nexus (FEW-N) dynamics. Climate change, human hostilities, an increase in poverty in several nations, and pandemics have all made the problem of FEW insecurity on a global scale worse. Many other factors contribute to this problem, including population growth, migration, economic development, and technological advancements (Ogbolumani & Nwulu, 2021). Figure 1.1 illustrates some of the stressors that negatively affect the FEW-N system, resulting in an increasing demand for these three resources.

The FEW-N describes the relationship between food, energy, and water. FEW-N is a conceptual framework that holistically recognises the interconnection between

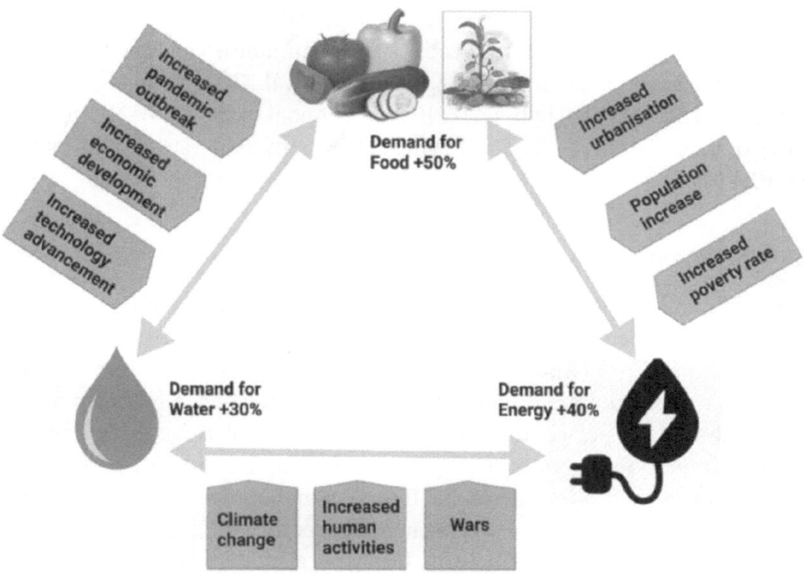

FIGURE 1.1 Overview of the FEW-N System with Stressors (Ogbolumani, 2021).

DOI: 10.1201/9781003633853-1

three equally vital sectors to reduce trade-offs and maximise synergies (Hoff, 2011). Therefore, the FEW-N refers to the interrelationships and synergies between these essential resources. FEW production, distribution, and consumption are interconnected, and policies and actions in one sector could have significant consequences in others. This interconnectedness of the three resources is evident across many sectors (Ogbolumani & Nwulu, 2020). For example, hydropower generation plants need water to generate electricity for the populace, which can harm downstream agricultural activities. This connection is also seen in food crop production and biofuel production, which simultaneously compete for water and land. The reliance on water and energy by irrigated agriculture further shows this interdependency. The manufacturing and food processing sectors also demonstrate this connection with solar panel production, which needs water for cleaning and cooling. Food processing companies use water and energy for cooking, cleaning, and refrigeration (Herrmann et al., 2018; Ogbolumani & Nwulu, 2022). These examples show how activities in all three sectors highlight their interconnected nature. They show how actions in one sector impact the availability of the others, confirming the need to manage the three critical resources holistically.

1.2 GOALS OF FOOD, ENERGY, AND WATER NEXUS

An integral part of the FEW-N would be introducing the concept to address the complex challenges arising from the interdependence of FEW systems (Albatayneh, 2023; Zhang et al., 2019). A key objective of FEW-N is to encourage collaboration across sectors and disciplines to ensure complex systems are managed sustainably and equitably (Estoque, 2023; Roidt & Avellán, 2019). It aims to identify solutions that will meet both the short-term needs of the communities and the long-term goals of sustainable development in the long run (David et al., 2023; Ghodsvali et al., 2019; Huntington et al., 2021; Olawuyi, 2020). Another goal of the FEW-N is to ensure the security of these resources. The following subsections elaborate further on the goals of the FEW-N.

1.2.1 Sectoral Collaboration

It is possible to achieve innovative and holistic approaches to addressing interconnected system challenges by collaborating across sectors and disciplines within the context of the FEW-N (Apeh & Nwulu, 2024). It is possible to develop integrated strategies for resource use, resource efficiency, and resilience by bringing together experts with various perspectives and knowledge from various disciplines, such as agriculture, energy, and water management, to combine diverse perspectives and expertise (Howarth & Monasterolo, 2017). A collaborative approach to managing FEW can also promote knowledge sharing, capacity building, and policy coordination, all of which can contribute to fostering a more sustainable and equitable FEW management system (Mantino et al., 2023). In this way, future generations will be able to benefit from long-term sustainable development, as well as present generations.

1.2.2 Sustainable and Equitable Resource Management

Equitable management within the context of the FEW-N means ensuring fair and just distribution of resources and benefits among different stakeholders, including communities, industries, and ecosystems. It recognises that access to FEW should not be determined by socioeconomic status or geographical location but based on principles of fairness and inclusion. By promoting equitable management, the FEW-N aims to address disparities and promotes sustainable development that benefits all (Sharma & Kumar, 2020; Wang et al., 2021).

1.2.3 Resource Security

The FEW-N focuses on three core elements: *food security, energy security, and water security*. These three core elements of the FEW-N are needed to meet the basic needs of the human race. The security of these critical resources is paramount to the nexus framework. Each of these concepts is examined in the following subsections.

a. **Food Security**

A widely accepted definition of food security is that from the World Economic Summit, held at the Food and Agriculture Organisation (FAO) headquarters in Rome in 1996, which states that: "Food security exists when all people, at all times, have physical and economic access to sufficient, safe and nutritious food that meets their dietary needs and food preferences for an active and healthy life" (Food and Agriculture Organization of the United Nations, 1996). According to the FAO report, this popular definition of food security has four dimensions: food availability, access, utilisation, and stability (FAO, 1996). Recently, a fifth dimension was added to food security, called sustainability. These five dimensions are briefly described as follows:

i. **Food Availability**: This is a key factor in food security. It refers to the physical availability of enough nutrient-dense and safe food at a specific place and time. This could be locally supplied or through importation (Gupta et al., 2021).
ii. **Food Access**: In simple terms, food access is when food resources can get to individuals, and the individuals have enough resources (money) to buy appropriate foods for a nutritious diet.
iii. **Utilisation**: This involves the ability of the individual to eat enough clean, safe, and nutritious food. The individual should also be able to drink adequate, clean, and safe water, have access to sanitation, and be physically healthy to be able to digest and utilise the food (Peng & Berry, 2019)
iv. **Stability**: For a population or community of people or person to be food secure, it means they must be able to deal with or withstand any risk caused by factors such as climate change, wars, earthquakes, and economic crises to the food supply chain system (Peng & Berry, 2019).

v. **Sustainability**: This is a newly added pillar in food security. It considers the economic, social, and environmental impact on food production, distribution, and consumption. It goes a step further to promote equitable food resource management, sectoral collaborations in food systems, conservation of the environment, afforestation, and biodiversity preservation. To meet the needs of current and future generations while preserving the ecosystem and natural resources and not being a threat to the environment and the wellbeing of the less privileged population (Opeyemi et al., 2024; Peng & Berry, 2019).

b. Energy Security

The definition of energy security and its scope keeps changing in response to changes in circumstances over time. For example, as new energy technologies evolve and awareness of climate change and sustainability issues increases, the critical aspects of energy security continue to evolve. Energy security is a concept that is needed to fulfil basic human needs. It is essential in fostering sustainable development (Bajracharya et al., 2022). Based on the definition given by the International Energy Agency, Energy security is the uninterrupted availability of energy resources at an affordable price (*What Is Energy Security? – University of Plymouth*, n.d.). According to research by the UK Parliament, Energy security is exposed to certain risks: high energy prices, fuel shortages, equipment failures, the effects of climate change, global unrest from wars in some parts of the world, and net-zero transition risks (Cabinet Office, 2022).

c. Water Security

Water security is the ability of people to have access to affordable, clean water to fulfil essential requirements, drive economic operations, and support overall health (Cosgrove & Loucks, 2015). There are many dimensions of water security, namely water quantity, water availability, water accessibility, water affordability, water quality, water safety, dependable water supply, and water sustainability (for current waste use practices). About 4 billion people are experiencing water scarcity for at least a month yearly (Mekonnen & Hoekstra, 2016; Young et al., 2021). Hundreds of millions of people are affected by coastal flooding due to excess rain and rising water bodies (Kulp & Strauss, 2019). Furthermore, biological and chemical contaminants like viruses, parasites, bacteria, heavy metals, pharmaceuticals, and plastic wastes exist in water bodies across the globe (Al-Hazmi et al., 2023). Some factors affecting water resources around the world include growing population, modernisation, climate change, ageing infrastructure, technological advancement, poor operations, and maintenance of water facilities, including wastewater (Cai et al., 2018). All these challenges are stressors to water resources.

The nexus encourages cross-sectoral dialogue to develop joint strategies for achieving sustainable outcomes and striking a balance between economic, social, and environmental objectives. To achieve this, nexus emphasises integrating social, economic, and ecological factors into decision-making processes. Considering the diverse needs and perspectives, the nexus can identify solutions that address immediate challenges and contribute to long-term resilience and sustainability. This holistic approach ensures adequate and equitable management of FEW systems for all stakeholders. The stakeholders in FEW-N are all those affected by production activities in the three sectors. These include food producers such as farmers in all areas of food production, energy producers, water utility companies, consumers of FEW, government bodies, non-governmental organisations, local community leaders, academics, and businesses involved in the production and distribution of these resources.

1.3 KEY COMPONENTS OF THE FOOD, ENERGY, AND WATER NEXUS

The critical components of the FEW-N are food production, energy production and water resource management. These three components are discussed briefly in the following subsections.

1.3.1 Food Production

Food production is growing crops, raising animals and aquatic organisms, and processing the raw forms of crops, animals, and aquatic organisms into consumable food. Food production is a complex process that includes agricultural activities, livestock and aquatic farming, and food processing (Oosting et al., 2022). Some examples of food production include crop cultivation, like grains, cereals, pulses, fruits, vegetables, herbs, and spices. Animals, birds, and dairy farming include cows, goats, sheep, rabbits, chickens, turkeys, and ducks. Other examples include fisheries and aquaculture, such as tilapia, catfish, salmon, oysters, and shrimp. Urban farming includes greenhouse, hydroponic, vertical, aquaponic, and aeroponics (Saad et al., 2021). Food processing and packaging are examples of food production transforming food by adding value and protecting food products (Ponnampalam & Holman, 2023; Shavanov, 2021).

1.3.2 Energy Production

Energy production turns raw energy sources into usable forms of energy. It is the transformation of primary energy sources into forms that people can use (Manzano-Agugliaro et al., 2013). There are two broad types of energy sources: non-renewable and renewable energy. Non-renewable sources include fossil fuel (oil and gas), coal, and nuclear fuel (which is in the form of uranium and thorium). The non-renewable category includes solar energy, which produces power via photovoltaic panels or concentrated solar power systems to intercept direct sunlight (Ajibola & Ogbolumani, 2024; Olorunfemi et al., 2022, 2023). Wind energy operates by using turbines to convert air movement into usable power. Flowing water generates hydropower, tidal power, and wave energy. Another form of renewable

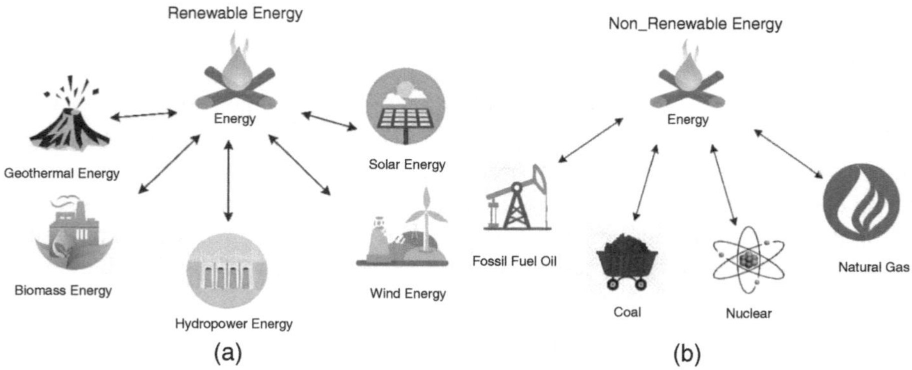

FIGURE 1.2 Renewable and Non-Renewable Energy Sources (Mastoi et al., 2022).

energy is derived from biomass. These are organic matters, including wood residues and animal, municipal, and agricultural waste (Seminario-Córdova & Rojas-Ortega, 2023). The various types of energy production have different strengths regarding availability and accessibility, environmental impact, cost of extraction and conversion, energy storage capabilities, and infrastructure requirements (Ogbolumani & Adekoya, 2025). Figure 1.2 shows the various types of renewable and non-renewable energy sources.

1.3.3 Water Management

Water management is the process of planning, developing, and managing water resources, in terms of quantity and quality, across all water uses. It is the coordinated use of different water sources: surface and groundwater, recycled water, advanced purified water, enhanced water storage, and water conservation. Examples include irrigation systems and water allocation, watershed management, groundwater extraction and monitoring, water treatment and distribution, wastewater management, and recycling. Figure 1.3 shows a framework for an urban water management cycle. This system has four main inputs: water, contaminants, energy, and chemicals. The water sources include surface water, groundwater, and precipitation. Wastewater is also part of the water system.

These three components (food, energy, and water) are all interconnected and interdependent. Some examples include the production of energy from food and water resources, the production of biofuels from plants, and the production of agricultural food based on energy and water resources. A water utility plant needs energy to pump, transport, and distribute water. Any changes to one component can significantly affect the others. Some interactions between FEW components in the FEW-N are illustrated in Table 1.1. For example, irrigation requires higher water use to produce more food, which can decrease water availability and cause water prices to rise (Ogbolumani & Nwulu, 2021, 2024).

Furthermore, energy production can affect water resources since power plants often use large quantities. Similarly, the use of pesticides to increase crop yield can

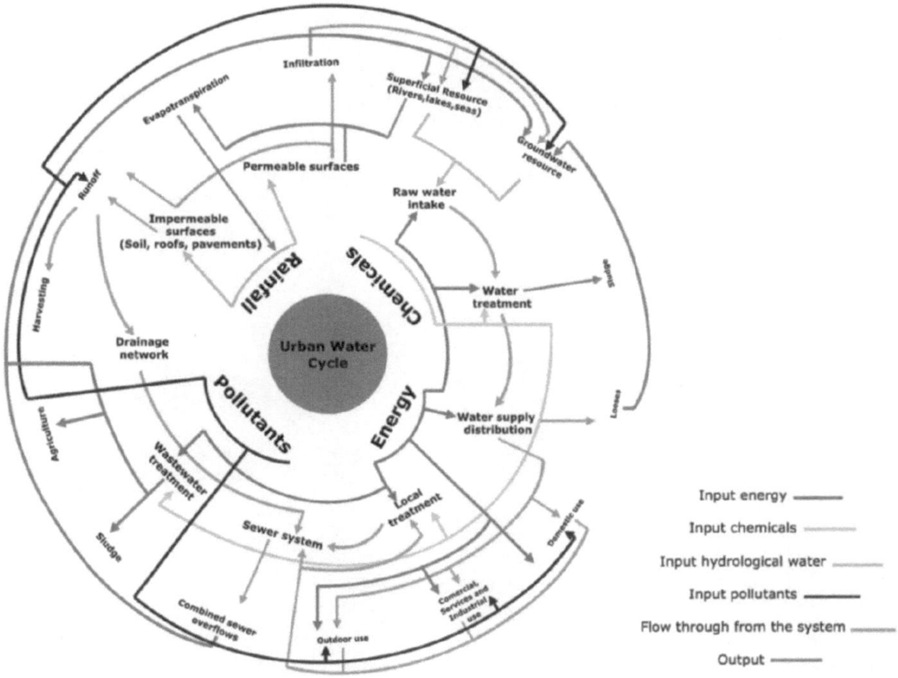

FIGURE 1.3 Water Resource Management (Peña-Guzmán et al., 2017).

TABLE 1.1
Potential Interaction of Resources → Subsectors in a FEW-N System (Ogbolumani, 2021)

FEW Resources	Subsectors		
	Food	**Energy**	**Water**
Food	Open field food production	Bioenergy production	
Energy	Greenhouse food production	Solar/Wind energy production	Desalination
	Fertiliser production	Bioenergy production	Waste water treatment
	Open field food production		
Water	Irrigated food production	Hydropower production	Pump water from reservoirs to the distribution network and final users

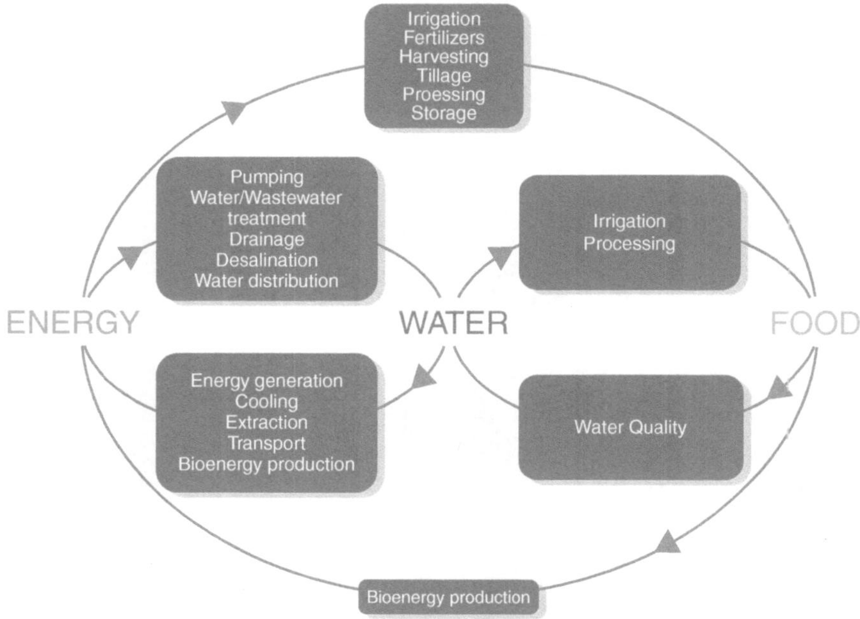

FIGURE 1.4 Food, Energy, and Water Nexus System as an Interacting System (IRENA, 2015).

lead to a decrease in biodiversity, contributing to the decline in the availability of natural resources. This can have a ripple effect, resulting in economic, social, and environmental impacts on the ecosystem. Figure 1.4 shows the FEW-N interacting system (IRENA, 2015). These interconnected components require careful and coordinated management to ensure sustainability. Figure 1.5 also illustrates how these three resources are interconnected, based on a case study in Refs. (Ogbolumani, 2021; Ogbolumani & Nwulu, 2021). Water is used for food production, with irrigation being an example. Water is used for energy production, such as hydropower production. Food for energy with biomass production is a relatable example. Renewable energy sources are used to provide energy for greenhouse food production (Ogbolumani & Nwulu, 2020)

1.4 SUMMARY

This chapter talks about the FEW-N, focusing on how it came to be and the interdependence of these resources. It explains how population growth, migration, and economic development, among many others, contribute to the inability to meet the demands for this resource. The chapter presents the interconnectedness of these resources and how understanding it is the first step to solve the problem. It also talks about the goals of FEW-N, fairness in the distribution of resources, and the core elements of the FEW-N system: food, water, and energy security. Examples such as energy production from food and water resources, biofuels from plants, and agricultural food production based on energy and water resources show how these resources are interconnected.

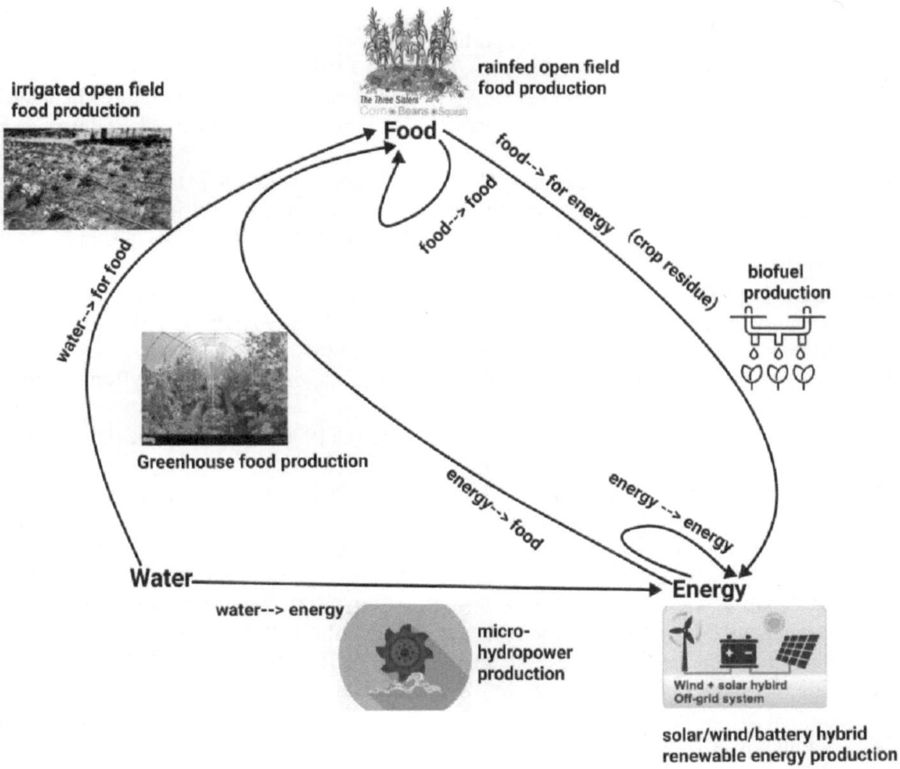

FIGURE 1.5 Intersectoral Interaction for a FEW-N Case Study in Refs. (Ogbolumani, 2021; Ogbolumani & Nwulu, 2021).

REFERENCES

Ajibola, O. A., & Ogbolumani, O. A. (2024). Smart battery storage integration in an IoT-based solar-powered waste management system. *Journal of Digital Food, Energy & Water Systems*, *5*(2), 17–42. https://doi.org/10.36615/HREWQ861

Albatayneh, A. (2023). Water energy food nexus to tackle climate change in the Eastern Mediterranean. *Air, Soil and Water Research*, *16*. https://doi.org/10.1177/11786221231170222

Al-Hazmi, H. E., Mohammadi, A., Hejna, A., Majtacz, J., Esmaeili, A., Habibzadeh, S., Saeb, M. R., Badawi, M., Lima, E. C., & Mąkinia, J. (2023). Wastewater reuse in agriculture: Prospects and challenges. *Environmental Research*, *236*, 116711. https://doi.org/10.1016/J.ENVRES.2023.116711

Apeh, O. O., & Nwulu, N. I. (2024). The water-energy-food-ecosystem nexus scenario in Africa: Perspective and policy implementations. *Energy Reports*, *11*, 5947–5962. https://doi.org/10.1016/J.EGYR.2024.05.060

Bajracharya, T. R., Shakya, S. R., & Sharma, A. (2022). Dynamics of energy security and its implications. *Handbook of Energy and Environmental Security*, 13–25. https://doi.org/10.1016/B978-0-12-824084-7.00019-9

Cabinet Office. (2022). *British Energy Security Strategy*. London, England: Cabinet Office. Retrieved from https://www.gov.uk/government/publications/british-energy-security-strategy

Cai, X., Wallington, K., Shafiee-Jood, M., & Marston, L. (2018). Understanding and managing the food-energy-water nexus – Opportunities for water resources research. *Advances in Water Resources*, *111*, 259–273. https://doi.org/10.1016/J.ADVWATRES.2017.11.014

Cosgrove, W. J., & Loucks, D. P. (2015). Water management: Current and future challenges and research directions. *Water Resources Research*, *51*(6), 4823–4839. https://doi.org/10.1002/2014WR016869

David, L. O., Nwulu, N. I., Aigbavboa, C. O., & Adepoju, O. O. (2023). Resource sustainability in the water, energy and food nexus: role of technological innovation. *Journal of Engineering, Design and Technology,* 15 April 2025; *23*(3), 725–743. https://doi.org/10.1108/JEDT-05-2023-0200

Estoque, R. C. (2023). Complexity and diversity of nexuses: A review of the nexus approach in the sustainability context. In *Science of the Total Environment* (Vol. 854, p. 158612). Elsevier. https://doi.org/10.1016/j.scitotenv.2022.158612

FAO. (1996, November 13). *World Food Summit – Final Report – Part 1*. FAO Rome. https://www.fao.org/4/w3548e/w3548e00.htm

Food and Agriculture Organization of the United Nations. (1996). *Rome Declaration on World Food Security and World Food Summit Plan of Action*. FAO.

Ghodsvali, M., Krishnamurthy, S., & de Vries, B. (2019). Review of transdisciplinary approaches to food-water-energy nexus: A guide towards sustainable development. *Environmental Science and Policy*, *101*(September), 266–278. https://doi.org/10.1016/j.envsci.2019.09.003

Gupta, P., Singh, J., Verma, S., Chandel, A. S., & Bhatla, R. (2021). Impact of climate change and water quality degradation on food security and agriculture. *Water Conservation in the Era of Global Climate Change*, 1–22. https://doi.org/10.1016/B978-0-12-820200-5.00006-3

Herrmann, R., Jumbe, C., Bruentrup, M., & Osabuohien, E. (2018). Competition between biofuel feedstock and food production: Empirical evidence from sugarcane outgrower settings in Malawi. *Biomass and Bioenergy*, *114*, 100–111. https://doi.org/10.1016/J.BIOMBIOE.2017.09.002

Hoff, H. (2011). Understanding the Nexus. *Background Paper for the Bonn2011 Conference: The Water, Energy and Food Security Nexus.*

Howarth, C., & Monasterolo, I. (2017). Opportunities for knowledge co-production across the energy-food-water nexus: Making interdisciplinary approaches work for better climate decision making. *Environmental Science & Policy*, *75*, 103–110. https://doi.org/10.1016/J.ENVSCI.2017.05.019

Huntington, H. P., Schmidt, J. I., Loring, P. A., Whitney, E., Aggarwal, S., Byrd, A. G., Dev, S., Dotson, A. D., Huang, D., Johnson, B., Karenzi, J., Penn, H. J. F., Salmon, A. A., Sambor, D. J., Schnabel, W. E., Wies, R. W., & Wilber, M. (2021). Applying the food–energy–water nexus concept at the local scale. *Nature Sustainability*, *4*(8), 672–679. https://doi.org/10.1038/s41893-021-00719-1

IRENA. (2015). Renewable energy in the water, energy and food nexus. *International Renewable Energy Agency* (Issue January 2015), 24.

Kulp, S., & Strauss, B. H. (2019). New elevation data triple estimates of global vulnerability to sea-level rise and coastal flooding. *Nature Communications*. https://www.nature.com/articles/s41467-019-12808-z

Mantino, F., Atterton, J., Nieto, E., Kong, X., & Durán-Díaz, P. (2023). Sustainable land governance for water–energy–food systems: A framework for rural and peri-urban revitalisation. *Land*, *12*(10), 1828. https://doi.org/10.3390/LAND12101828

Manzano-Agugliaro, F., Alcayde, A., Montoya, F. G., Zapata-Sierra, A., & Gil, C. (2013). Scientific production of renewable energies worldwide: An overview. *Renewable and Sustainable Energy Reviews*, *18*, 134–143. https://doi.org/10.1016/J.RSER.2012.10.020

Mastoi, M. S., Munir, H. M., Zhuang, S., Hassan, M., Usman, M., Alahmadi, A., & Alamri, B. (2022). A comprehensive analysis of the power demand–supply situation, electricity Usage Patterns, and the Recent Development of Renewable Energy in China. *Sustainability*, *14*(6), 3391. https://doi.org/10.3390/SU14063391

Mekonnen, M. M., & Hoekstra, A. Y. (2016). Sustainability: Four billion people facing severe water scarcity. *Science Advances*, *2*(2). https://doi.org/10.1126/SCIADV.1500323

Ogbolumani, O. A. (2021). *A Hybrid Decision-Making Framework for Optimal Resource Allocation in the Food-Energy-Water NEXUS* (Issue August). University of Johannesburg.

Ogbolumani, O. A., & Adekoya, M. (2025). Intelligent waste management optimization through machine learning analytics. *Journal of Science Research and Reviews*, *2*(1), 7–26. https://doi.org/10.70882/JOSRAR.2025.V2I1.25

Ogbolumani, O. A., & Nwulu, N. (2020). Integrated appliance scheduling and optimal sizing of an autonomous hybrid renewable energy system for agricultural food production. *Lecture Notes in Mechanical Engineering*, 651–660. https://doi.org/10.1007/978-981-15-5753-8_60

Ogbolumani, O. A., & Nwulu, N. I. (2021). Multi-objective optimisation of constrained food-energy-water-nexus systems for sustainable resource allocation. *Sustainable Energy Technologies and Assessments*, *44*(January), 100967. https://doi.org/10.1016/j.seta.2020.100967

Ogbolumani, O. A., & Nwulu, N. I. (2022). A food-energy-water nexus meta-model for food and energy security. *Sustainable Production and Consumption*, *30*, 438–453. https://doi.org/10.1016/J.SPC.2021.12.019

Ogbolumani, O. A., & Nwulu, N. I. (2024). Environmental impact assessment for a meta-model-based food-energy-water-nexus system. *Energy Reports*, *11*, 218–232. https://doi.org/10.1016/j.egyr.2023.11.033

Olawuyi, D. (2020). Sustainable development and the water-energy-food nexus: Legal challenges and emerging solutions. *Environmental Science and Policy*, *103*, 1–9. https://doi.org/10.1016/j.envsci.2019.10.009

Olorunfemi, B. O., Nwulu, N. I., & Ogbolumani, O. A. (2023). Solar panel surface dirt detection and removal based on arduino color recognition. *MethodsX*, *10*, 101967. https://doi.org/10.1016/J.MEX.2022.101967

Olorunfemi, B. O., Ogbolumani, O. A., & Nwulu, N. (2022). Solar panels dirt monitoring and cleaning for performance improvement: A systematic review on smart systems. *Sustainability*, *14*(17), 10920. https://doi.org/10.3390/SU141710920

Oosting, S., van der Lee, J., Verdegem, M., de Vries, M., Vernooij, A., Bonilla-Cedrez, C., & Kabir, K. (2022). Farmed animal production in tropical circular food systems. *Food Security*, *14*(1), 273–292. https://doi.org/10.1007/s12571-021-01205-4

Opeyemi, L., Clinton, D., Nnamdi, A., Omoseni, N., & Adepoju, O. (2024). *Sustainable Synergy: A Digital Framework for the Water-Energy-Food Nexus Project Delivery in Developing Economies*. Springer Nature. https://doi.org/10.1007/978-3-031-72833-4

Peña-Guzmán, C. A., Melgarejo, J., Prats, D., Torres, A., & Martínez, S. (2017). Urban water cycle simulation/management models: A review. *Water*, *9*(4), 285. https://doi.org/10.3390/W9040285

Peng, W., & Berry, E. M. (2019). The concept of food security. *Encyclopedia of Food Security and Sustainability*, 1–7. https://doi.org/10.1016/B978-0-08-100596-5.22314-7

Ponnampalam, E. N., & Holman, B. W. B. (2023). Sustainability II: Sustainable animal production and meat processing. *Lawrie's Meat Science*, 727–798. https://doi.org/10.1016/B978-0-323-85408-5.00001-7

Roidt, M., & Avellán, T. (2019). Learning from integrated management approaches to implement the Nexus. *Journal of Environmental Management*, *237*, 609–616. https://doi.org/10.1016/j.jenvman.2019.02.106

Saad, M. H. M., Hamdan, N. M., & Sarker, M. R. (2021). State of the art of urban smart vertical farming automation system: Advanced topologies, issues and recommendations. *Electronics*, *10*(12), 1422. https://doi.org/10.3390/ELECTRONICS10121422

Seminario-Córdova, R., & Rojas-Ortega, R. (2023). Renewable energy sources and energy production: A bibliometric analysis of the last five years. *Sustainability*, *15*(13), 10499. https://doi.org/10.3390/SU151310499

Sharma, P., & Kumar, S. N. (2020). The global governance of water, energy, and food nexus: allocation and access for competing demands. *International Environmental Agreements: Politics, Law and Economics*, *20*(2), 377–391. https://doi.org/10.1007/S10784-020-09488-2

Shavanov, M. V. (2021). The role of food crops within the Poaceae and Fabaceae families as nutritional plants. *IOP Conference Series: Earth and Environmental Science*, *624*(1), 012111. https://doi.org/10.1088/1755-1315/624/1/012111

Wang, X. C., Jiang, P., Yang, L., Fan, Y. Van, Klemeš, J. J., & Wang, Y. (2021). Extended water-energy nexus contribution to environmentally-related sustainable development goals. *Renewable and Sustainable Energy Reviews*, *150*, 111485. https://doi.org/10.1016/J.RSER.2021.111485

What Is Energy Security? – University of Plymouth. (n.d.). Retrieved February 8, 2025, from https://www.plymouth.ac.uk/discover/what-is-energy-security

Young, S. L., Frongillo, E. A., Jamaluddine, Z., Melgar-Quiñonez, H., Pérez-Escamilla, R., Ringler, C., & Rosinger, A. Y. (2021). Perspective: The importance of water security for ensuring food security, good nutrition, and well-being. *Advances in Nutrition*, *12*(4), 1058–1073. https://doi.org/10.1093/ADVANCES/NMAB003

Zhang, P., Zhang, L., Chang, Y., Xu, M., Hao, Y., Liang, S., Liu, G., Yang, Z., & Wang, C. (2019). Food-energy-water (FEW) nexus for urban sustainability: A comprehensive review. In *Resources, Conservation and Recycling* (pp. 215–224). Elsevier B.V. https://doi.org/10.1016/j.resconrec.2018.11.018

2 Food, Energy, and Water Nexus 4.0

2.1 THE FOURTH INDUSTRIAL REVOLUTION AND THE RISE OF FEW NEXUS 4.0

The first Industrial Revolution came at the end of the 18th century, using steam power and mechanisation. The second industrial revolution started at the beginning of the 20th century and brought about assembly lines, production in bulk, and electrical machinery. The third industrial revolution, in the early 1970s, brought information technology, automation, and machinery. Since 2011, cyber-physical systems/intelligent computers have been shaping the Fourth Industrial Revolution (4IR) (Bedi et al., 2024; Groumpos, 2021; Teker & Koc, 2019). The four revolutions are illustrated in Figures 2.1 and 2.2. The Fourth Industrial Revolution – also called Industry 4.0

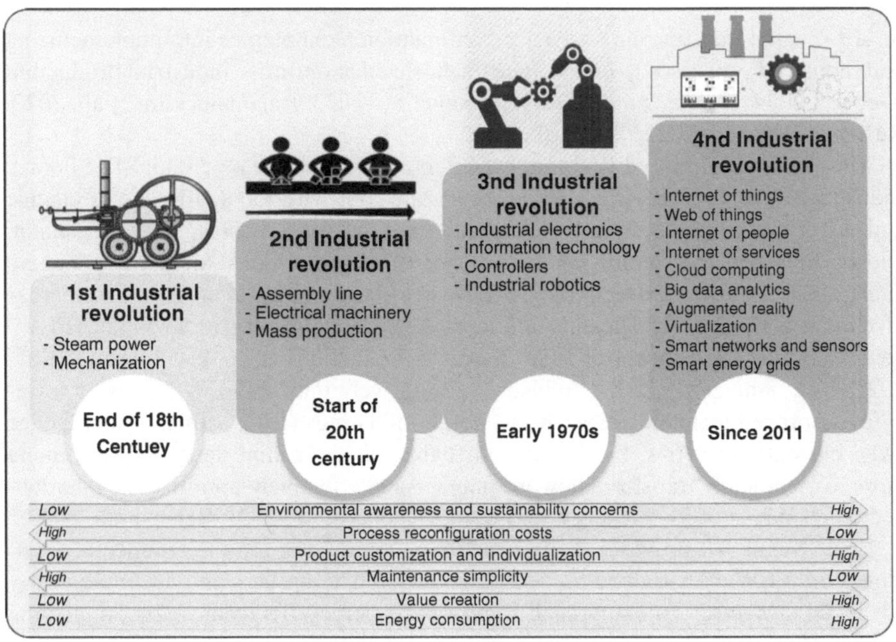

FIGURE 2.1 The Fourth Industrial Revolution in Picture (Ng et al., 2022).

DOI: 10.1201/9781003633853-2

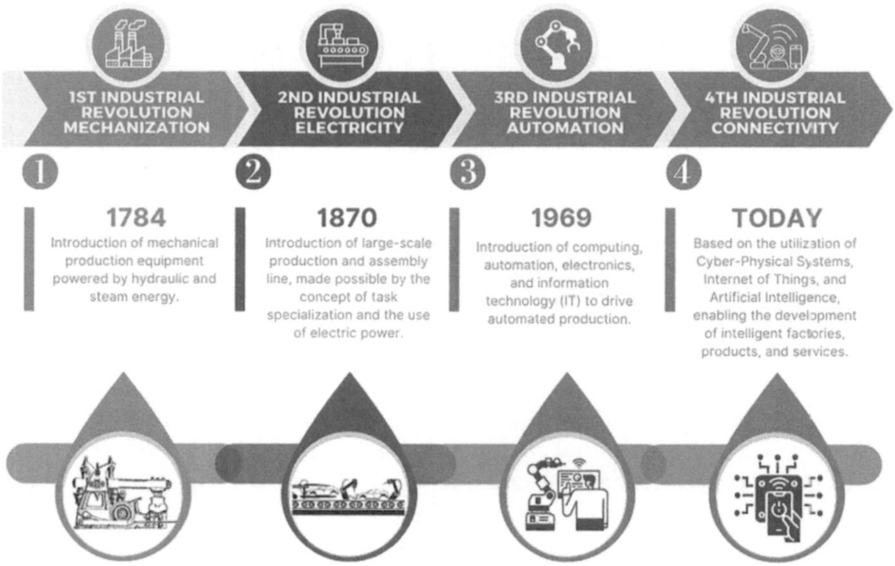

FIGURE 2.2 Rise of the Fourth Industrial Revolution (Filho et al., 2024).

or 4IR – is the interfusion of advanced automation technologies into manufacturing and industrial processes to create smart factories that optimise industrial production (George, 2024; Geuna et al., 2021; Hassoun et al., 2023; Rajamanickam et al., 2023; Rusko & Kosonen, 2022; Yin et al., 2018).

The 4IR is still in its early stages but offers transformative potential. "The Fourth Industrial Revolution is still in its nascent state. But with the swift pace of change and disruption to business and society, the time to join in is now," Gary Coleman, Global Industry and Senior Client Advisor of Deloitte Consulting, says (Zaman et al., 2023). Industry 4.0 technologies have led to technological progress in all areas of human activities, significantly influencing the Food, Energy, and Water (FEW) nexus, coining the concept of FEW Nexus 4.0 (Anandhi et al., 2023; David et al., 2022; Siekmann et al., 2023). Industry 4.0 positively impacts FEW Nexus by leveraging improved technologies such as Internet of Things (IoT), artificial intelligence (AI), big data analytics, blockchain, virtual reality, machine learning, and many more to completely transform how we manage and effectively optimise the interconnected resources of food, energy, and water (Dantas et al., 2021; Rejeb et al., 2022).

FEW Nexus 4.0 provides many benefits by addressing the interconnected challenges in FEW Nexus, and it does this through a unified and technologically advanced structure. This approach improves resource effectiveness by amplifying the use of resources and promoting circular economy principles (Tang, Chau, et al., 2022), thereby reducing waste and lowering the environmental impact. For instance, we can use big data technologies to assess and analyse the information in the food production industry that was previously too large or complex to handle, thus improving performance and production efficiency (Javaid et al., 2022b). A good description is shown in Figure 2.3 with smart practices and technologies for Agriculture 4.0. AI-driven energy management systems can monitor, analyse, and optimise energy usage within smart buildings. This maximises energy use and prevents waste (Muniandi et al., 2024). An example of the application of AI (reinforcement learning) in energy management systems is shown in Figure 2.4. The system displays a model for a reinforcement learning-based home energy management system consisting of consumer data, a weather station, and the utility company (Lee & Choi, 2019).

The use of IoT systems allows for the acquisition of data in real time, as well as the transmission and processing of data in a way that ensures efficiency (Adeleke et al., 2023; Ajibola & Ogbolumani, 2024; Ogbolumani & Mabaso, 2023). For example, Figure 2.5 shows an IoT-based integrated energy system consisting of the following major parts: demand-side services, regulation, market, utilities, and generation, and transmission and distribution (Motlagh et al., 2020). IoT has helped in the digitalisation of the energy system, transforming it from a "dumb grid" to a "smart grid" where information flows bi-directionally instead of the traditional unidirectional communication (Motlagh et al., 2020). Investments in FEW Nexus 4.0 can create new job opportunities in the food, energy, and water sectors, thereby contributing to wider economic development and leading to sustainability in food production, energy consumption, and water use (Ali et al., 2022; Dantas et al., 2021; Tang, Chau, et al., 2022). In addition, investing in technologies and practices that effectively help manage and make efficient use of these three resources can help with waste management (Rasul, 2016). Figure 2.6 shows that there is a link between food, energy, water, and waste: energy waste, food waste, and water waste.

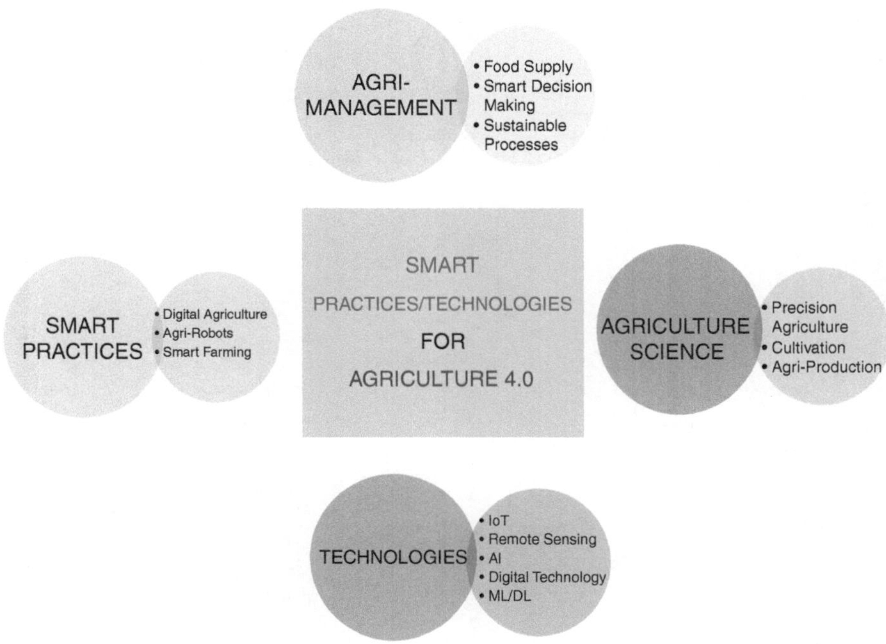

FIGURE 2.3 Smart Usage of Big Data Technologies (Javaid et al., 2022a).

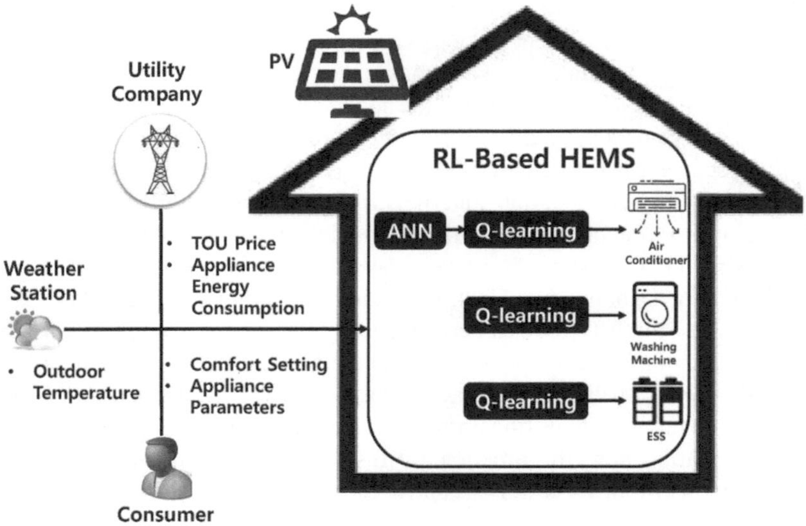

FIGURE 2.4 AI-Driven Management Systems (Lee & Choi, 2019).

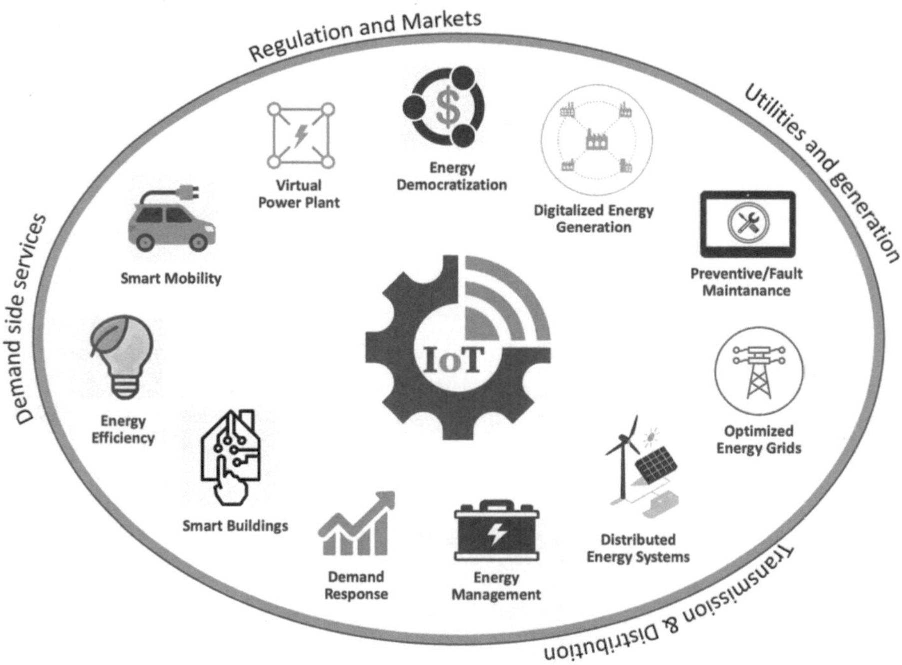

FIGURE 2.5 IoT Utilisation to Promote Energy Sector Efficiency (Motlagh et al., 2020).

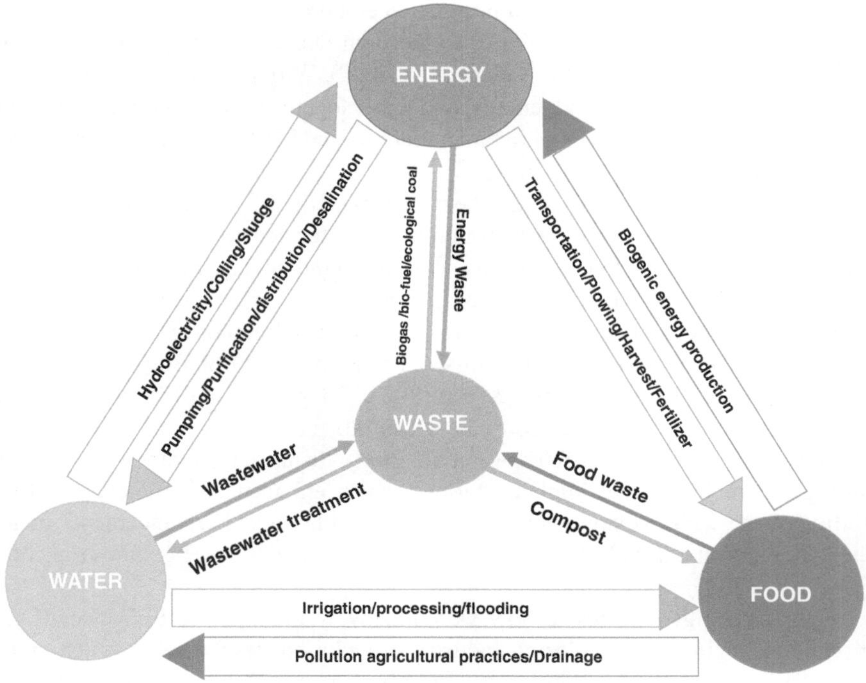

FIGURE 2.6 Management of Waste with the FEW Nexus 4.0 (Pemi et al., 2023).

The FEW Nexus, which involves the social framework needs, aims to understand how three basic resources – food, water, and energy – link together to meet our community needs. The evaluation explores multiple techniques that enhance sustainability and efficiency.

The resources connected through these innovations help strengthen systems and develop solutions that cover all three sector needs. This shows that problems resolved in one sector can also negatively impact the other sectors (Gain et al., 2015; Herrera-Franco et al., 2023). Through the FEW Nexus, we can see that the three sectors remain connected. The first connection demonstrates the specific quantity of water required for agricultural food production.

Agriculture is the largest water user, accounting for 85% of human water usage (D'Odorico et al., 2020) and 70% of global freshwater consumption (Ghosh et al., 2024). The second connection is between water and energy production, as water is not only for drinking and farming but is also crucial in energy production. In the energy sector, water is used to withdraw fuel from the ground, process it, and generate electricity (Bazilian et al., 2011; Chai et al., 2018; Shamoon et al., 2022). The third connection is like the second, as it is also between energy and water production. Just as water is required for energy production, energy is also required to pump water, circulate it, and, depending on the need, treat water for human consumption (Bazilian et al., 2011; Esmaeilion, 2020). Finally, the fourth connection is between energy and food production. Food production is a process that makes use of a lot of energy. From growing it to getting it to the final consumer, the process is energy-consuming (Ghosh et al., 2024; Naeem et al., 2022). With more people needing more of everything, looking at food, energy, and water together is not an option but a solution. Understanding and managing the interconnected FEW systems is quintessential to ensure sustainability and resilience.

Technology is critical to address the concerns of the FEW Nexus. Precision agriculture and smart water metres help maximise water consumption in food production. Renewable technologies, such as solar and wind energy, help in the long-term production of energy and are more sustainable.

Water management improvements result from desalination and water recycling procedures which minimise water loss (Abegaz et al., 2018; Bilgen et al., 2004; Hussain et al., 2017; Lalawmpuii & Rai, 2023; Odara et al., 2015; Sargentis et al., 2024). Industry 4.0, also referred to as 4IR or 4.0 (Frank et al., 2019; Ghobakhloo, 2020; Harahap & Rafika, 2020; Lasi et al., 2014), displays technology's importance through its technological implementations which enhance the efficiency and reliability of the FEW Nexus operations. These technological advancements include the IoT, blockchain, additive manufacturing (AM), and AI (Hassoun et al, 2023; Rane et al., 2023).

The 4IR brings changes to how we collectively manage the FEW Nexus. The discussion begins with exploring industrial revolution development which started with the first industrial revolution powered by steam and mechanisation. The second industrial revolution brought mass production and electrical power to industrial operations after steam-powered mechanisation, which was the first revolution, and automated technology, the third revolution. The present industrial revolution phase, known as Industry 4.0, implements advanced technologies such as the IoT, AI, and big data analytics, which perform better than those in the earlier industrial revolutionary stages.

Although this phase of the industrial revolution is still in its early stages, it can potentially transform the FEW Nexus by maximising the use of resources. Big data technologies process complex information in food production, thereby improving efficiency. AI-based systems reduce energy waste by maximising energy use in smart buildings, and IoT reduces water management costs through real-time monitoring and control of water quality. Applying these technologies improves the advantages of FEW Nexus 4.0 through the efficient use of resources. It aligns with circular economy concepts by minimising waste and reducing environmental effects. These technologies also support economic growth by creating employment opportunities and increasing sustainability in food production, energy supply, and water management.

2.2 THE EVOLUTION OF FEW NEXUS: HISTORICAL DEVELOPMENT OF THE FEW NEXUS CONCEPTS, FROM FEW NEXUS TO FEW NEXUS 4.0

The FEW Nexus was developed in the early 2000s when it was recognised that food, energy, and water could not be managed separately because changes in one sector affected the other sectors, and since then, it has improved dramatically (D'Odorico et al., 2018; Katz et al., 2020). The discussions initially revolved around the effects of energy generation on water resources and the consequences of water shortage on the agricultural output rate (Srigiri & Dombrowsky, 2022). Figure 2.7 shows some uses of water for energy; these include cooling thermoelectric power plants, generating hydropower, refining liquid fuel, and mining energy resources. Although the effects of one sector on the other were acknowledged, there was no discussion regarding strategies that would improve all three areas (Boas et al., 2016; Roggema et al., 2021). Figure 2.8 shows the social, economic, and environmental implications of water scarcity on agricultural productivity.

However, beginning in the 2010s, the FEW Nexus notion was acknowledged in academic study and policy-making circles (Simpson & Jewitt, 2019). Esteemed global forums, such as the World Economic Forum, highlighted the FEW Nexus as a crucial framework for tackling global sustainability issues (Basheer et al., 2018; Herrera-Franco et al., 2023). To assess the connections between the sectors, researchers developed various models and instruments (Cai et al., 2018b; Herrera-Franco et al., 2023; Kellner, 2023; Liu et al., 2018; Rasul, 2016; Rasul & Neupane, 2021). Figure 2.9 shows some of these models: FEW biophysical models (crop, soil, biomass, climate models), decision-making models (policy, agricultural land use, market

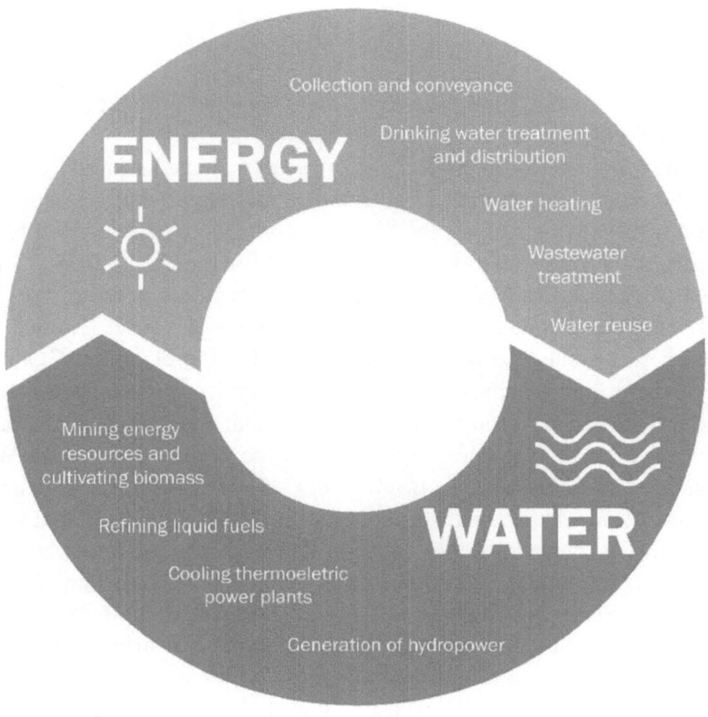

FIGURE 2.7 Impacts of Energy Production on Water (Fontenelle et al., 2022).

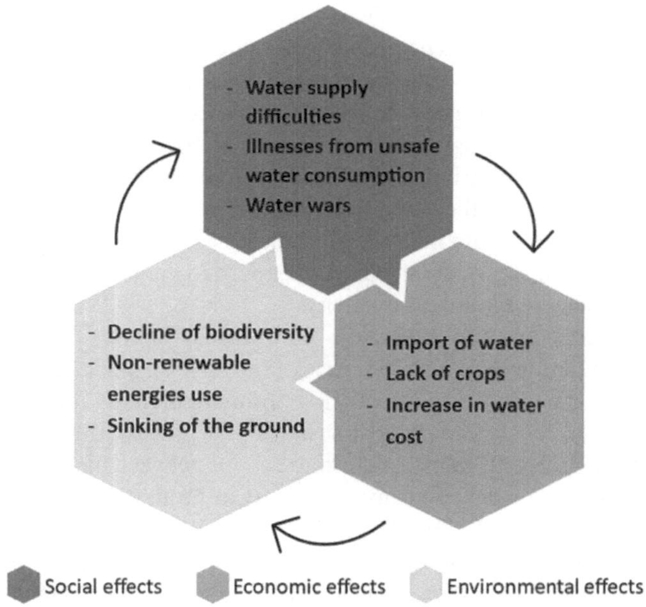

FIGURE 2.8 Implications of Water Scarcity on Agricultural Productivity (Ingrao et al., 2023).

signals), economic models (human system), and value and product models (value of food and fuel prices, value of ecosystem) (Nhamo et al., 2018).

Initially, the food, energy, and water sectors were managed separately until it was realised that the decisions made in one sector could significantly affect the other sectors. As a result of this insight, the "Nexus approach" was developed to manage resources across the three sectors (Ding et al., 2020; Han et al., 2021). "Nexus" is a term used to describe a network or series of connections between two or more things and was first introduced in the 1980s (Endo et al., 2017; Estoque, 2023b). This approach was created to address the Sustainable Development Goals (SDGs) and acknowledged that to achieve SDG objectives, collective management of food, energy, and water is crucial (Liu et al., 2018). Since that realisation, several advancements have been made. The most significant is the United Nations FEW Nexus partnerships (*Food, Energy-Water (FEW) NEXUS Partnerships | Department of Economic and Social Affairs*, n.d.; Kapucu et al., 2023). Due to global collaboration and examining the relationship between the food, energy, and water systems, advancements have increased to address the development challenges (Niet et al., 2021). The paper *Emerging Themes and Future Directions of Multi-sector Nexus Research* shows how the choices made in the energy sector affect water resources and how the integration of biofuels and bioenergy with carbon capture and storage affects the production of food and land use (Khan et al., 2022). Due to these discoveries, we better understand the FEW Nexus and make better decisions regarding SDGs. Figure 2.10 shows 17 United Nations SDGs.

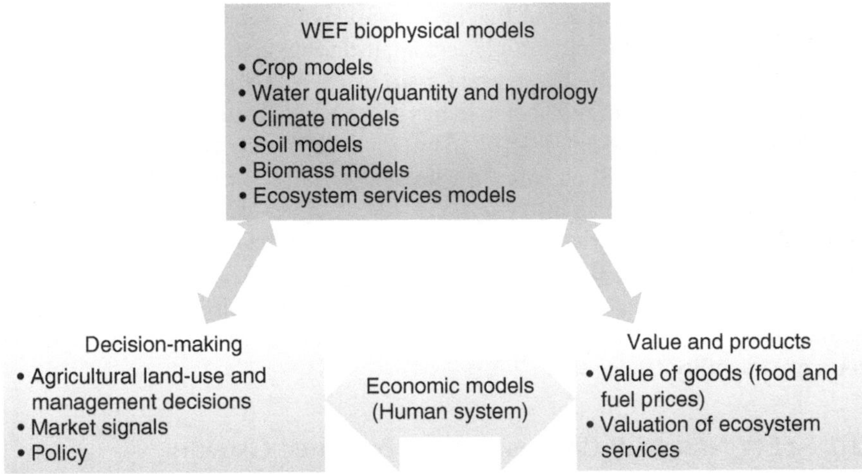

FIGURE 2.9 Models for Evaluating the Interconnectedness of the Food, Energy, and Water Sectors (Nhamo et al., 2018).

FIGURE 2.10 Sustainable Development Goals (Niet et al., 2021).

FEW Nexus evolved into FEW 4.0 by incorporating Industry 4.0 technologies, such as the IoT, advanced data analytics, and AI, among several other technologies. This led to a transformation in manufacturing processes and redefined industry and social value chains. An example of FEW 4.0 in action is the Phantom Edge, by Robert Bosch Engineering and Business Solutions Private Limited in May 2021, which leverages AI and IoT in multiple sectors, including manufacturing, healthcare, and agriculture (Do, 2022). The current state of FEW 4.0 is characterised by the increasing acceptance of digital technologies in the FEW Nexus. The global Industry 4.0 market size was valued at USD 114.55 billion in 2021 and, by 2029, it is projected to grow to USD 377.30 billion (*Industry 4.0 Market Size, Share | Growth Analysis [2022–2029]*, n.d.).Major industry players like Cognex Corporation, Schneider Electric SE, Emerson Electric Co., and many more are at the forefront, implementing innovative solutions such as digital twins and 3D printing to advance the technologies of Industry 4.0. Adopting Industry 4.0 technologies will further enable advanced manufacturing processes that can proactively respond to changes in demand and resource availability (Beier et al., 2021).

2.2.1 FEW Nexus 4.0 Contribution to Economic Growth

FEW Nexus 4.0 makes important contributions to economic growth and productivity; however, its technologies could impact the environment negatively (Chen et al., 2021; Li et al., 2020) and present some challenges, such as cybersecurity concerns and the need for a skilled workforce. These concerns are discussed in the following subsections.

2.2.2 FEW Nexus 4.0 Contribution to Productivity

FEW Nexus 4.0 enables the radical transformation of established techniques across FEW sectors to boost production levels subsequently. For example, the food sector battles three main challenges, including rising customer requirements, depleting resources, and the need to conserve the environment (Wang et al., 2021). FEW 4.0 creates solutions that solve these problems, resulting in productivity in the sector. Another example is that farmers can measure soil properties along with crop quality and meteorological data through precision agriculture methods that combine IoT tools and aerial vehicles with high-level measuring equipment in real time (Ajibola & Ogbolumani, 2024; RAkhter & Sofi, 2022). Likewise, deductive data management enables farmers to use water, fertiliser, and pesticides more accurately, generating higher yields at reduced operational costs (Dabić et al., 2023). Integrating automation and robotics in food processing leads to more efficient operations alongside uniform product results (Liberty et al., 2024). Computer systems perform sorting operations with packaging and quality assessment more quickly and precisely to decrease operational expenses and instances of human mistake. Through blockchain technology, users achieve supply chain optimisation by providing transparent food supply chain monitoring that reduces inventory waste and resource requirements for management (Perboli et al., 2018; Rejeb et al., 2019; Wang et al., 2019). This process optimises the distribution system to deliver improved conditions for products that move through supply chains to their endpoints (Bettiol et al., 2024; Calabrese & Falavigna, 2022).

AI algorithms fed with equipment data can detect equipment malfunctions before physical breakdown occurs. This proactive maintenance approach comes with the rewards of both time and long equipment durability and a reduction in maintenance expenses (Lee et al., 2020; Surucu et al., 2023). Through IoT-powered smart grids, customers can make real-time energy adjustments. At the same time, smart grids optimise network flow by minimising transmission waste. Digitalisation of real-time data has enabled the International Energy Agency to save 10% of their overall sectoral energy consumption (Abir et al., 2021; Rabie & Hassan, 2025). Organisation leaders are not left out; they now understand their energy patterns because advanced analytics allows them to develop sustainable energy-saving approaches. Organisations can maximise industrial production and reduce waste through streamlining across different industries (Bibri & Krogstie, 2020; Sanders et al., 2016). A combination of environmental monitoring and soil sensor data in the water sector produces automated irrigation technology, enabling exact watering at a specific time and producing enhanced water conservation and growing output (Adeleke et al., 2023; Gupta et al., 2021).

2.2.3 FEW 4.0 Contribution to Cybersecurity Concerns

Fourth industrial revolution technologies such as IoT and AI, together with large-scale data processing and automation, bring numerous benefits from Industry 4.0 yet introduce critical security threats to systems (Vaseashta, 2022). Several security weaknesses emerge from the FEW sectors' digital transformation process. The growing connection between IoT devices and systems creates many access routes that cyber attackers can exploit (Sasi et al., 2024). The interconnectedness of FEW systems means that a breach in a single area can cause a gradual decline throughout the entire nexus (Chen et al., 2021; D'Odorico et al., 2018). For example, a cyber attack on energy grids could disrupt water supply systems and food production, highlighting how vulnerable critical infrastructure is to digital threats (Sinha et al., 2023; T. Zhang et al., 2022). For example, the food industry uses Industry 4.0 technologies to enhance product quality by using live data, but this leaves security vulnerabilities if protective measures are not maintained. Numerous FEW sector infrastructures maintain outdated systems that were built without security concerns during development (Sasi et al., 2024). Modern technologies fail to detect all potential weaknesses when integrated with outdated systems (Farayola, 2024). The energy sector faces an increased cybersecurity threat because of new digital and sustainable technology transitions, which led to major incidents such as the attack on Colonial Pipeline in 2021(Saeed et al., 2024). The interconnected character of FEW sectors makes a single cyberattack against one component spread across the chain, leading to complete supply chain breakdowns. Attacks against fully connected ecosystems have the power to spread towards both customers and supply chain partners, thus enlarging the criminal advantage. Various data privacy issues emerge from the extensive data collection practices of Industry 4.0 because it incubates threats to consumer personal information together with corporate proprietary data (Aljohani & Almutairi, 2024; Cunico et al., 2024). To reduce these risks, robust cybersecurity measures must be implemented. This includes the development of secure communication protocols, regular security audits, and adopting advanced encryption technologies (David et al., 2022; Li, 2018).

2.2.4 FEW 4.0 Contribution to Skilled Workforce

Additionally, there is a need for live monitoring and response systems that can quickly identify and neutralise threats (Sundberg & Holmström, 2023). Another challenge is the growing demand for a highly skilled team to operate and maintain Industry 4.0 technologies within the FEW Nexus. Rapid technological progress requires skilled and adaptable workers. To meet this need, education and training programmes must be updated to focus on Industry 4.0 skills (Grybauskas et al., 2022; Velasco et al., 2024).

The strategy focuses on improving both STEM educational standards and vocational training alongside workforce skills retraining projects (Fernandez-Carames & Fraga-Lamas, 2018; Kılkış & Kılkış, 2017).

Workers need retraining through programmes established by Fernandez-Carames and Fraga-Lamas (2018) and Kılkış and Kılkış (2017). Industry, together with governmental bodies and academic institutions, must collaborate to develop educational programmes that correspond with industry needs.

The following part examines the developmental aspects of the FEW Nexus concept. It traces its evolution from an initial recognition of resource interconnectedness to the more advanced FEW 4.0 framework. The initial conceptual frameworks demonstrated recognition of connection points between sectors, but they did not feature comprehensive management approaches. During the 2010s, the research community, together with governments, emphasised the study of the FEW Nexus more frequently.

Global platforms have made FEW Nexus a central focus of their sustainability initiatives in relation to policy-making. The environmental management field adopted the usage of "nexus" as a term during the 1980s. The FEW Nexus framework exists as an infrastructure designed to back the SDGs by integrating food, energy, and water resource management. Recent technological advancements, particularly the incorporation of Industry 4.0 technologies like IoT and AI, have led to the evolution of this framework into FEW Nexus 4.0.

2.3 DISRUPTIVE CHANGES IN THE FEW NEXUS, DRIVEN BY INDUSTRY 4.0

The Internet of things, cloud computing, augmented reality, autonomous robots, simulation, cybersecurity, horizontal & vertical system integration, Additive Manufacturing and big data analytics when it comes to the high-tech advancements brought about by Industry 4.0, these nine technologies are considered its pillars. Figure 2.11 shows these nine pillars of Industry 4.0 (Alcácer & Cruz-Machado, 2019; Cohen et al., 2019; Pozzi et al., 2023). Figure 2.12 shows some enabling technologies of Industry 4.0; these technologies include industrial IoT, IoT, cyber-physical systems, simulation, cloud computing, augmented reality, cybersecurity, Additive Manufacturing Radio frequency identification (RFID), machine learning, AI, Big data, and blockchain. All these help in driving Industry 4.0.

The fourth industrial revolution, synonymous with smart manufacturing, represents the digitalisation of manufacturing and industrial practices, characterised by real-time decision-making, enhanced productivity, flexibility, and agility. This led to a revolutionary approach to how companies manufacture, improve, and distribute products – a "Smart Industry" (Morgan et al., 2021; Shao et al., 2021). Figure 2.13 shows a smart industry with its components, which include smart transport and

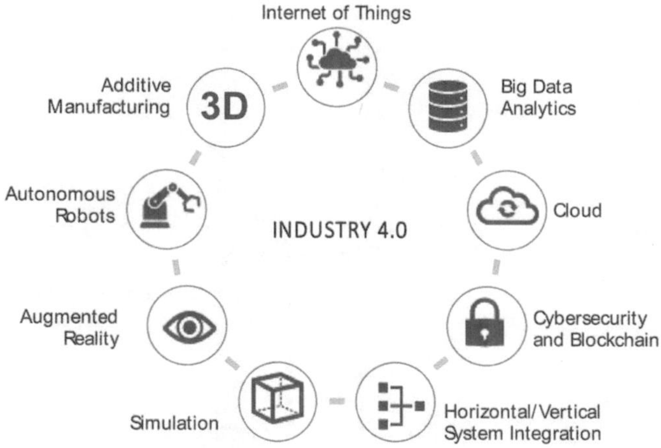

FIGURE 2.11 Pillars of Industry 4.0 (Cagno et al., 2021).

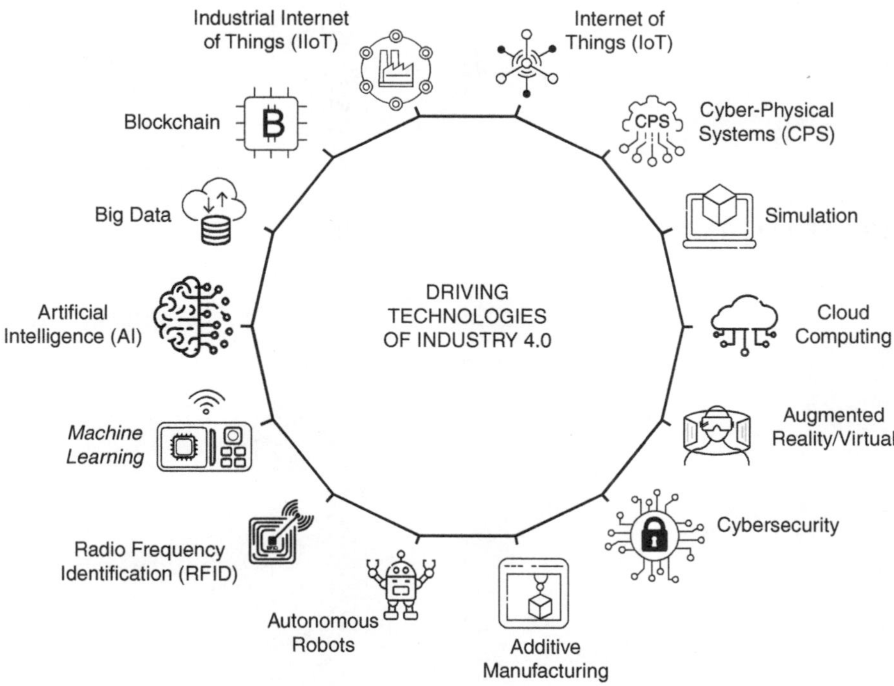

FIGURE 2.12 Enabling Technologies of Industry 4.0 (Filho et al., 2024).

logistics, smart industry servers, and Gateways. The technologies in the smart industry and those in the smart factory include smart IoT sensors for supply chain management, real-time monitoring, and cloud services. As mentioned, the connected factory, which uses advanced technologies to create an efficient production environment, plays a vital role in this revolution. A typical example of a smart factory is shown in Figure 2.14.

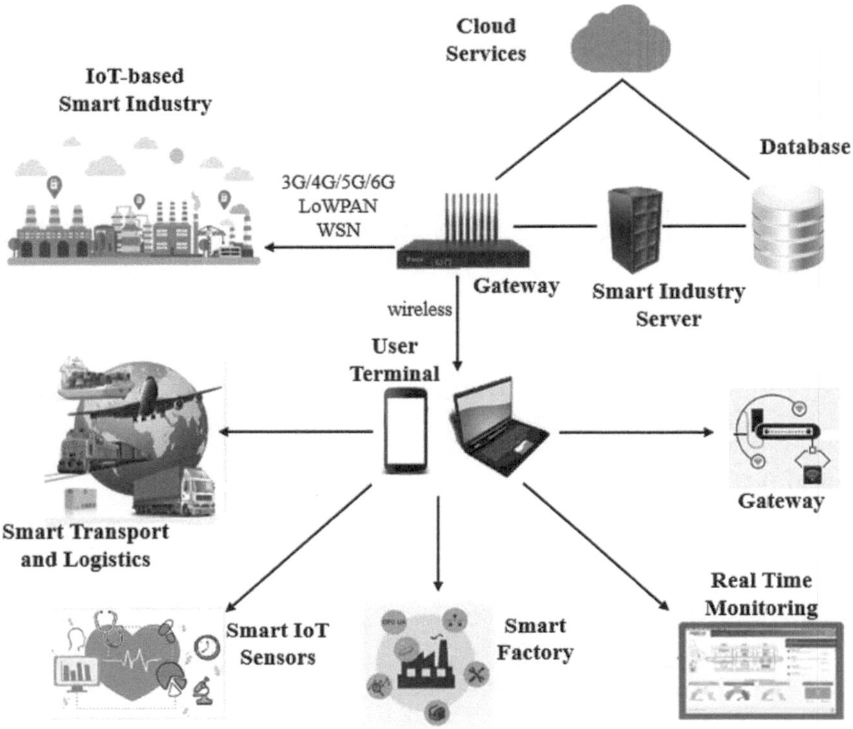

FIGURE 2.13 The Smart Industry (Farooq et al., 2023).

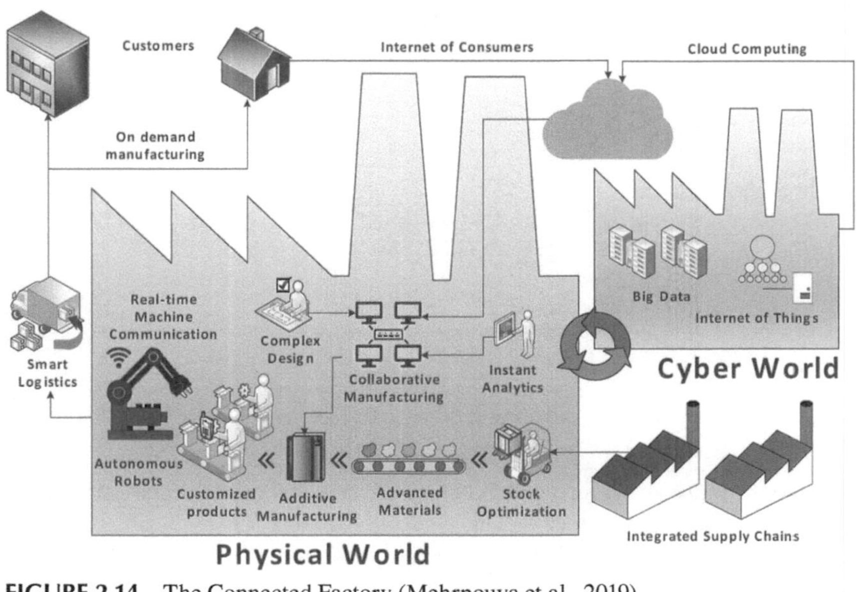

FIGURE 2.14 The Connected Factory (Mehrpouya et al., 2019).

2.4 USE CASES OF INDUSTRY 4.0-BASED PRODUCTION COMPANIES

Several companies have begun to gain significant value from implementing Industry 4.0 technologies (Rüßmann et al., 2015). For example, leading manufacturers have reduced machine downtime by 30%–50%, increased production by 10%–30%, and improved labour productivity by 15%–30% (S. Sharma et al., 2023). Renault's digital transformation journey began in 2015 and is a prime example of leveraging Industry 4.0 to enhance revenue, productivity, and customer relationship management, with an impact of approximately 300M euros per year (Yüksel, 2022). Some other examples of companies leveraging Industry 4.0 include BJC Healthcare, Alibaba, and GE Aerospace.

- **BJC Healthcare**: BJC Healthcare is a healthcare service provider that uses an integrated inventory management solution that leverages IoT to achieve cost savings in its supply chain. This company uses RFID technology (Dastres et al., 2022; B. Sarkar et al., 2022) (which uses radio waves to read and capture information stored on a tag attached to an object, such as medical supplies) to track and manage thousands of medical supplies. Since implementing the technology, BJC has reduced the stock kept onsite at each facility by 23%. Figure 2.15 shows an example of an RFID system. It consists of a reader, also called a transceiver, which is responsible for communicating with a computer system via a software application in charge of

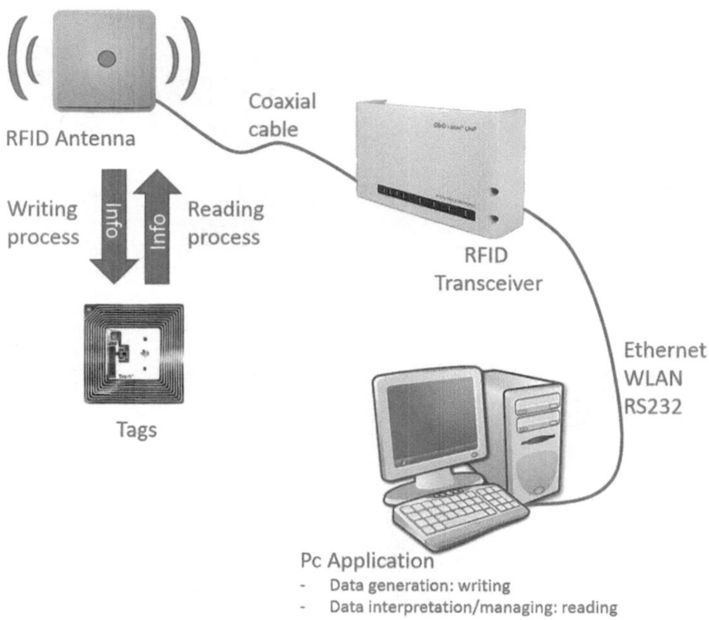

FIGURE 2.15 A Radio Frequency Identification (RFID) System (Valero et al., 2015).

managing data stored in the transponders. The reader is connected to the antenna, and some transponders (tags) are responsible for storing information. The antenna establishes a link between the reader and the tags (Valero et al., 2015).

- **Alibaba**: Alibaba's largest "smart warehouse" in China is managed by 60 high-tech robots. These self-maintaining robots are responsible for moving goods within the warehouse. They send the goods to manual labourers, who arrange for the items to be packed and posted to customers in their various locations globally. Through WIFI, the robots are given instructions and equipped with laser detection to ensure they do not cause harm to each other. After being powered to the maximum, they can work an entire eight-hour shift without a break, and as stated by Quicktron, their manufacturer, these robots have tripled the warehouse output in just one month. Doing this increases efficiency and reduces human labour (Huang et al., 2023). Figure 2.16 shows a digital twin model for a smart warehouse in a small-scale industry composed of IoT devices, an integrated management system, a real-time electronic tracking system, a smart logistics and tracking system, a smart perception system, and a smart delivery system (Tang, Ho, et al., 2022).
- **GE Aerospace**: GE Aerospace uses cloud-based analytics to monitor aircraft engines, predict maintenance needs, and optimise fuel efficiency. General Electric (GE) Aviation is also investing in data analytics to help determine how it makes engines and supports its customers (see Figure 2.17). In 2017, GE Aviation formed a strategic partnership with Teradata, a data and

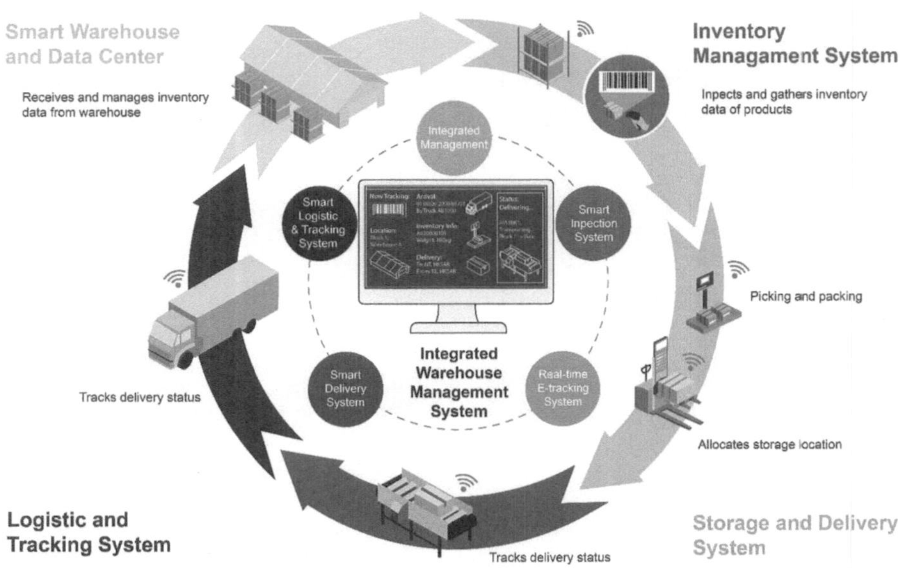

FIGURE 2.16 A Digital Twin Model of a Smart Warehouse Management System (Tang, Ho, et al., 2022).

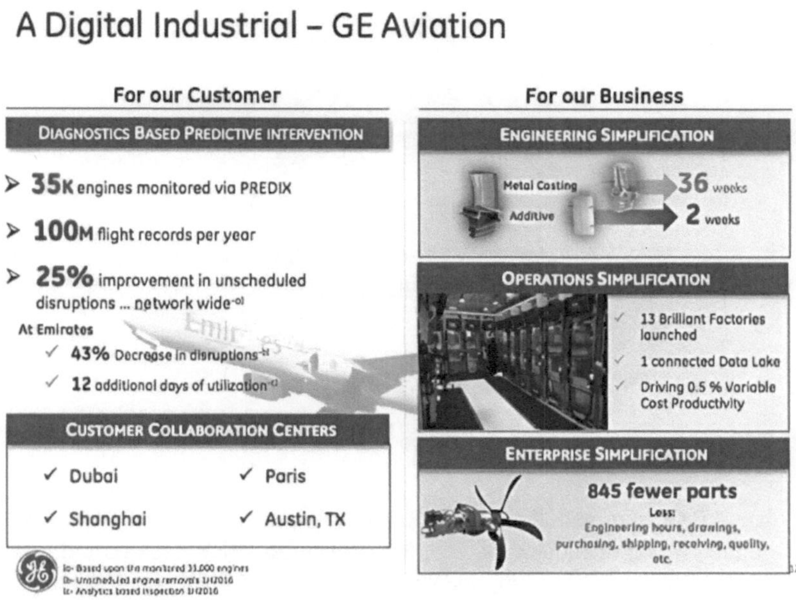

FIGURE 2.17 GE Aerospace ('Digital takeover', n.d.).

analytics company that uses analytics to improve the airline business. GE has also opened the GE Digital Centre for big data to give the company's different business units' access to 1,100 data scientists tasked with "marrying physics and analytics" to combine product knowledge with new digital capabilities. This enables GE Aviation to provide "personal healthcare" for the 35,000 aero engines currently flying worldwide (and also for a wide variety of other GE big data applications, including railways and street lighting). By using Predix, GE Aviation is now able to offer airlines and aircraft lessors predictive engine care and maintenance for their aircraft fleets, individual engines, and even individual components (Bouzidi, Terrissa, Zerhouni, & Ayad, 2020; 'Digital takeover', n.d.; Stanton, Munir, Ikram, & El-Bakry, 2023).

2.5 THE BUILDING BLOCKS OF FEW NEXUS 4.0

The Fourth Industrial Revolution, which is known as the 4IR, is changing production, organisational methods, and economic directions for all fields of human endeavours, making it necessary to examine the level to which its technologies affect the food, energy, and water sectors (David et al., 2022). The concept of FEW Nexus emerged when the interconnectedness of the three sectors was realised. Figure 2.18 illustrates this interconnected nature of food, energy and water systems. This necessitated addressing the issues in each sector together, rather than individually, and this is where Industry 4.0 came in with its technologies, giving way to a better and more sustainable way of problem-solving in these sectors, and leading to the concept of

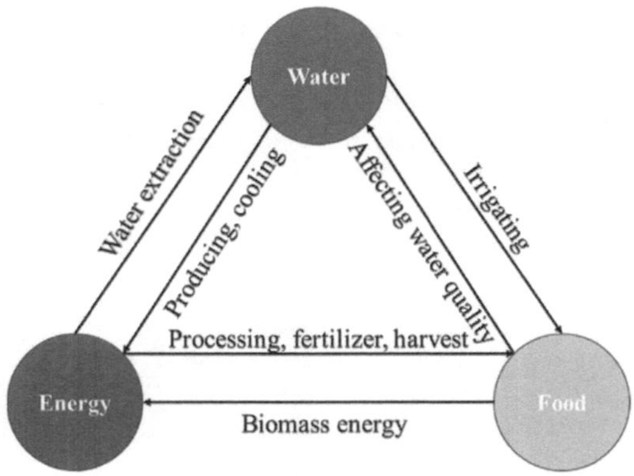

FIGURE 2.18 The Interconnectedness of the Food, Energy, and Water Sectors (Yu et al., 2021).

FEW Nexus 4.0 (Alhammadi et al., 2024; Grybauskas et al., 2022) FEW Nexus 4.0 is simply the integration of Industry 4.0 technologies in the FEW Nexus, and this concept is made up of three key components namely: Food 4.0, Energy 4.0, and Water 4.0. The following sections further explain these three components.

2.5.1 Food 4.0

The food sector is considered one of the costliest sectors for economic growth globally. This sector is responsible for a massive proportion of the total manufacturing value added in most countries, supplies the highest executives, and carries meaningful weight in the gross domestic product of almost all nations (Mauricio-Moreno, Miranda, Chavarria, Ramirez-Cadena, & Molina, 2015; Miranda et al., 2019; Weichhart, Molina, Chen, Whitman, & Vernadat, 2016). Food is one of the most essential resources for humans worldwide. Figure 2.19 shows a trend of global undernourishment between 2001 and 2022. There is a projection that the number of undernourished people will reach 840 million by 2030 (Wudil et al., 2022). Based on a United Nations estimate, the world population of 7.6 billion will increase to 8.6 billion in 2030, 9.8 billion in 2050, and 11.2 billion in 2100 (United Nations, n.d.). Regardless of whether fertility rates keep dropping, the increase in population is expected to continue, with about 83 million people added to the world population yearly (Watkins, 2006). There is a need to meet the increasing food demand for the growing population, and conventional crop management methods are inadequate to meet the soaring demands (Kadam et al., 2023).

In addition to issues related to the provision of food for the growing population, this sector also faces many problems that are obstacles to the sustainability of food in the future and the security of the Earth. These problems include, but are not limited to, climate change, food security control, environmental protection, increased levels of food waste and loss, and the possibility of new disease or pandemic outbreaks that threaten humanity (Hassoun et al., 2023; Miranda et al., 2019). These challenges need to be addressed urgently. For this reason, the food sector has taken advantage of Industry 4.0 technologies to solve these issues. Figure 2.20 shows some of these technologies, which

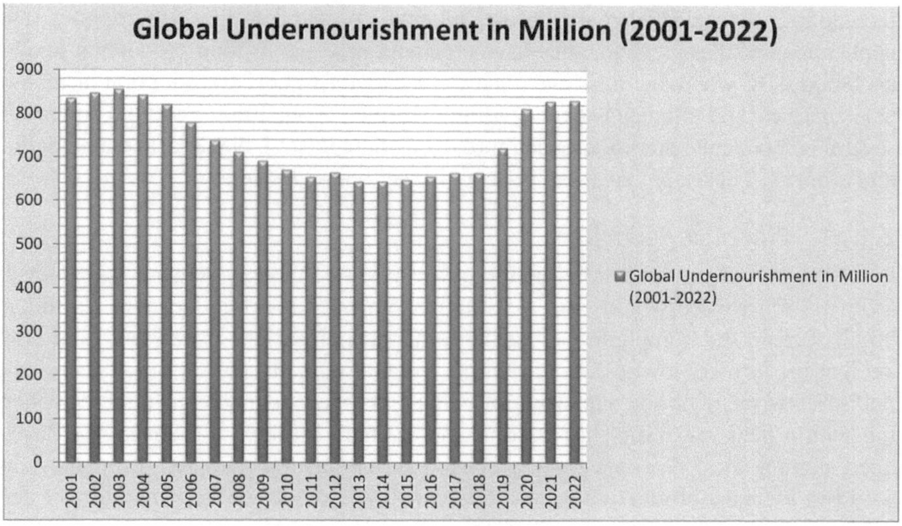

FIGURE 2.19 The Global Undernourishment in Millions (Hays & Fagan, 2016).

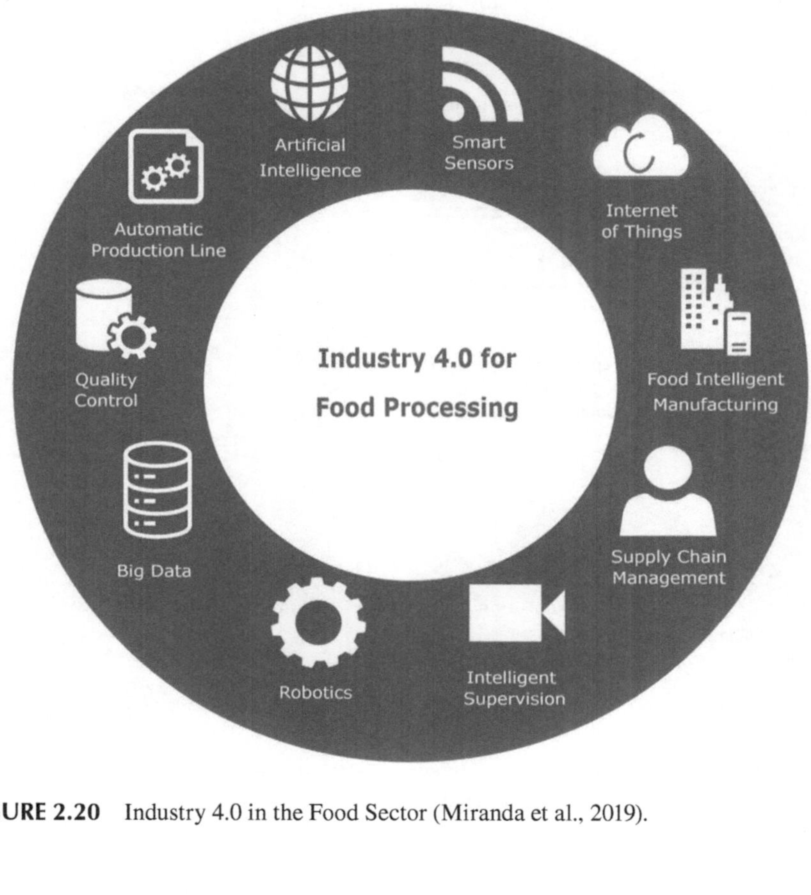

FIGURE 2.20 Industry 4.0 in the Food Sector (Miranda et al., 2019).

include IoT, AI, big data, smart sensors, automatic production lines, and robotics. The applications of these technologies have created a new age of food production known as "Food 4.0," where automation, connectivity, digitalisation, the adoption of renewable energies, and efficient use are leading (Miranda et al., 2019). Food 4.0 comprises several components: precision agriculture, biotechnology, AI, big data, and automation and robotics. These are explained in detail in the following section.

2.5.1.1 Precision Agriculture

Precision agriculture is defined as the use of high-technology sensors and analysis tools to improve crop yields and assist management in making decisions (Lynham, 2022). Before the completion of agricultural mechanisation, it was easy and convenient for farmers to consider variability manually due to the minimal size of the fields. However, with the expansion of fields and vigorous mechanisation, considering within-field variability has become increasingly difficult without highly developed technologies. The concept of precision agriculture was adopted throughout the world to increase productivity, reduce labour time, and effectively manage the fertilisers and irrigation processes (Brisco, Brown, Hirose, Mc Naim, & Staenz, 1998; Pierce & Nowak, 1999; Robert, 2002; Stafford, 2000; Zhang, Wang, & Wang, 2002). Precision agriculture uses several advanced technologies to help farmers enhance the productivity of their resources in their fields and support them in managing their businesses. These include IoT sensors and drones (Ogbolumani & Mabaso,

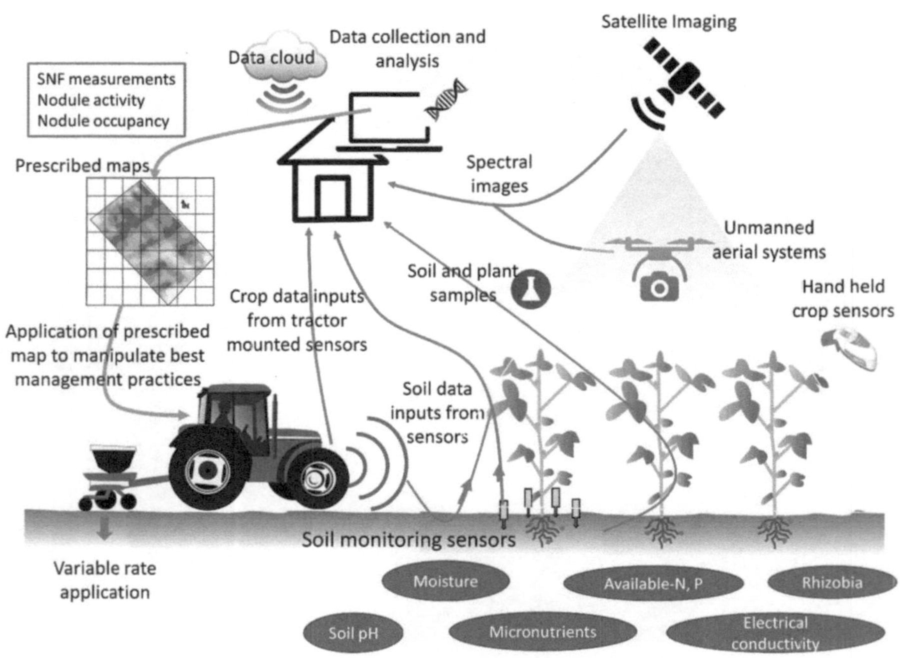

FIGURE 2.21 A Conceptual Model of Integrated Precision Agriculture Tools for Optimised Food Production (Thilakarathna & Raizada, 2018).

2023). Figure 2.21 shows a conceptual model of different precision agricultural tools like handheld sensors, soil monitoring sensors, satellite imaging, cloud servers, and mounted sensors for crop data are combined for optimised food production (Thilakarathna & Raizada, 2018)

2.5.1.2 IoT Sensors

The IoT is an innovative technology that creates an advanced environment around us with its systems that use trillions of sensor networks (Lee, Kim, Lee, & Kim, 2018). It is a connected network system that enables applications such as efficient irrigation for agriculture. Irrigation is a key process in agriculture. When done improperly, it results in water wastage, but proper irrigation can be done using the wireless sensor network (WSN) technology, a technology within IoT (Mat, Mohd Kassim, Harun, & Mat Yusoff, 2017; Sidhu, Kumar, Rana, & Jat, 2021) Figure 2.22 illustrates the use of IoT sensors to monitor and report in real time on some physical quantities within the system including humidity, temperature, wind speed and amount of rainfall (Machado et al., 2022). The WSN technology has improved monitoring and control applications. It has enabled efficient communication with numerous sensors and has tremendously enhanced monitoring and control applications. Figure 2.23 shows a typical WSN sensing for farming. This system comprises the user, sensor nodes, internet, base station, and farm sensing region. This application of WSN in agriculture has contributed to food security in many areas of the world.

The wireless moisture sensor network, another IoT technology, is a WSN with moisture sensors. The moisture sensors measure the soil's water level, enabling farmers to use the right water level for different soils as determined by the agronomist. It is vital because each type of crop requires different levels of water. Excess water in the soil damages the roots' nitrogen use and, simultaneously, makes the oxygen for the roots insufficient (Thompson, Gallardo, Valdez, & Fernández, 2007).

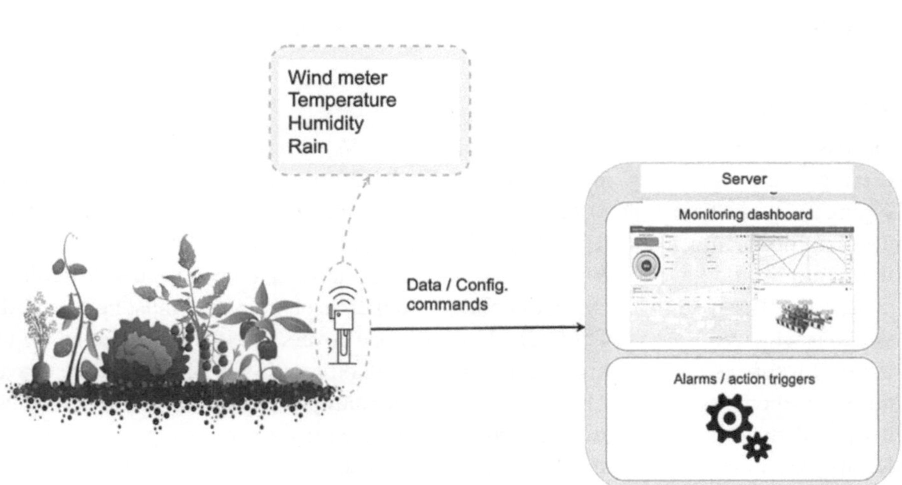

FIGURE 2.22 IoT Sensors (Machado et al., 2022).

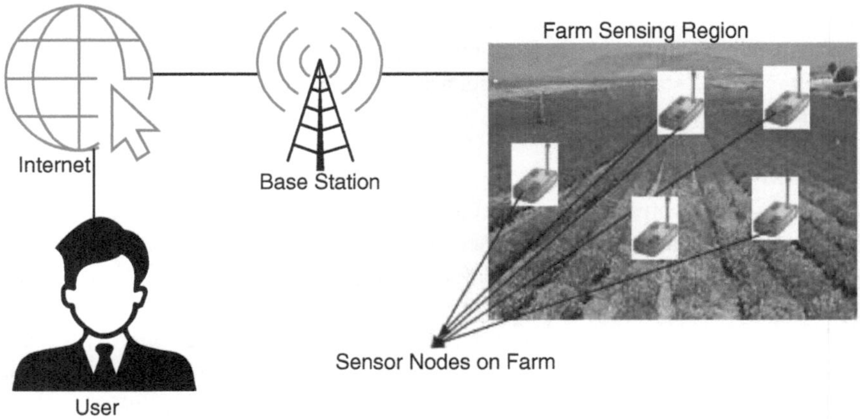

FIGURE 2.23 A Typical Wireless Sensor Network Technology for Farming (Ndjuluwa et al., 2023).

FIGURE 2.24 Use of Drones in Agri-Food Sector (Ivezić et al., 2023).

2.5.1.3 Drones

Drones, also called unmanned aerial vehicles (UAVs), are radio-controlled airplanes used as data collectors in precision agriculture to support many applications. Figure 2.24 shows a typical drone used for spraying insecticides on a farm where accessibility is minimal (Ivezić et al., 2023). Initially, if a farmer wanted to scout the field, they would have to do it on foot, but this is ineffective because it's difficult to cover the entire field. As farm sizes continue to increase, mole land must be watched, which is why UAVs are very efficient. Once a drone covers a field, it becomes easier to identify trouble spots and target them for scouting on foot (Stehr, 2015). Drones have numerous benefits and are of various types. For example, Figure 2.25 shows various uses of drones in agriculture, such as harvesting, spraying, mapping, and sensing. Drones are among the most crucial technologies to maximise the efficiency of the application of crop protection chemicals by reducing the requirement for manpower,

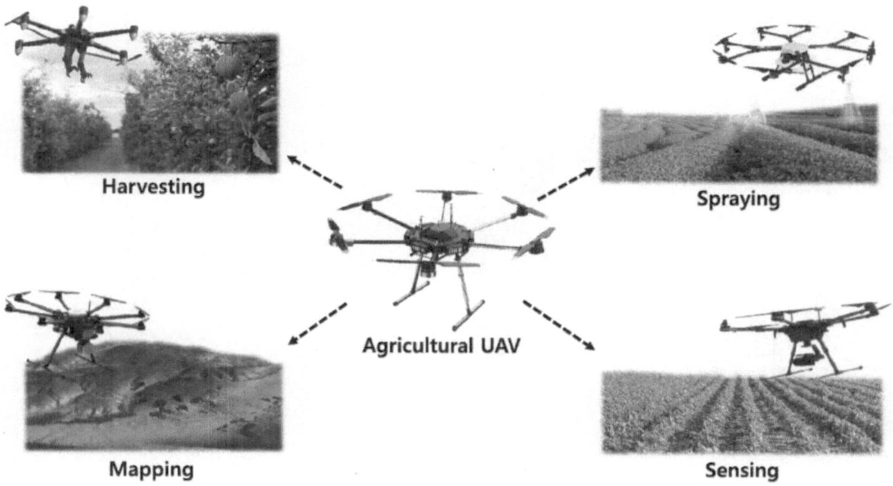

FIGURE 2.25 Types of Drones in Agri-Food Sector (Islam et al., 2021).

application time, water volume, amount of chemical usage, human exposure to health risks (Askerbekov et al., 2024; Das et al., 2024).

2.5.1.4 Biotechnology

According to Investopedia, biotechnology is a science-driven industry sector that uses living organisms and their derivatives to produce products and processes. As the global population continues to rise, and the probability of reduced yields of major food crops due to the changing climate increases, the development of genetically improved, stress-resilient crops has become a necessity (Steinwand & Ronald, 2020) Biotechnology in agriculture is simply the implementation of scientific techniques and apparatus to genetically modify plants and animals to improve food development and nutritional quality of food (Kumar et al., 2020). Figure 2.26 shows the various areas where plant biotechnology has contributed to crop improvement for resilient food systems and has also indirectly helped preserve the environment. These include soil biodiversity, crop improvement, water preservation, restoration of degraded soils, and net-zero emission in global food systems (Pathirana & Carimi, 2024).

Plant diseases and pests remain among the biggest challenges in agriculture despite all efforts to thwart them. Biotechnology helps to manage them through genetic modification and gene editing, i.e., by developing crops with greater resistance to diseases, pests, and harsh weather conditions (Kumar et al., 2020). These approaches expand the genetic tools available to crop breeders for combatting these plant pathogens (van Esse et al., 2020). There are two methods of genetic modification as shown in Figure 2.27. These are gene addition and gene silencing. Gene addition involves the provision of new traits to the plant, while gene silencing suppresses specific genes by blocking the expression of unwanted characteristics in the plant (Rodríguez et al., 2022).

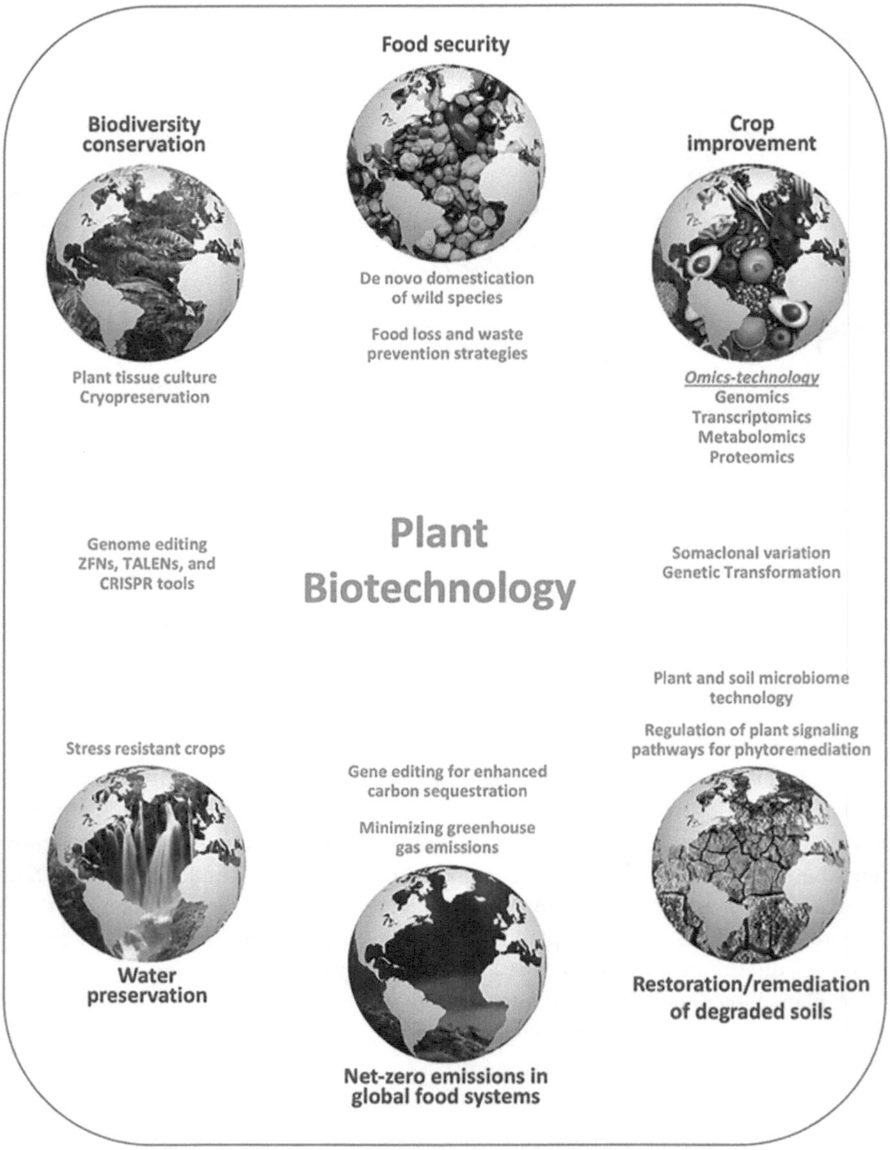

FIGURE 2.26 Applications of Biotechnology in Crop Improvement and Environmental Protection (Pathirana & Carimi, 2024).

2.5.1.5 Automation and Robotics

Robotics and automation systems help tackle improved production challenges with quality, safety, and security (Javaid et al., 2021). The robots used in the food industry specialise in handling tasks like loading, unloading, picking, placing, assembling,

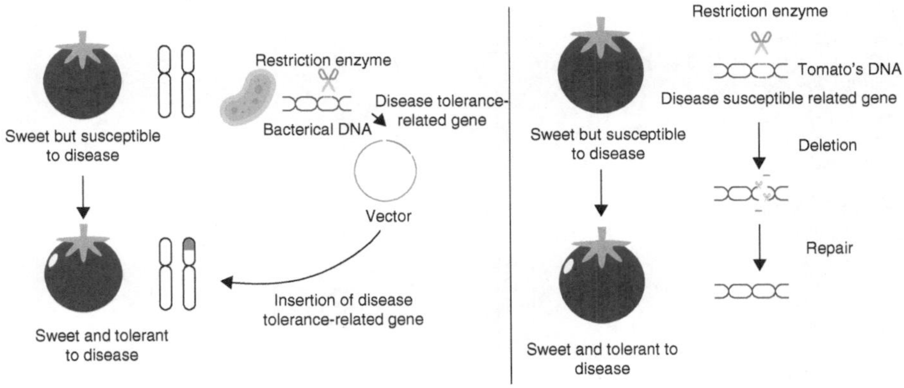

FIGURE 2.27 Genetic Modification of Food (Rodríguez et al., 2022).

palletising, and sorting at very high speeds (Javaid et al., 2021; Panda, Panigrahi, Mishra, & Kumar, 2023). They ensure that food safety and hygiene requirements are met, improving resource efficiency, simplifying maintenance, and preventing human injuries. The technology that is probably the most relevant in robotic applications is gripper technology. The gripper is a subsystem of equipment that meets an object and grips it before releasing it where needed. The gripper does not leave any visible marks on the items after gripping, thus meeting the hygiene standards and solving the problem of items not being delivered in optimal condition, enhancing customer satisfaction with the delivery process (Gulnergiz, Dilibal, Gormus, Danquah, & Emon, 2023; Hasnan, Yusoff, Lim, & Kamarudin, 2023; Zaidi, Maselli, Laschi, & Cianchetti, 2021) Figure 2.28 shows a piece of gripper equipment used in the food industry (Blanes et al., 2014)

2.6 ENERGY 4.0

The energy sector has reached a critical point. The shift to renewable energy is happening so quickly because we need to reduce pollution and stand against climate change. Still, our existing power systems were not designed for renewable energy, which is unpredictable and comes from different sources, making it harder to manage (Dey, Sreenivasulu, Veerendra, Rao, & Babu, 2022). Additionally, the management and distribution of electricity have become even more complicated due to the generation of energy from many local sources, such as rooftop solar panels, rather than a few large power plants (Gassar & Cha, 2021; Lakshmi, Rubanenko, Divya, & Lavanya, 2020; Olorunfemi et al., 2022). There is also an increasing shortage of labour that makes it hard to maintain and upgrade the energy infrastructures, along with the frequent and severe weather events likely caused by climate change (Jasiūnas, Lund, & Mikkola, 2021; Olorunfemi et al., 2023) All of these challenges are taking a toll on the ageing infrastructure leading to more failures and the need for constant costly repairs and upgrades. New technologies are

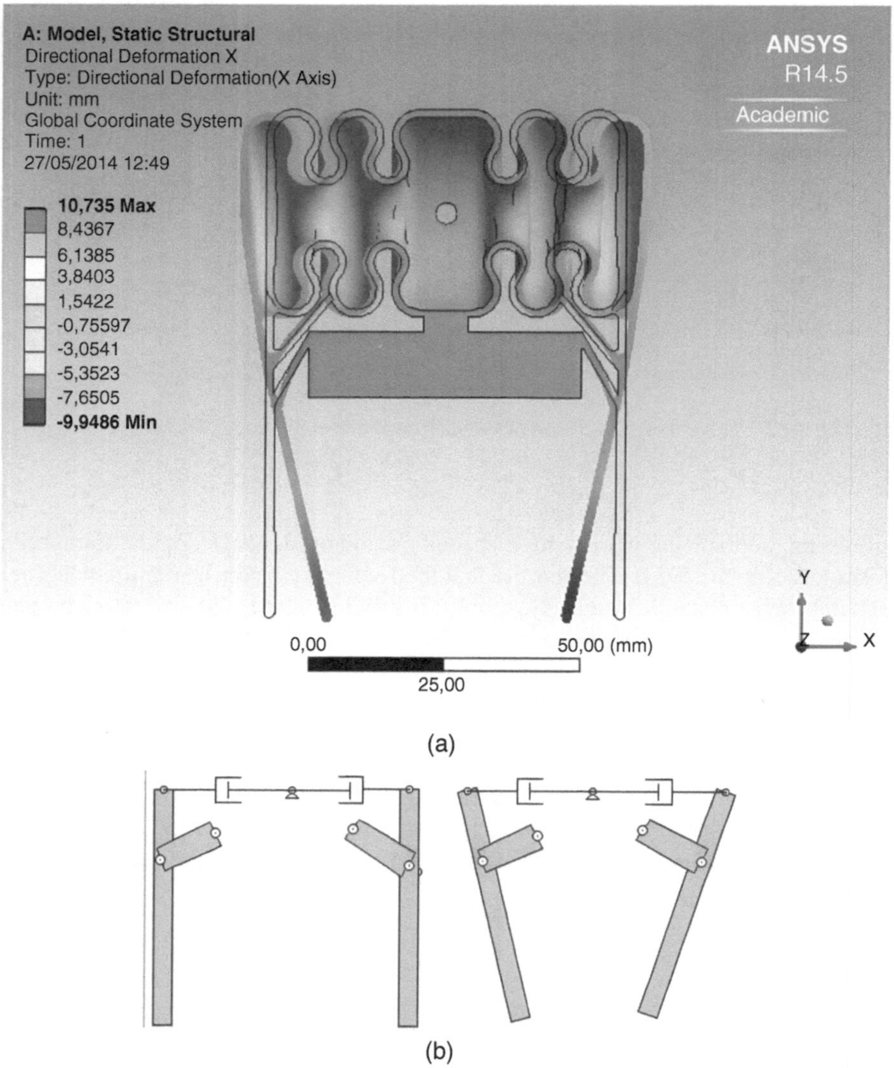

FIGURE 2.28 Gripper Equipment Used in the Food Industry (Blanes et al., 2014).

urgently needed to change grid conditions, optimise performance, and mitigate the risk of outages (Jasiūnas et al., 2021). Energy 4.0 is a concept related to a collection of utility hardware, software, and technologies that capitalise on Industry 4.0 technologies to increase efficiency, reliability, and sustainability in the energy sector. Figure 2.29 shows some of the technologies of Industry 4.0 with some useful in Energy 4.0. Some of its key technologies include cyber security, cloud computing, AI and the IoT (Khalid, 2024).

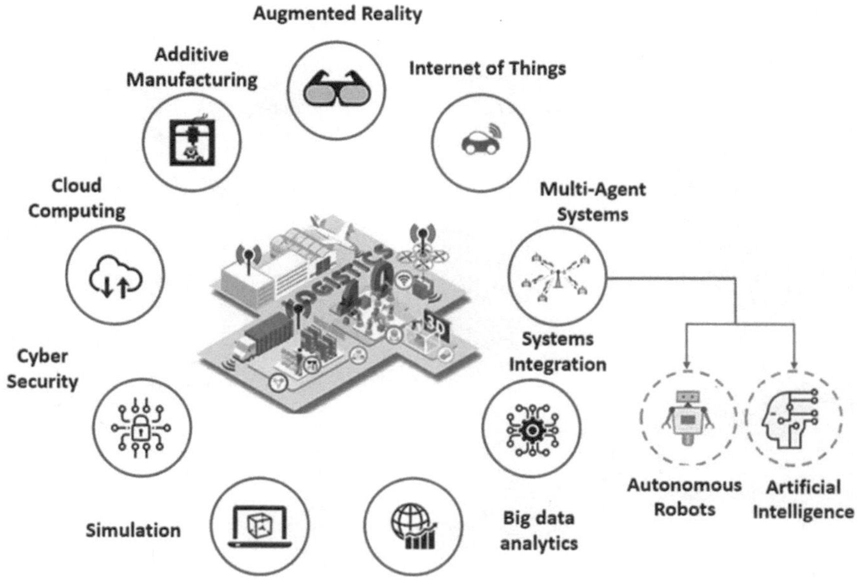

FIGURE 2.29 Industry 4.0 in the Energy Sector (Benotsmane et al., 2019).

2.6.1 Smart Grid System

Intelligent electrical power systems that use high-tech sensors, regulation systems, and communication technologies to improve the efficiency, reliability, and sustainability of the power grid are known as smart grids (Khalid, 2024). In simple terms, it is an electrical power system that makes use of improved technologies to observe and oversee the transportation of electricity from various energy sources to fulfil different user demands and allow them to participate in the optimisation of the operation of the system (Dileep, 2020; Majeed Butt, Zulqarnain, & Majeed Butt, 2021) It improves the reliability of the power grid by cutting down on quality disturbances and reduces the possibility of widespread blackouts (Khaleel, Abulifa, & Abulifa, 2023); reduces the prices that customers pay for electricity and promotes affordability, and reduces the loss and waste of energy (Ahmad & Zhang, 2021) Most importantly, it reduces environmental pollution through the decline of greenhouse gases and particulate matter, and facilitates cleaner power through the use of more reuseable distributed energy resources (DERs) (Adedeji, Ponnle, Abu-Mahfouz, & Kurien, 2022; Anjimoon et al., 2024; Kabeyi & Olanrewaju, 2023; Caballero-Peña, Cadena-Zarate, Parrado-Duque, & Osma-Pinto, 2022). A representation of the smart grid system is shown in Figure 2.30 (Onu et al., 2023).

2.6.2 Internet of Things in Energy

The IoT is a chain of interconnected systems that operate to perform specific works by enabling data transfer, analytics, and decision-making, thereby increasing

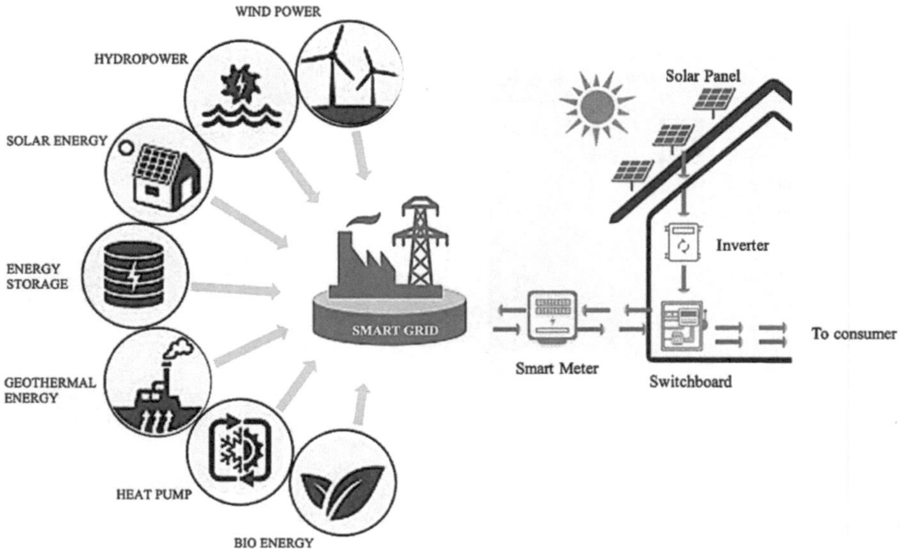

FIGURE 2.30 A Smart Grid System (Onu et al., 2023).

productivity and efficiency (Mohindru et al., 2020). IoT is increasingly used in the energy sector and is applied in various ways such as smart industry, smart harvesting, smart building, smart city, smart grid, and smart home. Figure 2.31 shows some applications of IoT in energy sectors (Motlagh et al., 2020). Some of these applications are discussed as follows:

- **Smart Industry –** This is also known as Industry 4.0, and it involves making use of improved systems like the IoT, AI, big data analytics, and automation to improve manufacturing and industrial operations. In the energy sector, it is employed for real-time monitoring, enhancing output, and effectively allocating resources (Li et al., 2021; Lu et al., 2022; Tang & Meng, 2020).
- **Smart Harvesting –** This is the use of advanced technologies such as IoT in agriculture to increase productivity, efficiency, and output yield from crops. It involves making use of sensors, drones, and automated machines to monitor the conditions of the crop, and predict the best harvest time to prevent waste and cost of labour while improving the yield (Sadowski & Spachos, 2020; Shah et al., 2021).
- **Smart Building –** Smart buildings were created by the Energy Performance Building Directive to improve energy flexibility, renewable energy production, and user interaction (Al Dakheel et al., 2020). Though several definitions exist, its concepts and features are still unclearly defined (Kumar et al., 2021; Yu et al., 2021). They are structures equipped with advanced systems

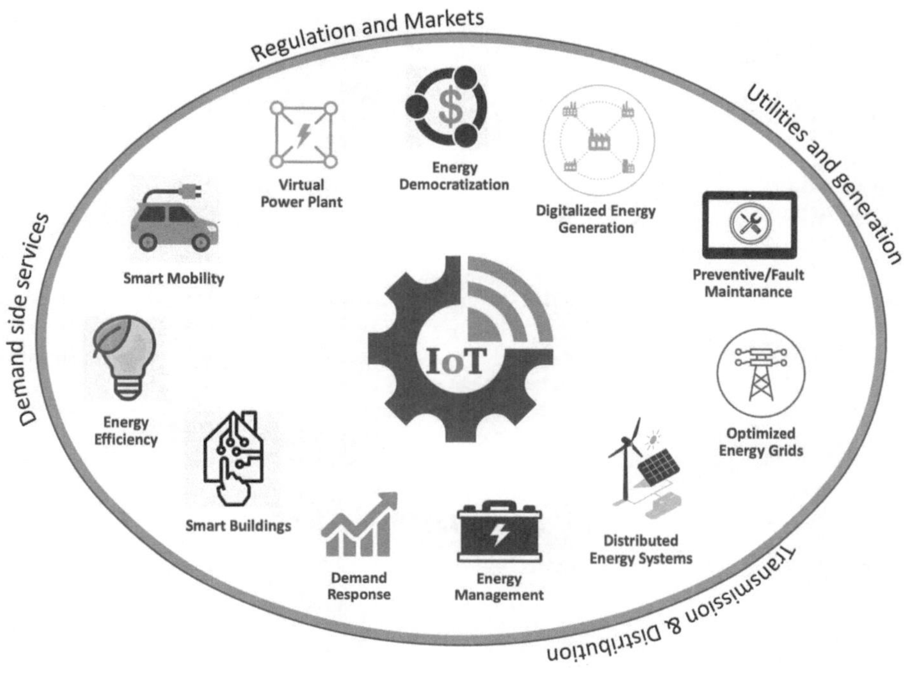

FIGURE 2.31 Ways in which IoT is used in the Energy Sector (Motlagh et al., 2020).

and technologies to automate building operations, including smart lighting, HVAC (heating, ventilation, and air conditioning) systems, security systems, and energy management (Himeur et al., 2023).

- **Smart City –** This refers to a technologically modern urban area that adopts and applies different types of electronic methods, sensing devices, and data processing systems to collect specific and advanced data used to improve the overall quality of life of the residents. These data management systems provide improvements in different aspects of operations and organisations, some of which include traffic regulation, sustainable resource use, quality of life, and smart city infrastructure development (Kirimtat, et al., 2020; Laufs et al., 2020; Sharif & Pokharel, 2022)
- **Smart Grid –** These electrical power systems incorporate digital technologies, sensors, and software to meet power supply and demand efficiently. At the same time, they minimise expenses and maintain the durability and reliability of the grid (Ahmed Abdulkadir & Al-Turjman, 2021).
- **Smart Home –** This is one of the major applications of IoT. It is the use of sensors, actuation technologies, and WSNs to connect various smart appliances inside homes to quickly resolve the interests of the occupants. Sensors in smart homes are used to enable smart automated services and ensure security and safety (Atlam & Wills, 2019).

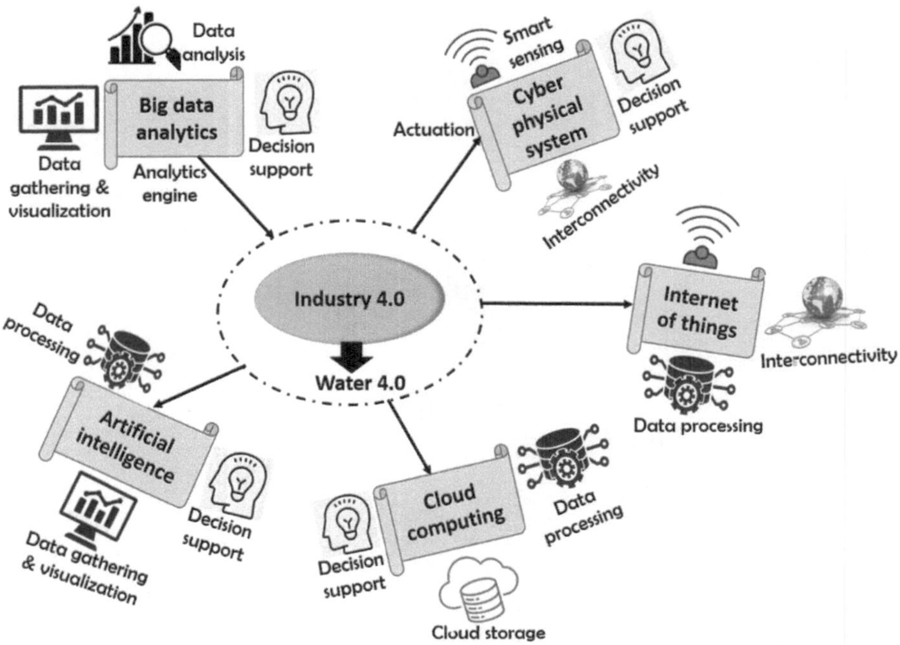

FIGURE 2.32 Industry 4.0 in the Water Sector (Adedeji et al., 2022).

2.7 WATER 4.0

Water is an essential resource to humans and is considered life itself. Still, its availability and quality are increasingly threatened by pollution, climate change, and overuse which necessitated leveraging Industry 4.0 technologies (Jain & Jain, 2020), which includes automation and robotics, IoT, big data analytics, and many more, to combat these challenges (Ahmed, Alam, Hoque, et al., 2023; Teh & Rana, 2023). This coined the concept of water 4.0. In simple terms, Water 4.0 is the next generation of water management that uses Industry 4.0 technologies (Figure 2.32) to effectively combat global water challenges due to the insufficiency of traditional methods to address them (Biazi & Marques, 2023; Mehmood, Liao, & Mahadeo, 2020; Stankovic et al., 2020). Furthermore, the world's demand for water keeps rising daily, making it crucial to employ a smart water management strategy to ensure sustainable access and distribution (Arfanuzzaman & Atiq Rahman, 2017).

2.7.1 Smart Water Management Systems

Smart water management is one of the essential requirements needed to meet the demand for water around the world (Sempere-Payá et al., 2013; Zekri et al., 2022). It involves the use of advanced technologies to allow live monitoring, predictive maintenance, and data-driven decision-making, which are necessary to make the most out of the water resources and reduce waste (Jasiulewicz-Kaczmarek, Legutko, & Kluk, 2020; Kamyab et al., 2023; Manish, Thakur, & Singh, 2023). Smart water management, as shown in Figure 2.33, is adopted in many areas like water distribution plants, water treatment plants, smart farming, water monitoring bodies, and standardisation of hydro-meteorological data format. In the area of the water distribution plant, we have a smart water system that involves leakage detection and a smart water grid. In contrast, smart metering is another subarea under the smart water system (Gupta et al., 2020).

Water 4.0 is essential in the transformation and upgrading of existing infrastructures; the implementation of smart metres (Lin, Tsao, & Lin, 2019) and automated control systems (Yuan et al., 2019) would make the process of billing, identifying inefficiencies, and responding to issues faster. AI and machine learning algorithms can predict future water needs, and this is used to plan and manage resources more effectively so as to ensure that water demands are met even when there is a scarcity (Kamyab et al., 2023; Xiang et al., 2021). Water 4.0, using advanced treatment technologies and live monitoring, facilitates better management of water and makes sure that the water is up to standard, thus reducing pollution, protecting the environment, and recycling and reusing water (Glasgow, Burkholder, Reed, Lewitus, & Kleinman, 2004; R. Sharma, Verma, Lugani, Kumar, & Asadnia, 2021). Water is a resource that must be sustained; hence, it is essential to make use of the solutions brought about by Water 4.0 so that water resources can be managed sustainably and build resilience against the future (Arlinghaus et al., 2017; Niemczynowicz, 1999).

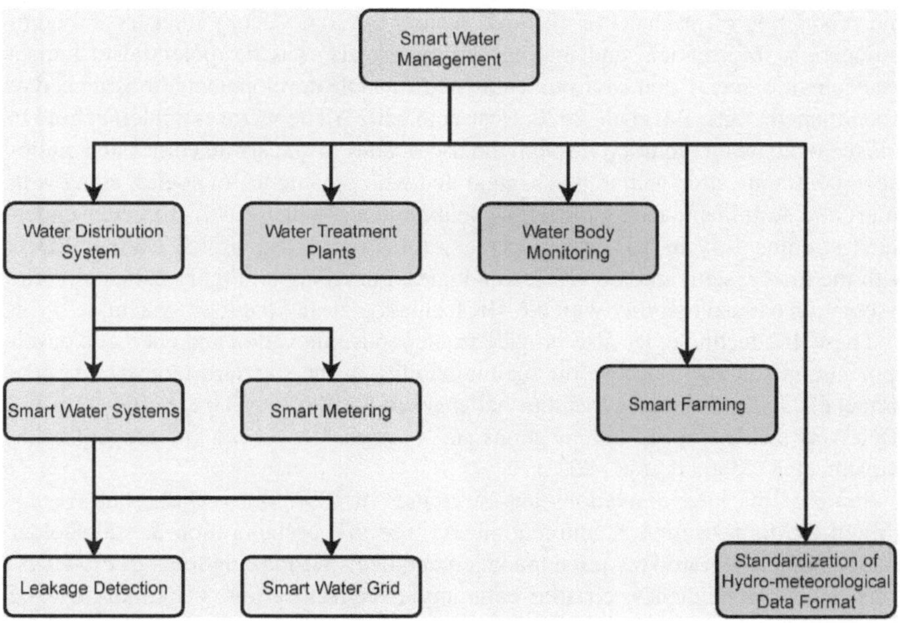

FIGURE 2.33 Smart Water Management System (A. D. Gupta et al., 2020).

The 4IR uses high-tech technologies to change the three sectors, leading to the 4.0 advancement of the FEW Nexus. This approach integrates Industry 4.0 technologies to address the challenges of these interconnected sectors more effectively and sustainably. Food 4.0 focuses on improving food production through technologies such as precision agriculture, biotechnology, and automation & robotics. Precision agriculture uses IoT sensors and drones to improve how crops are managed and irrigation. Biotechnology improves crops' resistance to pests and climate change through genetic modification. Automation and robotics simplify tasks in food production, increasing efficiency and maintaining hygiene. Energy 4.0 controls the difficulties of modern energy systems, especially with regard to the incorporation of sustainable energy sources and the management of decentralised generation. Key technologies include smart grids, which improve the effectiveness and dependability of power distribution, and IoT, which enhances many areas of energy management, including smart industries, harvesting, buildings, cities, grids, and homes. The goal of Water 4.0 is to overcome challenges related to water scarcity, pollution, and management by using Industry 4.0 technologies. Smart water management systems apply live monitoring and predictive analytics to maximise water use, reduce waste, and ensure water quality. This approach is important for long-term water resource management and environmental protection. FEW Nexus 4.0 uses these technologies to tackle global challenges in the food, energy, and water sectors, ensuring that these resources are better managed.

2.8 KEY INNOVATIONS AND STRATEGIES OF FEW NEXUS 4.0

Since the inception of FEW 4.0, there has been a higher level of flexibility, efficiency, and production of products in all three sectors through various Industry 4.0 communication, information, and intelligence, which also has the potential to impact the economic, social, and environmentally sustainable development (Abdelmajied & Abdelmajied, 2022; Bai et al., 2020; Höse et al., 2023) The sector is not left behind in this technological revolution through the use of start of the art machines and equipment, communication technologies integrated with sensing technologies, along with smart and sustainable technologies (Ogbolumani & Nwulu, 2020). To better understand sustainability in 4IR, it is necessary to associate the United Nations SDGs with the food system and the 4IR technologies. For example, Figure 2.34 shows the association of sustainability with the 4IR technologies and the food system.

These 4IR technologies also provide tremendous innovation and energetic development and can also improve the maintainability of the standard industrial system (Bai et al., 2020). To ensure that the challenges of FEW Nexus 4.0 are efficiently and effectively tackled, specific innovations and strategies have been and are still being implemented (Matheri et al., 2021).

Some of the key innovations and strategies of FEW 4.0 include, but are not limited to, digitalisation & automation, AI, the IoT, collaboration & stakeholder engagement, integrated resource management, decision-making tools in FEW systems, resource efficiency, circular economy, decentralisation, sustainability, and resilience & adaptability.

These key innovations and strategies of FEW Nexus 4.0 are discussed in the following section.

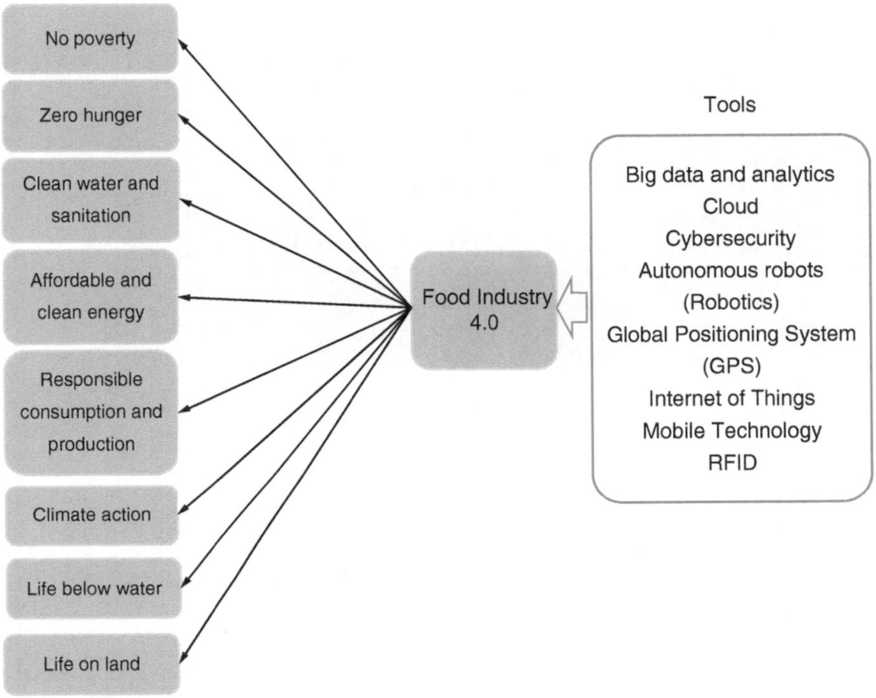

FIGURE 2.34 The Food Industry 4.0 and Sustainability (Jambrak et al., 2021).

2.8.1 Digitalisation and Automation

Digitalisation and automation are crucial to achieving efficiency and sustainability in the FEW Nexus 4.0. Figure 2.35 shows the importance of digitalisation in achieving sustainability. Digitalisation in FEW Nexus 4.0 is simply the collecting of vast amounts of data from the food sector, water systems, and energy grids and analysing it to provide an understanding of how the resources are being used, which in turn ensures the efficiency in the use of resources and provides multiple benefits. Figure 2.36 shows some of these benefits: Digitalisation drives enterprise innovation and enhances performance while considerably improving the performance of the capital market. It also gives a competitive advantage and enables high performance within the production system. Together, these benefits contribute to the industries' growth and success: Automation ensures that the processes are controlled efficiently to reduce waste and increase productivity (Kolberg & Zühlke, 2015). Automated processes monitor digitally and in real-time to ensure they can be adjusted as required ensure that faults and issues in the FEW Nexus 4.0 are swiftly identified and dealt with to prevent a hiccup in the production process (David et al., 2022; Javaid et al., 2022b). Figure 2.37 shows some of the advantages of industrial automation, which include but are not limited to the following: Improvement in productivity and quality, increased flexibility in production processes, and reduced cost of operation, which result in more competitiveness, security and assistance in gathering information within the production system

FIGURE 2.35 Digitalisation and Sustainability (Mondejar et al., 2021).

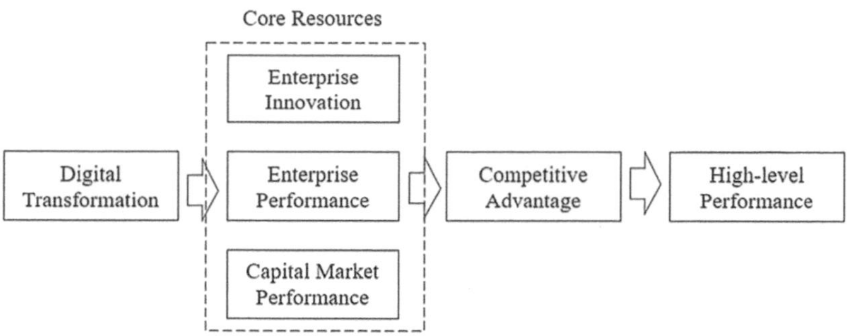

FIGURE 2.36 The Benefits of Digital Transformation (Liang & Tian, 2024)

FIGURE 2.37 Advantages of Industrial Automation (Al Shahrani et al., 2022)

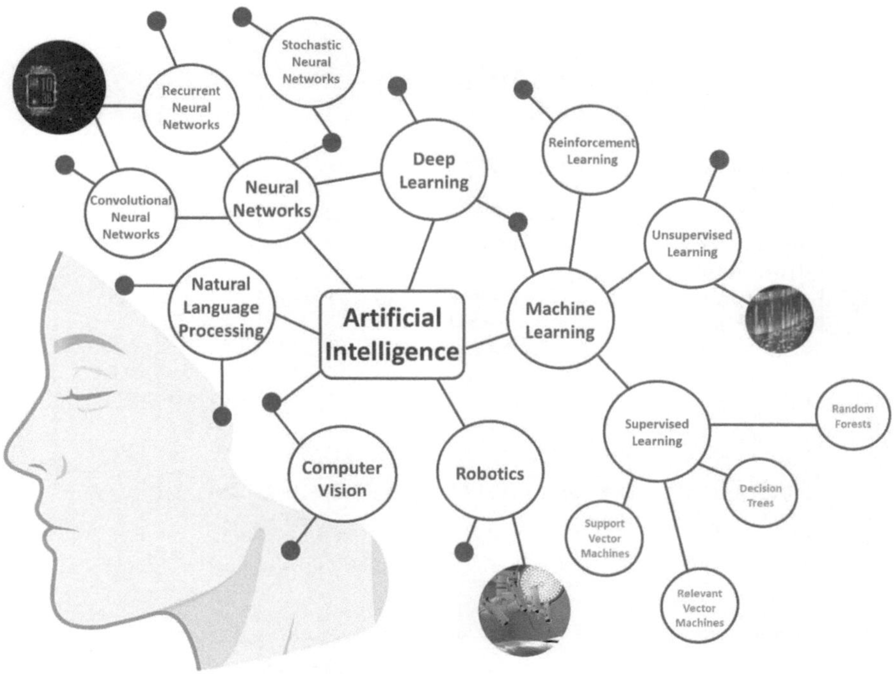

FIGURE 2.38 Artificial Intelligence (Athanasopoulou et al., 2022).

2.8.2 Artificial Intelligence (AI)

AI is a technology increasingly implemented in the FEW Nexus 4.0. It can process and correlate massive quantities of data and disclose information that would have remained secluded, thus enabling the production to be automated and energy-efficient (Nath et al., 2024). Still, AI isn't limited to data processing. It can recognise science-based results for environmental and climate deterioration issues. It supports multi-stakeholder decision-making procedures towards sustainability because, unlike individuals or groups, it's not biased and doesn't judge based on its interests (Ahmed et al., 2022). The capability of AI in the management of the FEW Nexus 4.0 has been severely underrated, and the same goes for its assistance to the governance of several interactions among the resources, sectors, and institutions that are composed of it, making it continue to grow and become more crucial in FEW 4.0 (D'amore et al., 2022). Figure 2.38 shows the six major areas in AI: machine learning, robotics, computer vision, natural language processing, neural networks, and deep learning.

2.8.3 Internet of Things (IoT)

The Internet of Things is a system of interconnected devices that perform specific works by enabling data transfer, analytics, and decision-making, thereby increasing productivity and efficiency (Mohindru et al., 2020). It is essential to solve the complex

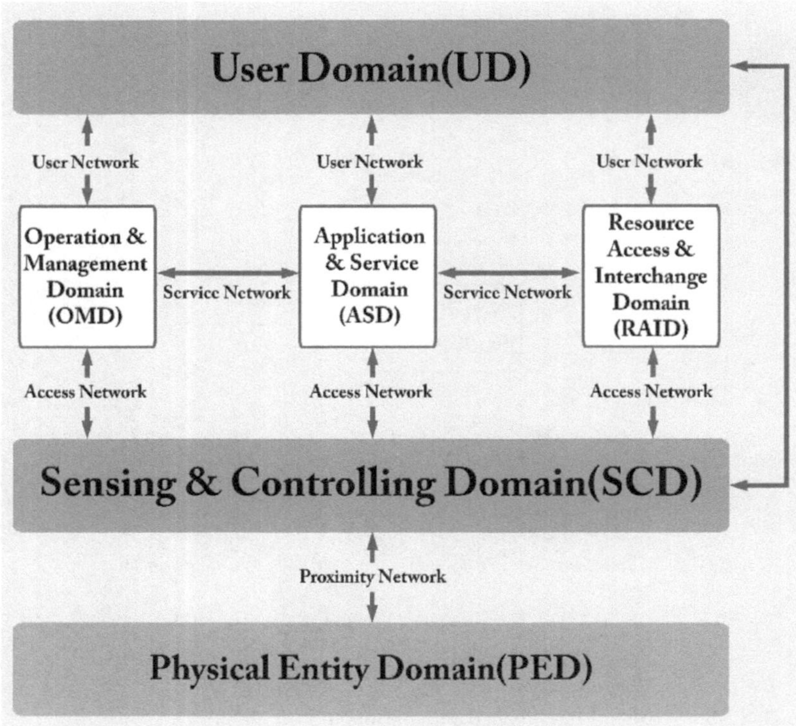

FIGURE 2.39 Application Domains of the Internet of Things (Bharathi et al., 2019).

issues that are overwhelming the sustainable life on this Earth, using its technologies to monitor and manage resources (Kumar et al., 2018). Figure 2.39 shows some application domains of IoT: smart home, smart factory, smart buildings, intelligent vehicle, security surveillance, telemedicine and health care, smart transport, water management, and agriculture automation are all areas where IoT can be applied.

2.8.4 Collaboration and Stakeholder Engagement

Collaboration, which is working together towards a common goal, and stakeholder engagement, which is the involvement of all relevant parties in decision-making, are both important for effectively managing resources in the FEW Nexus 4.0. Collaboration involves the alliance of industries in the food-energy-water sectors to form partnerships to enhance efficiency and sustainability through collective effort (Gebhardt et al., 2022; Morales et al., 2024). Figure 2.40 shows some of the benefits of collaboration, which include cooperation, coproduction, mutual support, and open innovation (Macedo et al., 2020). Engaging stakeholders in the development of tools and technologies for the FEW Nexus 4.0 not only benefits decision-making but also enables a better understanding of the alternative pathways for action under conditions

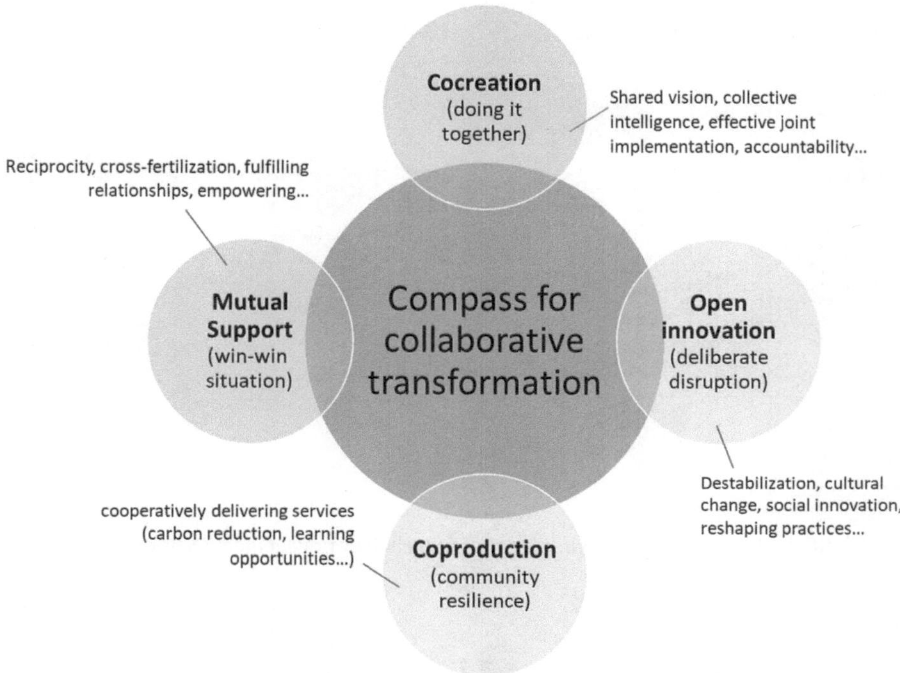

FIGURE 2.40 Benefits of Collaboration (Macedo et al., 2020).

of uncertainty that characterise the FEW Nexus 4.0 challenges (Hoolohan et al., 2018; MacHer et al., 2021; Namany et al., 2023; Yaroson et al., 2024). Figure 2.41 shows some of the processes and outcomes of stakeholder engagement. Game theory is among the most popular concepts in collaboration and stakeholder engagement. This is discussed in the following subsection.

2.8.4.1 Game Theory

Game theory, used as strategic analysis, has provided powerful tools and been applied to many fields, including the FEW Nexus 4.0. Its assumptions emphasise that rational players who pursue well-defined objectives and assume knowledge of others would form expectations of other decision-makers' behaviour (Guney & Ganjali, 2023). In simple terms, game theory predicts a party's behaviours towards fulfilling their interests during the interactive decision-making process with other parties, (David et al., 2022; Do, 2022) (Figure 2.42). Game theory concepts like leader/follower dynamic and hierarchy represent the strategic interactions and how decision-making works among different stakeholders.

- **Leader/Follower Dynamic**: This is also known as Stackelberg (Adil et al., 2021). It is when one party, which is the leader (government bodies or large corporations in the FEW Nexus 4.0), makes a decision first and the others,

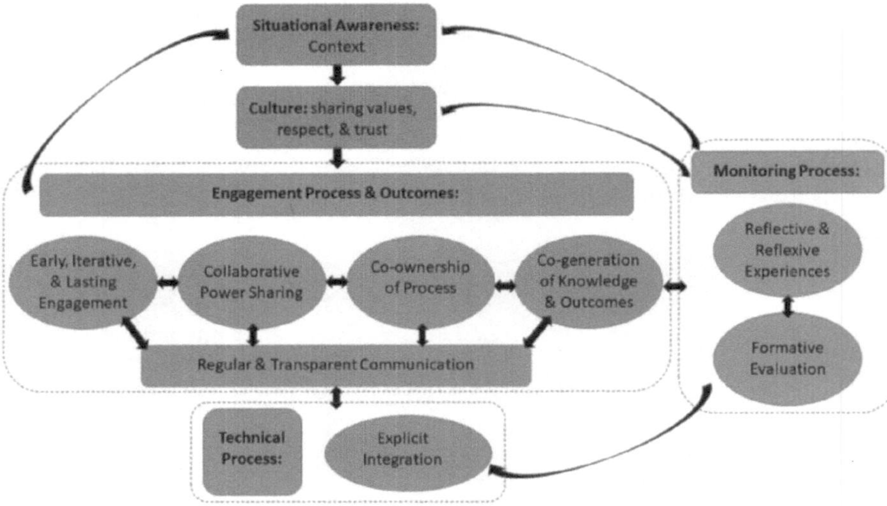

FIGURE 2.41 Processes and Outcomes of Stakeholder Engagement (Kliskey et al., 2021).

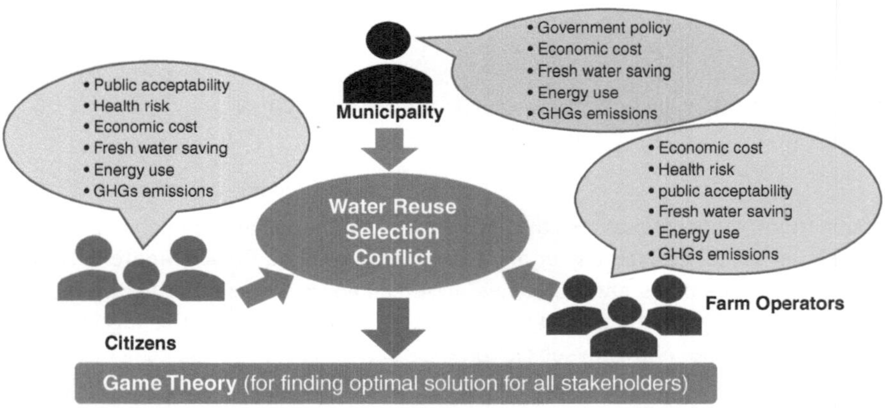

FIGURE 2.42 Game Theory (Huang & Li, 2022).

which are the followers (smaller companies or local communities in the FEW Nexus 4.0), make their decisions afterwards while taking their leader's decision into account (Ibrahim et al., 2021; Motlaghzadeh et al., 2020). Figure 2.43 shows a leader/follower dynamics in game theory.

- **Hierarchy**: Hierarchy in game theory represents the different levels or tiers among the stakeholders when it comes to influence, authority, control, and decision-making, from the international organisations to the individual farmers (Motlaghzadeh et al., 2020). Figure 2.44 illustrates the hierarchy in game theory showing from the needs of customers and stakeholders at the bottom of the pyramid with the senior management at the top most level.

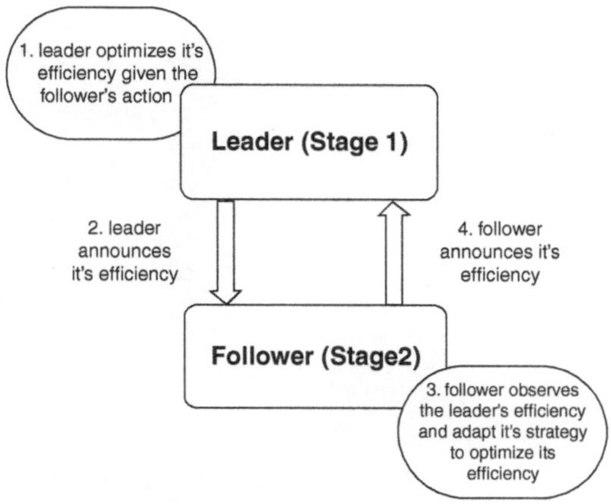

FIGURE 2.43 Leader/Follower Dynamic in Game Theory (Tatlari et al., 2025).

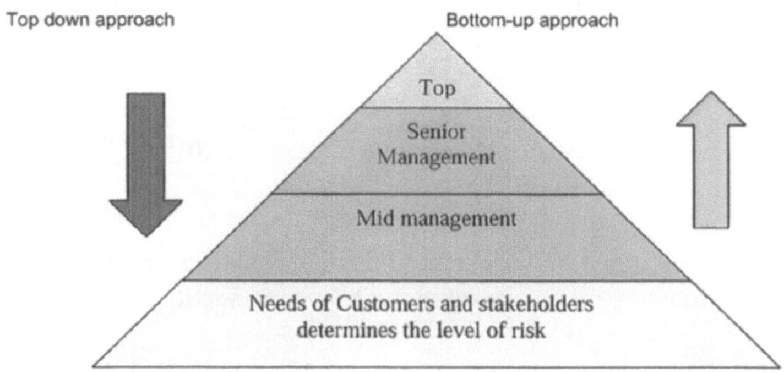

FIGURE 2.44 Hierarchy in Game Theory (Chibayambuya & Theron, 2007).

2.8.5 Integrated Resource Management

Rapid population growth and increased urbanisation contribute to the heightened presses on global food, water, and energy resource networks (Molajou et al., 2023) Therefore, there is a requirement for adaptable and efficient resource management that regards societal, environmental, and economic elements of sustainable development (Ahmad et al., 2023) Integrated resource management is simply the managing of the FEW Nexus 4.0 resources in a coordinated and sustainable way with the aim to balance various of its objectives, such as economic efficiency, and environmental sustainability by considering the interconnections between the different resource systems through a quantitative, qualitative approach and mixed method approach (Namany et al., 2019; Ogbolumani & Nwulu, 2021b; Simpson et al., 2020; Simpson et al., 2022; Stilwell & Harman, 2021). Figure 2.45 shows

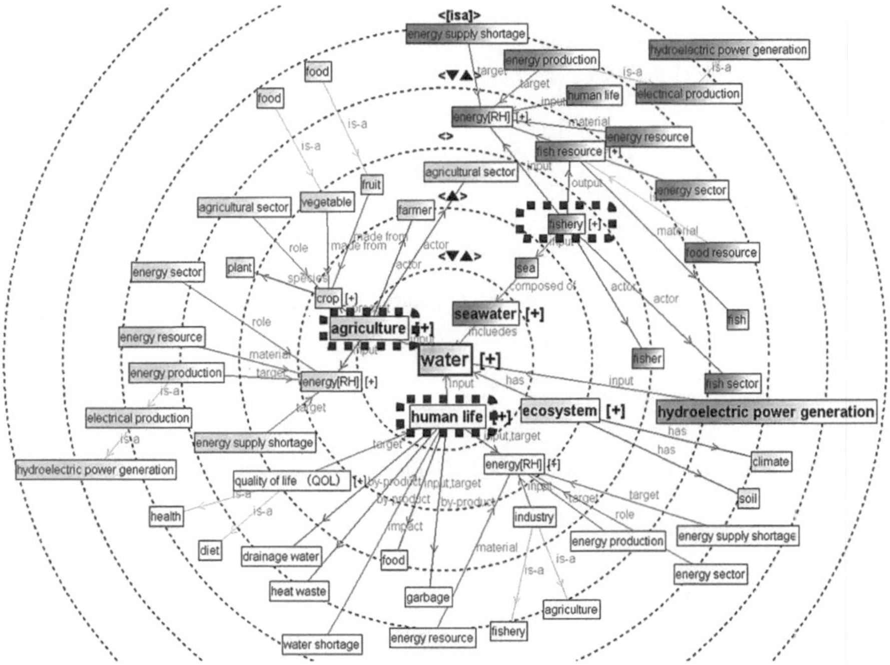

FIGURE 2.45 Integrated Resource Management (Endo et al., 2015).

a framework for integrated management of resources within a FEW Nexus system. These three approaches are discussed in the following section.

- **Quantitative**: These approaches are usually more data-intensive than qualitative approaches. Ali and Acquaye (2024) explain that structured frameworks allow researchers to analyse intricate FEW Nexus systems while assessing the interdependent effects between food and energy and water sectors. The FEW Nexus 4.0 system reveals hidden insights through the combination of statistical analyses and simulation techniques as quantitative approaches. Large datasets get processed efficiently by these methods which also enable the application of complex algorithms to perform objective evaluations on multiple scenarios (Ahmed, Alam, Hassan, et al., 2023; Moats & Tseng, 2024; Sahoo & Goswami, 2023). The approach turns out to be crucial when decisions need to be made under uncertain conditions or when managing multiple connected variables. Systems and trends are analysed through mathematical models which help measure variations and project results under varied environmental conditions using numerical data-driven analysis. The application of this method enables decision-makers to make the most of their resources by utilising the FEW Nexus 4.0 framework (Mao et al., 2021; Ogbolumani & Nwulu, 2021b, 2022, 2024).

- **Qualitative Method**: Evaluation of FEW-N system relations and exchanges through quantitative techniques generates workable resource use policies, and qualitative approaches augment them (Albrecht et al., 2018). Stakeholders participate in focus groups and workshops as well as interviews to generate insights through these qualitative system mapping approaches as described by (Henry et al., 2022). The analysis technique employs nodes for depicting system concepts alongside arcs that represent concept relationships with direction and strength information (Ringler et al., 2013) The model enables testing of potential actions on the system by analysing mutual element relationships (Basheer & Abdo, 2021).
- **Mixed Method**: The combination of qualitative and quantitative research approaches through mixed methods allows scientists to achieve better understanding of FEW systems. Premium research benefits from an integrated method that combines descriptive knowledge from qualitative research with quantitative data that produces numerical evidence through generalisation. The exploration of FEW systems through qualitative data delivers insights about individual experiences as well as motivational aspects within the system, while quantitative data generates statistical results according to Matović and Ovesni (2023) and Wallwey and Kajfez (2023). When researchers employ these methods together, they achieve an understanding of general frameworks as well as individual system behavioural components (Hendren et al., 2023; Mukumbang, 2023). Researchers can achieve better understanding of system connections through this combination technique since it results in improved diagnostic findings (Breit et al., 2023; *Human Insight AI: An Innovative Technology Bridging The Gap Between Humans And Machines For a Safe, Sustainable Future*, n.d.; Kharakhash, 2023). Researchers using mixed methods gain essential insights into FEW system complexities which leads them to deliver better recommendations.

2.8.6 Decision-Making Tools in FEW Systems

Decision tools evaluate FEW Nexus 4.0 systems by examining data from food, energy, and water sectors to maximise resource efficiency and enhance sustainability according to (David et al., 2022; Morales et al., 2024; Rising, 2020).The decision-making toolkit for FEW Nexus 4.0 systems includes Integrated Assessment Models (IAMs) as well as Geographic Information Systems and Multi-criteria Decision Analysis and System Dynamic Models and Life-Cycle Assessment.

- The methodology of IAMs provides analysis to demonstrate how shifts in any sector among food, energy, and water systems influence the other components. Scientists and policymakers can use these predictive tools due to their capability to handle extensive data through complex mathematical procedures which assess elements like energy consumption alongside food production and water reserves (Cheng et al., 2023; Nesje et al., 2023; Savin et al., 2023; Vahabzadeh et al., 2023). These modelling approaches provide critical insights into how decisions affect entire systems and direct strategic development towards resolving issues such as environmental degradation scarcity (Garver, 2023; Wood Hansen & Van Den Bergh, 2023). IAMs combine inter-sector information to produce resource sustainability policies with no adverse effects on sustainability (van Beek, Hajer, Pelzer, van Vuuren, & Cassen, 2020; Ogbolumani & Nwulu, 2024). Figure 2.46 shows an IAM for the sustainable allocation of food and energy resources for a FEW Nexus system (Ogbolumani & Nwulu, 2024).

- **Geographic Information Systems (GIS)**: GIS enables efficient management and analysis of location-specific data through its computerised system for understanding resource allocations within the FEW Nexus 4.0 framework. GIS technology helps display patterns in how resources get allocated and used, providing improved planning tools for management purposes (Ali et al., 2024). The combination of different data sources, including water availability, agricultural practices, and energy consumption, allows GIS to perform a complete analysis of resource relationships and their effects. Through this integrated approach, governments can generate specific policies to solve regional problems and enhance resource effectiveness and sustainability (Lakra et al., 2024; Azzam et al., 2023). GIS implements prediction forecast thanks to modelling and simulation which leads to active decision-making capabilities and strategic planning structures for FEW Nexus 4.0 (Robinson, 2020). Figure 2.47 shows a framework for the use of GIS technology for planning and forecasting purposes.
- **Multi-criteria Decision Analysis (MCDA)**: MCDA stands as an analytical approach for evaluating various contrasting decision factors throughout decision-making activities (Dean, 2022). Through a systematic process, stakeholders can make balanced decisions because the method provides structured methods to evaluate and measure requirements with varying significance levels (Baydaş et al., 2024; Mohamed Mouhoumed et al., 2023;

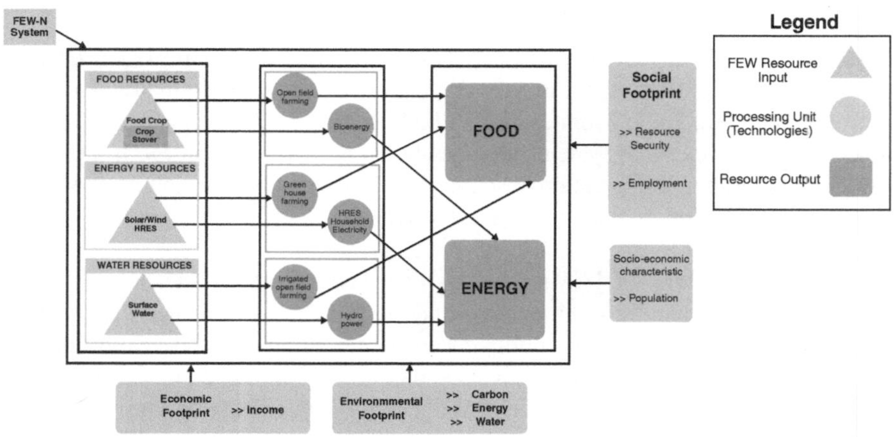

FIGURE 2.46 Integrated Assessment Model for Food, Energy and Water Nexus System (Ogbolumani & Nwulu, 2024).

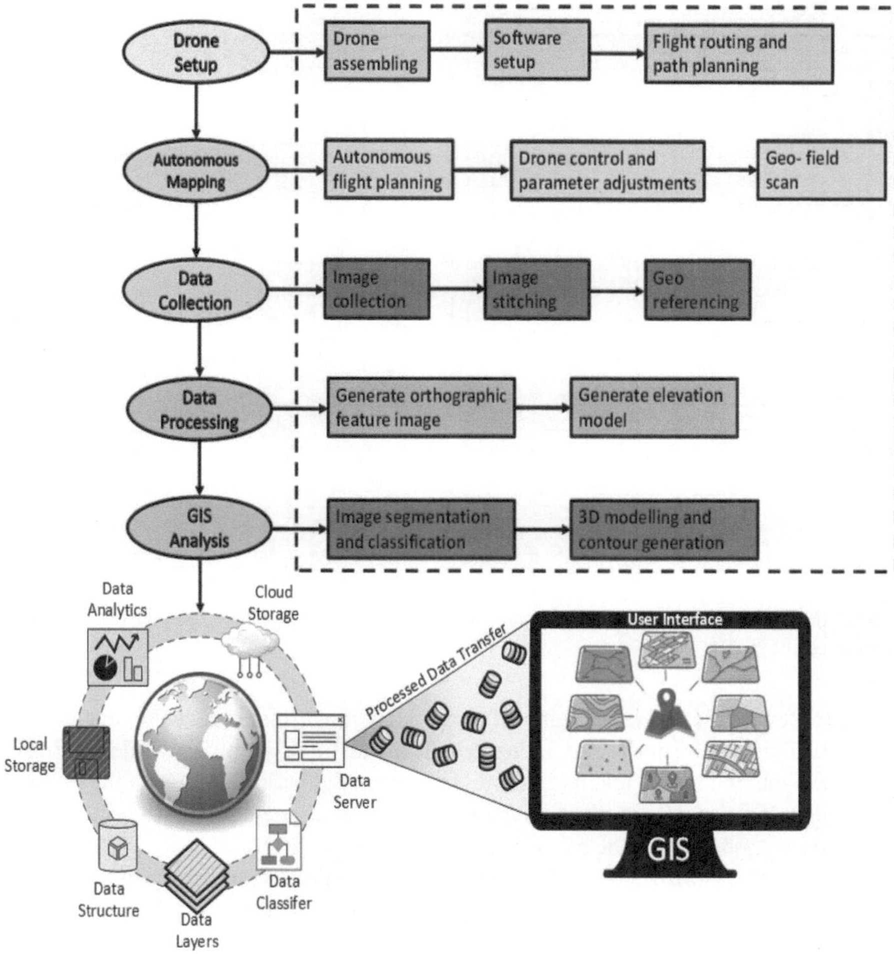

FIGURE 2.47 Geographic Information Systems (Quamar et al., 2023).

Pereira et al., 2024). The application of MCDA in resource management requires stakeholders to select essential decision factors followed by priority evaluations combined with performance assessments of different options against these criteria (Sahoo & Goswami, 2023). MCDA assesses resource management strategies by evaluating factors including expense, ecological effects, and societal advantages (Pascoe et al., 2023; Rihab et al., 2024). MCDA establishes a systematic evaluation process that helps decision-makers make balanced choices during complicated evaluations in the context of the FEW Nexus 4.0 framework. Figure 2.48 demonstrates the six MCDA evaluation steps for operation and system composition, including goal-setting and alternative sample generation alongside setting weightings and choosing methods. Aggregation and verification processes complete

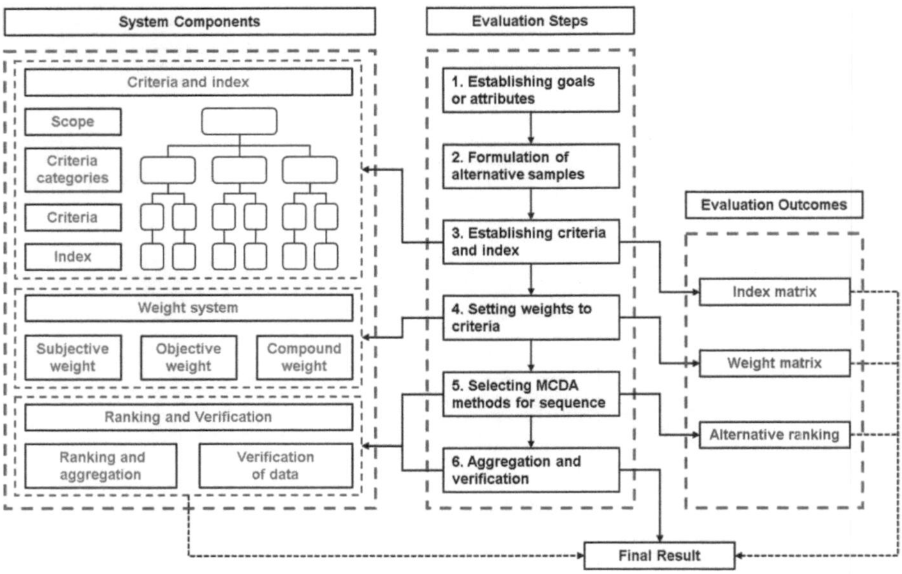

FIGURE 2.48 Operational Steps and Sub-systems of Multi-Criteria Decision Analysis (Yuan et al., 2022).

the evaluation stages. Some of these steps have outcomes. For instance, establishing criteria and index results in an index matrix, while setting weights to criteria results in a weight matrix. Selecting MCDA methods for sequences results in alternative ranking, while aggregation and verification are responsible for the overall result (Yuan et al., 2022).

- **System Dynamics Models (SDM)**: SDM functions as a time-based system analysis method for evaluating FEW Nexus 4.0 systems, according to Wang et al., (2024). The method requires the development of interactive models that represent system connections. A continuous exchange of feedback flows between diverse nexus components exists within the system. SDM provides stakeholders with insights about systematic changes between connected areas through which system outcomes are affected (Maghsoudi & Nezafati, 2023; Mhlongo, 2021; Zhang et al., 2023). SDM enables organisations to evaluate foreseeable impacts on the system before policy and technology implementation through scenario and strategy simulation. SDM provides tools to examine intervention results and identify obstacles alongside opportunities that lead to better strategic decisions in the complicated FEW Nexus 4.0 framework. The system dynamic model illustrated in Figure 2.49 presents a practical framework depicting how different elements contribute to water supply management in the Souss-Massa basin of Morocco. The analysis of water resource management used system dynamic models (Guemouria et al., 2023).
- **Life Cycle Assessment (LCA)**: LCA functions as a systematic evaluation method that tracks environmental effects across product and

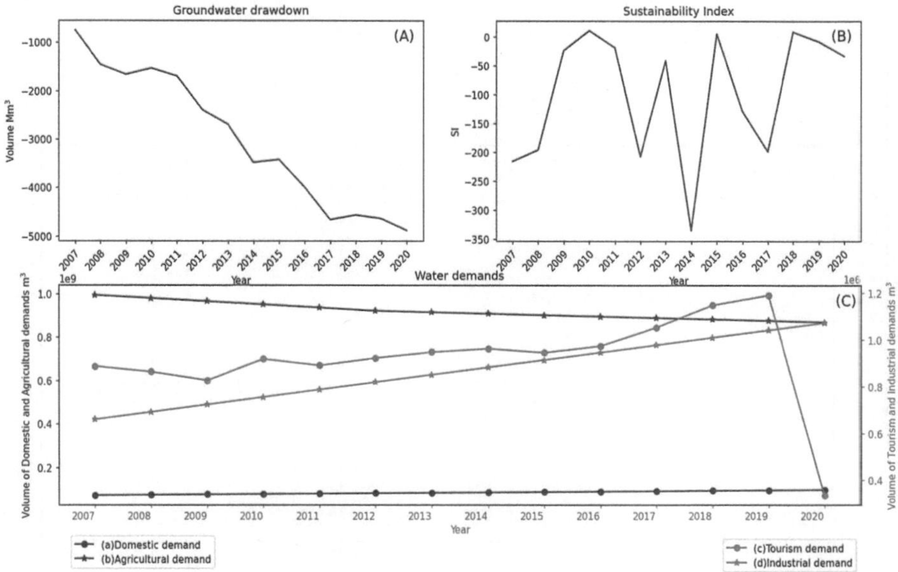

FIGURE 2.49 Diagram of A Causal Loop - A System Dynamic Model Illustration (Guemouria et al., 2023).

production stages, starting with initial resource extraction until final disposal (Corominas et al., 2020). Through phase-by-phase evaluation, LCA establishes the complete environmental impact of resource depletion, emission emissions, and waste production. The extensive review through LCA identifies particular sustainability weak spots while enabling better resource handle (Bhambhani et al., 2022; Sieverding et al., 2020). Through LCA, organisations can identify the life cycle stages which produce maximum environmental impacts, thus guiding their strategic impact mitigation plans (Croes & Vermeulen, 2021). Figure 2.50 displays an assessment framework for life cycle analysis. The LCA tools within FEW Nexus 4.0 enable better decisions concerning environmental resource management and sustainability by supporting ecologically friendly practices throughout combined food-energy-water systems.

2.8.7 Resource Efficiency

The Earth faces mounting difficulties in supplying sustainable FEW resources for human needs because of natural and human-made stresses. Resource efficiency functions as a main concept to tackle this problem by finding efficient resource usage throughout the FEW sectors that minimises waste while decreasing environmental effects and boosting production efficiency productivity (Ringler et al., 2013; Tian et al., 2018). Resource utilisation effectiveness increases from start to end of life cycle because each unit uses its resources optimally to deliver maximum

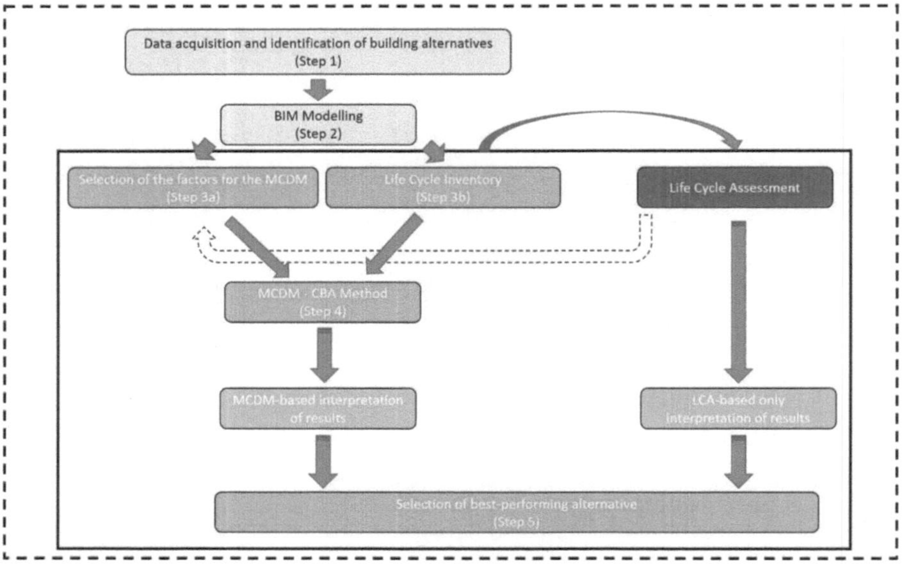

FIGURE 2.50 Life Cycle Assessment (Vázquez-Rowe et al., 2021).

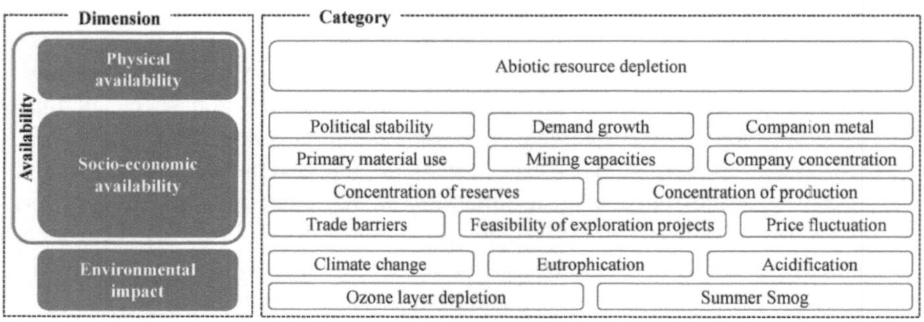

FIGURE 2.51 The ESSENZ Method's Resource Efficiency Dimensions and Categories *Henbler et al., 2016.*

outcomes while minimising environmental consequences. The FEW Nexus 4.0 system supports sustainable resource handling through practices that simultaneously meet human demands while protecting nature according to (Krishnan et al., 2020; Lim et al., 2022). The challenge involves sustaining present-day demands while protecting environmental vitality and future resource supplies to manage the coupled problems found in the FEW Nexus (Kremsa, 2021). The assessment framework for resource efficiency using ESSENZ method appears in Figure 2.51. The resource efficiency elements within ESSENZ method follow sustainable development principles (Henßler et al., 2016). The visual depiction showing assessment dimensions and categories appears in Figure 2.51.

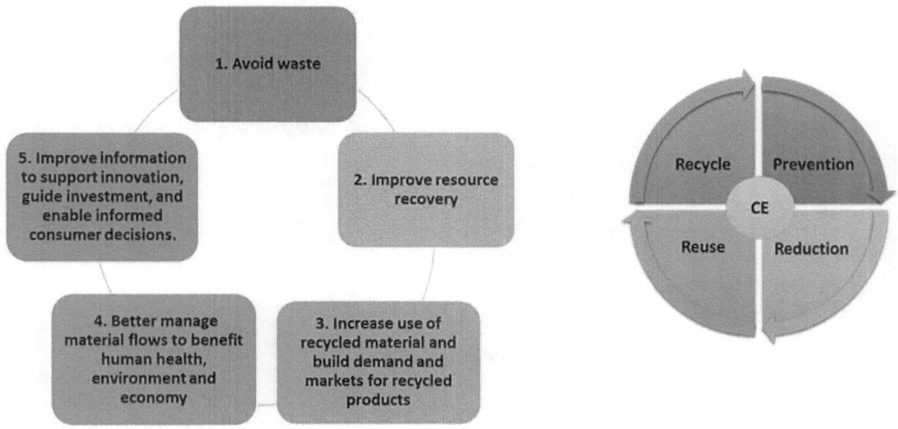

FIGURE 2.52 Principles of Circular Economy (Jahan et al., 2022).

2.8.8 Circular Economy

The circular economy presents an economic structure optimising waste elimination by employing circular resource flows for recycling operation and product resuse. Through continuous product and material recycling, the system helps lower fresh resource requirements because it operates unlike the linear "take-make-dispose" model. The resource loop remains in constant operation within a circular economy by transforming process outputs into following process inputs. This cycle reduces waste output (Brears, 2021; Del Borghi, Moreschi, & Gallo, 2020; Nautiyal & Goel, 2021; Slorach, Jeswani, Cuéllar-Franca, & Azapagic, 2020). Implementing this system decreases new resource consumption and cuts down production and disposal waste flows and greenhouse gas emissions (Yang et al., 2022). The FEW Nexus 4.0 model achieves sustainable resource management benefits in its food energy and water sectors through the implementation of the circular economy principle. A circular economy promotes sustainable resource management while decreasing its environmental impact, thus creating an economically sustainable system that meets FEW Nexus requirements (Lim et al., 2022; Zhou & Smulders, 2021). Figure 2.52 shows the principles of circular economy: The four core principles are: recycle, prevention, reduction and reuse, and these are broken down into 1. Avoid waste; 2. Improve resource recovery; 3. increase the use of recycled material and build demand and markets for recycled products; 4. Better management of material flows to benefit human health, environment, and economy; 5. Improve information to support innovation, guide investment, and enable informed consumer decisions.

2.8.9 Decentralisation

Governments today operate with a new structural approach where power shifts from headquarters to smaller locations in regions and cities. The decision-making process is improved through instantaneous data combination with local knowledge,

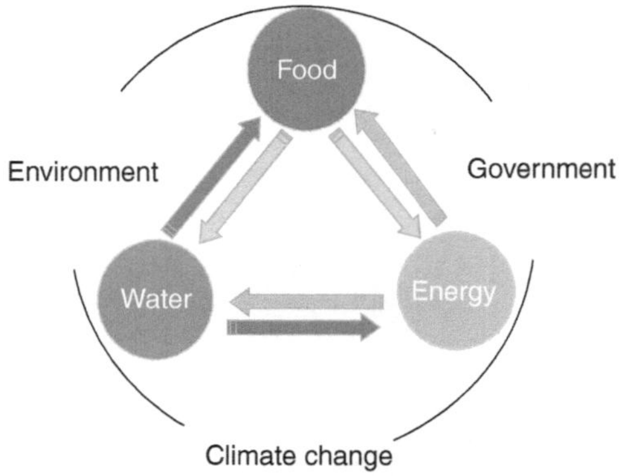

FIGURE 2.53 Decentralised FEW Nexus (Abdi et al., 2020).

which leads to specialised and more responsive resource management (Meir, Opfer, & Hernandez, 2022; Pahl-Wostl et al., 2021). Decentralisation brings government authority near communities, allowing decision-makers to deliver more efficient solutions for specific challenges. The localised control structure enhances policy accountability while reflecting community needs to the targeted populations. The governance system becomes more adaptive and flexible by enabling immediate changes to resource levels and community needs. Decentralisation enhances the managerial effectiveness of resource management combined with policy implementation by providing custom solutions adapted to regional and local circumstances (Figure 2.53).

2.8.10 Sustainable Practices

Sustainable practices in FEW Nexus 4.0 ensure that developments in food, energy, and water systems deliver positive economic, environmental, and social value (Dayioğlu & Türker, 2021). These practices encompass a range of strategies, from regenerative agriculture and renewable energy production to water conservation and waste reduction, all aimed at achieving long-term sustainability. By implementing these approaches, FEW Nexus 4.0 strives to meet current needs without compromising future generations' ability to meet their own. This balanced resource use and environmental preservation approach seeks to harmonise economic development with responsible resource management, supporting both present and future well-being (Hajian & Kashani, 2021; Henderson & Loreau, 2023; Muhie, 2022). Figure 2.54 shows sustainable practices in FEW Nexus. Ultimately, these sustainable practices serve as the foundation for a more robust and just food-energy-water system that can handle present issues while preserving resources for years to come (David et al., 2022).

2.8.11 Resilience and Adaptability

The FEW Nexus faces substantial risks from climate changes and socioeconomic barriers, which face potential breakdowns through droughts floods along with

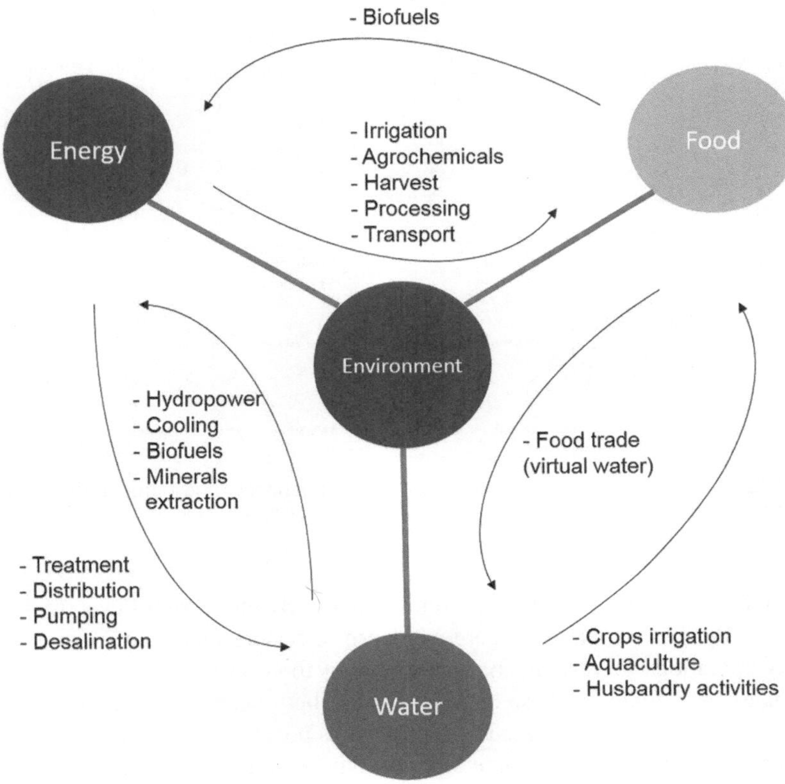

FIGURE 2.54 Sustainable Practices in the FEW Nexus (Salmoral et al., 2019).

extreme weather conditions while also experiencing market volatility that affects food, energy, and water operations.

Extreme weather phenomena, market fluctuations, population increase, and extreme weather manifestations pose challenges across the FEW Nexus. In this context, resilience and adaptability become crucial. The ability to persevere through adversity, along with the ability to bounce back after disasters, qualifies as resilience. Ending adaptations refer to the capability of modifying behaviour based on environmental transformations. Assets and capacities of communities link together as vulnerability forms a connection with resilience to adaptability, according to Zhai and Lee (2024)

People with vulnerability can predict hazards while maintaining control of destructive outcomes. The FEW Nexus requires these elements to successfully manage and reduce its related challenges. Systems utilise these capabilities to continue operations while disruptions arise and adapt to new challenges. Food, energy, along with energy and water resources, experience more substantial levels of sustainability because of their enhanced capacity to address new threats in the face of ongoing global changes (Birkmann & Pardoe, 2014; Li et al., 2023; Zhu et al., 2022). Figure 2.55 displays the relationship between vulnerability, resilience and adaptability.

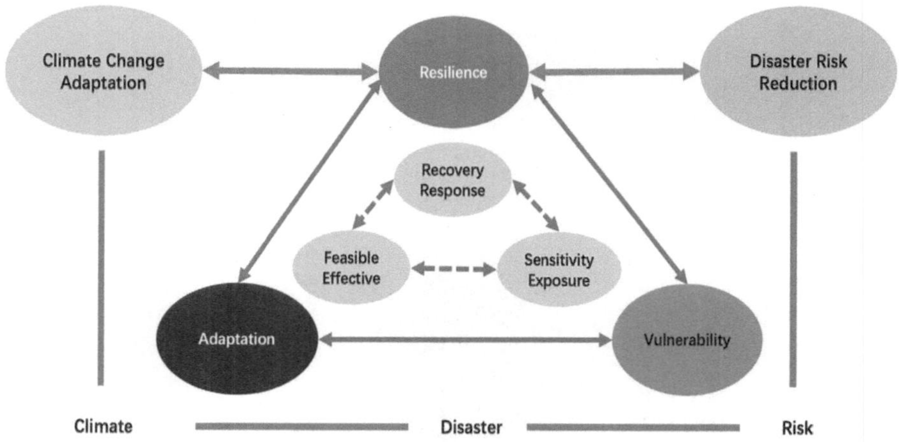

FIGURE 2.55 The relationship among adaptability, resilience, and vulnerability (Zhai & Lee, 2024).

The methods work to enhance community-level and system-level protective capabilities during trauma. Systems must learn to restore themselves while adapting their operations to various changes affecting food, energy, and water sectors. Stakeholders who focus on these qualities receive better capabilities to handle challenges affecting the FEW Nexus. The FEW Nexus benefits from such challenges by developing resilient and sustainable systems. These systems display the necessary attributes to survive while adapting to worldwide global changes.

2.9 GOALS OF FEW NEXUS 4.0

FEW 4.0 represents the concept that connects Industry 4.0 technologies with the integrated food, water, and energy sectors. The FEW Nexus includes interconnected components from the food, water, and energy sectors (Dantas et al., 2021; David, Adepoju, et al., 2024; David et al., 2022; Opeyemi et al., 2024). The implementation of modern technology serves the purpose of solving problems that led to the FEW Nexus (Dantas et al., 2021). FEW Nexus 4.0 aims to maintain resource security through its established set of goals. The goals of FEW Nexus 4.0 target sustainable production together with equitable allocation of resources and fair resource distribution for clean resources. The following section provides a brief discussion of the mentioned points:

2.9.1 Enhancing Resource Security

The economic paths of nations depend entirely on the availability of fundamental resources, which consist of food, water, and energy. These resources function as interconnected elements that produce sustainable economic results due to their essential role. The country adopted the FEW Nexus approach for promoting national economic development, according to Cansino-Loeza et al. (2020). Failure to address these interconnections could result in water, energy, and food insecurity (Bellezoni et al., 2022; Caiado Couto et al., 2021). As a result, all of these essential intakes could become less readily accessible and durable (David et al., 2024). A resource-efficient approach involves maximising productivity, minimising waste, reducing environmental impact, and using resources efficiently from the food, energy, and water sectors (Ringler et al., 2013; Tian et al., 2018).

2.9.2 Sustainable Resource Production

Changing climate, growing population, inefficient resource utilisation, environmental degradation, pollution, and degradation of natural resources are just some of the pressing issues that require a shift to a sustainable production process (Huo & Peng, 2023; Nguyen et al., 2023) Among other things, FEW Nexus 4.0 reduces waste, recycles, and minimises natural resource usage (Javaid et al., 2022b). A sustainable production process minimises the process's environmental impacts while being economically viable (Diaz et al., 2021).

2.9.3 Equity in Allocation of Resources

As a result of social injustice, economic disparities, and political instability, equity in resource allocation is becoming more critical (Smith et al., 2022). Resources must be allocated fairly among all members of society while also considering the diverse needs and circumstances of various groups (Li et al., 2022) To ensure that everyone benefits from economic and social progress, some barriers must be eliminated preventing them from gaining access to resources (Leclerc et al., 2012). For instance, equity and allocation of clean water are considered: when there are not enough clean resources, diseases and poor sanitation can occur (Ferreira et al., 2021). For instance, industrial pollution, burning fossil fuels for energy generation and lack of access to clean air could cause respiratory diseases, cardiovascular problems, or even premature deaths (Ofremu et al., 2024). Inadequate access to sanitary water causes water-borne diseases, which forces some families into purchasing bottled water, paper plates, and use laundromats more frequently for just daily activities, leading to high costs (Ferreira et al., 2021; Olatunde et al., 2024; Vizanko et al., 2024). This is why equitable access to clean resources is an important aspect of sustainable development in the FEW Nexus.

FEW Nexus 4.0 is an approach that aims to address the associated challenges of food security, energy supply, and water resources in a manner that affirms sustainability, resilience, and productivity across the entire production process and consumption processes. It highlights the need to understand how these three critical resources are intertwined to make them support each other efficiently. This nexus shows that our present and future needs can be met without harming the planet or its inhabitants. Basic food and nutrition availability is established through food security requirements. An adequate power supply is included in the energy supply concept. Different applications need these sources to support modern economies while enabling development and everyday life needs. The FEW system depends on the quality and amount of accessible drinking water as well as irrigation and industrial water needed for household needs. The improved connection between these systems helps decrease waste during human activities (Adeleke et al., 2023; du Plessis, 2019).

The future generations must have enough resource bases to sustain themselves. The achievement of any technological advancement needs immediate attention. Technological advancements function as the main driver for the advancement of FEW 4.0 goals. The evolution of technology provides better means to establish enhanced productive procedures. Water conservation occurs with enhanced energy utilisation and preserved clean usable water through efficient food production (Apeh & Nwulu, 2025). FEW Nexus 4.0 functions as a strategy to solve the linked problems of food security. This approach aims to achieve sustainable development by maintaining well-integrated food security, energy supply, and water resources (David, Aigbavboa, et al., 2024b).

The system focuses on boosting productivity and resiliency capabilities for the entire operational cycle, from production to consumption. The emphasis of these three vital resources lies in understanding their mutual relationship. The meaningful connection of water resources reminds everyone should remember their obligation to provide support to one another. Therefore, this nexus makes it clear if our current and future requirements are met without jeopardising the Earth's ecosystem or hurting humanity. Adequate food means everybody has access to enough food that is safe and nutritious (Apeh & Nwulu, 2024; Ogbolumani & Nwulu, 2020). This idea relates to various sources of power that are sufficient for the current use and development. A supply concerned with water entails the amount plus quality of drinking water, irrigation water, or industrial water used for household purposes. All these measures could help reduce waste from their operations, protect the environment, and ensure a resource base for future generations (Ajibola & Ogbolumani, 2024). Technical advancements play an essential part in achieving the FEW 4.0 goals. As technology develops, finding new ways to grow food with less water, use energy more efficiently, and ensure clean and drinkable water becomes easier.

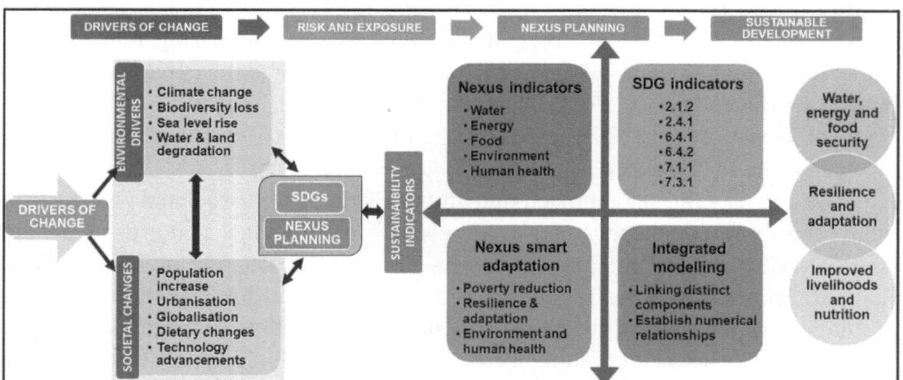

FIGURE 2.56 Balancing Relationships in the FEW Nexus Through Sustainable Development Goals (Mabhaudhi et al., 2021).

As technology advances, the FEW 4.0 approach will become more effective in achieving these goals. This means using new technologies and innovative practices to improve how food, energy, and water are used, addressing these issues in an integrated manner, and ensuring that improvements in one area do not negatively impact the others. FEW 4.0 balances food production, energy use, and water management. Figure 2.56 shows how SDGs help to balance relationships in the FEW Nexus system.

2.10 SUMMARY

This chapter focuses on the history and growth of the Fourth Industrial Revolution (or Industry 4.0), and how FEW 4.0 came to be. It explains the history and growth of the industrial Revolution from the first industrial revolution to the fourth. It also explains how FEW Nexus uses 4IR's high-tech technologies like AI and the IoT to address the problems of the food, energy, and water sectors more efficiently, leading to the rise of FEW 4.0. The chapter explores how FEW Nexus 4.0 harnesses Industry 4.0 technologies to enhance the flexibility, efficiency, and sustainability of food, energy, and water systems. At its core, digitalisation and automation drive innovation, optimising resource use through real-time data collection and automated processes. It also supports the management of the FEW resources in an interconnected manner. Encouraging collaboration and stakeholder engagement, involving diverse parties in decision-making, and employing game theory to strategise interactions. The chapter mentions the changes the food, energy, and water sectors experienced after these technologies were affected. It emphasises on how this integrated approach aims to create a more responsive, adaptable, and sustainable nexus of food, energy, and water resources.

REFERENCES

Abdelmajied, F. Y., & Abdelmajied, F. Y. (2022). Industry 4.0 and its implications: Concept, opportunities, and future directions. *Supply Chain – Recent Advances and New Perspectives in the Industry 4.0 Era*. https://doi.org/10.5772/INTECHOPEN.102520

Abdi, H., Shahbazitabar, M., & Mohammadi-Ivatloo, B. (2020). Food, energy and water nexus: A brief review of definitions, research, and challenges. *Inventions*, *5*(4), 56. https://doi.org/10.3390/INVENTIONS5040056

Abegaz, B. W., Datta, T., & Mahajan, S. M. (2018). Sensor technologies for the energy-water nexus – A review. *Applied Energy*, *210*, 451–466. https://doi.org/10.1016/J.APENERGY.2017.01.033

Abir, S. M. A. A., Anwar, A., Choi, J., & Kayes, A. S. M. (2021). Iot-enabled smart energy grid: Applications and challenges. *IEEE Access*, *9*, 50961–50981. https://doi.org/10.1109/ACCESS.2021.3067331

Adedeji, K. B., Ponnle, A. A., Abu-Mahfouz, A. M., & Kurien, A. M. (2022). Towards digitalization of water supply systems for sustainable smart city development—water 4.0. *Applied Sciences (Switzerland)*, *12*(18). https://doi.org/10.3390/APP12189174

Adeleke, I. A., Nwulu, N. I., & Ogbolumani, O. A. (2023). A hybrid machine learning and embedded IoT-based water quality monitoring system. *Internet of Things*, *22*, 100774. https://doi.org/10.1016/J.IOT.2023.100774

Adil, M., Mahmud, M. A. P., Kouzani, A. Z., & Khoo, S. Y. (2021). Energy trading among electric vehicles based on Stackelberg approaches: A review. *Sustainable Cities and Society*, *75*, 103199. https://doi.org/10.1016/J.SCS.2021.103199

Ahmad, M., Peng, T., Awan, A., & Ahmed, Z. (2023). Policy framework considering resource curse, renewable energy transition, and institutional issues: Fostering sustainable development and sustainable natural resource consumption practices. *Resources Policy*, *86*, 104173. https://doi.org/10.1016/J.RESOURPOL.2023.104173

Ahmad, T., & Zhang, D. (2021). Using the Internet of Things in smart energy systems and networks. *Sustainable Cities and Society*, *68*, 102783. https://doi.org/10.1016/J.SCS.2021.102783

Ahmed Abdulkadir, A., & Al-Turjman, F. (2021). Smart-grid and solar energy harvesting in the IoT era: An overview. *Concurrency and Computation: Practice and Experience*, *33*(4), e4896. https://doi.org/10.1002/CPE.4896

Ahmed, I., Jeon, G., & Piccialli, F. (2022). From artificial intelligence to explainable artificial intelligence in Industry 4.0: A survey on what, how, and where. *IEEE Transactions on Industrial Informatics*, *18*(8), 5031–5042. https://doi.org/10.1109/TII.2022.3146552

Ahmed, S. F., Alam, M. S. Bin, Hassan, M., Rozbu, M. R., Ishtiak, T., Rafa, N., Mofijur, M., Shawkat Ali, A. B. M., & Gandomi, A. H. (2023). Deep learning modelling techniques: Current progress, applications, advantages, and challenges. *Artificial Intelligence Review*, *56*(11), 13521–13617. https://doi.org/10.1007/S10462-023-10466-8

Ahmed, S. F., Alam, M. S. Bin, Hoque, M., Lameesa, A., Afrin, S., Farah, T., Kabir, M., Shafiullah, G. M., & Muyeen, S. M. (2023). Industrial internet of things enabled technologies, challenges, and future directions. *Computers and Electrical Engineering*, *110*, 108847. https://doi.org/10.1016/J.COMPELECENG.2023.108847

Ajibola, O. A., & Ogbolumani, O. A. (2024). Smart battery storage integration in an IoT-based solar-powered waste management system. *Journal of Digital Food, Energy & Water Systems*, *5*(2), 17–42. https://doi.org/10.36615/HREWQ861

Akhter, R., & Sofi, S. A. (2022). Precision agriculture using IoT data analytics and machine learning. *Journal of King Saud University – Computer and Information Sciences*, *34*(8), 5602–5618. https://doi.org/10.1016/J.JKSUCI.2021.05.013

Al Dakheel, J., Del Pero, C., Aste, N., & Leonforte, F. (2020). Smart buildings features and key performance indicators: A review. *Sustainable Cities and Society*, *61*, 102328. https://doi.org/10.1016/J.SCS.2020.102328

Al Shahrani, A. M., Alomar, M. A., Alqahtani, K. N., Basingab, M. S., Sharma, B., & Rizwan, A. (2022). Machine learning-enabled smart industrial automation systems using internet of things. *Sensors*, *23*(1), 324. https://doi.org/10.3390/S23010324

Albrecht, T. R., Crootof, A., & Scott, C. A. (2018). The water-energy-food nexus: A systematic review of methods for nexus assessment. *Environmental Research Letters*, *13*(4), 043002. https://doi.org/10.1088/1748-9326/AAA9C6

Alcácer, V., & Cruz-Machado, V. (2019). Scanning the Industry 4.0: A literature review on technologies for manufacturing systems. *Engineering Science and Technology, an International Journal*, *22*(3), 899–919. https://doi.org/10.1016/J.JESTCH.2019.01.006

Alhammadi, A., Alsyouf, I., Semeraro, C., & Obaideen, K. (2024). The role of Industry 4.0 in advancing sustainability development: A focus review in the United Arab Emirates. *Cleaner Engineering and Technology*, *18*, 100708. https://doi.org/10.1016/J.CLET.2023.100708

Ali, A., Ahmad, M., Nawaz, M., & Sattar, F. (2024). Spatial data infrastructure as the means to assemble geographic information necessary for effective agricultural policies in Pakistan. https://doi.org/10.1177/02666669241244503

Ali, Q., Parveen, S., Yaacob, H., & Zaini, Z. (2022). The management of Industry 4.0 technologies and environmental assets for optimal performance of industrial firms in Malaysia. *Environmental Science and Pollution Research*, *29*(35), 52964–52983. https://doi.org/10.1007/S11356-022-19666-1

Ali, S. M., & Acquaye, A. (2024). An examination of water-energy-food nexus: From theory to application. *Renewable and Sustainable Energy Reviews*, *202*, 114669. https://doi.org/10.1016/J.RSER.2024.114669

Aljohani, T., & Almutairi, A. (2024). A comprehensive survey of cyberattacks on EVs: Research domains, attacks, defensive mechanisms, and verification methods. *Defence Technology*, *42*, 31–58. https://doi.org/10.1016/J.DT.2024.06.009

Anandhi, A., Srivastava, P., Mohtar, R. H., Lawford, R. G., Sen, S., & Lamba, J. (2023). Methodologies and principles for developing nexus definitions and conceptualizations: Lessons from FEW nexus studies. *Journal of the ASABE*, *66*(2), 205–230. https://doi.org/10.13031/JA.14539

Anjimoon, S., Chandrashekar, R., Singh, N., Parmar, A., Sharma, N., & Mohammad, Q. (2024). Secure and sustainable energy distribution through blockchain technology in smart grids. *E3S Web of Conferences*, *505*, 02002. https://doi.org/10.1051/E3SCONF/202450502002

Apeh, O. O., & Nwulu, N. I. (2024). The water-energy-food-ecosystem nexus scenario in Africa: Perspective and policy implementations. *Energy Reports*, *11*, 5947–5962. https://doi.org/10.1016/J.EGYR.2024.05.060

Apeh, O. O., & Nwulu, N. I. (2025). Improving traceability and sustainability in the agri-food industry through blockchain technology: A bibliometric approach, benefits and challenges. *Energy Nexus*, *17*, 100388. https://doi.org/10.1016/J.NEXUS.2025.100388

Arfanuzzaman, M., & Atiq Rahman, A. (2017). Sustainable water demand management in the face of rapid urbanization and ground water depletion for social–ecological resilience building. *Global Ecology and Conservation*, *10*, 9–22. https://doi.org/10.1016/J.GECCO.2017.01.005

Arlinghaus, R., Alós, J., Beardmore, B., Daedlow, K., Dorow, M., Fujitani, M., Hühn, D., Haider, W., Hunt, L. M., Johnson, B. M., Johnston, F., Klefoth, T., Matsumura, S., Monk, C., Pagel, T., Post, J. R., Rapp, T., Riepe, C., Ward, H., & Wolter, C. (2017). Understanding and managing freshwater recreational fisheries as complex adaptive social-ecological systems. *Reviews in Fisheries Science & Aquaculture*, *25*(1), 1–41. https://doi.org/10.1080/23308249.2016.1209160

Askerbekov, D., Garza-Reyes, J. A., Roy Ghatak, R., Joshi, R., Kandasamy, J., & Luiz de Mattos Nascimento, D. (2024). Embracing drones and the Internet of drones systems in manufacturing – An exploration of obstacles. *Technology in Society*, *78*, 102648. https://doi.org/10.1016/J.TECHSOC.2024.102648

Athanasopoulou, K., Daneva, G. N., Adamopoulos, P. G., & Scorilas, A. (2022). Artificial intelligence: The milestone in modern biomedical research. *BioMedInformatics*, *2*(4), 727–744. https://doi.org/10.3390/BIOMEDINFORMATICS2040049

Atlam, H. F., & Wills, G. B. (2019). Technical aspects of blockchain and IoT. *Advances in Computers*, *115*, 1–39. https://doi.org/10.1016/BS.ADCOM.2018.10.006

Azzam, A., Samy, G., Hagras, M. A., & ElKholy, R. (2023). Geographic information systems-based framework for water–energy–food nexus assessments. *Ain Shams Engineering Journal*, *14*(12), 102224. https://doi.org/10.1016/J.ASEJ.2023.102224

Bai, C., Dallasega, P., Orzes, G., & Sarkis, J. (2020). Industry 4.0 technologies assessment: A sustainability perspective. *International Journal of Production Economics*, *229*, 107776. https://doi.org/10.1016/J.IJPE.2020.107776

Basheer, M., & Abdo, G. M. (2021). Implementing the water–energy–food–ecosystems nexus and achieving the sustainable development goals. *Implementing the Water–Energy–Food–Ecosystems Nexus and Achieving the Sustainable Development Goals*, 81–96. https://unesdoc.unesco.org/ark:/48223/pf0000379588

Basheer, M., Wheeler, K. G., Ribbe, L., Majdalawi, M., Abdo, G., & Zagona, E. A. (2018). Quantifying and evaluating the impacts of cooperation in transboundary river basins on the Water-Energy-Food nexus: The Blue Nile Basin. *Science of The Total Environment*, *630*, 1309–1323. https://doi.org/10.1016/J.SCITOTENV.2018.02.249

Baydaş, M., Elma, O. E., & Stević, Ž. (2024). Proposal of an innovative MCDA evaluation methodology: Knowledge discovery through rank reversal, standard deviation, and relationship with stock return. *Financial Innovation*, *10*(1), 1–35. https://doi.org/10.1186/S40854-023-00526-X

Bazilian, M., Rogner, H., Howells, M., Hermann, S., Arent, D., Gielen, D., Steduto, P., Mueller, A., Komor, P., Tol, R. S. J., & Yumkella, K. K. (2011). Considering the energy, water and food nexus: Towards an integrated modelling approach. *Energy Policy*, *39*(12), 7896–7906. https://doi.org/10.1016/J.ENPOL.2011.09.039

Bedi, P., Goyal, S. B., Rajawat, A. S., Kumar, J., Malik, S., & Radhakrishnan, L. C. (2024). *Industry Revolution 4.0: From Industrial Automation to Industrial Autonomy* (pp. 321–356). https://doi.org/10.1007/978-981-99-5354-7_17

Beier, G., Niehoff, S., & Hoffmann, M. (2021). Industry 4.0: A step towards achieving the SDGs? A critical literature review. *Discover Sustainability*, *2*(1), 1–21. https://doi.org/10.1007/S43621-021-00030-1

Bellezoni, R. A., Seto, K. C., & Puppim de Oliveira, J. A. (2022). What can cities do to enhance water-energy-food nexus as a sustainable development strategy? *Sustainable Development Goals Series, Part F*, *2735*, 39–57. https://doi.org/10.1007/978-3-031-05472-3_3

Benotsmane, R., Kovács, G., & Dudás, L. (2019). Economic, social impacts and operation of smart factories in Industry 4.0 focusing on simulation and artificial intelligence of collaborating robots. *Social Sciences*, *8*(5), 143. https://doi.org/10.3390/SOCSCI8050143

Bettiol, M., Capestro, M., Di Maria, E., & Ganau, R. (2024). Is this time different? How Industry 4.0 affects firms' labor productivity. *Small Business Economics*, *62*(4), 1449–1467. https://doi.org/10.1007/S11187-023-00825-8

Bhambhani, A., van der Hoek, J. P., & Kapelan, Z. (2022). Life cycle sustainability assessment framework for water sector resource recovery solutions: Strengths and weaknesses. *Resources, Conservation and Recycling*, *180*, 106151. https://doi.org/10.1016/J.RESCONREC.2021.106151

Bharathi, A., Balamurugan, B., Chokkanathan, K., Sathiyaraj, R., & Singh, A. (2019). Internet of things technologies. *Internet of Things in Biomedical Engineering*, 291–322. https://doi.org/10.1016/B978-0-12-817356-5.00015-2

Biazi, V., & Marques, C. (2023). Industry 4.0-based smart systems in aquaculture: A comprehensive review. *Aquacultural Engineering*, *103*, 102360. https://doi.org/10.1016/J.AQUAENG.2023.102360

Bibri, S. E., & Krogstie, J. (2020). Environmentally data-driven smart sustainable cities: Applied innovative solutions for energy efficiency, pollution reduction, and urban metabolism. *Energy Informatics*, *3*(1), 1–59. https://doi.org/10.1186/S42162-020-00130-8

Bilgen, S., Kaygusuz, K., & Sari, A. (2004). Renewable energy for a clean and sustainable future. *Energy Sources*, *26*(12), 1119–1129. https://doi.org/10.1080/00908310490441421

Birkmann, J., & Pardoe, J. (2014). Climate change adaptation and disaster risk reduction: Fundamentals, synergies and mismatches. *Adapting to Climate Change: Lessons from Natural Hazards Planning*, 41–56. https://doi.org/10.1007/978-94-017-8631-7_2

Blanes, C., Mellado, M., & Beltran, P. (2014). Novel additive manufacturing pneumatic actuators and mechanisms for food handling grippers. *Actuators*, *3*(3), 205–225. https://doi.org/10.3390/ACT3030205

Boas, I., Biermann, F., & Kanie, N. (2016). Cross-sectoral strategies in global sustainability governance: Towards a nexus approach. *International Environmental Agreements: Politics, Law and Economics*, *16*(3), 449–464. https://doi.org/10.1007/S10784-016-9321-1

Bouzidi, Z., Terrissa, L. S., Zerhouni, N., & Ayad, S. (2020). An efficient cloud prognostic approach for aircraft engines fleet trending. *International Journal of Computers and Applications*, *42*(5), 514–529. https://doi.org/10.1080/1206212X.2018.1488024

Brears, R. C. (2021). Circular economy and the water-food nexus. *The Palgrave Encyclopedia of Urban and Regional Futures*, 1–5. https://doi.org/10.1007/978-3-030-51812-7_98-1

Breit, A., Waltersdorfer, L., Ekaputra, F. J., Sabou, M., Ekelhart, A., Iana, A., Paulheim, H., Portisch, J., Revenko, A., Teije, A. Ten, & Van Harmelen, F. (2023). Combining machine learning and semantic web: A systematic mapping study. *ACM Computing Surveys*, *55*(14 S). https://doi.org/10.1145/3586163

Brisco, B., Brown, R. J., Hirose, T., Mc Naim, H., & Staenz, K. (1998). Precision agriculture and the role of remote sensing: A review. *Canadian Journal of Remote Sensing*, *24*(3), 315–327. https://doi.org/10.1080/07038992.1998.10855254

Caballero-Peña, J., Cadena-Zarate, C., Parrado-Duque, A., & Osma-Pinto, G. (2022). Distributed energy resources on distribution networks: A systematic review of modelling, simulation, metrics, and impacts. *International Journal of Electrical Power & Energy Systems*, *138*, 107900. https://doi.org/10.1016/J.IJEPES.2021.107900

Cagno, E., Neri, A., Negri, M., Bassani, C. A., & Lampertico, T. (2021). The role of digital technologies in operationalizing the circular economy transition: A systematic literature review. *Applied Sciences*, *11*(8), 3328. https://doi.org/10.3390/APP11083328

Cai, X., Wallington, K., Shafiee-Jood, M., & Marston, L. (2018). Understanding and managing the food-energy-water nexus – Opportunities for water resources research. *Advances in Water Resources*, *111*, 259–273. https://doi.org/10.1016/J.ADVWATRES.2017.11.014

Caiado Couto, L., Campos, L. C., da Fonseca-Zang, W., Zang, J., & Bleischwitz, R. (2021). Water, waste, energy and food nexus in Brazil: Identifying a resource interlinkage research agenda through a systematic review. *Renewable and Sustainable Energy Reviews*, *138*, 110554. https://doi.org/10.1016/J.RSER.2020.110554

Calabrese, G. G., & Falavigna, G. (2022). Does Industry 4.0 improve productivity? Evidence from the Italian automotive supply chain. *International Journal of Automotive Technology and Management*, *22*(4), 506–526. https://ideas.repec.org/a/ids/ijatma/v22y2022i4p506-526.html

Cansino-Loeza, B., Sánchez-Zarco, X. G., Mora-Jacobo, E. G., Saggiante-Mauro, F. E., González-Bravo, R., Mahlknecht, J., & Ponce-Ortega, J. M. (2020). Systematic approach for assessing the water-energy-food nexus for sustainable development in regions with resource scarcities. *ACS Sustainable Chemistry and Engineering*, *8*(36), 13734–13748. https://doi.org/10.1021/acssuschemeng.0c04333

Chai, L., Liao, X., Yang, L., & Yan, X. (2018). Assessing life cycle water use and pollution of coal-fired power generation in China using input-output analysis. *Applied Energy*, *231*, 951–958. https://doi.org/10.1016/J.APENERGY.2018.09.178

Chen, M., Sinha, A., Hu, K., & Shah, M. I. (2021). Impact of technological innovation on energy efficiency in Industry 4.0 era: Moderation of shadow economy in sustainable development. *Technological Forecasting and Social Change*, *164*, 120521. https://doi.org/10.1016/J.TECHFORE.2020.120521

Cheng, Y., Wang, J., & Shu, K. (2023). The coupling and coordination assessment of food-water-energy systems in China based on sustainable development goals. *Sustainable Production and Consumption*, *35*, 338–348. https://doi.org/10.1016/J.SPC.2022.11.011

Chibayambuya, J., & Theron, D. J. (2007). The application of holistic risk management in the banking industry. *Acta Commercii*, *7*(1). https://doi.org/10.4102/AC.V7I1.25

Cohen, Y., Naseraldin, H., Chaudhuri, A., & Pilati, F. (2019). Assembly systems in Industry 4.0 era: A road map to understand Assembly 4.0. *International Journal of Advanced Manufacturing Technology*, *105*(9), 4037–4054. https://doi.org/10.1007/s00170-019-04203-1

Corominas, L., Byrne, D. M., Guest, J. S., Hospido, A., Roux, P., Shaw, A., & Short, M. D. (2020). The application of life cycle assessment (LCA) to wastewater treatment: A best practice guide and critical review. *Water Research*, *184*, 116058. https://doi.org/10.1016/J.WATRES.2020.116058

Croes, P. R., & Vermeulen, W. J. V. (2021). The assessment of positive impacts in LCA of products. *International Journal of Life Cycle Assessment*, *26*(1), 143–156. https://doi.org/10.1007/S11367-020-01820-x

Cunico, F., Aldegheri, S., Avogaro, A., Boldo, M., Bombieri, N., Capogrosso, L., Caputo, A., Carra, D., Centomo, S., Cheng, D. S., Cinquetti, E., Cristani, M., Marchi, M. De,

Demrozi, F., Emporio, M., Fummi, F., Geretti, L., Germiniani, S., Giachetti, A., ... Quaglia, D. (2024). Enhancing safety and privacy in Industry 4.0: The ICE laboratory case study. *IEEE Access*. https://doi.org/10.1109/ACCESS.2024.3479411

Dabić, M., Maley, J. F., Črešnar, R., & Nedelko, Z. (2023). Unappreciated channel of manufacturing productivity under Industry 4.0: Leadership values and capabilities. *Journal of Business Research*, *162*, 113900. https://doi.org/10.1016/J.JBUSRES.2023.113900

D'amore, G., Di Vaio, A., Balsalobre-Lorente, D., & Boccia, F. (2022). Artificial intelligence in the water–energy–food model: A holistic approach towards sustainable development goals. *Sustainability*, *14*(2), 867. https://doi.org/10.3390/SU14020867

Dantas, T. E. T., de-Souza, E. D., Destro, I. R., Hammes, G., Rodriguez, C. M. T., & Soares, S. R. (2021). How the combination of Circular Economy and Industry 4.0 can contribute towards achieving the sustainable development goals. *Sustainable Production and Consumption*, *26*, 213–227. https://doi.org/10.1016/J.SPC.2020.10.005

Das, A., Kadawla, K., Nath, H., Chakraborty, S., Ali, H., Singh, S., & Dubey, V. K. (2024). Drone-based intelligent spraying of pesticides: Current challenges and its future prospects. *Applications of Computer Vision and Drone Technology in Agriculture 4.0*, 199–223. https://doi.org/10.1007/978-981-99-8684-2_12

Dastres, R., Soori, M., & Asamel, M. (2022). Radio frequency identification (RFID) based wireless manufacturing systems, a review. *Independent Journal of Management & Production, ISSN-e 2236-269X, Vol. 13, No. 1, 2022 (Ejemplar Dedicado a: Independent Journal of Management & Production), Págs. 258–290, 13*(1), 258–290. https://doi.org/10.14807/ijmp.v13i1.1497

David, L. O., Adepoju, O., Nwulu, N., & Aigbavboa, C. (2024). Determining the impact of economic indicators on water, energy and food nexus for sustainable resource security. *Clean Technologies and Environmental Policy*, *26*(3), 803–820. https://doi.org/10.1007/S10098-023-02651-8

David, L. O., Aigbavboa, C., Nwulu, N., & Adepoju, O. O. (2024). General description of water, energy, and food resources. *Green Energy and Technology, Part F3772*, 21–61. https://doi.org/10.1007/978-3-031-72833-4_2

David, L. O., Nwulu, N. I., Aigbavboa, C. O., & Adepoju, O. O. (2022). Integrating Fourth Industrial Revolution (4IR) technologies into the water, energy & food nexus for sustainable security: A bibliometric analysis. *Journal of Cleaner Production*, *363*, 132522. https://doi.org/10.1016/J.JCLEPRO.2022.132522

Dayioğlu, M. A., & Türker, U. (2021). Digital transformation for sustainable future – Agriculture 4.0: A review. *Journal of Agricultural Sciences*, *27*(4), 373–399. https://doi.org/10.15832/ANKUTBD.986431

Dean, M. (2022). Including multiple perspectives in participatory multi-criteria analysis: A framework for investigation. *Evaluation*, *28*(4), 505–539. https://doi.org/10.1177/13563890221123822

Del Borghi, A., Moreschi, L., & Gallo, M. (2020). Circular economy approach to reduce water–energy–food nexus. *Current Opinion in Environmental Science & Health*, *13*, 23–28. https://doi.org/10.1016/J.COESH.2019.10.002

Dey, S., Sreenivasulu, A., Veerendra, G. T. N., Rao, K. V., & Babu, P. S. S. A. (2022). Renewable energy present status and future potentials in India: An overview. *Innovation and Green Development*, *1*(1), 100006. https://doi.org/10.1016/J.IGD.2022.100006

Diaz, A., Schöggl, J. P., Reyes, T., & Baumgartner, R. J. (2021). Sustainable product development in a circular economy: Implications for products, actors, decision-making support and lifecycle information management. *Sustainable Production and Consumption*, *26*, 1031–1045. https://doi.org/10.1016/J.SPC.2020.12.044

'Digital takeover'. (n.d.). Retrieved July 15, 2024, from https://www.aerosociety.com/news/digital-takeover/

Dileep, G. (2020). A survey on smart grid technologies and applications. *Renewable Energy*, *146*, 2589–2625. https://doi.org/10.1016/J.RENENE.2019.08.092

Ding, T., Liang, L., Zhou, K., Yang, M., & Wei, Y. (2020). Water-energy nexus: The origin, development and prospect. *Ecological Modelling*, *419*, 108943. https://doi.org/10.1016/J.ECOLMODEL.2020.108943

Do, K. H. P. (2022). Water resource management: Challenges and opportunities with game theory approaches. *Oxford Research Encyclopedia of Environmental Science*. https://doi.org/10.1093/ACREFORE/9780199389414.013.763

D'Odorico, P., Chiarelli, D. D., Rosa, L., Bini, A., Zilberman, D., & Rulli, M. C. (2020). The global value of water in agriculture. *Proceedings of the National Academy of Sciences of the United States of America*, *117*(36), 21985–21993. https://doi.org/10.1073/PNAS.2005835117

D'Odorico, P., Davis, K. F., Rosa, L., Carr, J. A., Chiarelli, D., Dell'Angelo, J., Gephart, J., MacDonald, G. K., Seekell, D. A., Suweis, S., & Rulli, M. C. (2018). The global food-energy-water nexus. *Reviews of Geophysics*, *56*(3), 456–531. https://doi.org/10.1029/2017RG000591

dos Santos Filho, V. H., de Resende, L. M. M., & Pontes, J. (2024). Development of a theoretical model for digital risks arising from the implementation of Industry 4.0 (TMR-I4.0). *Future Internet*, *16*(6), 215. https://doi.org/10.3390/FI16060215

du Plessis, A. (2019). Primary water quality challenges, contaminants and the world's dirtiest places. *Springer Water*, 79–114. https://doi.org/10.1007/978-3-030-03186-2_5

Endo, A., Burnett, K., Orencio, P. M., Kumazawa, T., Wada, C. A., Ishii, A., Tsurita, I., & Taniguchi, M. (2015). Methods of the water-energy-food nexus. *Water*, *7*(10), 5806–5830. https://doi.org/10.3390/W7105806

Endo, A., Tsurita, I., Burnett, K., & Orencio, P. M. (2017). A review of the current state of research on the water, energy, and food nexus. *Journal of Hydrology: Regional Studies*, *11*, 20–30. https://doi.org/10.1016/J.EJRH.2015.11.010

Esmaeilion, F. (2020). Hybrid renewable energy systems for desalination. *Applied Water Science*, *10*(3), 1–47. https://doi.org/10.1007/S13201-020-1168-5

Estoque, R. C. (2023). Complexity and diversity of nexuses: A review of the nexus approach in the sustainability context. *Science of The Total Environment*, *854*, 158612. https://doi.org/10.1016/J.SCITOTENV.2022.158612

Farayola, O. A. (2024). Revolutionizing banking security: Integrating artificial intelligence, blockchain, and business intelligence for enhanced cybersecurity. *Finance & Accounting Research Journal*, *6*(4), 501–514. https://doi.org/10.51594/FARJ.V6I4.990

Farooq, M. S., Abdullah, M., Riaz, S., Alvi, A., Rustam, F., Flores, M. A. L., Galán, J. C., Samad, M. A., & Ashraf, I. (2023). A survey on the role of industrial IoT in manufacturing for implementation of smart industry. *Sensors*, *23*(21), 8958. https://doi.org/10.3390/S23218958

Fernandez-Carames, T. M., & Fraga-Lamas, P. (2018). A review on human-centered IoT-connected smart labels for the Industry 4.0. *IEEE Access*, *6*, 25939–25957. https://doi.org/10.1109/ACCESS.2018.2833501

Ferreira, D. C., Graziele, I., Marques, R. C., & Gonçalves, J. (2021a). Investment in drinking water and sanitation infrastructure and its impact on waterborne diseases dissemination: The Brazilian case. *Science of The Total Environment*, *779*, 146279. https://doi.org/10.1016/J.SCITOTENV.2021.146279

Fontenelle, A. L., Nilsson, E., Hidalgo, I. G., Uvo, C. B., & Peyerl, D. (2022). Temporal understanding of the water–energy nexus: A literature review. *Energies*, *15*(8), 2851. https://doi.org/10.3390/EN15082851

Food, Energy-Water (FEW) NEXUS Partnerships | Department of Economic and Social Affairs. (n.d.). Retrieved July 15, 2024, from https://sdgs.un.org/partnerships/food-energy-water-few-nexus-partnerships

Frank, A. G., Dalenogare, L. S., & Ayala, N. F. (2019). Industry 4.0 technologies: Implementation patterns in manufacturing companies. *International Journal of Production Economics*, *210*, 15–26. https://doi.org/10.1016/J.IJPE.2019.01.004

Gain, A. K., Giupponi, C., & Benson, D. (2015). The water–energy–food (WEF) security nexus: The policy perspective of Bangladesh. *Water International*, *40*(5–6), 895–910. https://doi.org/10.1080/02508060.2015.1087616

Garver, G. (2023). An ethical approach to assessments of harm from climate change and global heating: The problem with integrated assessment models and other Value-Laden tools. In *The Routledge Handbook of Applied Climate Change Ethics* (pp. 276–285). https://doi.org/10.4324/9781003039860-30

Gassar, A. A. A., & Cha, S. H. (2021). Review of geographic information systems-based rooftop solar photovoltaic potential estimation approaches at urban scales. *Applied Energy*, *291*, 116817. https://doi.org/10.1016/J.APENERGY.2021.116817

Gebhardt, M., Kopyto, M., Birkel, H., & Hartmann, E. (2022). Industry 4.0 technologies as enablers of collaboration in circular supply chains: A systematic literature review. *International Journal of Production Research*, *60*(23), 6967–6995. https://doi.org/10.1080/00207543.2021.1999521

George, A. S. (2024). The fourth industrial revolution: A primer on Industry 4.0 and its transformative impact. *Partners Universal Innovative Research Publication*, *2*(1). https://doi.org/10.5281/zenodo.10671872

Geuna, A., Guerzoni, M., Nuccio, M., Pammolli, F., & Rungi, A. (2021). *Digital Technologies and Industrial Transformations*. 7–38. https://doi.org/10.1007/978-3-030-85158-3_2

Ghobakhloo, M. (2020). Industry 4.0, digitization, and opportunities for sustainability. *Journal of Cleaner Production*, *252*, 119869. https://doi.org/10.1016/J.JCLEPRO.2019.119869

Ghosh, B., Gubareva, M., Ghosh, A., Paparas, D., & Vo, X. V. (2024). Food, energy, and water nexus: A study on interconnectedness and trade-offs. *Energy Economics*, *133*, 107521. https://doi.org/10.1016/J.ENECO.2024.107521

Glasgow, H. B., Burkholder, J. A. M., Reed, R. E., Lewitus, A. J., & Kleinman, J. E. (2004). Real-time remote monitoring of water quality: A review of current applications, and advancements in sensor, telemetry, and computing technologies. *Journal of Experimental Marine Biology and Ecology*, *300*(1–2), 409–448. https://doi.org/10.1016/J.JEMBE.2004.02.022

Groumpos, P. P. (2021). A critical historical and scientific overview of all industrial revolutions. *IFAC-PapersOnLine*, *54*(13), 464–471. https://doi.org/10.1016/J.IFACOL.2021.10.492

Grybauskas, A., Stefanini, A., & Ghobakhloo, M. (2022). Social sustainability in the age of digitalization: A systematic literature review on the social implications of Industry 4.0. *Technology in Society*, *70*, 101997. https://doi.org/10.1016/J.TECHSOC.2022.101997

Guemouria, A., Chehbouni, A., Belaqziz, S., Epule, T., Ait Brahim, Y., El Khalki, E. M., Dhiba, D., & Bouchaou, L. (2023). System dynamics approach for water resources management: A case study from the Souss-Massa Basin. *Water*, *15*(8), 1506. https://doi.org/10.3390/W15081506

Gulnergiz, E. T., Dilibal, S., Gormus, B., Danquah, J. O., & Emon, O. F. (2023). Additively manufactured soft pneumatic gripper integrated remotely operated underwater vehicle (ROV) for grasping archeological remains. *HORA 2023 – 2023 5th International Congress on Human-Computer Interaction, Optimization and Robotic Applications, Proceedings*. https://doi.org/10.1109/HORA58378.2023.10156774

Guney, C., & Ganjali, N. (2023). *Geodesigning A Game-Theory-Based Compromise Solution Approach in Determining the Cultivation Pattern*. https://doi.org/10.21203/RS.3.RS-2543128/V1

Gupta, A. D., Pandey, P., Feijóo, A., Yaseen, Z. M., & Bokde, N. D. (2020). Smart water technology for efficient water resource management: A review. *Energies*, *13*(23), 6268. https://doi.org/10.3390/EN13236268

Gupta, P., Singh, J., Verma, S., Chandel, A. S., & Bhatla, R. (2021). Impact of climate change and water quality degradation on food security and agriculture. In *Water Conservation in the Era of Global Climate Change* (pp. 1–22). https://doi.org/10.1016/B978-0-12-820200-5.00006-3

Hajian, M., & Kashani, S. J. (2021). Evolution of the concept of sustainability. From Brundtland report to sustainable development goals. In *Sustainable Resource Management: Modern Approaches and Contexts* (pp. 1–24). https://doi.org/10.1016/B978-0-12-824342-8.00018-3

Han, X., Zhao, Y., Gao, X., Wang, Y., Jiang, S., Zhu, Y., & An, T. (2021). The historical footprint and future challenges of water-energy-food nexus research: A bibliometric review towards sustainable development. *29*(2), 260–276. https://doi.org/10.1139/er-2020-0085

Harahap, N. J., & Rafika, M. (2020). Industrial Revolution 4.0: And the impact on human resources. *ECOBISMA (JURNAL EKONOMI, BISNIS DAN MANAJEMEN)*, *7*(1), 89–96. https://doi.org/10.36987/ECOBI.V7I1.1545

Hasnan, N. Z. N., Yusoff, Y. M., Lim, S. A. H., & Kamarudin, K. (2023). Potential of Industrial Internet of Things (IIoT) to improve inefficiencies in food manufacturing. *AIP Conference Proceedings*, *2907*(1). https://doi.org/10.1063/5.0171393

Hassoun, A., Aït-Kaddour, A., Abu-Mahfouz, A. M., Rathod, N. B., Bader, F., Barba, F. J., Biancolillo, A., Cropotova, J., Galanakis, C. M., Jambrak, A. R., Lorenzo, J. M., Måge, I., Ozogul, F., & Regenstein, J. (2023). The fourth industrial revolution in the food industry—Part I: Industry 4.0 technologies. *Critical Reviews in Food Science and Nutrition*, *63*(23), 6547–6563. https://doi.org/10.1080/10408398.2022.2034735

Henderson, K., & Loreau, M. (2023). A model of sustainable development goals: Challenges and opportunities in promoting human well-being and environmental sustainability. *Ecological Modelling*, *475*, 110164. https://doi.org/10.1016/J.ECOLMODEL.2022.110164

Hendren, K., Newcomer, K., Pandey, S. K., Smith, M., & Sumner, N. (2023). How qualitative research methods can be leveraged to strengthen mixed methods research in public policy and public administration? *Public Administration Review*, *83*(3), 468–485. https://doi.org/10.1111/PUAR.13528

Henry, S., Dujardin, S., Henriet, E., & Baltazar, S. C. S. (2022). *Qualitative Data and Approaches to Population–Environment Inquiry*. 139–163. https://doi.org/10.1007/978-3-030-76433-3_7

Henßler, M., Bach, V., Berger, M., Finkbeiner, M., & Ruhland, K. (2016). Resource efficiency assessment—Comparing a plug-in hybrid with a conventional combustion engine. *Resources*, *5*(1), 5. https://doi.org/10.3390/RESOURCES5010005

Herrera-Franco, G., Bollmann, H. A., Pasqual Lofhagen, J. C., Bravo-Montero, Lady, & Carrión-Mero, P. (2023). Approach on water-energy-food (WEF) nexus and climate change: A tool in decision-making processes. *Environmental Development*, *46*, 100858. https://doi.org/10.1016/J.ENVDEV.2023.100858

Himeur, Y., Elnour, M., Fadli, F., Meskin, N., Petri, I., Rezgui, Y., Bensaali, F., & Amira, A. (2023). AI-big data analytics for building automation and management systems: A survey, actual challenges and future perspectives. *Artificial Intelligence Review*, *56*(6), 4929–5021. https://doi.org/10.1007/S10462-022-10286-2

Hoolohan, C., Larkin, A., McLachlan, C., Falconer, R., Soutar, I., Suckling, J., Varga, L., Haltas, I., Druckman, A., Lumbroso, D., Scott, M., Gilmour, D., Ledbetter, R., McGrane, S., Mitchell, C., & Yu, D. (2018). Engaging stakeholders in research to address water–energy–food (WEF) nexus challenges. *Sustainability Science*, *13*(5), 1415–1426. https://doi.org/10.1007/S11625-018-0552-7

Höse, K., Amaral, A., Götze, U., & Peças, P. (2023). Manufacturing flexibility through Industry 4.0 technological concepts—impact and assessment. *Global Journal of Flexible Systems Management*, *24*(2), 271–289. https://doi.org/10.1007/S40171-023-00339-y

Huang, K., Wang, K., Lee, P. K., & Yeung, A. C. (2023). The impact of industry 4.0 on supply chain capability and supply chain resilience: A dynamic resource-based view. *International Journal of Production Economics*, *262*, 108913. https://doi.org/10.1016/j.ijpe.2023.108913

Huang, Wen & Li, Heng. (2022). Game theory applications in the electricity market and renewable energy trading: A critical survey. Frontiers in Energy Research. 10. 1009217. 10.3389/fenrg.2022.1009217. https://doi.org/10.3389/fenrg.2022.1009217

Human Insight AI: An Innovative Technology Bridging The Gap between Humans and Machines for a Safe, Sustainable Future. (n.d.). https://doi.org/10.5281/ZENODO.7723117

Huo, J., & Peng, C. (2023). Depletion of natural resources and environmental quality: Prospects of energy use, energy imports, and economic growth hindrances. *Resources Policy*, *86*, 104049. https://doi.org/10.1016/J.RESOURPOL.2023.104049

Hussain, A., Arif, S. M., & Aslam, M. (2017). Emerging renewable and sustainable energy technologies: State of the art. *Renewable and Sustainable Energy Reviews*, *71*, 12–28. https://doi.org/10.1016/J.RSER.2016.12.033

Ibrahim, M. A. R., Jaini, N. I., & Khalif, K. M. N. K. (2021). A comprehensive review of hybrid game theory techniques and multi-criteria decision-making methods. *Journal of Physics: Conference Series*, *1988*(1), 012056. https://doi.org/10.1088/1742-6596/1988/1/012056

Industry 4.0 Market Size, Share | Growth Analysis [2022-2029]. (n.d.). Retrieved July 15, 2024, from https://www.fortunebusinessinsights.com/industry-4-0-market-102375

Ingrao, C., Strippoli, R., Lagioia, G., & Huisingh, D. (2023). Water scarcity in agriculture: An overview of causes, impacts and approaches for reducing the risks. *Heliyon*, *9*(8), e18507. https://doi.org/10.1016/j.heliyon.2023.e18507

Islam, N., Rashid, M. M., Pasandideh, F., Ray, B., Moore, S., & Kadel, R. (2021). A review of applications and communication technologies for Internet of Things (IoT) and Unmanned Aerial Vehicle (UAV) based sustainable smart farming. *Sustainability (Switzerland)*, *13*(4), 1–20. https://doi.org/10.3390/SU13041821

Ivezić, A., Trudić, B., Stamenković, Z., Kuzmanović, B., Perić, S., Ivošević, B., Buđen, M., & Petrović, K. (2023). Drone-related agrotechnologies for precise plant protection in Western Balkans: Applications, possibilities, and legal framework limitations. *Agronomy*, *13*(10), 2615. https://doi.org/10.3390/AGRONOMY13102615

Jahan, I., Zhang, G., Bhuiyan, M., & Navaratnam, S. (2022). Circular economy of construction and demolition wood waste—A theoretical framework approach. *Sustainability (Switzerland)*, *14*(17). https://doi.org/10.3390/SU141710478

Jain, T. K., & Jain, N. (2020). *Fresh Water Needs and Conservation*. 257–268. https://doi.org/10.1007/978-3-319-95726-5_87

Jambrak, A. R., Nutrizio, M., Djekić, I., Pleslić, S., & Chemat, F. (2021). Internet of nonthermal food processing technologies (IoNTP): Food Industry 4.0 and sustainability. *Applied Sciences*, *11*(2), 686. https://doi.org/10.3390/APP11020686

Jasiulewicz-Kaczmarek, M., Legutko, S., & Kluk, P. (2020). Maintenance 4.0 technologies – New opportunities for sustainability driven maintenance. *Management and Production Engineering Review*, *11*(2), 74–87. https://doi.org/10.24425/mper.2020.133730

Jasiūnas, J., Lund, P. D., & Mikkola, J. (2021). Energy system resilience – A review. *Renewable and Sustainable Energy Reviews*, *150*, 111476. https://doi.org/10.1016/J.RSER.2021.111476

Javaid, M., Haleem, A., Singh, R. P., & Suman, R. (2021). Substantial capabilities of robotics in enhancing Industry 4.0 implementation. *Cognitive Robotics*, *1*, 58–75. https://doi.org/10.1016/J.COGR.2021.06.001

Javaid, M., Haleem, A., Singh, R. P., & Suman, R. (2022a). Enhancing smart farming through the applications of Agriculture 4.0 technologies. *International Journal of Intelligent Networks*, *3*, 150–164. https://doi.org/10.1016/J.IJIN.2022.09.004

Javaid, M., Haleem, A., Singh, R. P., Suman, R., & Gonzalez, E. S. (2022b). Understanding the adoption of Industry 4.0 technologies in improving environmental sustainability. *Sustainable Operations and Computers*, *3*, 203–217. https://doi.org/10.1016/J.SUSOC.2022.01.008

Kabeyi, M. J. B., & Olanrewaju, O. A. (2023). Smart grid technologies and application in the sustainable energy transition: A review. *International Journal of Sustainable Energy*, *42*(1), 685–758. https://doi.org/10.1080/14786451.2023.2222298

Kadam, S., Shinde, A. S., Bari, A. M., & Gujar, J. G. (2023). The role of artificial intelligence and the Internet of Things in smart agriculture towards green engineering. *Applied Computer Vision and Soft Computing with Interpretable AI*, 135–152. https://doi.org/10.1201/9781003359456

Kamyab, H., Khademi, T., Chelliapan, S., SaberiKamarposhti, M., Rezania, S., Yusuf, M., Farajnezhad, M., Abbas, M., Hun Jeon, B., & Ahn, Y. (2023). The latest innovative avenues for the utilization of Artificial Intelligence and Big Data Analytics in water resource management. *Results in Engineering*, *20*, 101566. https://doi.org/10.1016/J.RINENG.2023.101566

Kapucu, N., Beaudet, S., Chang, N. Bin, Qiu, J., & Peng, Z. R. (2023). Partnerships and network governance for urban food-energy-water (FEW) nexus. *International Journal of Public Administration*, *46*(1), 18–31. https://doi.org/10.1080/01900692.2021.1967981

Katz, S. L., Padowski, J. C., Goldsby, M., Brady, M. P., & Hampton, S. E. (2020). Defining the nature of the nexus: Specialization, connectedness, scarcity, and scale in food–energy–water management. *Water*, *12*(4), 972. https://doi.org/10.3390/W12040972

Kellner, E. (2023). Identifying leverage points for shifting water-energy-food nexus cases towards sustainability through the networks of action situations approach combined with systems thinking. *Sustainability Science*, *18*(1), 135–152. https://doi.org/10.1007/S11625-022-01170-7

Khaleel, M., Abulifa, S. A., & Abulifa, A. A. (2023). Artificial intelligent techniques for identifying the cause of disturbances in the power grid. *Brilliance: Research of Artificial Intelligence*, *3*(1), 19–31. https://doi.org/10.47709/BRILLIANCE.V3I1.2165

Khalid, M. (2024). Energy 4.0: AI-enabled digital transformation for sustainable power networks. *Computers & Industrial Engineering*, *193*, 110253. https://doi.org/10.1016/J.CIE.2024.110253

Khan, Z., Abraham, E., Aggarwal, S., Ahmad Khan, M., Arguello, R., Babbar-Sebens, M., Bereslawski, J. L., Bielicki, J. M., Campana, P. E., Silva Carrazzone, M. E., Castanier, H., Chang, F. J., Collins, P., Conchado, A., Dagani, K. R., Daher, B., Dekker, S. C., Delgado, R., Diuana, F. A., … Yue, Q. (2022). Emerging themes and future directions of multi-sector nexus research and implementation. *Frontiers in Environmental Science*, *10*, 918085. https://doi.org/10.3389/FENVS.2022.918085

Kharakhash, O. (2023). Data visualization: Transforming complex data into actionable insights. *Automation of Technological and Business Processes*, *15*(2), 4–12. https://doi.org/10.15673/ATBP.V15I2.2520

Kirimtat, A., Krejcar, O., Kertesz, A., & Tasgetiren, M. F. (2020). Future trends and current state of smart city concepts: A survey. *IEEE Access*, *8*, 86448–86467. https://doi.org/10.1109/ACCESS.2020.2992441

Kılkış, Ş., & Kılkış, B. (2017). Integrated circular economy and education model to address aspects of an energy-water-food nexus in a dairy facility and local contexts. *Journal of Cleaner Production*, *167*, 1084–1098. https://doi.org/10.1016/J.JCLEPRO.2017.03.178

Kliskey, A., Williams, P., Griffith, D. L., Dale, V. H., Schelly, C., Marshall, A. M., Gagnon, V. S., Eaton, W. M., & Floress, K. (2021). Thinking big and thinking small: A conceptual framework for best practices in community and stakeholder engagement in food, energy, and water systems. *Sustainability*, *13*(4), 1–19. https://doi.org/10.3390/SU13042160

Kolberg, D., & Zühlke, D. (2015). Lean automation enabled by Industry 4.0 technologies. *IFAC-PapersOnLine*, *48*(3), 1870–1875. https://doi.org/10.1016/J.IFACOL.2015.06.359

Kremsa, V. (2021). Sustainable management of agricultural resources (agricultural crops and animals). *Sustainable Resource Management: Modern Approaches and Contexts*, 99–145. https://doi.org/10.1016/B978-0-12-824342-8.00010-9

Krishnan, R., Agarwal, R., Bajada, C., & Arshinder, K. (2020). Redesigning a food supply chain for environmental sustainability – An analysis of resource use and recovery. *Journal of Cleaner Production*, *242*, 118374. https://doi.org/10.1016/J.JCLEPRO.2019.118374

Kumar, A., Sharma, S., Goyal, N., Singh, A., Cheng, X., & Singh, P. (2021). Secure and energy-efficient smart building architecture with emerging technology IoT. *Computer Communications*, *176*, 207–217. https://doi.org/10.1016/J.COMCOM.2021.06.003

Kumar, K., Gambhir, G., Dass, A., Tripathi, A. K., Singh, A., Jha, A. K., Yadava, P., Choudhary, M., & Rakshit, S. (2020). Genetically modified crops: Current status and future prospects. *Planta*, *251*(4), 1–27. https://doi.org/10.1007/S00425-020-03372-8

Kumar, N. M., Dash, A., & Singh, N. K. (2018). Internet of things (IoT): An opportunity for energy-food-water nexus. *2018 International Conference on Power Energy, Environment and Intelligent Control, PEEIC 2018*, 68–72. https://doi.org/10.1109/PEEIC.2018.8665632

Lakra, D., Singh, S. K., Gupta, S. K., & Kanga, S. (2024). Enhancing integrated resource management through remote sensing and GIS. *Journal of Geography and Cartography*, *7*(1), 4265. https://doi.org/10.24294/JGC.V7I1.4265

Lakshmi, G. S., Rubanenko, O., Divya, G., & Lavanya, V. (2020). Distribution energy generation using renewable energy sources. *Proceedings – 2020 IEEE India Council International Subsections Conference, INDISCON 2020*, 108–113. https://doi.org/10.1109/INDISCON50162.2020.00033

Lalawmpuii, & Rai, P. K. (2023). Role of water-energy-food nexus in environmental management and climate action. *Energy Nexus*, *11*, 100230. https://doi.org/10.1016/J.NEXUS.2023.100230

Lasi, H., Fettke, P., Kemper, H. G., Feld, T., & Hoffmann, M. (2014). Industry 4.0. *Business and Information Systems Engineering*, *6*(4), 239–242. https://doi.org/10.1007/S12599-014-0334-4

Laufs, J., Borrion, H., & Bradford, B. (2020). Security and the smart city: A systematic review. *Sustainable Cities and Society*, *55*, 102023. https://doi.org/10.1016/J.SCS.2020.102023

Leclerc, P. D., McLay, L. A., & Mayorga, M. E. (2012). Modeling equity for allocating public resources. *International Series in Operations Research and Management Science*, *167*, 97–118. https://doi.org/10.1007/978-1-4614-0806-2_4

Lee, H. J., Kim, M., Lee, H. J., & Kim, M. (2018). The internet of things in a smart connected world. *Internet of Things – Technology, Applications and Standardization*. https://doi.org/10.5772/INTECHOPEN.76128

Lee, J., Ni, J., Singh, J., Jiang, B., Azamfar, M., & Feng, J. (2020). Intelligent maintenance systems and predictive manufacturing. *Journal of Manufacturing Science and Engineering, Transactions of the ASME*, *142*(11). https://doi.org/10.1115/1.4047856

Lee, S., & Choi, D. H. (2019). Reinforcement learning-based energy management of smart home with rooftop solar photovoltaic system, energy storage system, and home appliances. *Sensors*, *19*(18), 3937. https://doi.org/10.3390/S19183937

Li, D., Zuo, Q., & Zhang, Z. (2022). A new assessment method of sustainable water resources utilization considering fairness-efficiency-security: A case study of 31 provinces and cities in China. *Sustainable Cities and Society*, *81*, 103839. https://doi.org/10.1016/J.SCS.2022.103839

Li, J., Dai, J., Issakhov, A., Almojil, S. F., & Souri, A. (2021). Towards decision support systems for energy management in the smart industry and Internet of Things. *Computers & Industrial Engineering*, *161*, 107671. https://doi.org/10.1016/J.CIE.2021.107671

Li, J. (2018). Cyber security meets artificial intelligence: A survey. *Frontiers of Information Technology and Electronic Engineering*, *19*(12), 1462–1474. https://doi.org/10.1631/FITEE.1800573

Li, Y., Dai, J., & Cui, L. (2020). The impact of digital technologies on economic and environmental performance in the context of Industry 4.0: A moderated mediation model. *International Journal of Production Economics*, *229*, 107777. https://doi.org/10.1016/J.IJPE.2020.107777

Liang, D., & Tian, J. (2024). The impact of digital transformation on the high-quality development of enterprises: An exploration based on meta-analysis. *Sustainability*, *16*(8), 3188. https://doi.org/10.3390/SU16083188

Liberty, J. T., Habanabakize, E., Adamu, P. I., & Bata, S. M. (2024). Advancing food manufacturing: Leveraging robotic solutions for enhanced quality assurance and traceability across global supply networks. *Trends in Food Science & Technology*, *153*, 104705. https://doi.org/10.1016/J.TIFS.2024.104705

Lim, M. K., Lai, M., Wang, C., & Lee, S. Y. (2022). Circular economy to ensure production operational sustainability: A green-lean approach. *Sustainable Production and Consumption*, *30*, 130–144. https://doi.org/10.1016/J.SPC.2021.12.001

Lin, Y. H., Tsao, S. Y., & Lin, Y. H. (2019). A smart home energy management system as an intelligent electricity energy audit based on AI-empowered non-intrusive load monitoring. *2019 IEEE International Conference on Consumer Electronics* – Taiwan, *ICCE-TW 2019*, 8991689. https://doi.org/10.1109/ICCE-TW46550.2019.8991689

Liu, J., Hull, V., Godfray, H. C. J., Tilman, D., Gleick, P., Hoff, H., Pahl-Wostl, C., Xu, Z., Chung, M. G., Sun, J., & Li, S. (2018). Nexus approaches to global sustainable development. *Nature Sustainability*, *1*(9), 466–476. https://doi.org/10.1038/s41893-018-0135-8

Lu, J., Wang, X., Cheng, X., Yang, J., Kwan, O., & Wang, X. (2022). Parallel factories for smart industrial operations: From big AI models to field foundational models and scenarios engineering. *IEEE/CAA Journal of Automatica Sinica*, *9*(12), 2079–2086. https://doi.org/10.1109/JAS.2022.106094

Lynham, P. (2022). From data to information. *Digital Agritechnology: Robotics and Systems for Agriculture and Livestock Production*, 11–47. https://doi.org/10.1016/B978-0-12-817634-4.00002-1

Mabhaudhi, T., Nhamo, L., Chibarabada, T. P., Mabaya, G., Mpandeli, S., Liphadzi, S., Senzanje, A., Naidoo, D., Modi, A. T., & Chivenge, P. P. (2021). Assessing progress towards sustainable development goals through nexus planning. *Water*, *13*(9), 1321. https://doi.org/10.3390/W13091321

Macedo, P., Huertas, A., Bottone, C., del Río, J., Hillary, N., Brazzini, T., Wittmayer, J. M., & Penha-Lopes, G. (2020). Learnings from local collaborative transformations: Setting a basis for a sustainability framework. *Sustainability*, *12*(3), 795. https://doi.org/10.3390/SU12030795

Machado, J., Antosz, K., Manupati, V. K., Ren, Y., Abdi, R. El, Mazurkiewicz, D., Ranga, M., Rea, P., Villani, E., Ottaviano, E., Palomar-Cosín, E., & García-Valls, M. (2022). Flexible IoT agriculture systems for irrigation control based on software services. *Sensors*, *22*(24), 9999. https://doi.org/10.3390/S22249999

MacHer, C., Steins, N. A., Ballesteros, M., Kraan, M., Frangoudes, K., Bailly, D., Bertignac, M., Colloca, F., Fitzpatrick, M., Garcia, D., Little, R., Mardle, S., Murillas, A., Pawlowski, L., Philippe, M., Prellezo, R., Sabatella, E., Thébaud, O., Ulrich, C., & Dankel, D. (2021). Towards transdisciplinary decision-support processes in fisheries: Experiences and recommendations from a multidisciplinary collective of researchers. *Aquatic Living Resources*, *34*, 13. https://doi.org/10.1051/alr/2021010

Maghsoudi, M., & Nezafati, N. (2023). Navigating the acceptance of implementing business intelligence in organizations: A system dynamics approach. *Telematics and Informatics Reports*, *11*, 100070. https://doi.org/10.1016/J.TELER.2023.100070

Majeed Butt, O., Zulqarnain, M., & Majeed Butt, T. (2021). Recent advancement in smart grid technology: Future prospects in the electrical power network. *Ain Shams Engineering Journal*, *12*(1), 687–695. https://doi.org/10.1016/J.ASEJ.2020.05.004

Manish, M., Thakur, P., & Singh, G. (2023). Water 4.0: The vision, potentials and simulation environments. *AIP Conference Proceedings*, *2495*(1). https://doi.org/10.1063/5.0123789

Mao, Y., Liu, Y., Zhuo, L., Wang, W., Li, M., Feng, B., & Wu, P. (2021). Quantitative evaluation of spatial scale effects on regional water footprint in crop production.

Resources, Conservation and Recycling, *173*, 105709. https://doi.org/10.1016/J.RESCONREC.2021.105709

Mat, I., Mohd Kassim, M. R., Harun, A. N., & Mat Yusoff, I. (2017). IoT in precision agriculture applications using wireless moisture sensor network. *ICOS 2016 – 2016 IEEE Conference on Open Systems*, 24–29. https://doi.org/10.1109/ICOS.2016.7881983

Matheri, A. N., Mohamed, B., & Ngila, J. C. (2021). Smart climate resilient and efficient integrated waste to clean energy system in a developing country: Industry 4.0. *African Handbook of Climate Change Adaptation: With 610 Figures and 361 Tables*, 1053–1080. https://doi.org/10.1007/978-3-030-45106-6_69

Matović, N., & Ovesni, K. (2023). Interaction of quantitative and qualitative methodology in mixed methods research: Integration and/or combination. *International Journal of Social Research Methodology*, *26*(1), 51–65. https://doi.org/10.1080/13645579.2021.1964857

Mauricio-Moreno, H., Miranda, J., Chavarria, D., Ramirez-Cadena, M., & Molina, A. (2015). Design S3-RF (Sustainable x smart x sensing – Reference framework) for the future manufacturing enterprise. *IFAC-PapersOnLine*, *48*(3), 58–63. https://doi.org/10.1016/J.IFACOL.2015.06.058

Mehmood, H., Liao, D., & Mahadeo, K. (2020). A review of artificial intelligence applications to achieve water-related sustainable development goals. *2020 IEEE/ITU International Conference on Artificial Intelligence for Good, AI4G 2020*, 135–141. https://doi.org/10.1109/AI4G50087.2020.9311018

Mehrpouya, M., Dehghanghadikolaei, A., Fotovvati, B., Vosooghnia, A., Emamian, S. S., & Gisario, A. (2019). The potential of additive manufacturing in the Smart Factory Industrial 4.0: A review. *Applied Sciences*, *9*(18), 3865. https://doi.org/10.3390/APP9183865

Meir, Y. Ben, Opfer, K., & Hernandez, E. (2022). Decentralized renewable energies and the water-energy-food nexus in rural Morocco. *Environmental Challenges*, *6*, 100432. https://doi.org/10.1016/J.ENVC.2021.100432

Mhlongo, T. (2021). *A Systems' Thinking Approach to Entrepreneurial Leadership : An Analysis of SMMES in the Gauteng Province*. https://doi.org/10.51415/10321/3925

Miranda, J., Ponce, P., Molina, A., & Wright, P. (2019). Sensing, smart and sustainable technologies for Agri-Food 4.0. *Computers in Industry*, *108*, 21–36. https://doi.org/10.1016/J.COMPIND.2019.02.002

Moats, D., & Tseng, Y. S. (2024). Sorting a public? Using quali-quantitative methods to interrogate the role of algorithms in digital democracy platforms. *Information, Communication & Society*, *27*(5), 973–1007. https://doi.org/10.1080/1369118X.2023.2230286

Mohamed Mouhoumed, R., Ekmekcioğlu, Ö., & Özger, M. (2023). A hybrid MCDA approach for delineating sites suitable for artificial groundwater recharge using drywells. *Journal of Hydrology*, *620*, 129387. https://doi.org/10.1016/J.JHYDROL.2023.129387

Mohindru, G., Mondal, K., & Banka, H. (2020). Internet of Things and data analytics: A current review. *Wiley Interdisciplinary Reviews: Data Mining and Knowledge Discovery*, *10*(3), e1341. https://doi.org/10.1002/WIDM.1341

Molajou, A., Afshar, A., Khosravi, M., Soleimanian, E., Vahabzadeh, M., & Variani, H. A. (2023). A new paradigm of water, food, and energy nexus. *Environmental Science and Pollution Research*, *30*(49), 107487–107497. https://doi.org/10.1007/S11356-021-13034-1

Mondejar, M. E., Avtar, R., Diaz, H. L. B., Dubey, R. K., Esteban, J., Gómez-Morales, A., Hallam, B., Mbungu, N. T., Okolo, C. C., Prasad, K. A., She, Q., & Garcia-Segura, S. (2021). Digitalization to achieve sustainable development goals: Steps towards a smart green planet. *Science of The Total Environment*, *794*, 148539. https://doi.org/10.1016/J.SCITOTENV.2021.148539

Morales, M., Nousala, S., Ghobakhloo, M., Morales, M., Nousala, S., & Ghobakhloo, M. (2024). The complexity of sustainable innovation, transitional impacts of Industry 4.0 to 5.0 for our societies: Circular society exploring the systemic nexus of socioeconomic transitions (pp. 31–56). https://doi.org/10.1007/978-981-99-9730-5_2

Morgan, J., Halton, M., Qiao, Y., & Breslin, J. G. (2021). Industry 4.0 smart reconfigurable manufacturing machines. *Journal of Manufacturing Systems*, *59*, 481–506. https://doi.org/10.1016/J.JMSY.2021.03.001

Motlagh, N. H., Mohammadrezaei, M., Hunt, J., & Zakeri, B. (2020a). Internet of Things (IoT) and the energy sector. *Energies*, *13*(2), 494. https://doi.org/10.3390/EN13020494

Motlaghzadeh, K., Kerachian, R., & Tavvafi, A. (2020). An evidential reasoning-based leader-follower game for hierarchical multi-agent decision making under uncertainty. *Journal of Hydrology*, *591*, 125294. https://doi.org/10.1016/J.JHYDROL.2020.125294

Muhie, S. H. (2022). Novel approaches and practices to sustainable agriculture. *Journal of Agriculture and Food Research*, *10*, 100446. https://doi.org/10.1016/J.JAFR.2022.100446

Mukumbang, F. C. (2023). Retroductive theorizing: A contribution of critical realism to mixed methods research. *Journal of Mixed Methods Research*, *17*(1), 93–114. https://doi.org/10.1177/15586898211049847

Muniandi, B., Kumar Maurya, P., Bhavani, C., Kulkarni, S., Reddy Yellu, R., Chauhan, N., & Author, C. (2024). AI-driven energy management systems for smart buildings. *Power System Technology*, *48*(1), 322–337. https://doi.org/10.52783/PST.280

Naeem, M. A., Karim, S., Hasan, M., & Kang, S. (2022). Nexus between oil shocks and agriculture commodities: Evidence from time and frequency domain. *SSRN Electronic Journal*. https://doi.org/10.2139/SSRN.4002236

Namany, S., Al-Ansari, T., & Govindan, R. (2019). Sustainable energy, water and food nexus systems: A focused review of decision-making tools for efficient resource management and governance. *Journal of Cleaner Production*, *225*, 610–626. https://doi.org/10.1016/J.JCLEPRO.2019.03.304

Namany, S., Govindan, R., & Al-Ansari, T. (2023). Operationalising transboundary cooperation through game theory: An energy water food nexus approach for the Middle East and North Africa. *Futures*, *152*, 103198. https://doi.org/10.1016/J.FUTURES.2023.103198

Nath, P. C., Mishra, A. K., Sharma, R., Bhunia, B., Mishra, B., Tiwari, A., Nayak, P. K., Sharma, M., Bhuyan, T., Kaushal, S., Mohanta, Y. K., & Sridhar, K. (2024). Recent advances in artificial intelligence towards the sustainable future of agri-food industry. *Food Chemistry*, *447*, 138945. https://doi.org/10.1016/J.FOODCHEM.2024.138945

Nautiyal, H., & Goel, V. (2021). Sustainability assessment: Metrics and methods. *Methods in Sustainability Science: Assessment, Prioritization, Improvement, Design and Optimization*, 27–46. https://doi.org/10.1016/B978-0-12-823987-2.00017-9

Ndjuluwa, L. N. P., Adebisi, J. A., & Dayoub, M. (2023). Internet of things for crop farming: A review of technologies and applications. *Commodities*, *2*(4), 367–381. https://doi.org/10.3390/COMMODITIES2040021

Nesje, F., Drupp, M. A., Freeman, M. C., & Groom, B. (2023). Philosophers and economists agree on climate policy paths but for different reasons. *Nature Climate Change*, *13*(6), 515–522. https://doi.org/10.1038/s41558-023-01681-w

Ng, T. C., Lau, S. Y., Ghobakhloo, M., Fathi, M., & Liang, M. S. (2022). The application of Industry 4.0 technological constituents for sustainable manufacturing: A content-centric review. *Sustainability*, *14*(7), 4327. https://doi.org/10.3390/SU14074327

Nguyen, T. T., Grote, U., Neubacher, F., Rahut, D. B., Do, M. H., & Paudel, G. P. (2023). Security risks from climate change and environmental degradation: Implications for sustainable land use transformation in the Global South. *Current Opinion in Environmental Sustainability*, *63*, 101322. https://doi.org/10.1016/J.COSUST.2023.101322

Nhamo, L., Ndlela, B., Nhemachena, C., Mabhaudhi, T., Mpandeli, S., & Matchaya, G. (2018). The water-energy-food nexus: Climate risks and opportunities in Southern Africa. *Water*, *10*(5), 567. https://doi.org/10.3390/W10050567

Niemczynowicz, J. (1999). Urban hydrology and water management – Present and future challenges. *Urban Water*, *1*(1), 1–14. https://doi.org/10.1016/S1462-0758(99)00009-6

Niet, T., Arianpoo, N., Kuling, K., & Wright, A. S. (2021). Embedding the United Nations sustainable development goals into energy systems analysis: Expanding the food–energy–water nexus. *Energy, Sustainability and Society*, *11*(1), 1–12. https://doi.org/10.1186/s13705-020-00275-0s

Odara, S., Khan, Z., & Ustun, T. S. (2015). Integration of precision agriculture and smart-grid technologies for sustainable development. *Proceedings* – 2015 *IEEE International Conference on Technological Innovations in ICT for Agriculture and Rural Development, TIAR 2015*, 84–89. https://doi.org/10.1109/TIAR.2015.7358536

Ofremu, G. O., Raimi, B. Y., Yusuf, S. O., Dziwornu, B. A., Nnabuife, S. G., Eze, A. M., & Nnajiofor, C. A. (2024). Exploring the relationship between climate change, air pollutants and human health: Impacts, adaptation, and Mmitigation strategies. *Green Energy and Resources*, 100074. https://doi.org/10.1016/J.GERR.2024.100074

Ogbolumani, O. A., & Mabaso, B. (2023). An IoT-based hydroponic monitoring and control system for sustainable food production. *Journal of Digital Food, Energy & Water Systems*, *4*(2), 106–140.

Ogbolumani, O. A., & Nwulu, N. I. (2020). Integrated appliance schedulingand optimal sizing of an autonomous hybrid renewable energy system for agricultural food production. In *Lecture Notes in Mechanical Engineering* (pp. 651–660). https://doi.org/10.1007/978-981-15-5753-8_60

Ogbolumani, O. A., & Nwulu, N. I. (2021). Multi-objective optimisation of constrained food-energy-water-nexus systems for sustainable resource allocation. *Sustainable Energy Technologies and Assessments*, *44*, 100967. https://doi.org/10.1016/j.seta.2020.100967

Ogbolumani, O. A., & Nwulu, N. I. (2022). A food-energy-water Nexus meta-model for food and energy security. *Sustainable Production and Consumption*, *30*, 438–453. https://doi.org/10.1016/J.SPC.2021.12.019

Ogbolumani, O. A., & Nwulu, N. I. (2024). Environmental impact assessment for a meta-model-based food-energy-water-nexus system. *Energy Reports*, *11*, 218–232. https://doi.org/10.1016/j.egyr.2023.11.033

Olatunde, T. M., Adelani, F. A., & Sikhakhane, Z. Q. (2024). A review of smart water management systems from Africa and the United States. *Engineering Science & Technology Journal*, *5*(4), 1231–1242. https://doi.org/10.51594/ESTJ.V5I4.1014

Olorunfemi, B. O., Nwulu, N. I., & Ogbolumani, O. A. (2023). Solar panel surface dirt detection and removal based on arduino color recognition. *MethodsX*, *10*, 101967. https://doi.org/10.1016/J.MEX.2022.101967

Olorunfemi, B. O., Ogbolumani, O. A., & Nwulu, N. (2022). Solar panels dirt monitoring and cleaning for performance improvement: A systematic review on smart systems. *Sustainability*, *14*(17), 10920. https://doi.org/10.3390/SU141710920

Onu, P., Pradhan, A., & Mbohwa, C. (2023). The potential of Industry 4.0 for renewable energy and materials development – The case of multinational energy companies. *Heliyon*, *9*(10). https://doi.org/10.1016/j.heliyon.2023.e20547

Opeyemi, L., Clinton, D., Nnamdi, A., Omoseni, N., & Adepoju, O. (2024). Sustainable synergy: A digital framework for the water-energy-food nexus project delivery in developing economies. *Springer Nature*. https://doi.org/https://doi.org/10.1007/978-3-031-72833-4

Pahl-Wostl, C., Gorris, P., Jager, N., Koch, L., Lebel, L., Stein, C., Venghaus, S., & Withanachchi, S. (2021). Scale-related governance challenges in the water–energy–food nexus: Toward a diagnostic approach. *Sustainability Science*, *16*(2), 615–629. https://doi.org/10.1007/s11625-020-00888-6

Panda, B. K., Panigrahi, S. S., Mishra, G., & Kumar, V. (2023). Robotics for general material handling machines in food plants. *Transporting Operations of Food Materials within Food Factories: Unit Operations and Processing Equipment in the Food Industry*, 341–372. https://doi.org/10.1016/B978-0-12-818585-8.00005-2

Pascoe, S., Cannard, T., Dowling, N. A., Dichmont, C. M., Asche, F., & Little, L. R. (2023). Use of data envelopment analysis (DEA) to assess management alternatives in the presence of multiple objectives. *Marine Policy*, *148*, 105444. https://doi.org/10.1016/J.MARPOL.2022.105444

Pathirana, R., & Carimi, F. (2024). Plant biotechnology—An indispensable tool for crop improvement. *Plants*, *13*(8), 1133. https://doi.org/10.3390/PLANTS13081133

Pemi, B. A. P., Njomo, D., Tchinda, R., Seutche, J. C., Kamta Legue, D. R., Babikir, M. H., & Chara-Dackou, V. S. (2023). Modeling and quantitative analysis in the energy–food–water–waste nexus (EF2W): Case study in cameroon. *Sustainability (Switzerland)*, *15*(11), 8483. https://doi.org/10.3390/su15118483

Perboli, G., Musso, S., & Rosano, M. (2018). Blockchain in logistics and supply chain: A lean approach for designing real-world use cases. *IEEE Access*, *6*, 62018–62028. https://doi.org/10.1109/ACCESS.2018.2875782

Pereira, V., Basilio, M. P., & Santos, C. H. T. S. H. T. (2024). *Enhancing Decision Analysis with a Large Language Model: pyDecision a Comprehensive Library of MCDA Methods in Python*. https://arxiv.org/abs/2404.06370v1

Pierce, F. J., & Nowak, P. (1999). Aspects of precision agriculture. *Advances in Agronomy*, *67*(C), 1–85. https://doi.org/10.1016/S0065-2113(08)60513-1

Pozzi, R., Rossi, T., & Secchi, R. (2023). Industry 4.0 technologies: Critical success factors for implementation and improvements in manufacturing companies. *Production Planning & Control*, *34*(2), 139–158. https://doi.org/10.1080/09537287.2021.1891481

Quamar, M. M., Al-Ramadan, B., Khan, K., Shafiullah, M., & El Ferik, S. (2023). Advancements and applications of drone-integrated geographic information system technology—A review. *Remote Sensing*, *15*(20), 5039. https://doi.org/10.3390/RS15205039

Rabie, K., & Hassan, F. U. (2025). IoT for critical big data applications. *Empowering IoT with Big Data Analytics*, 235–264. https://doi.org/10.1016/B978-0-443-21640-4.00009-0

Rajamanickam, M., John Gerard Royan, E. N., Ramaswamy, G., Rajendran, M., & Vadivelu, V. (2023). Fourth industrial revolution. *Integration of Mechanical and Manufacturing Engineering with IoT: A Digital Transformation*, 41–84. https://doi.org/10.1002/9781119865391.CH2

Rane, N., Choudhary, S., & Rane, J. (2023). Leading-edge artificial intelligence (AI), machine learning (ML), blockchain, and internet of things (IoT) technologies for enhanced wastewater treatment systems. *SSRN Electronic Journal*. https://doi.org/10.2139/SSRN.4641557

Rasul, G. (2016). Managing the food, water, and energy nexus for achieving the sustainable development goals in South Asia. *Environmental Development*, *18*, 14–25. https://doi.org/10.1016/J.ENVDEV.2015.12.001

Rasul, G., & Neupane, N. (2021). Improving policy coordination across the water, energy, and food, sectors in South Asia: A framework. *Frontiers in Sustainable Food Systems*, *5*, 602475. https://doi.org/10.3389/fsufs.2021.602475

Rejeb, A., Keogh, J. G., & Treiblmaier, H. (2019). Leveraging the Internet of Things and blockchain technology in supply chain management. *Future Internet*, *11*(7), 161. https://doi.org/10.3390/FI11070161

Rejeb, A., Rejeb, K., Zailani, S., Keogh, J. G., & Appolloni, A. (2022). Examining the interplay between artificial intelligence and the agri-food industry. *Artificial Intelligence in Agriculture*, *6*, 111–128. https://doi.org/10.1016/J.AIIA.2022.08.002

Rihab, R., Gannouni, S., & Rebai, N. (2024). A comparative study of different methods of multi-criteria decision analysis (MCDA) with spatial reference to select of suitable sites for treated wastewater of irrigated perimeters. *Sustainable Water Resources Management*, *10*(4), 1–21. https://doi.org/10.1007/s40899-024-01120-4

Ringler, C., Bhaduri, A., & Lawford, R. (2013). The nexus across water, energy, land and food (WELF): Potential for improved resource use efficiency? *Current Opinion in Environmental Sustainability*, *5*(6), 617–624. https://doi.org/10.1016/J.COSUST.2013.11.002

Rising, J. (2020). Decision-making and integrated assessment models of the water-energy-food nexus. *Water Security*, *9*, 100056. https://doi.org/10.1016/J.WASEC.2019.100056

Robert, P. C. (2002). Precision agriculture: A challenge for crop nutrition management. In *Progress in Plant Nutrition: Plenary Lectures of the XIV International Plant Nutrition Colloquium* (pp. 143–149). https://doi.org/10.1007/978-94-017-2789-1_11

Robinson, A. C. (2020). Geographic information systems (GIS) for disaster management. *Geographic Information Systems (GIS) for Disaster Management*. https://doi.org/10.4324/9781351034869

Rodríguez, A. V., Rodríguez-Oramas, C., Velázquez, E. S., de la Torre, A. H., Armendáriz, C. R., & Iruzubieta, C. C. (2022). Myths and realities about genetically modified food: A risk-benefit analysis. *Applied Sciences*, *12*(6), 2861. https://doi.org/10.3390/APP12062861

Roggema, R., Yan, W., & Keeffe, G. (2021). A moveable nexus: Framework for FEW-design and planning. *Contemporary Urban Design Thinking*, *Part F3*, 9–37. https://doi.org/10.1007/978-3-030-61977-0_2

Rusko, R., & Kosonen, M. (2022). Emerging new technologies and industrial revolution. *Encyclopedia of Data Science and Machine Learning*, 2161–2181. https://doi.org/10.4018/978-1-7998-9220-5.CH129

Rüßmann, M., Lorenz, M., Gerbert, P., Waldner, M., Justus, J., Engel, P., & Harnisch, M. (2015). *Industry 4.0: The Future of Productivity and Growth in Manufacturing Industries. Boston consulting group*, *9*(1), 54-89

Sadowski, S., & Spachos, P. (2020). Wireless technologies for smart agricultural monitoring using Internet of Things devices with energy harvesting capabilities. *Computers and Electronics in Agriculture*, *172*, 105338. https://doi.org/10.1016/J.COMPAG.2020.105338

Saeed, S., Gull, H., Aldossary, M. M., Altamimi, A. F., Alshahrani, M. S., Saqib, M., Zafar Iqbal, S., & Almuhaideb, A. M. (2024). Digital transformation in energy sector: Cybersecurity challenges and implications. *Information*, *15*(12), 764. https://doi.org/10.3390/INFO15120764

Sahoo, S. K., & Goswami, S. S. (2023). A comprehensive review of multiple criteria decision-making (MCDM) methods: Advancements, applications, and future directions. *Decision Making Advances*, *1*(1), 25–48. https://doi.org/10.31181/DMA1120237

Salmoral, G., Schaap, N. C. E., Walschebauer, J., & Alhajaj, A. (2019). Water diplomacy and nexus governance in a transboundary context: In the search for complementarities. *Science of The Total Environment*, *690*, 85–96. https://doi.org/10.1016/J.SCITOTENV.2019.06.513

Sanders, A., Elangeswaran, C., & Wulfsberg, J. (2016). Industry 4.0 implies lean manufacturing: Research activities in Industry 4.0 function as enablers for lean manufacturing. *Journal of Industrial Engineering and Management*, *9*(3), 811–833. https://doi.org/10.3926/JIEM.1940

Sargentis, G. F., Mamassis, N., Kitsou, O., & Koutsoyiannis, D. (2024). The role of technology in the water–energy–food nexus. A case study: Kerinthos, North Euboea, Greece. *Frontiers in Water*, *6*, 1343344. https://doi.org/10.3389/frwa.2024.1343344

Sarkar, B., Takeyeva, D., Guchhait, R., & Sarkar, M. (2022). Optimized radio-frequency identification system for different warehouse shapes. *Knowledge-Based Systems*, *258*, 109811. https://doi.org/10.1016/J.KNOSYS.2022.109811

Sasi, T., Lashkari, A. H., Lu, R., Xiong, P., & Iqbal, S. (2024). A comprehensive survey on IoT attacks: Taxonomy, detection mechanisms and challenges. *Journal of Information and Intelligence*, *2*(6), 455–513. https://doi.org/10.1016/J.JIIXD.2023.12.001

Savin, I., Creutzig, F., Filatova, T., Foramitti, J., Konc, T., Niamir, L., Safarzynska, K., & van den Bergh, J. (2023). Agent-based modeling to integrate elements from different disciplines for ambitious climate policy. *Wiley Interdisciplinary Reviews: Climate Change*, *14*(2), e811. https://doi.org/10.1002/WCC.811

Schneider, U. A., Havlík, P., Schmid, E., Valin, H., Mosnier, A., Obersteiner, M., Böttcher, H., Skalský, R., Balkovič, J., Sauer, T., & Fritz, S. (2011). Impacts of population growth, economic development, and technical change on global food production and consumption. *Agricultural Systems*, *104*(2), 204–215. https://doi.org/10.1016/J.AGSY.2010.11.003

Sempere-Payá, V., Todolí-Ferrandis, D., & Santonja-Climent, S. (2013). ICT as an enabler to smart water management. *Smart Sensors, Measurement and Instrumentation*, *4*, 239–258. https://doi.org/10.1007/978-3-642-37006-9_11

Shah, N., Kamdar, L., Gokalgandhi, D., & Mehendale, N. (2021). Walking pattern analysis using deep learning for energy harvesting smart shoes with IoT. *Neural Computing and Applications*, *33*(18), 11617–11625. https://doi.org/10.1007/s00521-021-05864-4

Shamoon, A., Haleem, A., Bahl, S., Javaid, M., Bala Garg, S., Chandmal Sharma, R., & Garg, J. (2022). Environmental impact of energy production and extraction of materials – A review. *Materials Today: Proceedings*, *57*, 936–941. https://doi.org/10.1016/J.MATPR.2022.03.159

Shao, X. F., Liu, W., Li, Y., Chaudhry, H. R., & Yue, X. G. (2021). Multistage implementation framework for smart supply chain management under Industry 4.0. *Technological Forecasting and Social Change*, *162*, 120354. https://doi.org/10.1016/J.TECHFORE.2020.120354

Sharif, R. Al, & Pokharel, S. (2022). Smart city dimensions and associated risks: Review of literature. *Sustainable Cities and Society*, *77*, 103542. https://doi.org/10.1016/J.SCS.2021.103542

Sharma, R., Verma, N., Lugani, Y., Kumar, S., & Asadnia, M. (2021). Conventional and advanced techniques of wastewater monitoring and treatment. In *Green Sustainable Process for Chemical and Environmental Engineering and Science: Analytical Techniques for Environmental and Industrial Analysis* (pp. 1–48). https://doi.org/10.1016/B978-0-12-821883-9.00009-6

Sharma, S., Malik, A., Sharma, C., Batra, I., Kaswan, M. S., & Garza-Reyes, J. A. (2023). Adoption of Industry 4.0 in different sectors: A structural review using natural language processing. *International Journal on Interactive Design and Manufacturing*, 1–23. https://doi.org/10.1007/s12008-023-01550-y

Sidhu, R. K., Kumar, R., Rana, P. S., & Jat, M. L. (2021). Automation in drip irrigation for enhancing water use efficiency in cereal systems of South Asia: Status and prospects. *Advances in Agronomy*, *167*, 247–300. https://doi.org/10.1016/BS.AGRON.2021.01.002

Siekmann, F., Schlör, H., & Venghaus, S. (2023). Linking sustainability and the fourth industrial revolution: A monitoring framework accounting for technological development. *Energy, Sustainability and Society*, *13*(1), 1–18. https://doi.org/10.1186/s13705-023-00405-4

Sieverding, H., Kebreab, E., Johnson, J. M. F., Xu, H., Wang, M., Grosso, S. J. D., Bruggeman, S., Stewart, C. E., Westhoff, S., Ristau, J., Kumar, S., & Stone, J. J. (2020). A life cycle analysis (LCA) primer for the agricultural community. *Agronomy Journal*, *112*(5), 3788–3807. https://doi.org/10.1002/AGJ2.20279

Simpson, G. B., & Jewitt, G. P. W. (2019). The development of the water-energy-food nexus as a framework for achieving resource security: A review. *Frontiers in Environmental Science*, *7*(FEB). https://doi.org/10.3389/fenvs.2019.00008

Simpson, G. B., Jewitt, G. P. W., Becker, W., Badenhorst, J., Masia, S., Neves, A. R., Rovira, P., & Pascual, V. (2022). The water-energy-food nexus index: A tool to support integrated resource planning, management and security. *Frontiers in Water*, *4*, 825854.https://doi.org/10.3389/FRWA.2022.825854/BIBTEX

Simpson, G., Jewitt, G., Becker, W., Badenhorst, J., Neves, A., Rovira, P., & Pascual, V. (2020). *The Water-Energy-Food Nexus Index: A Tool for Integrated Resource Management and Sustainable Development*. https://doi.org/10.31219/OSF.IO/TDHW5

Sinha, S. K., Davis, C., Gardoni, P., Babbar-Sebens, M., Stuhr, M., Huston, D., Cauffman, S., Williams, W. D., Alanis, L. G., Anand, H., & Vishwakarma, A. (2023). Water sector infrastructure systems resilience: A social–ecological–technical system-of-systems and whole-life approach. *Cambridge Prisms: Water*, *1*, e4. https://doi.org/10.1017/WAT.2023.3

Slorach, P. C., Jeswani, H. K., Cuéllar-Franca, R., & Azapagic, A. (2020). Environmental sustainability in the food-energy-water-health nexus: A new methodology and an application to food waste in a circular economy. *Waste Management*, *113*, 359–368. https://doi.org/10.1016/J.WASMAN.2020.06.012

Smith, G. S., Anjum, E., Francis, C., Deanes, L., & Acey, C. (2022). Climate change, environmental disasters, and health inequities: The underlying role of structural inequalities. *Current Environmental Health Reports*, *9*(1), 80–89. https://doi.org/10.1007/s40572-022-00336-w

Srigiri, S. R., & Dombrowsky, I. (2022). Analysing the water-energy-food nexus from a polycentric governance perspective: Conceptual and methodological framework. *Frontiers in Environmental Science*, *10*, 725116. https://doi.org/10.3389/fenvs.2022.725116

Stafford, J. V. (2000). Implementing precision agriculture in the 21st century. *Journal of Agricultural Engineering Research*, *76*(3), 267–275. https://doi.org/10.1006/JAER.2000.0577

Stankovic, M., Hasanbeigi, A., Neftenov, M. N., Basani, M., Núñez, A., & Ortiz, R. (2020). *Use of 4IR Technologies in Water and Sanitation in Latin America and the Caribbean*. https://doi.org/10.18235/0002343

Stanton, I., Munir, K., Ikram, A., & El-Bakry, M. (2023). Predictive maintenance analytics and implementation for aircraft: Challenges and opportunities. *Systems Engineering*, *26*(2), 216–237. https://doi.org/10.1002/SYS.21651

Staupe-Delgado, R. (2020). The water–energy–food–environmental security nexus: Moving the debate forward. *Environment, Development and Sustainability*, *22*(7), 6131–6147. https://doi.org/10.1007/s10668-019-00467-5

Stehr, N. J. (2015). Drones: The newest technology for precision agriculture. *Natural Sciences Education*, *44*(1), 89–91. https://doi.org/10.4195/NSE2015.04.0772

Steinwand, M. A., & Ronald, P. C. (2020). Crop biotechnology and the future of food. *Nature Food*, *1*(5), 273–283. https://doi.org/10.1038/s43016-020-0072-3

Stilwell, P., & Harman, K. (2021). Phenomenological research needs to be renewed: Time to integrate enactivism as a flexible resource. *International Journal of Qualitative Methods*, *20*. https://doi.org/10.1177/1609406921995299

Sundberg, L., & Holmström, J. (2023). Democratizing artificial intelligence: How no-code AI can leverage machine learning operations. *Business Horizons*, *66*(6), 777–788. https://doi.org/10.1016/j.bushor.2023.04.003

Surucu, O., Gadsden, S. A., & Yawney, J. (2023). Condition monitoring using machine learning: A review of theory, applications, and recent advances. *Expert Systems with Applications*, *221*, 119738. https://doi.org/10.1016/J.ESWA.2023.119738

Tang, L., & Meng, Y. (2020). Data analytics and optimization for smart industry. *Frontiers of Engineering Management*, *8*(2), 157–171. https://doi.org/10.1007/S42524-020-0126-0

Tang, Y. M., Chau, K. Y., Fatima, A., & Waqas, M. (2022). Industry 4.0 technology and circular economy practices: Business management strategies for environmental sustainability. *Environmental Science and Pollution Research*, *29*(33), 49752–49769. https://doi.org/10.1007/S11356-022-19081-6/METRICS

Tang, Y. M., Ho, G. T. S., Lau, Y. Y., & Tsui, S. Y. (2022). Integrated smart warehouse and manufacturing management with demand forecasting in small-scale cyclical industries. *Machines*, *10*(6), 472. https://doi.org/10.3390/MACHINES10060472

Tatlari, A. A., Perchkolaei, B. R., & Lotfi, F. H. (2025). Innovative efficiency measurement using stackelberg games and neutrosophic numbers. *Neutrosophic Sets and Systems*, *80*, 387–409. https://doi.org/10.5281/ZENODO.14708763

Teh, D., & Rana, T. (2023). The use of Internet of Things, Big Data Analytics and artificial intelligence for attaining UN's SDGs. In *Handbook of Big Data and Analytics in Accounting and Auditing* (pp. 235–253). https://doi.org/10.1007/978-981-19-4460-4_11

Teker, S., & Koc, T. C. (2019). Industrial revolutions and its effects on quality of life. *Pressacademia*, *9*(9), 304–311. https://doi.org/10.17261/pressacademia.2019.1109

Thilakarathna, M. S., & Raizada, M. N. (2018). Challenges in using precision agriculture to optimize symbiotic nitrogen fixation in legumes: Progress, limitations, and future improvements needed in diagnostic testing. *Agronomy*, *8*(5), 78. https://doi.org/10.3390/AGRONOMY8050078

Thompson, R. B., Gallardo, M., Valdez, L. C., & Fernández, M. D. (2007). Using plant water status to define threshold values for irrigation management of vegetable crops using soil moisture sensors. *Agricultural Water Management*, *88*(1–3), 147–158. https://doi.org/10.1016/J.AGWAT.2006.10.007

Tian, H., Lu, C., Pan, S., Yang, J., Miao, R., Ren, W., Yu, Q., Fu, B., Jin, F. F., Lu, Y., Melillo, J., Ouyang, Z., Palm, C., & Reilly, J. (2018). Optimizing resource use efficiencies in the food–energy–water nexus for sustainable agriculture: From conceptual model to decision support system. *Current Opinion in Environmental Sustainability*, *33*, 104–113. https://doi.org/10.1016/J.COSUST.2018.04.003

United Nations. (n.d.). *World Population Projected to Reach 9.8 Billion in 2050, and 11.2 Billion in 2100 | United Nations*. United Nations. Retrieved January 30, 2025, from https://www.un.org/en/desa/world-population-projected-reach-98-billion-2050-and-112-billion-2100

Vahabzadeh, M., Afshar, A., & Molajou, A. (2023). Framing a novel holistic energy subsystem structure for water-energy-food nexus based on existing literature (basic concepts). *Scientific Reports*, *13*(1), 1–21. https://doi.org/10.1038/s41598-023-33385-8

Valero, E., Adán, A., & Cerrada, C. (2015). Evolution of RFID applications in construction: A literature review. *Sensors*, *15*(7), 15988–16008. https://doi.org/10.3390/S150715988

van Beek, L., Hajer, M., Pelzer, P., van Vuuren, D., & Cassen, C. (2020). Anticipating futures through models: The rise of integrated assessment modelling in the climate science-policy interface since 1970. *Global Environmental Change*, *65*, 102191. https://doi.org/10.1016/J.GLOENVCHA.2020.102191

van Esse, H. P., Reuber, T. L., & van der Does, D. (2020). Genetic modification to improve disease resistance in crops. *New Phytologist*, *225*(1), 70–86. https://doi.org/10.1111/NPH.15967

Vaseashta, A. (2022). *Nexus of Advanced Technology Platforms for Strengthening Cyber-Defense Capabilities*. 14–31. https://doi.org/10.3233/NHSDP220003

Vázquez-Rowe, I., Córdova-Arias, C., Brioso, X., & Santa-Cruz, S. (2021). A method to include life cycle assessment results in choosing by advantage (CBA) multicriteria decision analysis. A case study for seismic retrofit in peruvian primary schools. *Sustainability*, *13*(15), 8139. https://doi.org/10.3390/SU13158139

Velasco, J. S., Arredondo-Soto, K. C., & Miranda-Ackerman, M. A. (2024). Review of the challenges in implementing Industry 4.0 technologies in the context of sustainable supply chains. In *Automation and Innovation with Computational Techniques for Futuristic Smart, Safe and Sustainable Manufacturing Processes* (pp. 37–65). https://doi.org/10.1007/978-3-031-46708-0_2

Vizanko, B., Shmaya, T., Pankaj Boindala, S., Ostfeld, A., & Berglund, E. (2024). Operating water distribution systems for equitable access to clean water. *World Environmental and Water Resources Congress* 2024*: Climate Change Impacts on the World We Live In – Proceedings of the World Environmental and Water Resources Congress* 2024, 1229–1235. https://doi.org/10.1061/9780784485477.109

Wallwey, C., & Kajfez, R. L. (2023). Quantitative research artifacts as qualitative data collection techniques in a mixed methods research study. *Methods in Psychology*, *8*, 100115. https://doi.org/10.1016/J.METIP.2023.100115

Wang, Y., Singgih, M., Wang, J., & Rit, M. (2019). Making sense of blockchain technology: How will it transform supply chains? *International Journal of Production Economics*, *211*, 221–236. https://doi.org/10.1016/J.IJPE.2019.02.002

Wang, Y., Yuan, Z., & Tang, Y. (2021). Enhancing food security and environmental sustainability: A critical review of food loss and waste management. *Resources, Environment and Sustainability*, *4*, 100023. https://doi.org/10.1016/J.RESENV.2021.100023

Wang, Z., Cao, X., Ren, X., & Taghizadeh-Hesary, F. (2024). Can digital transformation affect coal utilization efficiency in China? Evidence from spatial econometric analyses. *Resources Policy*, *91*, 104940. https://doi.org/10.1016/J.RESOURPOL.2024.104940

Watkins, K. (2006). *Human Development Report 2006 – Beyond Scarcity: Power, Poverty and the Global Water Crisis*. https://papers.ssrn.com/abstract=2294691

Weichhart, G., Molina, A., Chen, D., Whitman, L. E., & Vernadat, F. (2016). Challenges and current developments for Sensing, Smart and Sustainable Enterprise Systems. *Computers in Industry*, *79*, 34–46. https://doi.org/10.1016/J.COMPIND.2015.07.002

Wood Hansen, O., & Van Den Bergh, J. (2023). Environmental problem shifting from climate change mitigation: A mapping review. *PNAS Nexus*, *3*(1). https://doi.org/10.1093/PNASNEXUS/PGAD448

Wudil, A. H., Usman, M., Rosak-Szyrocka, J., Pilař, L., & Boye, M. (2022). Reversing years for global food security: A review of the food security situation in Sub-Saharan Africa (SSA). *International Journal of Environmental Research and Public Health*, *19*(22), 14836. https://doi.org/10.3390/IJERPH192214836

Xiang, X., Li, Q., Khan, S., & Khalaf, O. I. (2021). Urban water resource management for sustainable environment planning using artificial intelligence techniques. *Environmental Impact Assessment Review*, *86*, 106515. https://doi.org/10.1016/J.EIAR.2020.106515

Yang, M., Chen, L., Wang, J., Msigwa, G., Osman, A. I., Fawzy, S., Rooney, D. W., & Yap, P. S. (2022). Circular economy strategies for combating climate change and other environmental issues. *Environmental Chemistry Letters*, *21*(1), 55–80. https://doi.org/10.1007/S10311-022-01499-6

Yaroson, E. V., Chowdhury, S., Mangla, S. K., & Dey, P. K. (2024). Unearthing the interplay between organisational resources, knowledge and Industry 4.0 analytical decision support tools to achieve sustainability and supply chain wellbeing. *Annals of Operations Research 2024*, 1–48. https://doi.org/10.1007/S10479-024-05845-5

Yin, Y., Stecke, K. E., & Li, D. (2018). The evolution of production systems from Industry 2.0 through Industry 4.0. *International Journal of Production Research*, *56*(1–2), 848–861. https://doi.org/10.1080/00207543.2017.1403664

Yu, L., Qin, S., Zhang, M., Shen, C., Jiang, T., & Guan, X. (2021). A review of deep reinforcement learning for smart building energy management. *IEEE Internet of Things Journal*, *8*(15), 12046–12063. https://doi.org/10.1109/JIOT.2021.3078462

Yu, X., Shan, L., & Wu, Y. (2021). Land use optimization in a resource-exhausted city based on simulation of the F-E-W nexus. *Land*, *10*(10), 1013. https://doi.org/10.3390/LAND10101013

Yuan, Z., Olsson, G., Cardell-Oliver, R., van Schagen, K., Marchi, A., Deletic, A., Urich, C., Rauch, W., Liu, Y., & Jiang, G. (2019). Sweating the assets – The role of instrumentation, control and automation in urban water systems. *Water Research, 155*, 381–402. https://doi.org/10.1016/J.WATRES.2019.02.034

Yuan, Z., Wen, B., He, C., Zhou, J., Zhou, Z., & Xu, F. (2022). Application of multi-criteria decision-making analysis to rural spatial sustainability evaluation: A systematic review. *International Journal of Environmental Research and Public Health, 19*(11), 6572. https://doi.org/10.3390/IJERPH19116572

Yüksel, H. (2022). Industry 4.0 transformation: Factors affecting adoption and impacts on companies. *International Journal of Industrial Engineering and Operations Management, 4*(3), 63–89. https://doi.org/10.1108/IJIEOM-06-2022-0020

Zaidi, S., Maselli, M., Laschi, C., & Cianchetti, M. (2021). Actuation technologies for soft robot grippers and manipulators: A review. *Current Robotics Reports, 2*(3), 355–369. https://doi.org/10.1007/S43154-021-00054-5

Zaman, S. A. A., Dawood, H. M., Zehra, S. N., & Saeed, S. Z. (2023). Blockchain driven supply chain and Industry 4.0 technologies. *Management for Professionals, Part F539*, 219–238. https://doi.org/10.1007/978-981-99-0699-4_13

Zekri, S., Jabeur, N., & Gharrad, H. (2022). Smart water management using intelligent digital twins. *Computing and Informatics, 41*(1), 135–153. https://doi.org/10.31577/CAI_2022_1_135

Zhai, L., & Lee, J. E. (2024). Investigating vulnerability, adaptation, and resilience: A comprehensive review within the context of climate change. *Atmosphere, 15*(4), 474. https://doi.org/10.3390/ATMOS15040474

Zhang, N., Wang, M., & Wang, N. (2002). Precision agriculture—a worldwide overview. *Computers and Electronics in Agriculture, 36*(2–3), 113–132. https://doi.org/10.1016/S0168-1699(02)00096-0

Zhang, P., Liu, L., Yang, L., Zhao, J., Li, Y., Qi, Y., Ma, X., & Cao, L. (2023). Exploring the response of ecosystem service value to land use changes under multiple scenarios coupling a mixed-cell cellular automata model and system dynamics model in Xi'an, China. *Ecological Indicators, 147*, 110009. https://doi.org/10.1016/J.ECOLIND.2023.110009

Zhang, P., Zhang, L., Chang, Y., Xu, M., Hao, Y., Liang, S., Liu, G., Yang, Z., & Wang, C. (2019). Food-energy-water (FEW) nexus for urban sustainability: A comprehensive review. *Resources, Conservation and Recycling, 142*, 215–224. https://doi.org/10.1016/J.RESCONREC.2018.11.018

Zhang, T., Tan, Q., Wang, S., Zhang, T., Hu, K., & Zhang, S. (2022). Assessment and management of composite risk in irrigated agriculture under water-food-energy nexus and uncertainty. *Agricultural Water Management, 262*, 107322. https://doi.org/10.1016/J.AGWAT.2021.107322

Zhou, S., & Smulders, S. (2021). Closing the loop in a circular economy: Saving resources or suffocating innovations? *European Economic Review, 139*, 103857. https://doi.org/10.1016/J.EUROECOREV.2021.103857

Zhu, Y., Zhang, C., Fang, J., & Miao, Y. (2022). Paths and strategies for a resilient megacity based on the water-energy-food nexus. *Sustainable Cities and Society, 82*, 103892. https://doi.org/10.1016/J.SCS.2022.103892

3 Emerging Technologies and Innovations in FEW Nexus 4.0

3.1 EMERGING TECHNOLOGIES AND INNOVATIONS IN FEW 4.0 APPLICATIONS

Recent discussions on emerging technologies are gaining momentum in the research, education, and policy spheres; accordingly, this translates to a surge in publications and news stories (Rotoloet al., 2015). Behind this attention is the fact that coming up with a definition of emerging technologies is quite ambiguous, making the conceptualisation not uniform about researchers (Litvinski, 2018). While some definitions closely rely on the associated social and economic benefits or how long the technology is new, others focus on unexplored risks and how new the technology is (Haessler et al., 2023). Also, perceptions towards emerging technologies might vary depending on individualistic worldviews and areas of specialisation. Having this in mind, classifying technologies under a broad label such as "nanotechnology" or "synthetic biology" tends to obscure the individual distinctiveness of the technologies, which further challenges the definition (Boon & Moors, 2008; Parry & Battista, 2019; Small et al., 2014).

3.1.1 Definition of Emerging Technologies

The term emerging technologies is described in the present context as new products or innovations based on existing technologies. Its meaning can be defined for different fields such as media/business, science, or education (Winston & Strawn LLP, 2023). Regarding food, energy, and water (FEW) Nexus 4.0, emerging technology is linked to future technologies already in their early stages of development or are estimated to be available between five and ten years from now and will impact societies or economies in the FEW nexus space.

In general terms, some of the emerging technologies that are already being harnessed for creative and destructive purposes include but are not limited to the following: Internet of Things (IoT), artificial intelligence (AI), blockchain, data analytics, machine learning (ML), virtual reality (VR), and sensor and monitoring technologies. Figure 3.1 shows a pictorial representation of some of these technologies (Erkenntnis, 2021). However, with FEW Nexus 4.0, some emerging technologies include precision farming, gene editing, second-generation biofuels, and agrivoltaics (Miao & Khanna, 2020). Figure 3.2 shows a pictorial view of these

DOI: 10.1201/9781003633853-3

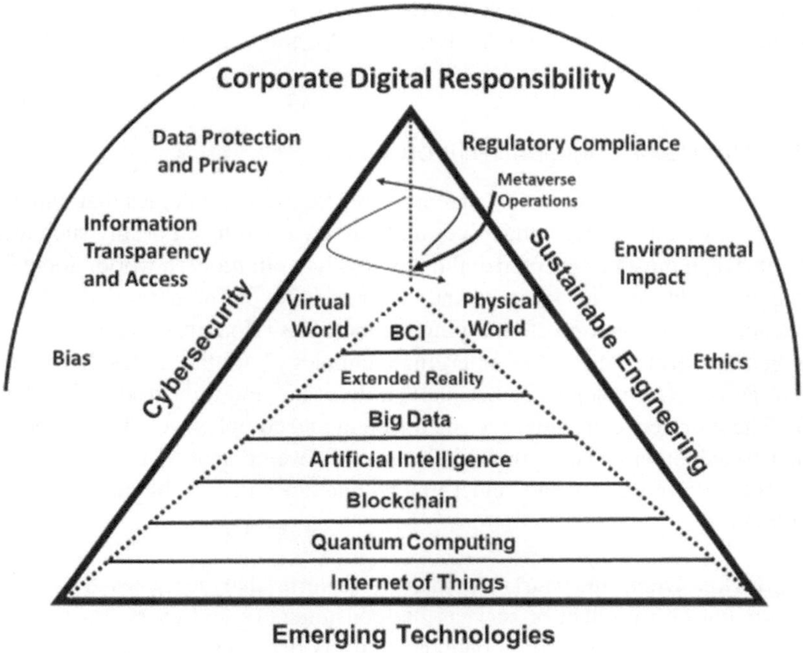

FIGURE 3.1 Emerging Technologies (Wynn & Bechkoum, 2024).

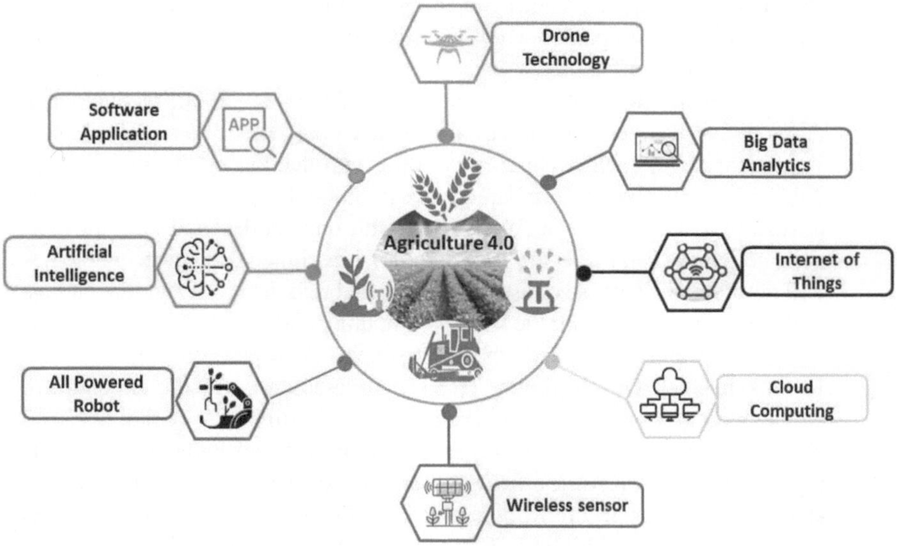

FIGURE 3.2 Emerging Technologies in FEW Nexus 4.0 (Factors et al., 2023).

emerging technologies applicable in Agriculture 4.0. They include the IoT, Cloud Computing (CC), AI, Big Data Analytics (BDA), and wireless sensor technologies (Factors et al., 2023).

3.1.2 Drivers of Emerging Technologies

With the coming of Industry 4.0 came new innovative technologies that may revolutionise how we engage with machines and our surroundings. For any such technology to thrive, it must be accepted and integrated into our modern world. Some factors drive them so they can be accepted and integrated into our modern world and way of life, hence the term drivers of emerging technologies (Hopkins, 2021).

In general terms, "drivers of emerging technology" are the forces that create and encourage new technologies to accomplish domestic and industrial tasks (Truman, 2019). These drivers can sometimes be different and complex, yet they can positively influence each other and determine the course of development of future technologies. Some of the drivers of new and emerging technologies include but are not limited to the following:

1. **Market Demand**: Market demand is a term that describes the precise amount of a product or service that consumers or end users can purchase at a specific price. Market demand impacts firms and consumers by influencing production and shaping competitive potential in the marketplace (Dwivedi, Ismagilova, et al., 2021). Businesses must possess knowledge of market demand to effectively design, develop, and promote products and services to suit consumer needs (Iheanachor et al., 2021). Identified consumer needs and preferences can influence the advancement of emerging technologies as companies work towards providing satisfying products and services to consumers.
2. **Technological Advances**: Since the onset of the year 2000, technological development has become the key driver of globalisation. Some of the most significant technological advances are the introduction of the Internet and Android mobile phones. Compared to the traditional era, these inventions have drastically changed how people interact with one another and plan for their daily activities (Fakunle et al., 2023). Thus, it is assumed that new technologies will develop whenever scientists achieve some breakthroughs and innovations in their existing ideas with the help of computing and electronics, new materials, telecommunications, and other related fields (Betz et al., 2023).
3. **Research and Development (R&D) Investments**: R&D expenditure agrees with an organisation's overall strategic readiness to market new and improved products, services, etc. Traditionally, such investments are undertaken internally within the firm, but enhanced local investment is associated with improved innovation and organisational performance. Firms seek innovation outside the R&D space to achieve results (Wadhwa, 2016). Public and private investment for R&D activities from the government and other private bodies, as well as those from institutions, can be put forward to enhance the speed of identifying and improving sophisticated technologies.

4. **Regulatory Environment**: There must be some form of regulation on certain activities carried out regarding the influx of new technologies, such as compatibility, amount of data being transferred, data security issues, and individual information security (Lee, 2019). For this regulation to occur, numerous interested parties must actively participate, state and private. There is also a need to launch programmes that encourage the acceptance of new technologies; some of these programmes include subsidies that help facilitate activities that minimise the resource constraint for small and medium-scale enterprises and also make them benefit from digitalisation (Mohiuddin et al., 2023). It is crucial to note that government policies, regulations, and standards play a significant role in encouraging or discouraging the adoption of new technologies by offering incentives, laying down rules and regulations, and ensuring that the latest technologies meet the set safety and regulatory requirements.
5. **Globalisation and Interconnectedness**: Globalisation refers to the constant rising interaction or interconnectedness between entities across different borders: people, ideas, goods, and money. This phenomenon has transformed the structure of international relations throughout the past few decades, hence the increased intertwined networks of trade and finance. As the world becomes a global village primarily due to the ease of information flow and sharing of knowledge and resources, emerging technologies stand the chance to spread across the world and even within regions (Cassimon et al., 2019).
6. **Societal and Cultural Factors**: Social attitudes and other cultural factors that are prevailing or emerging can define people's attitudes towards new technologies and their adoption. They can also affect the development and evolution of these new technologies. The faster the society adopts technologies, the faster the growth and development of such technologies.
7. **Economic Factors**: Funding and development decisions in new and emerging technologies are based on assessing market competition, cost to business, and evolution of opportunities for higher revenues. If the latest technology is worth it, it can be funded and developed by the government or any organisation that needs it.

3.1.3 Technologies Driving FEW Nexus 4.0

The fourth industrial revolution, or Industry 4.0, also known as 4IR, has significantly impacted the global economy in different areas of life due to its productivity, performance, and adaptability. These technologies remain useful in our contemporary world, especially in making efficient use of resources to meet the three pillars of sustainable development (Ogedengbe et al., 2023). Technologies used in 4IR have helped change the nature of production from manual to mechanised over time. Advances in science and technological innovation define the concept of the Industrial Revolution. Likewise, the current revolution, the fourth industrial revolution and the digital revolution, has its activities digitally controlled or regulated by information and communication technology, cyber-physical systems (CPSs) and the IoT (David et al., 2022).

This constant technological revolution is changing the structure of production, business, and economic prospects in nations and thus calls for a strategic assessment of how these technologies affect fundamental resources, namely, water, energy, and food, in the human economy. These resources are important for socioeconomic development. Other authors such as Li et al. (2021) and Qi et al. (2021) highlight the importance of these resources in solving the global challenges affecting food security, water and energy crises, human health, and sustainable economic growth. There has been a lot of research going on lately on how to effectively use technology to manage these vital life resources, food, energy, and water, in such a way as to see them as one and make them so connected that the output of one can serve as input to the other. This approach has birthed a new research field known as FEW nexus or WEF nexus.

The production and consumption of water, energy, and food are tightly connected with corresponding interdependencies (Ogbolumani & Nwulu, 2021; Olawuyi, 2020), as well as (Endo et al., 2015); the concept of WEF nexus manages one resource for several sectors and seeks efficiency without rivalry (Pueppke, 2021).

The FEW nexus is a systematic approach to managing limited resources for sustainable resource security and development (Ogbolumani & Nwulu, 2022). Some areas of the FEW sectors that affect the human economy are their connections and relations, which provoke the stimulation of economic processes and the possibility of managing the impacts resulting from the sectors with the help of the nexus approach (Torres et al., 2022).

The technologies driving FEW Nexus 4.0 include, but are not limited to, *IoT, AI, data analytics, Machine Learning (ML), blockchain, geographic information system (GIS), precision agriculture (PA) technologies, smart cities (SCs), green energy, Virtual Reality (VR), and, renewable energy*, Their importance and the technique of integrating them into the FEW 4.0 are explained under several headings in the following sections.

3.2 INTEGRATING IOT AND AI IN FEW NEXUS 4.0 SYSTEMS

The IoT refers to devices with sensors, processing ability, software, and other technologies that enable them to connect with other devices and systems through the Internet or other available communication networks. This area extends to electronics, communication, computer science, and mechatronic engineering. Devices in this network do not necessarily need to have a direct connection to the public internet; instead, they will be connected to a dedicated network and be addressable with protocols to communicate and share data effectively (Bakar et al., 2023; Shafiq et al., 2024).

In a nutshell, IoT enables the networking of diverse devices and systems to collect and exchange data smoothly, enhance efficiency, and automate functions across various applications and industries (Dey et al., 2018; Swamy & Kota, 2020). The idea of IoT is to enable two devices to communicate without any human intervention.

The concept of putting sensors and intelligence into physical objects, commonly known as the IoT, dates back to the 1980s, when university students hacked a Coca-Cola vending machine to report its contents remotely. However, the technology was awkward and somewhat cumbersome because of its nature; few people paid attention to the innovation, which wasn't widespread (Chin et al., 2019).

Computer scientist Kevin Ashton, who worked for Procter and Gamble, first coined the term "Internet of Things" in 1999. Ashton's idea used small radio frequency identification (S-radio frequency identification [RFID]) chips embedded in products to track them as they moved along a supply chain. He included the buzzword "internet" in his suggestion to draw the attention of executives and other notable personnel in attendance (Manglani et al., 2019). The next decade began to witness a surge of interest in IoT as more devices connected to the internet. Milestones include LG announcing the world's first smart refrigerator in 2000 and Apple launching the first iPhone in 2007. By 2008, the number of devices connected to the internet had surpassed the number of people on the planet (Rayes & Salam, 2016). Major innovations followed those of LG and Apple in 2009 as Google began testing autonomous vehicles on roads and highways and launched the Google Nest in 2011, a smart thermostat people can use to remotely control with their smartphones to adjust the central heating in a house. It is fair to say that these innovations have been milestones in the spread of IoT technologies (Marchant, 2021).

In addition to what was said earlier, IoT-connected devices are split into three broad domains. These include consumer IoT, enterprise IoT, and public spaces IoT (Siow et al., 2018). Consumer IoT includes devices like wearables, i.e., devices worn on human bodies. Enterprise IoT applications include those made in smart factories and PA, used for large-scale activities and production. Public spaces IoT consists of waste management systems and waste management systems, among others; they are commonly used for monitoring and control (Ajibola & Ogbolumani, 2024; Ogbolumani & Adekoya, 2025).

Enterprises use IoT technology to improve and boost supply chain optimisation, inventory management, and customer experience; they are commonly used to enhance efficiency and improve productivity. On the other hand, smart consumer devices like the smart speaker, Amazon Echo and others have become everyday household items. Most new home appliances are now smarter than before and enable remote operations, thanks to the ready enable remote operations and availability of cheap and energy-efficient sensors.

The past decade has witnessed a surge in the application of IoT technology in most aspects of city structures in the United States, the United Kingdom, and China. This deployment aims to simplify operations ranging from collecting water metre readings to traffic management. This clearly shows how IoT solutions can be effectively used in cities in terms of versatility. Their use has brought about the simplification, ease, and convenience of daily activities (Marchant, 2021).

The focus of technologies such as IoT on the FEW nexus is primarily driven by the goal of mitigating climate change through a targeted approach to efficient natural resource management (Kumar et al., 2018).

The dreams of sustainable development goals are recently fuelling significant increases and improvements in various technologies and concepts, both within national and international frameworks. Such efforts foster long-term sustainability practices across various sectors, including energy, food, and water, while supporting economic stability and promoting sustainability (Biggs et al., 2015). The Food- Energy-Water

(FEW) nexus has emerged as a primary idea with good-sized implications for sustainable improvement. The FEW nexus has attracted significant international interest, mainly in reaction to the profound and adverse changes witnessed in climate and society. These changes include urbanisation, industrialisation, globalisation, an increase in population, financial crises in several countries, depletion of resources, and disconnections among communities (Endo et al., 2015). Important essential issues several nations are dealing with include these challenges. Several nations are dealing with other important issues that revolve around energy, food, and water security. The intricacies of the FEW nexus gifts are a powerful venture for many countries, compounded by various factors, including uncontrollable human population increases and climate change impacts. Meeting the growing needs for energy, food, and water while curtailing greenhouse gas (GHG) emissions poses a large, enormous task for a few nations. Therefore, attaining sustainability hinges on adopting efficient practices within the FEW nexus (Leck et al., 2015; Rasul & Sharma, 2016). IoT technologies facilitate how we interact with technology and the world by connecting objects to perform specific tasks, increasing productivity and efficiency through data transfer, analysis, and quick decision-making. IoT should be strategically placed in addressing the complex issues of the FEW nexus that affect livelihood sustainability. There are higher threats to water resources due to water scarcity, pollution, and climate change, and responses to IoT are needed for efficient and effective water protection, improved water management efficiency, and enhanced relief efforts. Agriculture is a significant food industry and requires an efficient electrical response to water generation. IoT provides a comprehensive response to these challenges, making it very easy to effectively and conveniently address the challenges associated with the FEW nexus. Since many industries and governments face FEW nexus challenges, IoT adoption will be critical in solving these overlapping problems in the FEW nexus, and this technology will prove very effective in mitigating these challenges (Kumar et al., 2018; Lin et al., 2017).

3.2.1 IoT Architecture

The IoT is a technology that enables the interconnection of devices to the virtual realm through the Internet, thereby enhancing effective decision-making and computational capabilities. The structure of IoT lacks a defined pattern and can range across diverse applications. IoT operates on a three-layered pyramid, which is commonly referred to as its architecture. The layers are the perception layer, network layer, and application layer. At the same time, some other structures of IoT depict the technology as having a five-layer architecture, including a notion layer, a delivery layer, a processing layer, a utility layer, and a commercial enterprise layer (Kumar et al., 2018; Lin et al., 2017; Sethi & Sarangi, 2017). Often, the IoT architecture is defined through the technologies applied in its development, comprising things like sensors and devices for the collection of data, gateways for providing connectivity and control for the transfer of data, community infrastructure including internet and routers, and cloud infrastructure, aiding in decision-making, analytics, and computation (Ray, 2018). Figure 3.3 shows the most common IoT architectures. They are the three-layer architecture, SoA-based architecture, and middleware-based architecture.

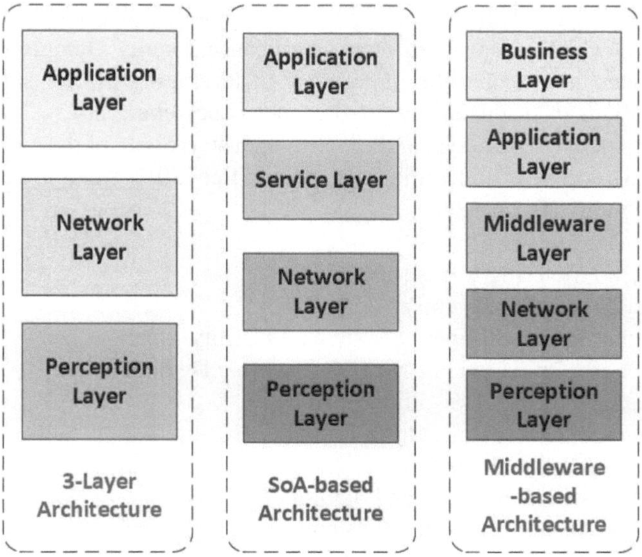

FIGURE 3.3 Common IoT Architectures (Lombardi et al., 2021).

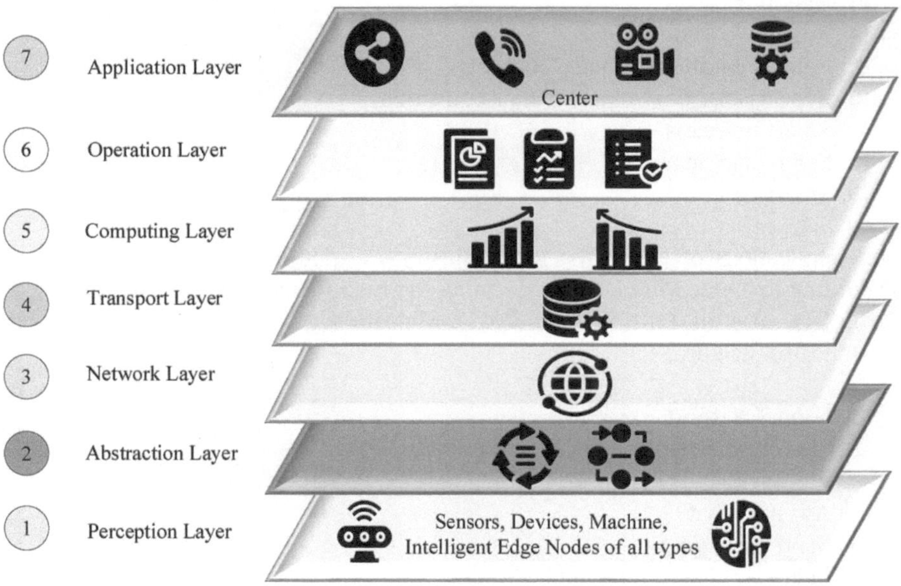

FIGURE 3.4 Schematics of the Seven-layer IoT architecture (Krishna et al., 2021).

Figure 3.4 shows a seven-layer structure in IoT for the FEW nexus, which has been proposed by Krishna et al. (2021). The seven layered architecture includes the perception layer, abstraction layer, network layer, transport layer, computing layer, operation layer, and application layer.

Understanding IoT within the scope of sustainability is critical, mainly in the FEW nexus in which energy holds great significance. Energy's interdependence with food and water underscores the importance of leveraging IoT for sustainable and sensible solutions in sustainability programmes (Leck et al., 2015). The technology of IoT has opened opportunities in the energy industry. Some of the emerging opportunities brought up by IoT according to Choi and Lee (2017; Kul and Şen (2017; Orsi and Nesmachnow (2017) include:

1. Advanced Metering Infrastructure
2. Battery Energy Management
3. Control of Electrical Energy System and Utility
4. Control and Operation of Energy Consuming Devices
5. Decrease in Energy Downtime
6. Energy and Performance Optimisation
7. Energy Storage and Analytics
8. Energy System Reliability and Stability
9. Fault Maintenance in Energy Systems
10. Hybrid Energy System Intelligent Control
11. Hybrid Electric Vehicle Intelligent Control
12. Micro-Grid
13. Nano-Grid
14. Predictive Maintenance
15. Remote Monitoring and Reliability
16. Supervisory Control and Data Acquisition
17. Smart Inverters
18. Safety and Security
19. Smart Grid.

The applications of IoT in the food industry are quite promising, some of them, according to Three Way IoT transforming the food industry (n.d.), (Abraham et al. (2017), Desai et al. (2017), Dolci (2017); Maksimović et al. (2015), and Ramundo et al. (2016) include:

1. Food Transport
2. Storage and Handling of Food Products
3. Freshness and Quality Monitoring of Food Products from Plant Species and Animal Meat
4. Food Packaging
5. Intelligent Packaging and Delivery
6. Vegetable Washing System
7. Removing Dried and Damaged Leaves from Food Materials
8. Food Delivery Monitoring and Alerts
9. Sorting of Vegetables, Fruits, Packed Food Products
10. Cereals and Pulses Drying in Controlled Temperatures
11. Meat and Other Food Storage in Cold Conditions
12. Online Payment of Food Bills
13. Food Safety and Security.

In the area of water, IoT technology plays a prominent role in tackling challenges confronting this field. Some of the applications, according to Narendran et al. (2017), Predescu et al. (2017), and Ramesh et al. (2017), include:

1. Water Quality Monitoring
2. Water Safety Monitoring
3. Wastewater Management
4. Quality Control on Water Reserves
5. Monitoring Drought Locations
6. Water Transport
7. Leakage Detection in Water Flow Paths
8. Monitoring Water Consumption Patterns
9. Efficient and Systemic Water Management
10. Demand Side Water Price Monitoring.

Realising the synergies involving the FEW nexus and the potential of IoT, we can see that IoT presents numerous possibilities for optimising properties, resource consumption, and organisational features of the interrelated connections regarding the FEW nexus. IoT can be applied in the FEW nexus in the following ways:

1. **Smart Agriculture**: Such agriculture uses digital devices which digital devices that are competent in capturing data pertaining to soil humidity, climatic conditions, the status of the plants, and the level of nutrients available in the plants in very short periods. It means that the farmer can control the amount of water used for watering the crops, the amount of fertiliser used, or whether there is a need to apply pest control. Doing so adds to the yield from the crop but does so in a responsible and sustainable manner (Azadi et al., 2021; Ogbolumani & Nwulu, 2020; Ogbolumani & Nwulu, 2021).
2. **Water Management**: Smart gadgets using IoT technology can help regulate water quality usage and delivery schemes in urban and horticulture applications. On the practical level, IoT-based smart metres minimise water wastage and help detect leaks in an integrated farm system; they also help forecast demand and, therefore, support conservation while helping to reduce losses in water supply systems.
3. **Energy Optimisation**: The deployment and use of smart grid infrastructure and advanced energy analysis software via IoT can become a potential tool for power systems technology in distributing and utilising resources in the energy food and water nexus. These include green-energy products, gadgets, integration of renewable electricity, and control structures specifically designed and tailored to guarantee that energy supply for water and food production processes is always available (Kumar et al., 2020; Ogbolumani & Nwulu 2020).
4. **Supply Chain Transparency**: IoT devices keep monitoring delivery channels of consumables such as foodstuffs, power, and water right from production, through the supply channel, down to the consumer. It also helps in better tracking and controlling the quality and safety aspects that deal with food theft or even diversion, energy theft, and water wastage or contamination (Madumidha et al., 2019).

5. **Predictive Maintenance**: Integrating IoT in structures supporting FEW Nexus 4.0 offers the benefit of predictive maintenance. When deployed, IoT technologies offer higher monitoring and tracking of the fitness and overall performance of machinery utilised in food processing industries, water treatment plants, and energy generation plants. Significantly, in this manner, the chances of downtime are greatly reduced, and operational effectiveness and efficiency are expanded via spotting faults and deviations from normal operating styles of machines earlier and reporting these faults to the maintenance team. They also forecast maintenance before a breakdown happens.
6. **Cross-Sector Collaboration**: This includes hubs for the collection, transfer and integration of data in the FEW systems space. This principle works effectively when combined or integrated into the decision-making process, utilisation of resources and management of risks related to challenges like climate change, population growth, and resources scarcity.

The concept of IoT within the FEW nexus under Industry 4.0 holds significant property. As can be seen from our earlier discussions, IoT provides transformative prospects to achieve sustainable resource utilisation, build the resilience of our systems, and prepare for the world's future in light of new challenges.

3.2.2 Integrating Artificial Intelligence in FEW 4.0 Systems

Over the years, various authors have highlighted the key role of innovation in reaching sustainable goals and performance (Silvestre & Ţîrcă, 2019). Data overlapping, limited knowledge of human resources, and unpredictable climate events, for example, may make decision-making cumbersome, leading to poor sustainability plans (Albrecht et al., 2018). Therefore, AI technologies serve as powerful tools in handling a large chunk of data; AI also helps identify unseen insights and provide scientific and technically based solutions related to environmental and climate problems (Nishant et al., 2020). AI supports the decision-making processes of multi-stakeholders to attain sustainability goals through the different industries responsible for the UN 2030 Agenda (D'amore et al., 2022). However, the potential of AI in the WEF nexus has been underestimated from existing literature, which mostly relates to other stakeholders, especially governments and communities (Di Vaio, Boccia et al., 2020; Di Vaio et al., 2022; Di Vaio, Palladino, et al., 2020). Although digital technologies are recognised for having similar productive potentials in the FEW nexus, the potential of AI in managing the FEW nexus and improving governance across resources, sectors, and institutions is almost overlooked (Caprani, 2016; Duan et al., 2019; Dwivedi, Hughes, et al., 2021). To delve more into AI and its potential to transform the FEW Nexus 4.0 positively, we need to understand the meaning of AI.

AI is defined as a machine's ability to mimic human skills in reasoning, learning, and planning for the solution to a problem. ML models and algorithms enable AI to analyse and learn from massive datasets acquired from an external environment through sensors and devices such as satellites, motion sensors, cameras, etc.

(D'amore et al., 2022). Such interaction enables AI to interpret the data correctly and pursue specific objectives through flexible configurations (Davenport et al., 2020; Haenlein et al., 2019; Nawaz et al., 2024). The symbiosis between AI and human intelligence through algorithms may support managers in making quality decisions that will be fruitful for their organisations. The emergence of AI brings a cultural change wherein data, connectivity, and interactivity are interwoven in standard management practices. The algorithms used in AI commonly outperform human decisions. Therefore, AI is a powerful tool to support advanced decision-making processes within the WEF nexus.

There are several types of AI based on technologies, functionalities and capabilities; they are briefly described below (Tuomi, 2018)

Based on Technologies, we have the following branches of AI:

a. **ML**: AI architectures that can learn independently from previous experience and do not require explicit programming. This concerns the creation of self-learning software that acquires data and utilises it on its own (Ogbolumani & Adekoya, 2025).
b. **Deep Learning (DL)**: A kind of ML that goes beyond a single neural network layer. It is usually employed in learning large volumes of data, and it is foundational to voice control in consumer electronics, image recognition and many other uses (Matsui et al., 2022; Soori et al., 2023).
c. **Natural Language Processing**: This technique of AI enables a computer to read and understand any natural spoken language. It is used in interactive operations within the chatbots, translation services, and consideration of attitudes within the applications (Matsui et al., 2022).
d. **Robotics**: This is a term that refers to the managerial way of handling robots as well as other computer systems that are used to manage the robots, their built-in sensors, and data processing units (Mishra & Gupta, 2017; Soori et al., 2023).
e. **Computer Vision**: It is concerned with enabling machines to perceive and interpret the world in ways similar to humans, and with the capacity for these computerised worldviews to be understood. This technology has the following applications: computer tomography, image analysis, watchful eye, and production.
f. **Expert Systems**: These are other AI systems that answer questions and engage with issues that occur in a specific area of specialisation using if-then programming.

Based on functionalities, we have the following types of AI:

a. **Reactive Machines**: Unlike human beings while preparing a course of action for AI systems of this nature, there are no memories or past encounters that are remembered. They also categorise and/or address various states of affairs within the targeted culture. IBM's Deep Blue is an example of a computer system that beat Garry Kasparov in chess.

b. **Limited Memory**: AI of this nature functions in collecting historical data. The gathered historical data enables these AI systems to make optimum and wise decisions as the possible outcomes have already been considered. Most of the AI embodied in today's society can be grouped under this class, including chatbots, virtual assistants and self-driving cars.
c. **Theory of Mind**: It is a bit more developed from the previous type, and researchers are still studying it. It would take recognition and memorisation of emotions, beliefs, needs, and depending on them to make the right decision. It implies that the machine must be able to comprehend people.
d. **Self-aware AI**: This captures the future of AI, which means that computing systems will have consciousness, sentient, and self-awareness. This sort of AI is yet to be and exists only in the realm of distant possibility; yet such an intelligent entity would be able to learn and feel, and thus, it would form opinions and set objectives.

Based on capabilities, we have the following types of AI:

a. **Narrow AI**: This type of AI is also called weak AI. It is the type of AI that has limited power in what it can search for, such as facial recognition, web search, and driving a car. They comprise a large percentage of AI systems under development today and include algorithms playing games like Go and Chess. They are made to operate in a restricted field under certain conditions.
b. **General or true AI.** This kind of AI can manage new and emergent tasks independently because it contains all the functionalities in people. This type of AI is fully autonomous, and there is no limit to the capability to analyse and understand the problems that this AI framework can solve.
c. **Superintelligent AI**: This kind of AI can represent the next level of intelligence that can even outperform human intelligence complemented through creativity, general wisdom, and strategy. Currently, superintelligent AI remains a theory and concept that has not been actualised or empirically proven.

AI can be applied in the FEW Nexus 4.0 to solve the complexities within these integrated systems. AI can improve efficiency, sustainability, and resilience within the FEW nexus by:

1. **Predictive Analytics**: AI works like a mechanism whereby the models can capture raw data feeds from different sources or suck data and can forecast, predict demands and even scarcities of stocks of food, energy, and water.
2. **Optimising Resources**: AI models can be deployed to optimise water use in agriculture irrigation and trim down energy consumption levels for food and water production, with perfect management of these resources.
3. **Smart Monitoring and Control**: The control and monitoring of various parameters like environmental conditions, water quality, energy use, and

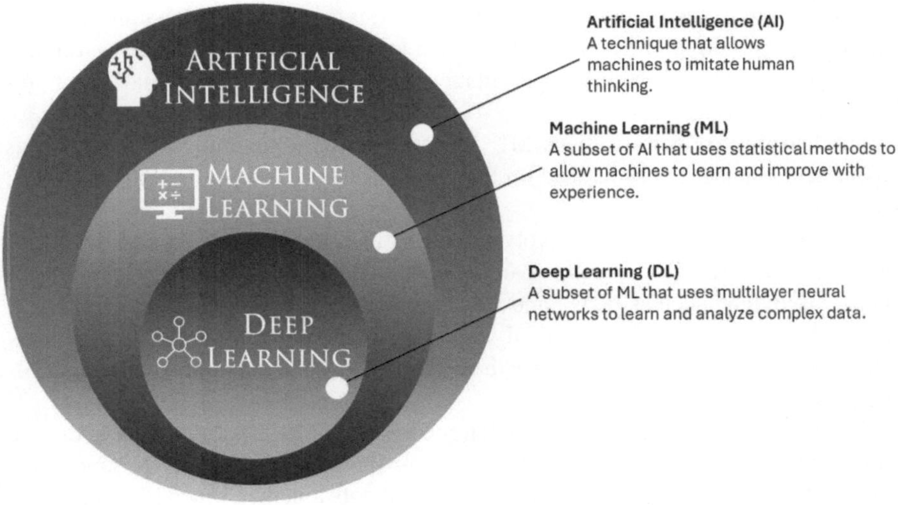

FIGURE 3.5 Artificial Intelligence and its Subcategories (Mondillo et al., 2024).

crop conditions have become quite approachable in real time and hence allow effective and timely remedial measures using the sensor networking of IoT-based smart gadgets.

4. **Decision Support Systems**: This is also known as DSS; they use advanced computing techniques and formulations to present policy makers, planners, and stakeholders with well-reasoned information in endeavouring to analyse and make informed decisions on the prudent management, allocation, policies, and investments in FEW resources.
5. **Resilience Planning**: Several approaches can help AI build resilience for the FEW nexus by identifying risks, predicting an outbreak of an event, and formulating coping mechanisms against the effects of climate changes, water shortages, and energy breakdowns.

AI has some other technologies under it. They are referred to as the subsets of AI. For the scope of FEW Nexus 4.0, we will only elaborate on ML and profound learning aspects of AI. Figure 3.5 shows a pictorial representation of AI and its subsets, ML, and DL.

3.2.3 Deep Learning vs. Machine Learning

Many people treat "Deep Learning" and "Machine Learning" as the same thing in AI. This is untrue. Therefore, anyone interested in exploring AI must understand the differences between these terms. The good news is that these differences are not difficult to understand, as evidenced by several published articles. Both ML and DL are subsets of AI (Soori et al., 2023; Wolfewicz, 2023).

ML: ML is a subfield of AI that algorithms and statistical models to boost the intelligence of computerised machines, so that computers can learn from data and make decisions, all without being programmed. It is essential to understand that ML algorithms are taught on large extensive data and are used to find patterns, relative connections, and trends between variants for the subsequent decision-making or prediction (David et al., 2024).

The learning process of these ML algorithms would generally be classified as supervised or unsupervised based on the data used in their training (Sarker, 2021b; Wolfewicz, 2023).

DL: DL is a form of ML that involves using artificial neural networks (ANNs) with more than one layer. DL algorithms are more suitable for tasks with large amounts of unprocessed data, such as image and voice identification and visual, speech, natural language, and driving automation. DL models can automatically draw out important characteristics from raw data and use these features to solve complex problems accurately (Kumar et al., 2024).

DL can also be considered a supercharged version of ML that has been under the spotlight lately since it can do things we never thought possible. ANNs, modelled after the brain, work by putting a lot of connected dots together, much like our brains do. This setup is used by DL, which allows them to learn and understand much more complex patterns than regular ML methods (Kufel et al., 2023).

3.2.4 Differences between Machine Learning and Deep Learning

Deep Learning is defined as a subfield of ML because it has some peculiarities that make it stand out from the rest of the scope of this field. Its characteristics are best summed up as the architecture of the ANNs, low level of intervention, and increased demands for data (Taye et al., 2023).

Unlike others called traditional ML algorithms, which regularly have a simpler model, for example, linear regression trees or DL makes use of complex ANN resembling the human brain. Besides, DL algorithms contain less human interference than conventional ML algorithms. The difference between DL and more traditional approaches is that feature selection and classification are generally done by software engineers, but DL algorithms perform feature extraction on their own and improve all their decisions based on the data they receive (Dargan et al., 2020). DL also requires significantly larger amounts of data for learning and fine-tuning weights appropriately. Whereas, with thousands of samples, the traditional ML algorithms may suffice, in DL architectures, millions are sometimes needed due to the multiple layers of the model (Jamwal et al., 2022; *What Is Deep Learning? | IBM*, n.d.). This large variety of data helps DL systems to reduce the variation and provide higher correct interpretation (Sarker, 2021a).

3.2.5 Deep Learning and Machine Learning in FEW 4.0

The idea of using ML methods to model the interconnections between energy and water or energy and food or food energy and water has not been explored in recent times. Irrespective of this, conventional methods such as ANNs and support vector machines have applied and isolated water and energy as external resource systems (Han et al., 2018). In a survey by Zaidi et al. (2018), they depicted two related organisations. The first organisation defines and categorises the ML approaches, while the second determines and categorises other types of target problems within the energy-water nexus domain.

The use of ML, more specifically recurrent neural networks (RNNs) has the following advantages in terms of their method: robustness, ability to solve nonlinear problems, suitability for application in the case of small sample size, statistical accuracy, and high level of generalisation (Akhter et al., 2019; Guo et al., 2021; Şahin et al., 2021). However, RNNs suffer from vanishing gradient problem, thereby being constrained in their capacity to handle long-term dependencies in sequences of information, for instance, temporal data (Williams & Zipser, 1989). This problem led to the development of the long short-term memory (LSTM) network (Hochreiter & Schmidhuber, 1997), with gated recurrent units (GRUs) as a less complex version than LSTM (Cho et al., 2014). For time series applications, both LSTM and GRU have been widely used in several domains like solar radiation (photovoltaic energy), electricity load, water usage, and air pollution (Rajagukguk et al., 2020; Salloom et al., 2021; Sarkar et al., 2022; Torres et al., 2022).

In the energy-water nexus, ML techniques have seldom been used, with most modelling of water-energy interactions done through various techniques such as ANNs and support vector machines. The use of these techniques generally considers the systems of water and energy as separate resources. There are various types of ML techniques utilised in the energy-water nexus domain. These techniques are categorised into three classes: supervised, unsupervised, and reinforcement learning. ML can be used to address various problems that exist in the FEW nexus. The various target problems relevant to the domain of interest in the FEW nexus relating to ML include energy generation, quantity or intensity of energy generation from various sources, non-renewable sources like fossil fuels such as natural gas, coal, petroleum, and renewable sources like biomass, solar energy, wind energy, and hydropower. These specific problems can be described below:

- Energy generation: Modelling energy consumption across the residential, industrial, or commercial sectors.
- Water Usage: Model water usage across residential, industrial, and commercial sectors.
- Energy for Water: Simulating the flow of energy involved in the extraction, supply, treatment, or utilisation of water.
- Water for Energy: Modelling the use of water used in energy (Zaidi et al., 2018).

Challenges in using ML for PA include (1) spatial-temporal resolutions that are variable and (2) the presence of missing data resulting from various factors such as malfunctioning IoT or failures in communication (Kussul et al., 2017). DL models, especially convolutional neural networks and unmanned aerial vehicles (UAVs), have recently been increasingly used in agricultural remote sensing applications. Low adaptation of UAV-based technology in specialty crops is attributed to complexities in data preprocessing and analysis, along with some limitations of the availability of commercial tools (Krizhevsky et al., 2012). BDA and CC have become very important in terms of efficiently processing data and handling other issues such as security, scalability, and reduction of cost for UAVs. UAVs generate and handle volumes upon volumes of unstructured data, which makes them useful for applications involving the extraction of large datasets (Kim et al., 2019). Cloud applications combined with analytic tools could help perform tasks such as the detection of certain characteristics of trees and crop estimation using UAV imagery (Ghatrehsamani et al., 2020). Figure 3.6 shows a pictorial concept of the integration of AI into FEW Nexus 4.0.

ML and DL have become important in the context of the FEW Nexus 4.0, i.e., food, energy, and water, integrated with the Fourth Industrial Revolution. These computational tools tend to offer insightful decisions in meeting complex challenges that arise from issues of resource management, sustainability, and efficiency within the integrated fields of food, energy, and water resources. ML algorithms can parse large quantities of data to spot trends. This ability, among other things, is useful for resource optimisation and predictive analysis related to supply and demand dynamics in the FEW nexus. It also helps optimise decisions along other dimensions. DL is a subdomain of ML, which furthers its capabilities using deep neural network structures to extract complex features and relationships from high-dimensional data. This allows for far more accurate prediction capabilities and the creation of automated decision support system. ML and DL technological integration in FEW Nexus 4.0

FIGURE 3.6 Integrating Artificial Intelligence (AI) into FEW Nexus 4.0 (Gbashi & Njobeh, 2024).

will offer very strong potential to address certain issues associated with the sustainability and resilience goals of the FEW Nexus 4.0.

3.3 APPLICATIONS OF BLOCKCHAIN IN FOOD, ENERGY, AND WATER SYSTEMS

Blockchain has invaded the digital space by changing the entire setup of contracts, transactions, and records that make up the basic unit of an economy, polity, society, and legal system around the globe (Ertz & Boily, 2019). Primarily, the birth of blockchain can be ascribed to the presentation of bitcoin by the pseudonym Satoshi Nakamoto in 2008 and is regarded as the first generation of blockchain. This generation has witnessed the rise of cryptocurrencies in financial applications and changed the whole perspective of digital payment systems (Nerurkar et al., 2021). The second generation of blockchain introduced smart contracts, thereby paving the way for cashless transactions and even lending services in the form of stocks, loans, and smart properties. Further, the third generation of blockchain has shifted the terrain from banking to new applications such as healthcare, government, science, and so on. In less than a decade, blockchain has undergone three generations and is on the threshold of entering the fourth generation with AI and digital intelligence (Tripathi et al., 2023). Figure 3.7 shows the working principles of blockchain technology.

Blockchain technology and its system of decentralised storage and control of records have changed business, industries, and commercial segments. Instead of belonging to one owner, blockchain consists of time-stamped and unalterable data chunks known as blocks populated by a network of nodes or computers. Each block is connected to the next one in a secure manner, using cryptographic hash functions. This unalterable and distributed architecture has transformed traits and frameworks of trust, possession, personality, and monetary systems into something safe, clear, fleeting, and anonymous (Tripathi et al., 2023).

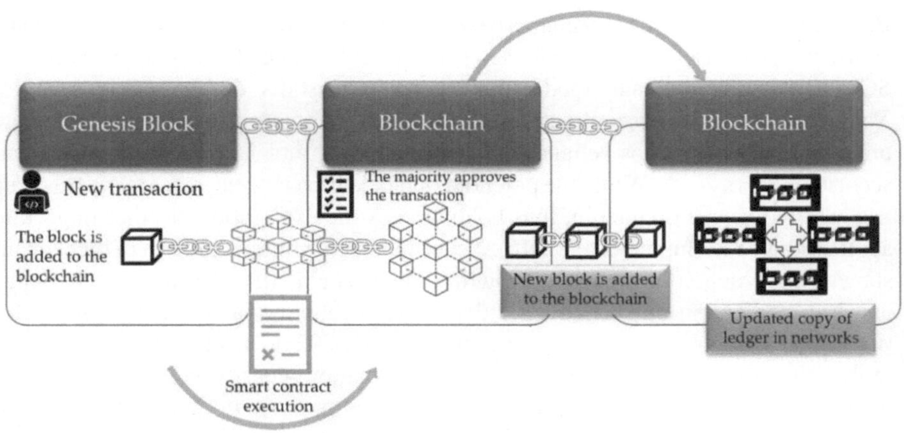

FIGURE 3.7 Basic Working Operation of Blockchain Technology (Juszczyk & Shahzad, 2022).

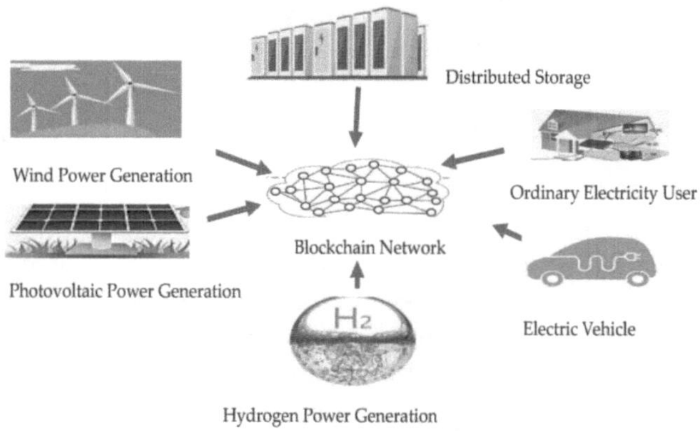

FIGURE 3.8 Application of Blockchain Technology in a Micro Grid (Wang et al., 2022).

Blockchain is a distributed database that differentiates it from other databases because it stores data in linked blocks using cryptography. Blockchain can hold many forms of data, the most common form involves using blockchain as a ledger for transactions (Tripathi et al., 2023). It is somewhat like a record book containing data in blocks that are connected by chains. Each block contains data, and computers view the system or arrangement to assist in the chain's composition. Once the data set has been added, it is "sealed," highly secure and resistant to any alteration due to the cryptographic principles built into the systems security structure. Blockchain means transparency and offers a way of organising records and checks of transactions or data without the use of a specific centre. Figure 3.8 shows an application of blockchain technology in a microgrid system.

3.3.1 Applications of Blockchain Technology in FEW Nexus 4.0

Beginning with bitcoin, blockchains have progressed to transparent and trusted smart contracts for a variety of applications in various fields, some of which include finance, SCs, healthcare, and agri-food value chains (Christidis & Devetsikiotis, 2016; Nakamoto, n.d.). Smart contracts usually execute on consensus nodes to automate processes, which lessens reliance on intermediaries and increases efficiency and security (Szabo, 2018). While the potential of blockchain is great, research in this area has mostly focused on bitcoin, thus holding back the exploration of other important applications (Yli-Huumo et al., 2016). Nevertheless, the market growth of blockchain shows huge potential for industry growth (Lucena et al., 2018). However, a challenge exists in perceiving blockchain as a disruptive technology and a converging solution in various business processes (Toufaily et al., 2021). Blockchains are increasingly being used in agri-food value chains, information security, manufacturing, and management of water sustainably. They show great promise in improving both supply chain management and sustainability (Tripathi et al., 2023). Similar to the idea of Tse et al. (2017), the use of blockchain is recommended in the agri-food value chain to address market needs. Kumar and Mallick (2018) have further provided

details of the use of blockchain in different agri-food chain activities. Tian (2017) has developed a traceability system integrating RFID and blockchain for Chinese agri-food markets. To overcome the challenges associated with RFID, Boehm et al. (2018) combined blockchain with Near Field Communication and suggested that only authorised users should be allowed while conducting his studies. Papa (2017) notes that blockchain could improve transparency and customers' trust. Nonetheless, the scalability problem persists; recent advancements aim to solve it (Croman et al., 2016; Gupta, 2017). However, the centralisation of blockchain solutions brings privacy and security issues into discussion (Yu et al., 2018). For example, approaches based on a public blockchain can provide secure and safe communication between IoT devices with no possibility cyberattacks (Hammi et al., 2018).

Tian (2017) considers a traceability system that integrates HACCP, blockchain, and IoT to enhance safety and productivity. According to Badzar and Sternberg (2016), blockchain has features that make transactions transparent and trustworthy. The position of blockchain in agri-food with a focus on the ability to provide a secure record. Biswas et al. (2017) put forward a wine traceability system using blockchain technology. Maouchi et al. (2018) argue that the existing proposals for using blockchain in agri-businesses are not feasible and applicable. Even with certain drawbacks, some researchers and companies have continued exploring this technology's potential in other sectors of the economy, especially the agricultural sector. For example, Walmart's investment in blockchain makes many people realise its potential positive impact (Kamath, 2018; Yiannas, 2018). Lucena et al. (2018) agreed that blockchain technology should be adopted for grain value chain traceability to minimise disputes. Kshetri (2018) mentioned that there are very few systematic studies regarding blockchain applications in agri-food chains. Hence, more research needs to be done in this area to improve the usability of this technology in agriculture.

From the perspective of Industry 4.0 and distributed computing, the information systems in the agri-food industry have undergone many transformations. A few notable advantages of blockchain technology include real-time transaction recording, no intermediary, and the ability to have a distributed ledger that is immutable and increases information security within the agri-food value chains (Mukri, 2008; Nwulu & Damisa, 2023).

Leng et al. (2018) investigated the agricultural value chain and developed a double-chain structure with a public blockchain for a secure, private, believable, and efficient system provided an analysis of data accountability and history employing a blockchain technology that seems very appropriate for the sharing of dynamic data in the agri-food value chain. As more participants in the agri-food chain adopt IoT, this makes them use electronic devices that are interconnected to make complex data collection and analysis in a secure way (Reyna et al., 2018; Neisse et al., 2017) Thus, it is possible to conclude that the use of the blockchain helps to solve the issue of information security and improve the quality of the agri-food value chains with the help of IoT developments.

Managing large volumes of information and services in today's knowledge-based economy is vital for food manufacturers. Manufacturer collaboration is needed to maintain competitiveness (Ersoy et al., 2022). To tackle this problem, some researchers have incorporated blockchain into manufacturing processes. Smart contracts and edge computing were used by Z. Li et al. (2018) to propose a framework for the decentralised management of knowledge and services for enterprises. This framework provides security and facilitates connections and updates of transaction records in the blockchain, making them scalable and flexible while improving quality, efficiency, and effectiveness (Li et al., 2018). Zhao et al. (2019) design a model for manufacturing service information based on blockchain and CC that includes resource sharing, data recall, and quality assurance. According to Kumar and Mallick (2018), the use of blockchain in manufacturing includes management, production data, and product assurance. However, blockchain's ability to enhance robustness, performance, security, and self-sufficiency of manufacturing services has been discussed less (Da Xu et al., 2018).

The productivity of agriculture depends on effective water resource management. Water scarcity and sustainable agri-food development are very critical areas of consideration. Civic Ledger, in 2017, put up a blockchain-enabled platform, Water Ledger, which granted the power of peer-to-peer water trading amongst irrigators, ensuring efficiency in water resources use and ensuring water resources are managed securely and transparently. While incurring upfront costs, using blockchain in water trading, coupled with smart contracts, allows water companies and governments to access real-time data (Apeh & Nwulu, 2025; Tripathi et al., 2023).

Blockchain is also a useful technology in the energy sector, and it has proven promising. Blockchain can optimise the use of energy sources and infrastructure management and spreads across the energy generation system, which touches how power is generated and distributed from bottom to top. It also offers the basic framework of a later energy generation system that allows for adept management of both ends of the spectrum and a smooth integration of multiple energy sources (David et al., 2025; Rejeb et al., 2024).

The energy industry consists of a range of subsectors; most of them have intricate supply chains that range from the extraction of fossil fuels to electricity production and, finally, the distribution of energy. Global and regional restrictions are in place to make things even more complicated. Considering this situation, assistance from the energy industry for sustainability and renewable energy programmes is essential. Blockchain technology fulfils such requirements through enhanced traceability, transparency, and reliable coordination in supply chains. It finds a perfect fit in the energy sector regarding compliance with regulations and the integration of renewable energy solutions (Friedman & Ormiston, 2022).

Some key benefits of blockchain technology that would benefit the energy industry include reduced costs, increased transparency, and a decentralised process (Khatoon et al., 2019). The organisation of complex networks and supply chains makes communication and other processes more effective and generally more efficient, reducing inefficiency and saving on costs across the supply chain from exploration to distribution. At the same time, blockchain increases transparency through verifiable information on a ledger (Nwulu & Damisa, 2023). The immutability of this technology assures that records will never be changed, and this will further increase regulatory compliance without compromising sensitive data. This technology operates on a decentralised and programmable ledger system, meaning that peer-to-peer networks can function without oversight from some central authority

(Nwulu & Damisa, 2023). It is also regulated through smart contracts, so transactions become much safer and generate more energy innovation. Lastly, blockchain can digitise physical assets via smart contracts so that tokens represent assets like energy commodities or carbon credits (C. S. Han et al., 2023). In this regard, digitalisation changes energy trading and facilitates new ways of trading excess energy and other resources (Kirli et al., 2022; Xu et al., 2022).

3.4 SMART CITIES AND FEW NEXUS 4.0: A SUSTAINABLE FUTURE

SCs are created by leveraging the technologies of Industry 4.0 and many organisational changes to improve management in cities and the quality of life (QoL) of citizens. Technology in SCs relies on data and digital technologies to make better decisions and increase the general life quality of people living in a specific city where such technology is applied (Gracias et al., 2023). Technology in SCs relies on three interrelated levels, specifically:

1. **Technological Infrastructure**: This is the most basic layer of this technology; in essence, it comprises smartphones, sensors, and high-speed communication networks, which will be activated through 5G and high-speed internet. All these components pick up and transmit data.
2. **Applications**: This represents the second layer of SC technology. This layer uses specific tools and professional software developed by technology suppliers and application developers. Such tools are used for processing raw data and transforming them into valuable insights and alerts.
3. **End-usage**: This is the final level of SC technology. This third layer concerns how cities, companies, and the public use multiple applications developed to support this technology (Woetzel et al., 2018).

Wide acceptance and a change of people's attitudes to apply these technologies successfully are required for it to fulfil its potential benefits. In this aspect, culture and religious beliefs are dominant factors, but embracing and utilising the SC technology may bring about long-term benefits associated with the optimised use of transportation facilities, energy, and water conservation, as well as the promotion of preventive healthcare systems. These layers interact smoothly with each other to give rise to real-time monitoring capabilities that respond accordingly, hence enabling more effective and adaptive urban management. The purpose of SCs is to plan and develop optimum, structural, and linked infrastructures in cities to enhance the QoL for their inhabitants and encourage economic development (Talebkhah et al., 2023). The advent of advanced computing from standard desktop computing and the increase in connected devices and sensors have led to the emergence of SCs. There are smart environments, such as smart homes (Pal et al., 2018; Yassine et al., 2019), smart grids (Lu et al., 2018), smart transportation systems (Pundir et al., 2022), smart healthcare facilities, and even SCs (Silva et al., 2017). Governments are adopting big data or BD applications to help SCs attain sustainability and standards (Brous et al., 2020; Silva et al., 2017). SCs utilise different technologies for enhancing health services (Pal et al., 2018), transportation (Pundir et al., 2022), energy (Lu et al., 2018), education (Mishra & Tyagi, 2022), and water supply. All these efforts are geared towards optimising costs and resource usage and encouraging greater citizenship activity.

Technologies and innovations such as BDA hold significant promise for enhancing SC services (Brous et al., 2020). Integrating BDA technologies will enable better management and utilisation of large chunks of data obtained from cities. CC infrastructure also supports the management and implementation of BD applications in SCs, thus providing robust platforms for data processing and analysis, which will yield results that can be used to make accurate decisions. Figure 3.9 shows the layout of the smart area consisting of a smart environment and SCs.

The concepts of SCs depend on IoT and innovative technologies and applications for their operations. These applications cut across various domains, from optimising city operations, such as real-time traffic control, to improving entertainment and tourism using augmented reality (AR) (Sánchez-Corcuera et al., 2019). The applications are expected to meet specific criteria by the cities, including sustainable development, adaptation changes within the cities, giving solutions that foster collaboration between various stakeholders, and offering privacy when handling citizens' data. SCs provide a good platform for efficiently transferring large volumes of data between different sections of cities (Dai et al., 2024). These data flows are used in developing SC applications that impact a wide range of aspects such as those of economic, environmental, energy, water, waste, and public engagement

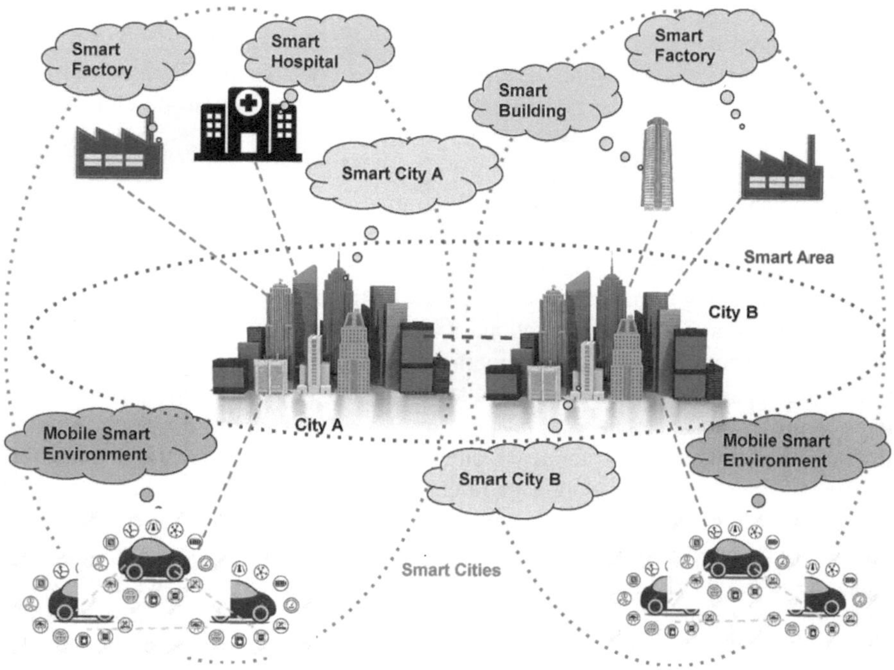

FIGURE 3.9 An Example Layout of a Smart Area (Tricomi et al., 2024).

and intellectual endowment, lifestyle improvements, building efficiency, transportation, and public space improvement (Bastidas et al., 2022; Cowley & Caprotti, 2019; Julieta Cedillo-Elias et al., 2018; Mishra & Tyagi, 2022).

SCs try to bring together applications of IoT in terms of better usage of public resources with enhanced Quality of Services while at reduced costs. Though not officially defined, the term SCs has become very popular. SCs aim to utilise the increasing technological capabilities to solve urban problems and improve life (Bibri & Krogstie, 2017; Trindade et al., 2017). Since cities significantly influence the economy, society, and the environment, they bring forth the practical solutions needed in transport, land use, and the quality of public services, among others. For example, an efficient transport system shall contribute to the economy while effectively providing mobility. Increasingly, cities are adopting innovative solutions, usually based on Information and Communication Technology (ICT), to optimise municipal services and infrastructure, thus aligning with SC goals (Talebkhah et al., 2023).

The adoption of an SC concept drastically increases the data volume involved in the process, and BD is at the heart of IoT-rendered services (Hashem et al., 2016). BD involves volume, velocity, and variety, with data streams emerging at frantic speeds (Hashem et al., 2016). This phenomenon will help provide opportunities for meaningful extraction from vast volumes of data collected from several sources, mainly non-structured features (Gil et al., 2019).

The SC term has become more famous in recent times; this is mainly due to computing devices and wireless, mobile, and wearable sensing technologies. The trend will likely continue, with rapid development expected over the next few years. The efficient use of real-world data in the context of IT infrastructures is necessary to improve citizens' QoL through smart computing (Talebkhah et al., 2023). The concept of SCs arose owing to urbanisation, where more than 54% of the world's population lives in urban areas today. This number is likely to increase to 66% by 2050 (Chang et al., 2019), and by this, the world's urban centres will be hosting 2.5 billion more people by that year (Wang et al., 2018). Although cities occupy only 3% of Earth's land, they are major consumers of natural resources and account for 60%–80% of GHG emissions (Chamoso et al., 2018). As the world population grows, so does the demand for energy, water, food, and transport, causing severe environmental problems such as depletion of resources and global warming. These challenges must be addressed by creating economic, environmental, and social sustainability. The concept of SC has, therefore, gained much attention as a way of achieving the above, as cities strongly influence social, economic, and environmental considerations (Albino et al., 2015). Figure 3.10 shows the basic components of a smart city.

3.4.1 Applications of Smart Cities

SCs use different sensor technologies besides platforms to deliver effective and intelligent services to the public and the authorities. Each of these sensors, including positional, pressure, airflow, motion, and sound, are networked by the IoT, and together they form the core of data collection and monitoring and management

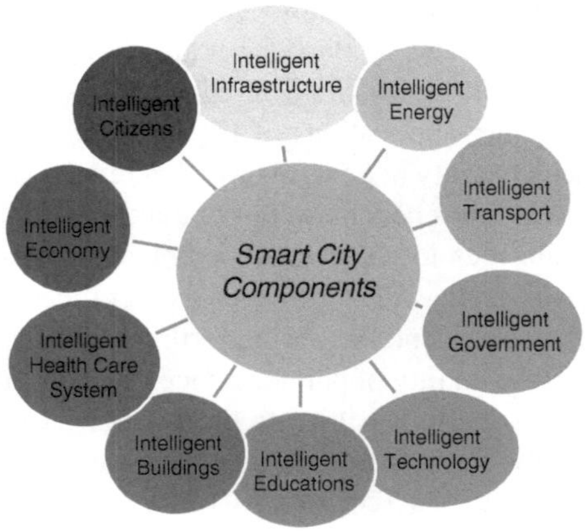

FIGURE 3.10 Basic Components of a Smart City (Issa Zadeh & Garay-Rondero, 2023).

of urban activities in real time (Ajibola & Ogbolumani, 2024; Ogbolumani & Mabaso, 2023; Talebkhah et al., 2023). Wearable sensors are becoming very popular, particularly smartphones, because they provide a lot of information regarding the activities and behaviour of users. SCs use data from these wearable sensors to improve several services, such as traffic and parking, to decrease driving stress, fuel consumption, and pollution while increasing productivity in the economy (Ramírez-Moreno et al., 2021). The SC also incorporates technologies such as smart public lighting equipped with motion sensors and energy-efficient LED light bulbs and smart buildings equipped with sensors to optimise energy consumption in the buildings within the city. All these are urban infrastructures that pertain to sustainability and resource efficiency. Other sectors in which sensor technology has been used in the city include Smart Healthcare, Smart Transportation, Smart Governance, Smart Grid, and Smart Home. This technology has proven to improve the QoL in cities and make urban environments liveable while supporting (Kim et al., 2022; Mofidi & Akbari, 2020).

The concept of a smart city and FEW Nexus 4.0 are, two correlated ideas that aim to deal with urban sustainability problems. Advanced technologies and data analytics put into practice make urban areas efficient, sustainable, and more liveable. Integrating ICT, IoT devices, and data analytics optimises several factors that characterise city life, such as transportation, energy management, waste disposal, public safety, and health care. SCs create that are resilient, resource-efficient, and responsive to the requirements of the people (Okonta & Vukovic, 2024; Sukhwani et al., 2020). When SCs become integrated within the FEW nexus, it allows for holistic solutions to the problems plaguing cities holistically (Batty et al., 2012; Lee et al., 2023). Digital insights gained from data in a smart city environment are useful in improving efficiency in food, energy, and water systems. This information will be useful in promoting efficient and effective resource allocation and waste reduction, thus building resiliency to environmental and socioeconomic changes, and in the long run, making the environment sustainable. Integration of the FEW nexus strategy into smart city design and development would mean increasing the sustainability and QoL for citizens in cities around the world.

In this section, blockchain technology was discussed with emphasis on its applications in the FEW Nexus 4.0. The following points are worth noting about how blockchain technology influences the FEW 4.0:

1. **Supply Chain Transparency**: With the help of the distributed ledger enabled by blockchain, it is possible to verify the status of each transaction that takes place in the supply chain, starting from the farm down to the plate of a consumer. This also helps combat fraud and contamination issues with food products in an effort to develop quality and original food products.
2. **Water Management**: It can also be used in water supply management; here, it entails controlling use of water, analysing water quality, and allowing the users to sell water to one another. Smart contracts may also assist in engaging the systems in the way and manner of handling water and governing water usage and scarcity.
3. **Energy Trading**: Blockchain enables P2P trading between various microgrids, where electricity consumers and producers are also customers in this setting. Smart contracts are helpful in energy commerce transactions. After all, consumers may be able to sell or buy the energy at a particular time concerning the surplus energy produced, to maximise the efficiency and versatility of the energy resources.
4. **Waste Management**: Through the solutions based on the application of the blockchain platform, we can locate the waste streams from any of the components of the FEW nexus in real time, at the same time, also enhance the processes of recycling and the development of the necessary reward system in the form of token which will be beneficial to minimise the valuable potentials of wastes generated by the citizens of a particular region. This assists in passing sustainable waste practices and, at the same time, guarantees the enhancement of circular economy advocacy and implementation.
5. **Decentralised Governance**: Decentralised governance platforms using blockchain can help in collaborative decision-making and management of resources for strengthening the FEW nexus. Introducing innovations that increase trust and agree on fair solutions to problems and disputes regarding resources is an essential solution that blockchain contributes to improving the situation.
6. **Data Sharing and Collaboration**: Using blockchain solutions such as security and access to data for collaboration among all players potentially benefit the FEW nexus. They create a platform for information sharing, networking, cooperation in R&D, and encouraging adoption of improved practices in the sustainable use of resources.

In general, the potential of blockchain technologies is to contribute practical solutions for solving the problems of interdisciplinary interactions involved in the global food, energy, and water systems within Industry 4.0, thus implying that organisations would develop more sustainable and less prone to failure.

3.5 PRECISION AGRICULTURE AND FEW NEXUS 4.0: ADVANCEMENTS AND CHALLENGES

The world's agricultural industry recently witnessed significant and mechanisation in the 20th century; the result of this technological transformation created a revolution which forced the substitution of labour or human workforce with machines. This consequently led to an improvement in land productivity and the attainment of large economies of scale (Martín-Retortillo & Pinilla, 2015). The transition from labour-intensive agriculture to capital-intensive agriculture enabled farmers to effectively manage larger fields and farms. The Green Revolution, which started in the mid-20th century, resulted in increased production using genetically modified plants and animals, synthetic chemical fertilisers, and pesticides. The effect of this revolution reduced crop losses to the barest minimum (Jordan, 2013). These advancements promoted the growth of larger and more consistently controlled agricultural areas in various regions across the globe. Before the introduction of agricultural mechanisation, farmers could adapt their management techniques within their fields to accommodate variations in yield potentials, topography, soil characteristics, nutrient requirements, and both abiotic (such as weather and climatic factors) and biotic (such as pests and weed infestation, and other biological factors) stresses. These adjustments were primarily made through manual practices (Zhang et al., 2002). However, in acquiring economies of scale from mechanisation and moving to standard techniques, farmers compromised the ability to manage the variety of farm fields efficiently. The advent of mechanisation in farming and advances in information technology opened a new field in agriculture known as PA.

PA is a novel and continuous agricultural revolution in information technology. The idea of PA came into existence in the 1980s. PA technology was commercialised during the early days of the 1990s. PA tackles the difficulty of customising management to site, crop, and environmental features (Lowenberg-DeBoer, 2015a, 2015b; Swinton & Lowenberg-DeBoer, 1998). It also supports using new technology and data to address the heterogeneities of farms (Zhang et al., 2002). Thus, PA uses standardised techniques to reduce the unknowns connected to the information base for farm management decisions (Liaghat & Balasundram, 2010) and enables temporal and site-specific farm management even for agricultural systems that become mechanised and big scale. PA is not restricted to specific farms but could be suitable and valuable for all farms, ranging from subsistence to commercial, organic to conventional, and traditional to developing rural farms. According to the works of Berckmans (2014) and Busse et al. (2015), PA can be split into two categories: precision farming and precision livestock farming. Adoption of PA technologies to date varies geographically and by technology type. Although various components of PA became mature technologies in developed countries (e.g., geo-referencing technologies and guidance systems), the overall picture is that PA has not yet been taken up widely in the agricultural sector at large (Bramley, 2009; Tey & Brindal, 2012).

However, the continual growth of many PA technologies indicates lower costs and extended applications in ways that indicate more extensive acceptance. Recent advancements in information technology provide a finer and more accurate degree of precision than previously achievable (Aubert et al., 2012), and the costs for sensor technologies are falling rapidly, further contributing to this technology's usability. Advances currently follow technological breakthroughs connected to the digitalisation of the agricultural sector in data processing and robotics (Walter et al., 2017; Weersink et al., 2018; Wolfert et al., 2017). These discoveries have led some to claim that PA will promote the sustainability of agriculture in the long run (Walter et al., 2017).

PA seeks to optimise agricultural management practices in a unified manner. It aims to actualise this by building on the high spatial and temporal variability of crops and the environmental features (Lowenberg-DeBoer et al., 2020). The idea behind this approach is to address certain qualities of the environment like the variability in the potential of crop yield, topography, soil properties, nutrient requirements, and factors such as weather conditions, pests, and weed infestations.

There are some technologies associated with PA; they include precision farming technology paired with BD and ML, gene editing technologies, second-generation biofuels, and agrivoltaics. These technologies will be briefly explained below (Miao & Khanna, 2017)

- **Precision Farming Technologies**: With the development of technologies such as UAVs, robotics, internet infrastructure, wireless technologies, remote sensing, sensor technologies, and ML, precision farming technologies, available to farmers for some 30 years, have advanced in the past few years. These advanced precision farming techniques and ushered agriculture into the era of BD and the digital revolution (Khanna, 2021; Miao & Khanna, 2020). Precision farming technologies offer farmers new tools for increasing agroproduce while reducing the adverse impacts associated with mechanised agriculture and heavy chemical use (Gallardo & Sauer, 2018; Weersink et al., 2018). For example, variable rate technology, together with soil and yield sensors and smart fertilisers, enables farmers to apply optimal amounts of input like seeds, water, fertilisers, and pesticides at the subfield level or even at the plant level in farms (Calabi-Floody et al., 2018; Faulkner et al., 2014; Gallardo & Sauer, 2018). In contrast to this, conventional farming practices treat fields as uniform and homogenous units. Doing this results in over-application of inputs in certain field areas (Oliver et al., 2013). In addition, precision farming technologies permit farmers to capture large amounts of high-resolution data associated with on-site farming activities. These large chunks of data are usually referred to as BD. Advanced internet infrastructure, wireless technologies, and cloud storage enable real-time access to this data, as well as with real-time market and weather information. All these allow farmers to control and manage production from anywhere at any time. Moreover, ML and robotics can bring greater levels of automation that will help manage agricultural production. This digitalisation in agriculture can help control pollution problems by precisely identifying the location and timing of discharge. This can help government to design environmental policies that encourage agricultural practices among the citizens (Weersink et al., 2018).

- **Gene Editing Technologies**: Breakthroughs in gene editing technologies over the past decade have revolutionised the breeding of new crop varieties with desirable traits (Jaganathan et al., 2018). Traditional transgenesis methods, which involve introducing foreign genes from related or unrelated organisms, are labour-intensive, they are also not precise, and they also raise concerns about potential health and environmental hazards (Rosi-Marshall et al., 2007; Sedeek et al., 2019).

New gene editing technologies such as zinc finger nucleases and transcription activator–like effector nucleases and a clustered regularly interspaced short palindromic repeats (CRISPR)-associated protein technologies offer precise and efficient alternatives. These technologies can knock out or replace genes in the original genome to confer desired traits without introducing foreign DNA (Sedeek et al., 2019; Wolt et al., 2016).

These emerging gene editing tools have already been employed to develop new crop varieties with improved yields, nutritional properties, and good levels of nitrogen efficiency. For example, rice and wheat yields have been enhanced by knocking out yield constraining genes while drought tolerant traits in corn have been improved using CRISPR/Cas9 technology (Li et al., 2016; Osakabe et al., 2016; Xu et al., 2016; Zhang et al., 2016). In addition to this, efforts have been made to enhance nitrogen use efficiency in crops to reduce dependence on nitrogen fertilizer and mitigate nitrogen pollution (Beatty & Good, 2011; Fan et al., 2017).

The idea of gene editing technologies being applied to improve the yield of bioenergy crops such as switchgrass, miscanthus, willow, and poplar, and as well as biodiesel crops, holds promise for addressing global food security challenges and enhancing agricultural sustainability. Gene editing technologies also help mitigate environmental impacts associated with conventional breeding methods (Clifton-Brown et al., 2019; Karavolias et al., 2021).

- Second-Generation Biofuels: Liquid fuels are projected to maintain a significant role in transportation and logistics over the next 30 years. This is driven by the need to reduce greenhouse gas emissions and transition towards more renewable and less carbon-intensive fuels (Debnath et al., 2019; Robertson et al., 2017). First-generation biofuels are derived from food crops, and there have been significant concerns about competition with food production and indirect land use changes (Khanna & Crago, 2012). Second-generation biofuels differ from first-generation fuels in the following ways: they are sourced from non-food feedstocks like perennial energy crops, agricultural and forestry residues. They offer a solution by avoiding competition with food crops.

Perennial energy crops have shown promise in providing higher biofuel yields per unit of land compared to conventional crops like corn ethanol. These crops also offer environmental benefits such as improved soil carbon sequestration, reduced soil erosion, and lower nitrogen leaching (Dwivedi et al., 2015; Hudiburg et al., 2016; Langholtz et al., 2016; Liu & Rajagopal, 2019). Studies have documented that the carbon intensity of cellulosic biofuels, which is a type of second generation biofuel, can be up to 90% lower than that of gasoline (Farrell et al., 2006; Morales et al., 2015).

When compared with first-generation biofuels, second-generation biofuels give a more sustainable and environmentally friendly alternative to first-generation biofuels, and they offer a pathway to meet transportation and energy needs while reducing reliance on fossil fuels, thus mitigating climate change.

Energy from the sun can also be harnessed and used in the agro space for energy generation. Agrivoltaics is the term used to describe the involvement of solar panels in the agro space to generate electricity. This concept is explained below in detail:

Agrivoltaics: The deployment of photovoltaic panels has seen meaningful growth mainly due to adequate insurance (Bolinger et al., 2018; Timilsina et al., 2012). However, deploying solar energy, which requires mounting photovoltaic panels on the ground, has raised concerns about space for tillable land that could be used for food production (Moore et al., 2022).

Agrivoltaic systems create a situation where crop yield and photovoltaic panels coexist on the same land. Agrivoltaic systems can harmonise the reign between energy and food yield for land by allowing both activities to take place simultaneously (Weselek et al., 2019).

The idea of agrovoltaics has recently gained momentum in Europe and Asia, with Japan leading in acceptance since the composition of the first agrivoltaic transcription in 2004 (Movellan, 2013). Studies on agrivoltaics have mainly focused on their potential impacts on biology and economic benefits without adequate consideration of the land's inhabitants.

Research on drought has shown increased crop yields, reduced drought for plants as well as lower heat for photovoltaic panels in agrivoltaic systems (Barron-Gafford et al., 2019). Some studies have reported a slight increment in row crop yields under agrivoltaic conditions. For example, experiments have shown that corn yields could be improved by up to 5% in agrivoltaic systems compared to formal monocrop fields without photovoltaic panels (Sekiyama & Nagashima, 2019). Some researchers also suggested that agrivoltaic systems could improve total land yields by combining crop yields and power production. However, experience shows that this system usually results in slightly reduced crop yields compared to monocrop production (Dupraz et al., 2011).

While agrivoltaic systems are still in the empirical stage in the United States, their effectiveness in allowing diversified income for farmers, conservation of water, and simultaneous production of food and energy makes them a perfect choice of agro-practice in the future (*A Tech Revolution Will Help Farmers Harvest Sunshine with Their Crops | WIRED*, n.d.).

Figure 3.11 illustrates an agrivoltaic system in which photovoltaic panels and agricultural activities coexist on the same land. Such an arrangement simultaneously offers the benefit of food and energy production, which brings about diversified income generation for the farmers. The diagram also shows that water conservation is further attained through the use of solar panels and crops on the same plot of land (Miao & Khanna, 2020).

PA, together with the FEW nexus, is a duo strategy that is focused on targeting sustainability issues within the agricultural sector of the economy. This farming

FIGURE 3.11 An Agri-Voltaic System (Carreño-Ortega et al., 2017).

technique uses practices more advanced than the actual farming techniques. It works with the help of technologies like GPS, sensors, drones, data analytics, and many others. The targeted conditions include soil type, weather, water, nutrient applications, pest infestations, etc. The main goal of PA is to yield minimising high-quality crops while minimising the use of resources and its adverse effects on the environment. The insight from this topic shows that incorporating PA with the FEW nexus is a nouvelle model for farming, which is sustainable. Awareness has been raised among relevant stakeholders on how PA and all the technologies related to it can help improve resource efficiency and production capacity at the farm level. This awareness considers the totality of the food energy and water nexus.

3.6 DATA ANALYTICS AND MACHINE LEARNING IN FEW NEXUS 4.0

As the world continues to get digitised due to the emergence of computers and internet-enabled devices, information about the users of these devices continues to get transmitted to various data sinks. These sinks need to process these data to make efficient use of it. Some data sources exist in multiple fields, such as demography, climate, research, medicine, and energy use (Borkar & Thakur, 2021). This large quantity of data represents the behaviours, interests, consumption patterns, and activities of the users and the usage of infrastructure and machines. As the population of internet users and mobile phone users continues to rise, the amount of data keeps growing at an exponential rate (Morley et al., 2018). This entire process represents a possible transition in society towards an Informational Society, otherwise known as the digital age, and by extension, a move towards a Knowledge-Based Society. In an Information Society, data is central to the economy, and all other activities and decisions revolve around it (Hannan et al., 2018). The Information Society would give importance to the access and application of information to create knowledge to improve society (Database Systems Journal, 2012; Riahi & Riahi, 2018).

The term BD comprises developing and deploying technologies delivering to deliver well-timed data to users from vast and exponentially growing datasets

(Rao et al., 2019). BD is a multifaceted concept with varying definitions depending on the user's view. Precisely defining BD is challenging due to the unreliable information about what constitutes big in terms of data volume, which varies across various domains (Sivarajah et al., 2017). Rather than being defined by appropriate technologies, BD represents a family of techniques and technologies to handle large and compound datasets.

As an emerging field, the definition of BD continues to develop as we hunt its concepts and draw its value. This unreliable information about the whole concept of BD underscores the ongoing strategies to leverage BD effectively (Riahi & Riahi, 2012, 2018). In simple terms, BD is a novel concept that describes a vast array of enormous and complex datasets that are very strenuous to analyse, analyse, and process by making conventional data analysis methods. BD mostly involves datasets from numerous sources, which require some advanced tools and techniques to analyse and draw insights effectively.

3.6.1 Characteristics of Big Data

BD usually refers to datasets that are so huge in volume, variety, and speed (velocity) of arrival compared to conventional sources of data. The terms volume, variety and velocity are the main features of Bigdata, and they are commonly called the 3Vs of Bigdata (Gewirtz, 2018). Figure 3.12 gives a pictorial description of this 3V concept.

These concepts are discussed:

- **Volume**: In Bigdata volume represents the actual quantity of data generated, collected, and analysed within the system under consideration (Iqbal et al., 2020). The growth in volume is caused mainly by the increase in the

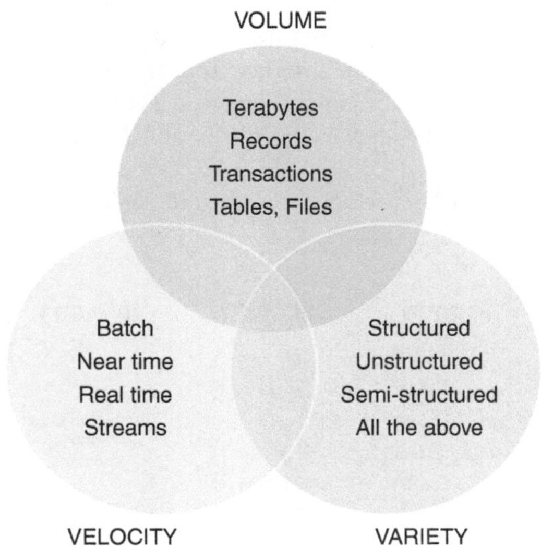

FIGURE 3.12 The 3V Concept (Riahi & Riahi, 2018).

amount of data that has been produced and collected over the years, as well as the need to make use of this data efficiently and effectively.

- **Variety**: Variety is the heterogeneity of data that an information system process. This comprises structured data, semistructured datacomprises structured, semistructured, and unstructured data, thus giving rise to matters like relations and connections of the dissimilar data types (Mostajabi et al., 2021).
- **Velocity**: Velocity in BD refers to the relative rate at which data is produced, collected, processed, and used in decision-making processes (Mohammadpoor & Torabi, 2020). In BD, the analysis must happen on a real-time basis because the generation and transmission of data are often continuous and in real-time. The idea of velocity in BD relates to the speed of processing of data, the rate at which data analysis is done and also the time it takes to gain value from data.

Another form of classification of the features of BD exists in 5Vs; they are pictorially described in Figure 3.13.

The remaining two concepts added are discussed:

A more recent way to characterise Bigdata as against using the previous 3Vs is by adding another 2V, as shown in the picture above. These two Vs are described below:

- **Veracity**: Veracity in BD relates to the quality, correctness, and doubtfulness connected with the data and its sources. It covers characteristics such as data integrity, the consistency as well as and reliability, and the level of confidence and reliability, and the confidence level of the sources delivering the data. It is vital to ensure data validity, as this is crucial for making good decisions and generating relevant insights from BD (Lozano et al., 2020).
- **Value**: Value represents the importance and effectiveness derived from the data. It also reflects the secondary attributes, insights and actions enabled by analysing and leveraging BD. Maximising the value of BD involves extracting a large number of insights from the data, optimising concern processes, improving decision-making, and eventually making a driving base and competitor advantage (Riahi & Riahi, 2018).

Recently, the features of BD have been extended to 10 due to advancements in technology. Figure 3.14 illustrates this.

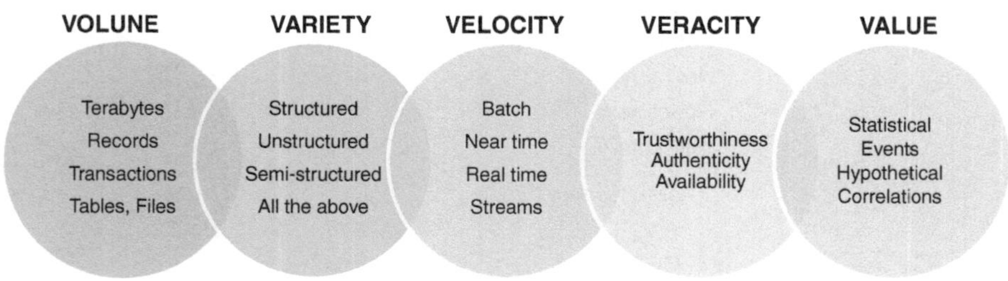

FIGURE 3.13 5V Description of the Features of Big Data (Riahi & Riahi, 2018).

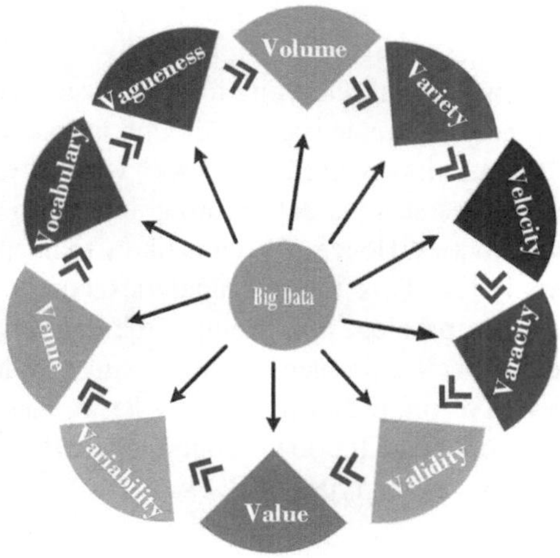

FIGURE 3.14 The 10Vs of Big data (Bhat & Huang, 2021).

BDA is a significant component that goes hand in hand with BD. BD could also be defined as an extremely large volume of data, which cannot be suitably handled by our conventional data warehouse and analytical tools our conventional data warehouse and analytical tools cannot suitably handle. Over the years, the results that the analysis of structured data yields have become detailed as a result of the unpredictable rate at which the data is produced and transmitted, hence there is a need for the use of data analysis tools and not traditional reporting tools to analyse datasets of this nature (Ikegwu et al., 2022). This implies that the patterns or links within the fast-changing data have to be identified automatically to allow for the data to be analysed and exploited. BDA is, therefore, the method of collecting, handling, and analysing large volumes of data with the main motivation of finding distinctive patterns and other necessary characteristics that are otherwise hidden in data (Mohamed et al., 2020). The process of finding patterns in large and complex data sets using the help of modern technology and techniques can be seen as part of the concept of BDA. BDA is a technique designed to meet the new or current challenges facing data processing more efficiently and effectively, which means that it solves problems differently.

3.6.2 Types of Big Data Analytics

According to Balali et al. (2020; Kibria et al. (2018), there are four common types of BDA. They are described:

a. Descriptive Analytics: This type of analytics entails analysing historical data; hence, it focuses on the question, "What is going on?" Descriptive analytics merely structures and mines data to reveal patterns so that one has a glimpse of specific probabilities of such incidences in the future.

b. Diagnostic Analytics: This type of analytics investigates the cause of an event. The typical question for this stage is "Why did it happen?" since diagnostic analytics is about searching for the causes behind events.
c. Predictive Analytics: Future activity prediction is based on what happened in the past. This question is used to answer what is expected to happen considering techniques such as data mining and AI to infer forecasts and potential future activities. Hence, "What is likely to happen?" is asked here.
d. Prescriptive Analytics: This type of analytics comes last when ranked because it solves the problem that prompts the data analysis. Prescriptive analytics solves a specific problem by answering, "What should be done next?" To achieve a result, prescriptive analytics utilises historical and predictive data to recommend the ideal course of action. Figure 3.15 shows a pictorial description of the types of BDA.

A good percentage of farmers in the world tend to reside and practice their profession in rural areas; this has spurred researchers to explore the potential of this technology to influence the dynamics of the food system and provide connections between farmers and large corporations (Bronson & Knezevic, 2016). For example, Canada's agricultural research departments in the 1920s gathered reliable yield statistics for crop and weather circumstances. This data was used to enhance wheat production, although the procedures were done manually.

However, BD in agriculture is a new idea regarding the volume and analysis of the data being accumulated differently from the traditional way. BD technologies are very effective in accurately analysing agro data. Technological restrictions made this difficult to attain earlier (Lokers et al., 2016). For instance, farmers in the past use notebooks to record agricultural performance, whereas employing digital applications today boosts logging operations as they help develop features like speech recognition. BD programmes also supply farmers with vast data sets and powerful data analytical tools, both result in formulating policies other than resolving individual farm needs (Wolfert et al., 2017).

The use of BD as a decision-making tool in agriculture also comes with some drawbacks. However, its positive prospects outweigh the negative. BD provides

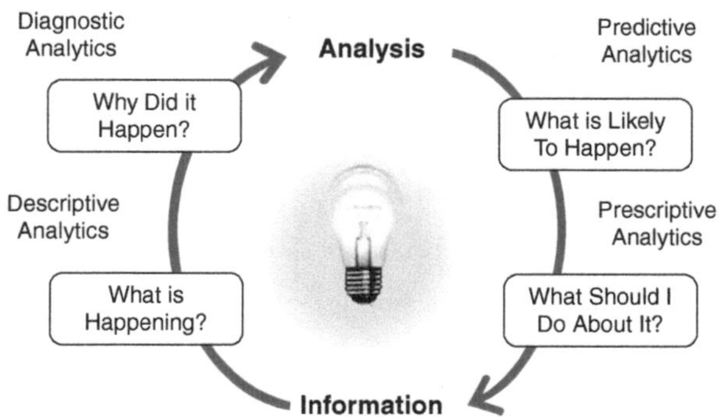

FIGURE 3.15 Types of Big Data Analytics (Perwej, 2017).

rural practitioners and related industries with quantitative insights into single factors affecting rural production, enabling them to make informed decisions in their daily farming activities (Wally, 2021). By leveraging BDA, farmers could stay informed about foodstuff prices, gain valuable knowledge about the demands of certain crops, and stay updated on new technologies emerging in the agro space of animal husbandry (Weersink et al., n.d.). According to Bhat and Huang (2021), using BDa will empower farmers to optimise their agro produce, boost efficiency and adapt effectively to changing patterns in agro cultivation practices.

In agriculture, BD is effective in mitigating the challenges associated with crop yield farming patterns and farming patterns and getting adequate data about soil and climatic conditions, which will aid farmers make decisions regarding farming. This technology can tell soil quality, observe diseases and pests, predict water requirements, and forcast crop draw times (Bhat & Huang, 2021).

The application of BD in the agriculture and food sector is still in its early stage. It has a great potential to improve the agro space, thus helping to increase the efficiency and the effectiveness of the supply chain as well as making sure food is in adequate supply. While BD was initially a phenomenon mostly associated with the primary phase of the agricultural supply chain, its use is now increasingly spread along the entire agro-value chain (Wolfert et al., 2017). Large and popular applications of BD in agriculture are apparent in countries such as America, Canada, European countries, and China (Beaulieu et al., 2015; Dakshayini & Balaji Prabhu, 2020; Tantalaki et al., 2019; Tarik & Mohammed, 2020)Misra et al., 2020; Sarker et al., 2019). However, more recent trends show growing interest in countries situated in Australia and Africa, for example in Australia and Africa, such as Morocco (Cui & Gao, 2020; Tarik & Mohammed, 2020). Today, BD stands as the basis of contemporary business solution and between accuracy in timing and decision-making. The agro space has taken into consideration predictive analytics, the use of IoT sensors in the farms, and improved analytics. All these variables foster significant changes in agricultural and business models (Majumdar et al., 2017). There is fear that experience will be replaced by algorithms in the long run; however, the BD application, to which significant improvements are expected to be brought to farm operation/management, may also be a new experience on farmers (Pham & Stack, 2018; Stubbs, 2016) and as such, the notion that algorithm will replace experience will not hold because these algorithms will certain need agro inclined men to either write new algorithms or modify existing ones. BD thus means enhancement in various areas pertaining to real-time prediction, production tracing, and business operations. However, the scale-up of such applications creates potential risks that can influence farm organisations and supply chains (Devlin, 2012; Bhat & Huang, 2021).

As depicted in Figure 3.16, some large-scale industrial farms are technologically advanced (smart farms). The farm management practices used in these smart farms do not resemble the practices used in traditional farming at all. In these smart farms, the management systems rely on BD technologies to process real-time farm data (Said Mohamed et al., 2021). The advantage of this is that it boosts productivity. Relevant farm data in smart farms is collected by IoT devices, remote sensing technologies, and other sensor networks (A. Ali & Alshmrany, 2020). The data regarding the soil conditions, crop status, weather, and ambient conditions obtained from

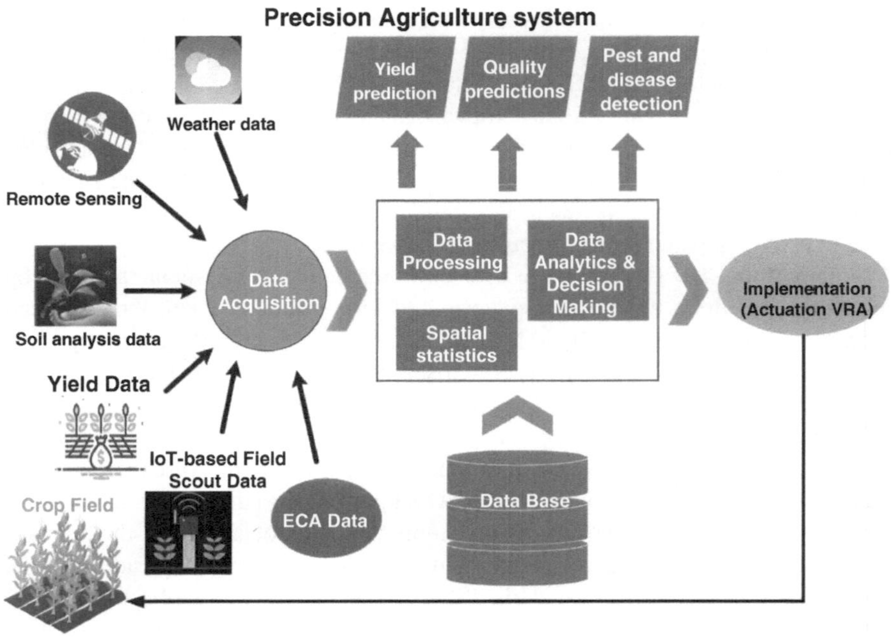

FIGURE 3.16 Big Data-Based Precision Agriculture System Representation (Bhat & Huang, 2021).

these IoT sensor networks are stored locally or in cloud storage. Critical insights are extracted from this data through ML BD algorithms to assist in making the right decisions for the farmer (Bayih et al., 2022). The smart farm uses these technologies to form a decision system, which then physically carries out suggested actions through an intelligent control system aided by advanced machinery (Bhat & Huang, 2021). This process is repeated continuously in a systematic cycle in a cycle until the time to harvest comes. Figure 3.16 illustrates this.

The 20th century ushered in of different wireless technologies, ranging from smartphones to wireless technologies as small as a grain of rice, for example, RFID chips. These technologies generate massive amounts of data, which can be used for various purposes ranging from tracking to making accurate decisions. Handling this voluminous amount of data is a big challenge since useful information contained in them may be corrupted noise or any other form of random data (García-Gil et al., 2019). When wireless technologies are used for farm management, the information embedded in the data they generate should be presented in a meaningful manner so that the end user can comprehend the various processes taking place in the far (Cui & Gao, 2020). To make the utilisation of BD easier in managing farms, mapping the farm is essential. Mapping provides a platform for grouping homogeneous farms into zones for agro data collection (García-Gil et al., 2019). Maps give the various zones with relevant parameters for applying farm practices (Saiz-Rubio & Rovira-Más, 2020). These maps work by interpolation methods. Kriging is the standard interpolation method to manage these farm zones (Nawar et al., 2017). Due to the voluminous data that smart agriculture generates, numerous software applications handle data interpolation.

Software-based farm management solutions, such as GIS, facilitate the automation of data collection and analysis, supervision, planning, record-keeping, decision-making, and farm operation management (Tummers et al., 2021). These tools also assist in basic tasks such as recording produce harvests and yields, scheduling farm tasks, tracking soil nutrients, weather prediction, and farm mapping, as well as even more advanced functions related to farm management automation. There are some specific GIS data management systems, such as the farm management information system (FMIS), which have been designed for many PA applications (Liu et al., 2019; Liu, Zhu, Yue, & Guo, 2020). FMIS aids farmers in operational planning, record-keeping, implementation, and evaluation of any farm project to reduce production costs, maintain quality and safety of produce, and enable informed decision-making. Many commercial agriculture information management systems are available, including ADAPT, WinGIS, SpiderWeb GIS, AGERmetrix, FieldView, SST Software, AgVerdict Inc., Trimble, etc. Some of them are designed to cater to farmers and other stakeholders within the agricultural industry and those in the agriculture supply chain (Bhat & Huang, 2021). However, the effectiveness of the recommendations from these software tools depends on the parameters included in the algorithms of each of the specific software platforms. For instance, the Decision Support System for Agrotechnology Transfer generates outputs based on experimental data used for crop model evaluations. It allows users or farmers to compare experimental results with observed results. This is important if modelled results form the basis of real-world decisions (Sarkar, 2012) are a lot of BD analysis software tools. Many BD analysis software tools are available in agriculture, each serving specific purposes and requirements (Kamilaris et al., 2017). Figure 3.17 illustrates some of these tools.

Figure 3.18 shows a pictorial depiction of BD decision support system.

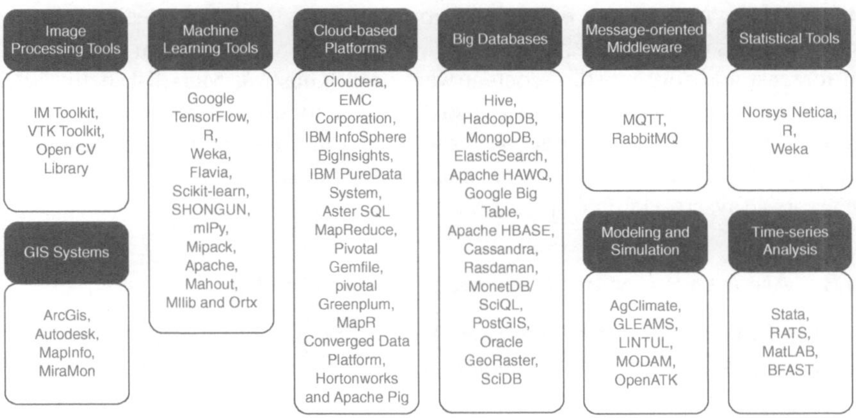

FIGURE 3.17 Software Tools of Big Data Analysis Employed in Agriculture (Bhat & Huang, 2021).

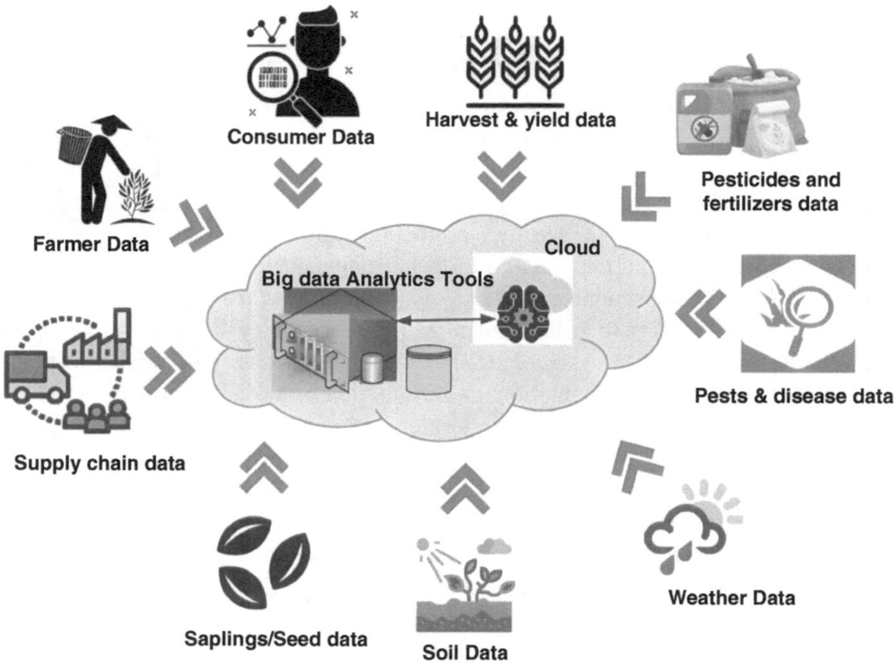

FIGURE 3.18 Big Data-Based Decision Support System (Bhat & Huang, 2021).

PA relies on complex data analysis to support effective decision-making in farm management in order to get better yields (Sishodia et al., 2020). Due to the multiple parameters involved in the farm, it becomes crucial to apply AI techniques such as DL, genetic algorithms, ML, and expert systems. These technologies help evaluate and model complex agriculture data, thus making precision farming more achievable. For instance, an example of applying fuzzy logic in a decision support system includes crop management for kiwi, potato and corn where variables such as rain forecast and soil moisture content were used (Giusti & Marsili-Libelli, 2015). Another decision support system was also developed for estimating weekly irrigation requirements for citrus orchards based on the soil moisture and climate data, with data input of real-time soil parameter measurement control to minimise the estimation errors (Navarro-Hellín et al., 2016).

3.6.3 Applications of Big Data Analytics in FEW 4.0

1. **Big Data-driven Crop Selection Decision Support System**: This system is proposed to assist in decision-making in different agro-production and supply chains. It provides farmers and other stakeholders with data relevant to soil quality, seed characteristics, weather patterns, marketability, and the potential hazards of crops. It supplies these data from cloud sources and local databases. They inform and advise about the quality of crops, market

demand, and performance. It will also help distribute the crops optimally so as not to harvest too much in one place (Bhat & Huang, 2021).

2. **Management of Crops, Growth Monitoring, and Assessment of Produce Quality**: Effective and efficient processing of farm data will be needed to turn numerical and visual data into valuable insights. Image data from farms can be applied to crop growth and quality monitoring. Examples of tools for image processing are OpenCV, Matplotlib, and Scikit-Image in Python. From these tools, it will be possible to measure plant height and width, and it will be possible to know crop quality based on leaf colour and crop that can be extracted from images. Besides, IoT devices collect environmental parameters and store them in the cloud. Networks in the food industry gather data at every stage of food production, from quality evaluation of raw materials to the ingredients in the final product. The initial analysis defines and applies essential features that train ML or DL models; these models extract required features from large datasets accumulated in the field. Crop yield estimation is supported by all the factors influencing production, from irrigation to soil composition and structure, topography, climate, crop stress, diseases, and pests. This helps in the efficient management of resources by providing decision-makers with timely and accurate estimates to know the scarcity or excess of the crop and the action to be taken (Bhat & Huang, 2021).
3. **Sustainability**: the emergence of novel technologies such as AI, IoT, and drones promote sustainability in agriculture using BD. Over the years, the size of land available for agricultural practices has not witnessed significant expansion; the same applies to the amount of groundwater sipping down the soil; the soil quality over time keeps deteriorating due to environmental and industrial factors. Using technologies will ensure that the available land and water will be used optimally for agricultural purposes (Abiri et al., 2023). Using BD will enable optimising farming practices to meet global demands and save for upcoming generations (Saiz-Rubio & Rovira-Más, 2020). The solutions offered by BD show that this technology can sustain agriculture and conserve natural resources.
4. **Reduced Chemical Application**: new technologies such as AI can be employed along with other technologies like ML, Robotics, and computer vision in fighting weeds. Data obtained from IoT devices used in farms can be integrated with AI to detect and analyse weed-infested areas on the farm. This will enable farmers to spray chemicals strategically to limit their growth. Targeted spraying prevents blanket spraying, which involves spraying across the whole farm. With targeted spraying, fewer chemicals are now sprayed on the farm, which will have little or no effect on the growth of crops (Bhat & Huang, 2021).
5. **Early Disease Detection**: Plant diseases pose a significant threat to the quality of agricultural produce. While DL algorithms based on neural networks have emerged for identifying such diseases, they typically only work when diseases are well-developed. For this reason, such algorithms have a limited impact on improving produce quality. Plant diseases are hard to

detect at an early stage; they are difficult to monitor on the farm and costly to implement. Improving the quality of farm produce requires DL algorithms because they can perform timely recognition and classification of plant diseases. However, DL techniques will work here if pathological datasets of plants containing disease symptoms in their early stages are created (Bhat & Huang, 2021).

6. **Risk Management System in Agriculture**: Another critical aspect of PA is managing risks caused by farm location, soil type, and heat stress or frost. Climatic variability plays a crucial role in cultivation conditions. An interesting approach to interpreting data in this context involves the integration of a variety of datasets. Climate patterns are mixed with world datasets and historical meteorological data to provide climate information on smaller units than spatial, hence enabling the environment to adapt to changes immediately. For example, Kashiwao et al. (2017) examined the potential of BD for forecasting rainfall by using meteorological datasets. The outcomes show that there is great potential in using data fusion within the domain of PA.
7. **Agriculture Management System**: BD facilitates data sharing among farmers. The interactions of agricultural practitioners form a network. This network, when managed with data-driven software, will, in turn, lead to software-defined management. These management systems come in various forms, whether rural, urban, or even between different geographical regions. Rural management systems are designed to deliver accounting services, connect all stakeholders involved in agriculture, and facilitate benchmarking for farmers. BD assists farm managers and agribusinesses worldwide by integrating and analysing vast amounts of data from real-time sources to enable them to make informed business decisions. Such systems offer intelligent solutions to the farming business. Smart farming practices extend PA by adding farm management tasks based on data availability (Kamilaris et al., 2017). Studies conducted on small farms in the developing world reveal that farmers cannot trade their harvest due to oversupply or lack of information. Tools for improved productivity can help integrate crops into the global supply chain (Bhat & Huang, 2021; Giusti & Marsili-Libelli, 2015).

In summary, BD offers a lot of improvements in the FEW nexus. Components of the nexus can be analysed using BD, and stakeholders can then develop various data-driven decisions to improve the existing structure. BD's improvements in the nexus include real-time tracking, control, sustainability, and resource optimisation.

3.7 APPLICATIONS GEOGRAPHIC INFORMATION SYSTEM (GIS) IN FEW NEXUS 4.0 SYSTEMS

GIS is an integrated system that comprises a computer running advanced software used to manage, store, and handle geographical data. GIS allows managers to analyse and visualise geographic data with a view of making relevant decisions. A spatial database is commonly used in GIS but is not needed in the definition. A typical GIS infrastructure includes human users, which consists of some level of support staff, a detailed procedure and defined workflows, and relevant knowledge. In addition to this, some institutional organisations are also involved in its operations (*Geographic Information System (GIS): Definition, Development, Applications & Components*, 2022)

Modern agricultural systems have evolved beyond the traditional farming principles; farmlands today are no longer domesticated; they have grown to cover most parts of the world. For agricultural institutions to meet up with today's agro-practices that will guarantee food security across the globe, technological innovations have to come in, especially in the aspect of maintaining healthy and productive soil and crop. GIS finds suitable application in this aspect of agriculture (Giller et al., 2021). It facilitates regular observation of the field remotely and provides relevant information that will help stakeholders to solve problems associated with the field. The application of GIS in our contemporary agricultural practices helps farmers to create maps of field data, and process it for use in monitoring their crops remotely. Figure 3.19 shows the various applications of GIS technology in the agric-food sector.

Montealegre et al. (2022) applied the use of GIS to estimate the energy performance of buildings in the city of Ferrara. Göçer et al. (2016) made a detailed plan for repositioning historic campus buildings in their country. In addition, Johansson et al. (2017) developed an energy chart for a building in Sweden. They used this to analyse

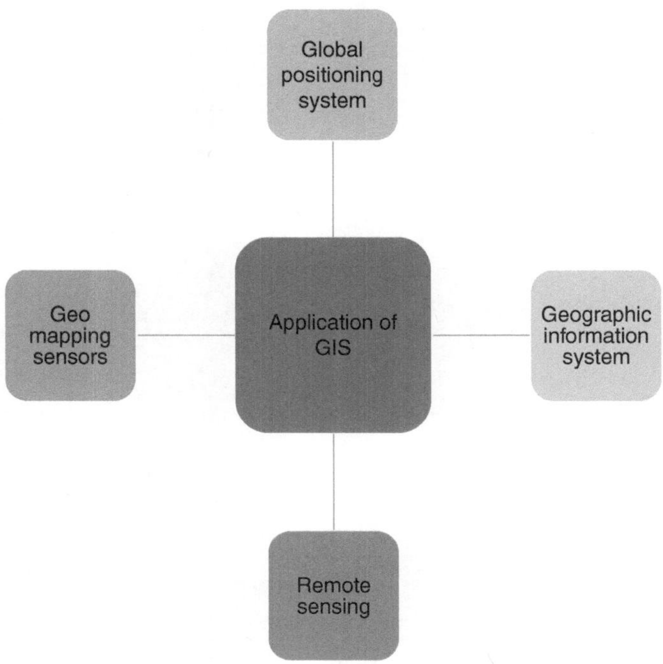

FIGURE 3.19 Application of GIS Technologies in the Food Sector (Bharti et al., 2021).

the energy consumption and detect if there was a need for renovation and if the need was urgent. The idea of evaluating potential urban productivity has mainly focused on food, energy, and water, which are the three main resources of urban metabolism.

La Rosa et al. (2014) introduced a system to determine the use of urban areas for commercial agriculture, utilising GIS and multicriteria conclusion analysis. This analysis considered factors like closeness to residential areas as well as closeness to farmlands. Kadhim et al. (2016) used GIS technology for sensing data to justify urban-rural capabilities in Barcelona. They focused on rooftop areas in non-residential urban zones. In contrast to this, Saha and Eckelman (2017) assessed ground and rooftop effectiveness in Boston, they also considered empty spaces and underutilised areas. The outcome of these studies highlighted the effectiveness of geospatial analysis in identifying appropriate locations for urban agriculture, albeit with considerations for data acquisition methods and economic viability.

Figure 3.20 shows a pictorial representation of GIS in sensing and planning residential homes.

In the energy sector, GIS enhances the efficiency of renewable energy. Recent developments in estimating the solar energy potential of urban regions have moved towards building-by-building approaches to reduce uncertainties that occur with unplanned installations of solar-powered stations, and improve accuracy (Formolli et al., 2023). The traditional methods used so far, like regional or national-level GIS-based assessments, have, over time, proven to be very helpful. Relevant studies,

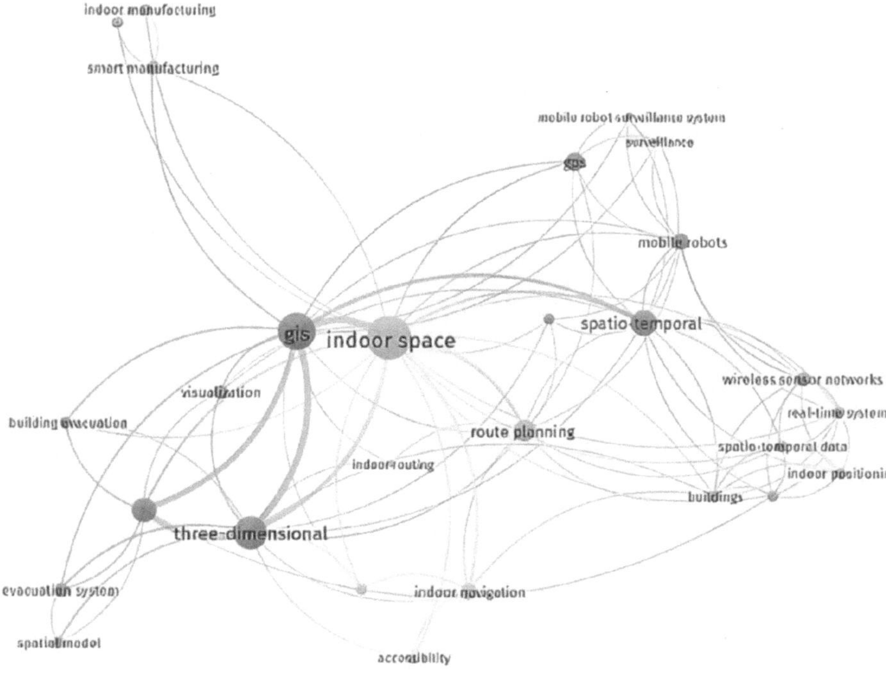

FIGURE 3.20 GIS in Residential Urban Areas (Teixeira et al., 2021).

including those from Bódis et al. (2019) for the European Union and Gagnon et al. (2016) for the United States, have shown that rooftop solar photovoltaic systems could significantly contribute to electricity generation. However, to improve accuracy in installation, the research has shifted towards building-specific analysis. Despite these advances, challenges still exist, especially in integrating models of the building stock and accounting for the entire shading effects. Due to these challenges, further work is needed to refine the approaches used so far and to develop standard procedures for assessing urban solar potential, which is an important aspect of the FEW nexus. Figure 3.21 shows a framework for using GIS for energy use planning of an urban building.

In the water aspect of the FEW Nexus 4.0, GIS plays a critical role, especially in rainwater harvesting. In recent years, there has been a growing interest in evaluating the possibilities of rainwater harvesting in urban areas (AL-Shammari et al., 2021).

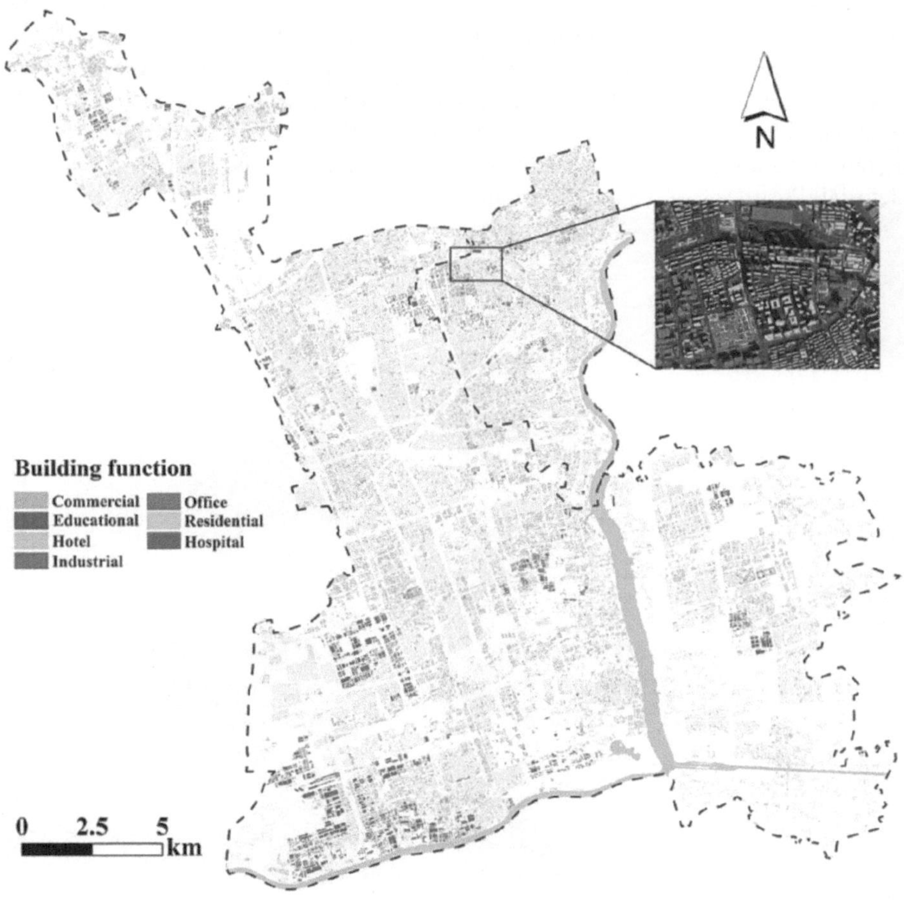

FIGURE 3.21 Geographical Information System (GIS) in Planning Energy Use of Urban Building (L. Chen et al., 2024).

Bottom-up approaches have been used to estimate the possibilities of using rainwater for domestic purposes other than consumption. As regards the assessment of rainwater harvesting potential at a city level, GIS technology and characteristics of buildings have been used by researchers, such as Lúcio et al. (2020), who devised methods for rating the possibilities of rainwater harvesting in Lisbon, Portugal. A. Ferreira et al. (2023) applied GIS and LiDAR data to estimate potable water savings for households in Escambia County, Florida.

Based on the location of collected rainwater, some works are focused on using it for urban gardening on rooftops. Salvador et al. (2019) evaluated the ability of technology to support urban metabolism based on rooftops within the FEW nexus systems. According to Rufí-Salís et al. (2020) and Sanjuan-Delmás et al. (2018), there are high percentages of rainwater usable in crop irrigation. Other researchers López-Felices et al. (2023) also addressed the feasibility of rainwater harvesting for urban agriculture in planted gardens on the ground in different areas. Figure 3.22 shows a pictorial view of the ongoing research in the application of GIS in water

While most studies on the deployment of GIS in the FEW nexus have focused on individual uses, there is an increasing need to quantify the FEW potentials at the level of buildings (Montealegre et al., 2022). Accepting that roofs in urban areas come in different physical layouts, uses, and characteristics, such as roof type, orientation, and shading, means that the FEW systems have to be considered when designing buildings and planning cities (Todeschi et al., 2020). Figure 3.22 shows a pictorial illustration of the use of GIS in the FEW Nexus 4.0.

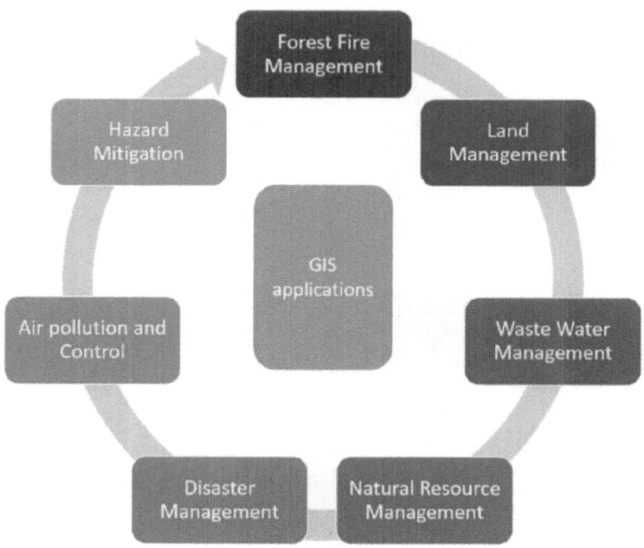

FIGURE 3.22 Geographic Information System (GIS) in Environment Studies (Duarte & Teodoro, 2021).

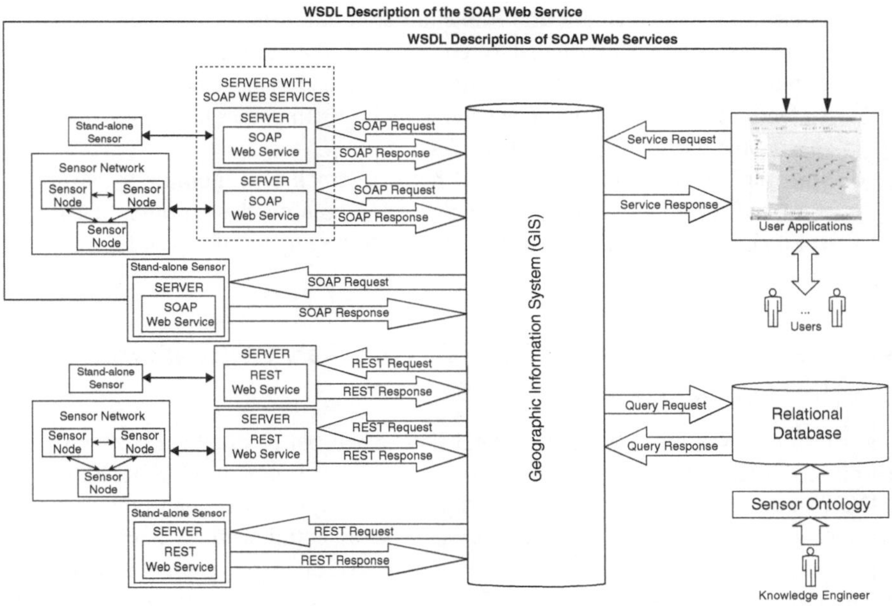

FIGURE 3.23 GIS Sensor Management in FEW Nexus (Borge-Diez et al., 2021).

In summary, regarding the FEW nexus, GIS has proven to be important in planning and the creation of effective strategies used to manage resources efficiently. Some of the important applications of GIS in the FEW nexus to note include:

1. **Data Integration and Visualisation**: GIS is a tool that integrates spatial datasets associated with food, energy, and water resources on a conceptual level. It also shows the availability and location in a certain geographical region. The mapping of these datasets and the use of spatial analysis to analyse the FEW nexus is what makes GIS important in the FEW system.
2. **Spatial Analysis for Resource Management**: GIS is helpful in detecting areas where the extraction, production, distribution, and consumption of these resources in the FEW nexus are ideal.
3. **Scenario Modelling and Planning**: Modelling done with the aid of GIS enables policy makers to view the consequences of their policies on the FEW resources.
4. **Risk Assessment and Resilience Planning**: When used in a creative way, GIS can be used to evaluate risks influenced by natural disasters, fluctuating climate patterns, and human activities on FEW resources. Adding layers of hazard maps, critical infrastructure and resources, decision-makers are in a position of being able to make informed decisions about which communities to invest in resilience and in which they should prepare the emergency services to respond.

5. **Community Engagement and Stakeholder Collaboration**: GIS applications create an enabling platform that fosters effective communication between communities and stakeholders for effective planning and management of the FEW resources. It also makes the people living in the communities to be aware of the idea behind the decisions of stakeholders.
6. **Monitoring and Evaluation**: GIS helps to track and monitor the availability of resources in the FEW nexus when combined with other sensing technologies. It can also be used to evaluate the impact of resources on the environment.

3.8 GREEN ENERGY AND THE INTEGRATION OF RENEWABLES IN FEW NEXUS 4.0 SYSTEMS

The FEW nexus is an important concept that, when put to practical use, will help actualise the SDG 2030 agenda and three of its goals, namely: food (SDG 2), energy (SDG 7), and water (SDG 6). The importance of this concept is due to the interconnection and reliance of the human race on these essential sectors. There has been exponential growth in the deployment of renewable energy systems, especially in the period between 2008 and 2013. This growth is seen in the added capacity and production of biofuels. Figure 3.25 shows a graph depicting the growth in renewable energy. In the power sector alone, 120 gigawatts of renewable energy capacity were installed in 2013. This, for perspective, is equivalent to the entire electricity generation capacity of Brazil. Additionally, this capacity addition represents more than half, 58%, of all new capacity installed globally during the period (Jaeger, 2021).

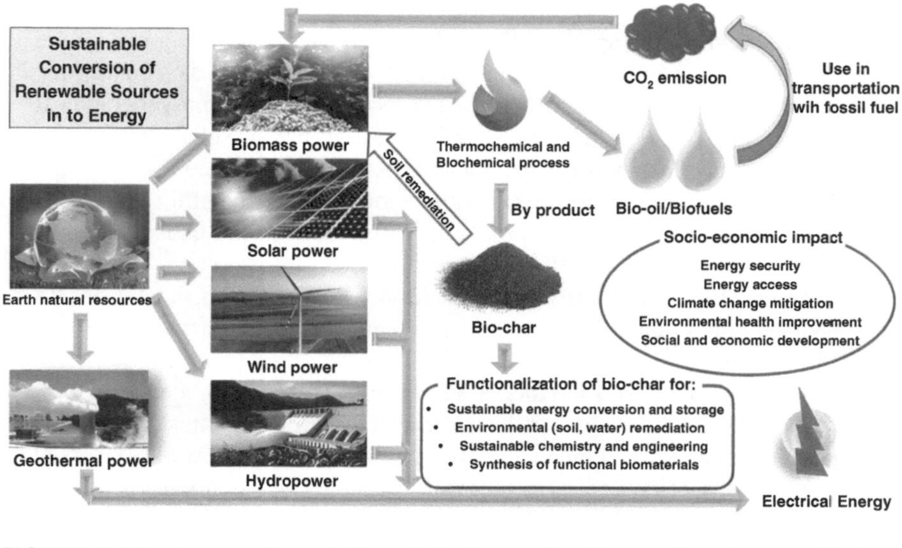

FIGURE 3.24 Average Annual Growth in Renewable Energy Capacity and Biofuels Production (IRENA, 2014).

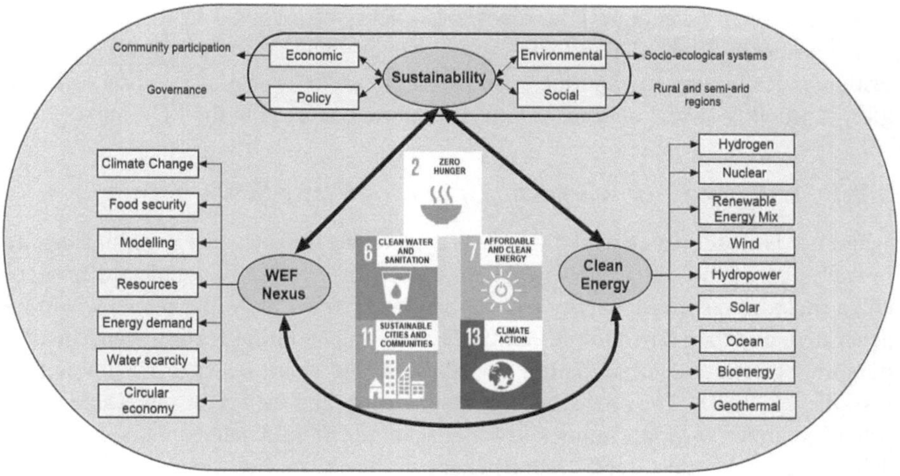

FIGURE 3.25 A Sustainability Approach to Integrated Renewable Energy FEW Nexus System (Jaiswal et al., 2022).

Renewable energy's share of global electricity generation reached 22.1% in 2013, with marked growth in the heat and transport sectors (Androniceanu & Sabie, 2022). Liquid biofuels supplied 0.8% of global transport fuels in the same year. Renewable energy's share of global electricity generation reached 22.1% in 2013, with growth in the manufacturing and transport sectors (Ahmad & Zhang, 2020). The growing recognition of the advantages of renewable energy use and a rising need to address issues associated with climate change led the United Nations to formulate a policy to double the share of renewables in the energy mix by the year 2030 in order to achieve sustainable energy for all (Gielen et al., 2019). Analysis by IRENA's Renewable Energy map (REmap) 2030 shows that renewables will only reach 21% of the global energy mix by 2030 (Gielen et al., 2019). However, this could exceed 30% if efforts are focused on accelerating deployment in buildings, transport, and industry. Energy efficiency measures and access to clean energy further provide an opportunity to increase the share of renewables to 36% (Mohsin et al., 2021). In the FEW nexus, the relationship that exists between food, energy, and water systems is further linked to renewable energy development.

Renewable resources such as solar, wind, and tidal energy consume very negligible amounts of water resources throughout their lifecycle. In this regard, renewable energy reduces environmental impacts and improves energy security; this makes it a vital source of energy in the 21st century and also in the FEW nexus (Hemeida et al., 2022). While concerns have been raised regarding bioenergy's impacts on land use, food, and water security, sustainable practices and other protective measures can address such impacts. Reducing the risks and maximising benefits associated with renewable energy requires a good understanding of the energy mix, the technology involved and the relevant standards and regulations (Ang et al., 2022). Figure 3.25 shows the framework sustainability approach to the integrated renewable energy-based FEW nexus system.

Further sections explore how renewable energy is involved in the energy-water and food-energy nexus. The FEW nexus creates security for the three most vital resources for human life and the economy. Renewable energy generation technologies, if properly used, could resolve most of these concerns in the FEW nexus.

3.8.1 Deployment of Renewable Energy for Water Supply

Water use is closely intertwined with energy delivery because energy input is needed in each part of the water supply system. Most of the time, the whole water supply chain includes the entire energy source, beginning from the water plants where it is generated, and then down to the end-use. Extracting and transporting water involves pumping, treating, or desalinating, distributing, and returning it to its source (Abu Hasan et al., 2020). The process is energy-intensive, and the use of renewable energy provides numerous opportunities to act at several different stages. Although renewable energy does not directly reduce energy intensity, it can reduce environmental impact, and this is valuable in rural areas because it opens up access to reliable water supplies. Renewable energy interventions throughout the water supply chain can improve water security and reduce the pressure on the water-energy nexus. Good opportunities for renewable energy implementation are found in water pumping, water distribution, desalination, and heating, as illustrated in Figure 3.26. Another good part of renewable energy is that it can also be applied in wastewater treatment. Water supply systems often involve setting up treatment plants in large areas, particularly in regions worldwide where water is scarce. After treatment, it is a common practice to transfer large volumes of treated water from one basin to another at both a national and regional level to meet demand in areas where clean water is scarce (Purvis & Dinar, 2020). This is done to meet the growing water needs in those areas. One challenge associated with this water transfer system is that it requires a significant amount of energy input, which, in the long run, increases the energy component of the FEW nexus at the expense of clean water availability (Javan et al., 2024).

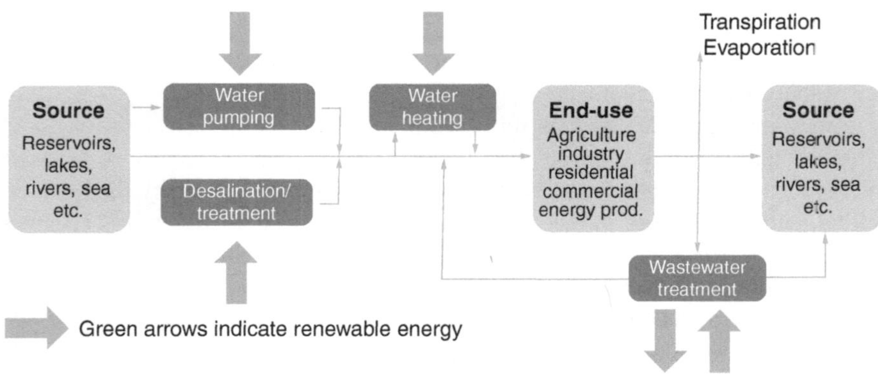

FIGURE 3.26 Renewable Energy across the Water Supply Chain (Renewable Energy Agency, 2015).

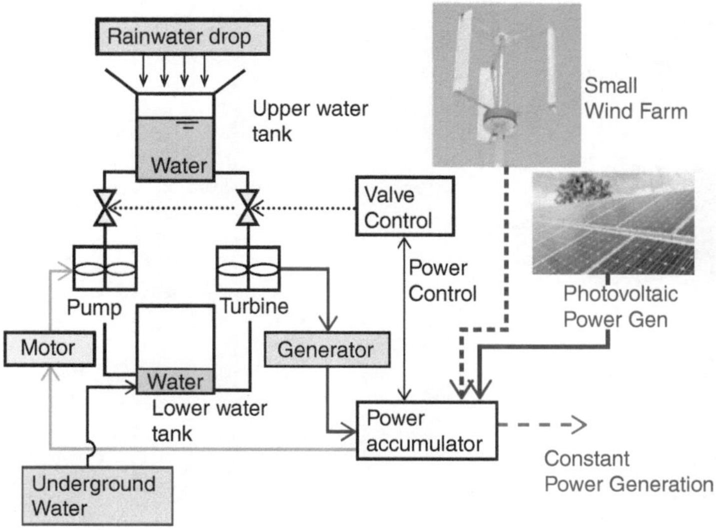

FIGURE 3.27 Pumped Hydro (PHS) Energy Storage for a Hybrid Renewable Energy System (Ali & Jang, 2020).

Water can also be used to store energy. Figure 3.27 shows a hybrid renewable energy system with a pumped hydro storage unit used to store energy.

Roughly 1.2 billion people live in areas of the world with physical water scarcity. Also, an additional 1.6 billion people live in areas of economic water shortage with inadequate man-made water infrastructure to meet their water needs due to a lack of extraction capacity from water sources. (UNDESA, n.d.). Groundwater plays an important role in the sustainability of life. It is a primary source of fresh water for both household and agricultural use. However, the extraction of groundwater consumes significant amounts of energy through electricity which sometimes is generated using fossil fuel. In countries such as India and Nigeria, the use of groundwater accounts for about 80% of all fresh water used for domestic and commercial purposes (Gonzalez Sanchez et al., 2020). Countries such as India and Nigeria depend on highly subsidised electricity for agricultural purposes. These countries use electrically powered pumps to generate water for watering their farms (Durga et al., 2024). As discussed earlier, this system is energy-intensive and as such, water is generated at the expense of energy. A suitable solution to this is the use of renewables, especially solar-powered systems for water generation. This is less energy-intensive, and the energy used here is often gotten via natural sources, which are constantly being replenished. Also, we don't need to mount pressure on the grid because these systems are mostly off-grid energy solutions. Figure 3.28 shows a solar pumping system for drinking water.

Solar pump solutions are pumping machines which are powered by energy from the sun. They have been deployed successfully in India. The governments of many nations, especially in Asia and Africa, have launched programmes that promote the use of solar technologies not only for agricultural water generation purposes but also

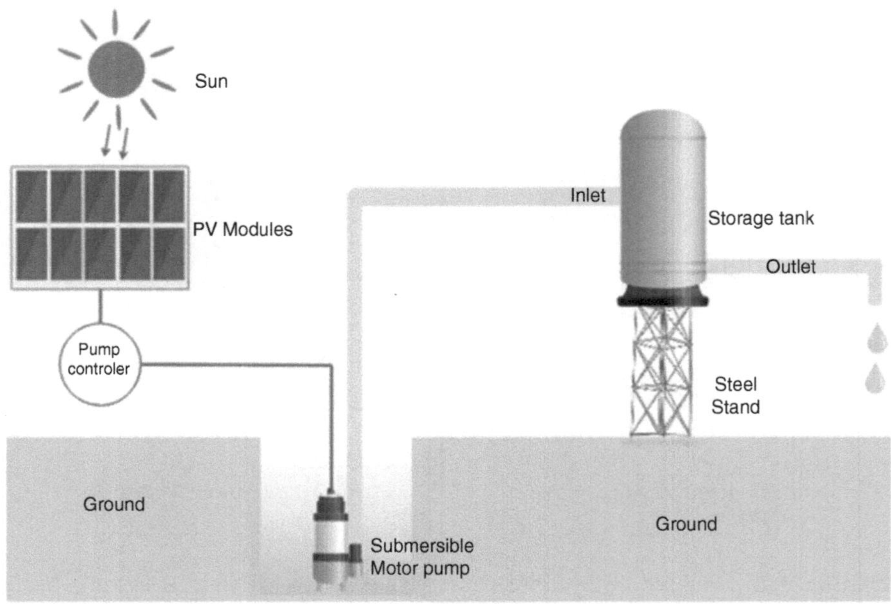

FIGURE 3.28 A Solar Pumping System for Drinking Water (Phiri et al., 2023).

for electricity generation (Makinde & Obikoya, 2024). One of such programmes is the Regional Solar Programme (RSP), which was initiated in the Sahel in 1986. This programme illustrates the potential of solar pumps in addressing the water and energy challenges. RSP has deployed around 995 solar pumping stations and 649 community systems that have provided improved water and electricity access to 2 million people. The interventions of RSP have been effective in substantially reducing the population without safe drinking water in the Sahel countries (Wolf et al., 2022). Realising these benefits, several countries have initiated programmes to promote solar pumping solutions to implement renewable energy for sustainable water provision and energy access (Zakariazadeh et al., 2024).

Several countries have launched programmes to convert to solar-based water pumping systems because of its significant environmental and economic benefits. For example, India has announced plans to replace most diesel pumpsets with solar pumps. This is believed to save 18.7 GW and 23.3 billion units of electricity, 10 billion litres of diesel, and 26 million tonnes of CO_2 emissions (Bhatt, 2019). The government of India will also deploy 20,000 solar pumps in selected districts to expand the water supply. In Morocco, the government-owned bank, Crédit Agricole USD 300 million grant for installing 100,000 solar pumps in the year 2020. South Africa also gave out USD 195 million for the supply of solar pumps; 10% were used to convert traditional pump sets (Renewable Energy Agency, 2015). In Tunisia, the government developed a Renewable Energy plan which includes developing renewable energy system in the agricultural sector. This plan installed 200 large water-pumping stations powered by hybrid technologies in strategic locations across the country (*Tunisia – Electrical Power Systems and Renewable Energy*, 2024).

However, the conversion to solar-based water pumps is facing significant barriers. The initial high capital costs on access to some form of financial support from international organisations (Ukoba et al., 2024). Other challenges include people's resistance to new technology, the difficulty in establishing distribution channels, and the provision of adequate training for installers and operators (Aghimien et al., 2022). Concerns have also been raised that solar pumping is leading to excessive water use, particularly in regions such as India and China. To address this challenge, many programmes coupled with financial support have been launched, although there is a limit which these programmes can reach.

Some other opportunities are still available for renewable energy deployment in the water sector via wastewater treatment. When water is used, it is either treated or discharged to the environment. Countries with high-income and prosperous economies treat about 70% of their wasterwater. Upper-middle-income countries treat 38%, lower-middle-income countries treat 28%, while low-income countries treat only 8% of their wastewater (Obaideen et al., 2022).

Wastewater treatment requires considerable energy in its collection, treatment, and disposal. Treatment operations in the United States and other developed countries account for about 3% of total electricity consumption, mainly for water pumping and aerobic treatment processes (Liu et al., 2020). Some of the by-products of treated wastewater include solids and decomposing matter. These solids recovered after treatment can also be treated and recycled. When this is done, the product can be used for fertiliser production, while the waste is disposed of in landfills (Kolawole et al., 2024). However, the greater the solids produced, the more the amount of energy needed for disposing or incinerating.

In developing countries with erratic power supply, wastewater treatment plants usually fail or are abandoned. A critical aspect of wastewater is that it has significant energy value itself, which is ten times the energy needed for treatment (Obaideen et al., 2022). Energy obtained from biogas in anaerobic treatment and also bio-solids obtained from wastewater treatment plants are significant sources of energy which can be used for energy generation.

Water utilities in developed countries are not only trying to keep energy use low but also to increase the physical security of the distribution systems. One of the best strategies known to lower energy costs is the conservation of water by reducing delivery leakages. Energy cost takes up a good chunk of the budgets of water utility organisations, ranging between 30% and 55% in most cases (Babel et al., 2021). Fluctuations in energy costs at present are related to water utility's cost structure, especially where the water and energy pricing policy-making processes are different. A way of mitigating this challenge is to reduce the reliance on foreign energy sources and integrate renewable energy systems across the entire water-energy supply chain (Bogdanov et al., 2021).

Desalination is another aspect of water production that involves considerable energy usage. It is the technique of separating salt and other impure substances from brackish water or seawater to produce fresh water (Zhang et al., 2021). This technology is essential in regions where fresh water is scarce but there is an abundance of seawater. Desalination technologies have become highly significant in solving water scarcity. Currently, there are about 16,000 small-scale desalination plants across the world capable of treating 70 million cubic metres of seawater daily. However, desalination is a costly process that consumes about 75% of its energy as electricity (Fawzy, 2020). Most electricity used in powering up desalination plants is generated with fossil fuels. As the growth of desalination continues, the use of fossil fuel is likely to increase. The energy-water nexus is against this increment because it comes at the expense of water generation to mitigate this challenge, the power for the desalination process has to be modified to use renewable energy (Panagopoulos, 2021). Doing this will save cost, conserve energy resources and reduce the impacts of fossil fuels on the environment and the economy. Some of the renewable energy technologies used in supplying power for the desalination process comprise thermal energy (use of thermal power or heat for generating electricity. It is only the membrane processes that require electricity), solar energy (use of energy from the sun to generate electricity), and wind reverse osmosis (RO) (combination of wind energy with RO, Rashidi et al., 2022). Out of the renewable-based technologies used for powering up the desalination process, solar energy and RO technology top the list. But other technologies are used for seawater desalination based on renewable energy (Sonar, 2020).

Another way to conserve energy in the FEW Nexus 4.0 is to make the water utilities renewable. Some scenarios have been observed where this technique was successfully used. A few of these scenarios are briefly explained:

In California, United States, the Valley Center Municipal Water District implemented a 1.1 MW of solar power systems in 2009 to generate about 2.1 GWh of electricity yearly. This saves up to 20% of the electricity consumption of its largest pumping station (Bukhary et al., 2019). The Idyllwild Water District is another town in Riverside County, California, that implemented solar solutions. This district implemented a 44.1 kW solar PV system to power water pumps so as to ensure reliable water supply across the city. This system addresses all problems related to frequent power outages due to high wind speeds (*Solar – Idyllwild Water District*, n.d.).

South Africa implemented mini-hydropower plants in her eThekwini Water and Sanitation to generate electricity using excess pressure from flowing water. This project is believed to reduce GHG emissions, optimise infrastructural returns, and leave a foundation for managers of water projects to follow (Climate Links, 2020).

Another technique for carrying out desalination is through the use of concentrated solar power (CSP). CSP-based desalination can be a very promising solution to mitigate water stress in the Middle East and North Africa region and could provide a large share of water desalination by 2050. CSP offers the following advantages: integration with water desalination techniques such as RO and thermal options. However, there are issues involved in integrating the existing desalination technology with CSP plants in order to provide a cost-effective solution (Omar et al., 2020). Despite the high costs of desalinating salt water by using renewable energy sources, long-term efforts have to be made to reduce costs and make the process more efficient in terms of its ability to meet the global water and energy demands both now and in the future (Shokri & Sanavi Fard, 2022).

Water heating for domestic use is another energy-intensive activity in the water extraction and utilisation process. Water heating itself consumes a considerable amount of energy, ranging from 15% to 30% of the overall energy consumption in developed countries such as Europe and the United States (Tow et al., 2021).

Most of the energy used for heating water in these regions is generated using fossil fuels. This will pollute the environment with carbon emissions and damage the health of the masses. (Chen et al., 2023).

Renewable sources of energy used for water heating include solar and geothermal power. According to the International Energy Agency, the global solar water heating capacity was 326 GWh in 2013. Solar thermal systems are widely installed in urban centres, accounting for 40%–80% of the hot water required for household use and community heating systems (Ang et al., 2022). Solar thermal systems, when combined with other technologies, are used for water heating in Europe (Herrando et al., 2019)

3.8.1.1 Hydro Energy Generation

Among the many water applications, energy generation is one prominent feature of the water-energy relationship matrix. In our contemporary world, energy facilities consume roughly 15% of global freshwater sources annually. Next to agriculture, this is the second highest use (Mahlknecht et al., 2020). Current energy space is based on oil, coal, and natural gas. This means that much water will be used to explore, produce, transport, convert, and treat these energy sources. As energy generation increases to meet current demands, water availability and accessibility become major challenges due to its limited use.

Proper water resources management policies are essential when it comes to preventing problems that may arise with the current water supply as input to existing energy generation equipment, which runs on fossil fuels (Meran et al., 2021). The need for the further development of energy generation equipment that requires less water is important if we must conserve water. Figure 3.29 shows the interaction between energy and water—the use of water for energy, and energy for water treatment, distribution, and wastewater management. Likewise, in Figure 3.30, it is shown that the supply chain of energy consists of three basic stages: fuel extraction, processing, and transportation.

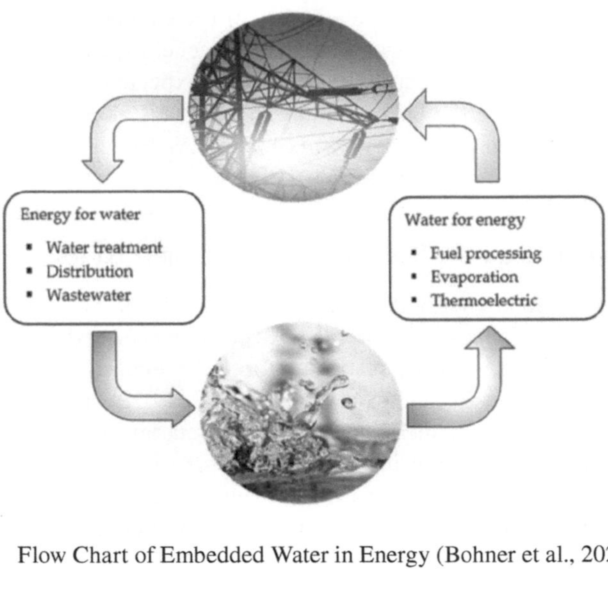

FIGURE 3.29 Flow Chart of Embedded Water in Energy (Bohner et al., 2024).

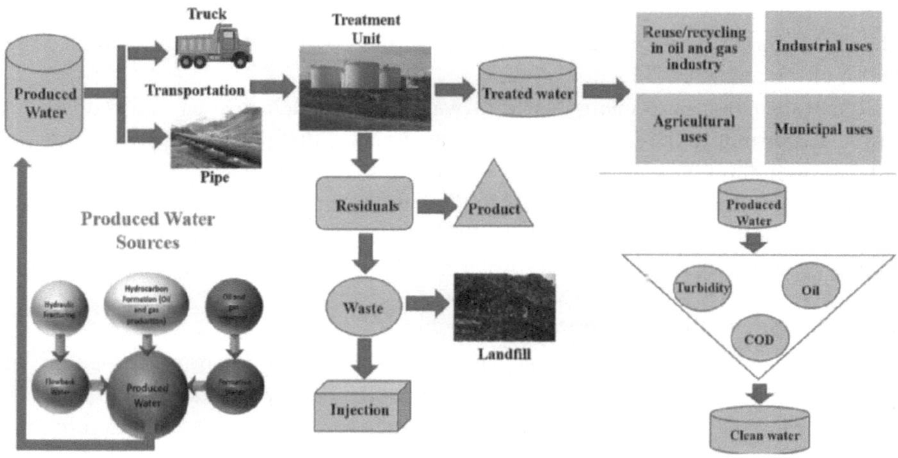

FIGURE 3.30 Withdrawal and Water Consumption during Primary Fuel Extraction, Processing, and Transportation (Nath et al., 2023).

According to global predictions, renewables in the energy sector are one way of reducing water usage (Zakariazadeh et al., 2024). In different cases, for instance, under the New Policies Scenario (NPS) and the 450 Scenario, an enhancement of the share of renewable energy sources decreases water withdrawals as against business-as-usual estimates (Zakariazadeh et al., 2024). European Union and GCC regions are examples of regional-level projections, which also support the positive impact of renewables on water resources (Renewable Energy Agency, 2015).

According to renewable energy plans, the GCC region, where water resources are scarce, is expected to reduce water withdrawal and consumption by 31% by 2030 (Renewable Energy Agency, 2015). Realisation of these plans could yield significant water savings, mainly through lower water use for generating electricity from crude oil. In addition, the change in focus to renewables is expected to decrease water consumption in countries like the United States and China (Zakariazadeh et al., 2024). Figure 3.31 shows a supply side energy mix for a FEW Nexus system. It consists of both renewable and non-renewable energy sources.

3.8.2 Renewable Energy's Role in the Food-Energy Nexus

Components of the food-energy nexus provide frameworks for understanding ways that the contribution of renewables in this nexus can be analysed. The core elements of the food-energy nexus provide platforms for understanding the potential role of renewable energy sources in the nexus (Lazaro et al., 2022). These options include the use of renewable energy technology to satisfy energy demand throughout the entire food value chain and the increase in the share of bioenergy in the world's total energy supply. Figure 3.32 shows various entry points through which renewable energy integrates into the conventional energy supply system (Renewable Energy Agency, 2015).

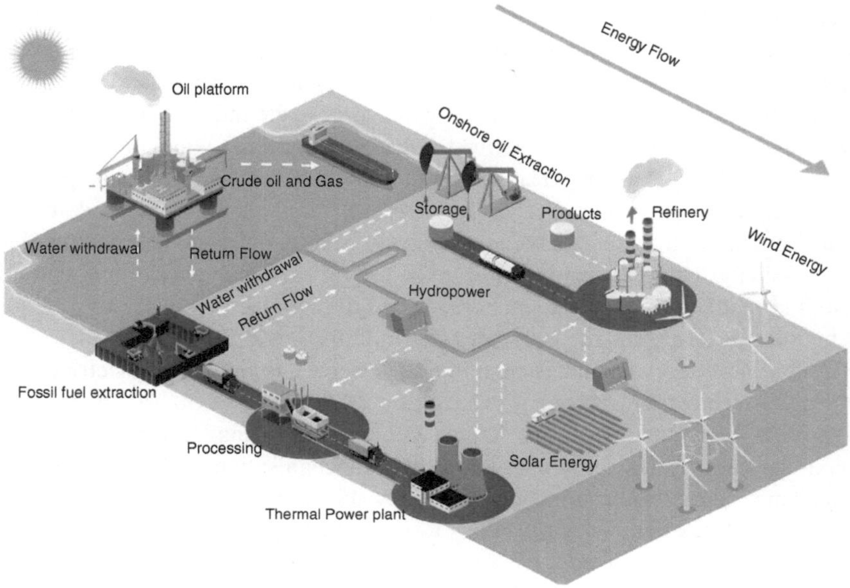

FIGURE 3.31 Supply-side Energy Mix of a FEW Nexus System (Vahabzadeh et al., 2023b).

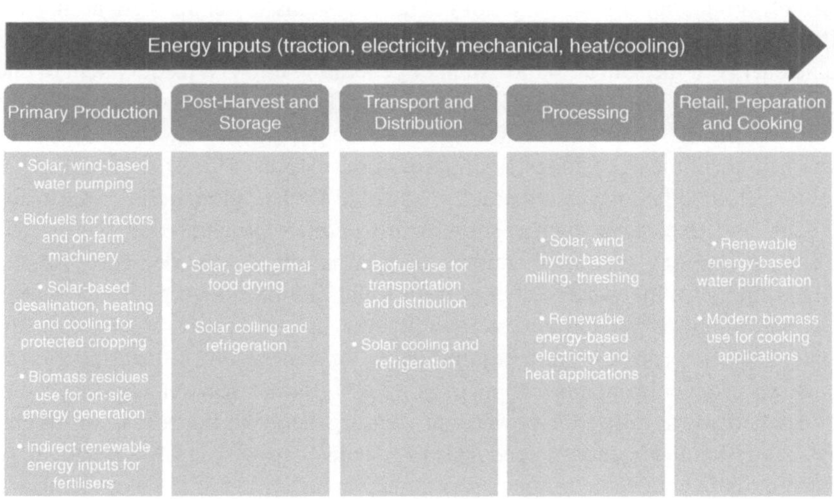

FIGURE 3.32 Illustration of Different Entry Points for Renewable Energy into Conventional Energy Supply Systems (Renewable Energy Agency, 2015, p. 74).

One way to understand how renewable energy can be integrated into the agri-food system is to break the system into simpler units that include processing, transport, distribution, sale, production, and cooking. The following few paragraphs discuss the role of renewable energy at certain stages where its effect is felt, namely in the production, processing and cooking of the supplied food.

During the primary production phase, there is a requirement for direct energy input for activities such as land preparation, farming, and irrigation. The energy use level differs depending on the type of farming being conducted. Small-scale subsistence farming uses human and animal labour, while large-scale industrialised farming use power and machinery and are inclined to be capital-intensive and energy-intensive (Aćin et al., 2023). The extent of mechanisation varies within the developing area. For example, in sub-Saharan Africa, where draft animals continue predominating in land preparation, tractors have progressively been withdrawn. On the other hand, South Asia has witnessed an increase in the use of tractors in agricultural activities. Also, indirect energy requirements are needed to manufacture agricultural requirements such as fertilisers and pesticides (Kirui, 2019).

Renewable energy resources can provide power requirements for electricity, heat, cooling, and bioenergy needs in agriculture (Babatunde et al., 2019). Regarding energy, there are challenges that come with the increasing degree of mechanisation in agriculture. This growing high level of mechanisation comes with a growing dependence on external energy sources. Renewable energy can meet electricity, heat, or traction energy demand for primary food production directly or by providing an offset to fossil sources of energy. An example is the increasing use of local biomass resources by households or small industries for the generation of electricity and/or heat using small-scale digesters (Logan et al., 2020). For instance, in the United States, there were approximately 144 systems for anaerobic digestion in commercial livestock production by January 2014. These systems differ in their ability to use different types of agricultural crop residues and animal and food wastes to produce usable energy at the source to the extent of generating 840 GWh equivalent of energy in the fiscal year 2013 (icct, 2023). Conventionally, it is employed in combined heat and power plants. However, it is indispensable to pay attention to the other possible uses of lignocellulosic residues essentially in developing nations, where it might be used for better soil management practices, animal feed, or biomaterials (Kirui, 2019).

Weak and expensive energy sources are another problem that works against the food processing sector. Therefore, the goal is introducing decrease the energy demand, increase the ratio of energy utilisation, and introducing renewable energy to support the improvement in sustainability and competitiveness. Current food processing factories can apply biomass products for generation of heat and electricity for use in production plants (Antar et al., 2021). India uses wet bagasse, which is produced in sugar mills, for power and steam generation purposes, and therefore plays a key input for energy production in India. Likewise, in Mauritius, bagasse cogeneration is noticeable in the proportionate contribution to the overall electricity generation in the country, with resultant gains in the extent of GHG emissions and imported coal (Kirui, 2019).

A large share of the total energy inputs is associated with food losses that occur during harvesting and storage, and consequently, a lot of emphasis has been placed on renewable energy in devising decentralised processing and storage facilities (Majeed et al., 2023). Solar energy and biomass are two examples that have been effective for both dry and cold storage. Drying is a preservation method that can be applied to a wide variety of foodstuffs and uses low-temperature sources of heat (EL-Mesery et al., 2022). The dried food forms are light and have an extended shelf life. The use of wood biomass for spice drying has broadened income flows and increased the revenue of small operators in the spice market chain in Sri Lanka. Food drying using geothermal energy is also an immense opportunity to reduce food loss and strengthen food security initiatives (Kirui, 2019).

Cooking, especially in developing countries, is also a very energy-intensive activity, and there is little adherence to energy-efficient practices. Approximately 2.7 billion persons rely on local biomass fuels such as fuelwood, crop residues, and animal dung for cooking worldwide (Joardder & Islam, 2022). However, these traditional biomass sources are sometimes unsustainable and generate emissions of smoke and carbon monoxide which constitute huge health and safety hazards (Mawusi et al., 2023). Also, collecting traditional biomass could be time-consuming and a lot of effort could be exerted.

To address these challenges, efforts have been made to guarantee energy efficiency in cooking through an upgrade of stoves. These stoves can reduce the need for fuelwood by up to 50% compared to that required by the open fires. Another a long-term sustainable approach to deal with the problem is to increase the availability of improved cooking fuels like LPG and biogas (Adams et al., 2023). Local bioenergy resources are complemented by supplying clean cooking fuels to rural users in a sustainable manner. This is critical in delivering the goal of universal access to modern energy (Kirui, 2019).

Another approach towards improving and sustaining clean and renewable energy is the deployment of bioenergy in the global energy mix. Traditional uses like cooking and heating dominate bioenergy utilisation at the moment, with more than 60% of the world's biomass demand in 2010 accounted for by the residential and commercial building sectors (Duarah et al., 2022). However, modern bioenergy can increase its share of the world energy mix because it has many applications in the electricity, heat, and transportation fuels sectors. According to the analysis of IRENA's REmap 2030, by 2030, the demand for modern biomass can double from 60% of total final renewable energy use (Renewable Energy Agency, 2015). Bioenergy development must be effectively and sustainably managed. The potentials it opens towards a more secure and localised energy, on the one hand, are matched, on the other hand, by the concerns over its impact on food security and the environment. Large-scale agriculture production and competition for natural resources are also considered some negative implications of bioenergy development (Gomiero, 2018). For example, the impacts of liquid biofuels on food prices, economic growth, land use, deforestation, and climate change are complex, multi-dimensional, and influenced by factors such as feedstocks used, production methods, and geography (Barr et al., 2021). Evidence shows that biomass energy can contribute to food security if it is sustainably produced and managed. Several approaches, tools, and frameworks for sustainability are currently under development and application to provide decision-makers in bioenergy development with the necessary tools for making better decisions (Abdel-Basset et al., 2021). Such initiatives include the Sustainability Indicators for Bioenergy developed by the Global Bioenergy Partnership, the European Commission, and the FAO Sustainable Bioenergy Support Package. All these tools are designed to assist in quantifying trade-offs and allow for more comprehensive cost-benefit analysis in the context of decision-making concerning bioenergy development (Vera et al., 2022).

It can be observed that bioenergy is associated with the supply and demand of food because both energy crops and food crops compete for resources such as land. Strategies to minimise impacts of this competition, according to Ale et al. (2019), include:

Intensify production by improving the yields of energy and food crops, improving land use efficiency, and implementing integrated food energy systems.

Cultivating crops on available or depleted land and areas, using energy crops that can grow under unfavourable conditions.

The agricultural waste streams and residues can be used to generate bioenergy and reduce the additional pressure on the land while bringing added value to the whole supply chain.

Generate by-products, including wastes that livestock can consume, reduce the utilisation of new land and water for producing the feed.

Because first-generation fuels (biofuels or conventional fuels, mostly obtained from agricultural sources) hurt world food supplies, second- and third-generation biofuels, derived from non-food cellulosic feedstocks, are the focus of research. These feedstocks need elaborate processing, although they hold the prospects of sustainability advantages, and they are insensitive to feedstock fluctuations (Kirui, 2019).

Within the energy area, a variation of the energy types demands land input energy sources like fossil fuels, fossil fuels, and nuclear energy. Land-use impacts must be evaluated based on the area of effect, type, and duration of change for each energy lifecycle stage (Ang et al., 2022). There are measures and methods to assess land intensity and determine the possibility of post-use land restoration.

3.8.3 Dual Use of Land in FEW Nexus

It is important to note that land can be dualised for both energy generation and agriculture. A case study of this has been carried out in some countries, including Germany (Trommsdorff et al., 2021). As the world has advanced technologically, farm areas have been used for renewable energy production and crop production. This multiple purpose is well illustrated by wind farms built on pastures across countries, biogas plants that process animals' manure and recycle nutrients back to the soil, firing of straw for heat and electricity, and growing energy crops for production of bioenergy crops, and feed and fibre from bioenergy crops used in biofuels, electricity generation, etc. (Noureldeen Mohamed, 2020). Some other ways this can be done are:

Leasing to renewable energy developers, which wind power companies can be a part of, greatly benefits farmers who are vulnerable to rising fossil fuel prices and could barely make profits from product sales (Cunningham & Seidman, 2024).

Farmer ownership of renewable energy: In Germany, for instance, about 11% of renewable energy capacity is farmer-owned as of 2012 (Mosher & Corscadden, 2019). Solar farms can also improve the capacity of the ground in the way that bad soil production areas can be made productive again through the improvement of grazing land (Guerin, 2019). Therefore, the agreements governing solar farms must incorporate a grazing framework that allows farmers continued access to the solar farmlands.

Agricultural land management is another concept embraced by many developers under solar farm development, with consideration for the ecosystem and economic benefits as a result of the two related ventures (Goldberg, 2023). Most plans for solar farms feature a wildlife management plan that allows small animals to graze beneath and between the rows of PV modules. Examples include sheep and free-range poultry for grazing in solar farms. This provides proof of the two-fold use of land serving solar energy production and also grazing by the animals (Agrivoltalics, 2023).

3.8.4 Comparison of Renewables and Conventional Energy Generation in the FEW Nexus

Previous research examining land use intensities of renewable and conventional energy technologies has provided mixed outcomes depending on the specific region or the type of technology. In comparing total land usage by renewable and traditional power technologies, Ang et al. (2022) concluded that of all renewable energy technologies, solar PV uses the minimum amount of land, particularly in regions with high levels of solar radiation. Onshore wind occupies a more significant amount of space due to the size of wind farms; however, it also uses very little floor area in the process, thus allowing other forms of utilisation of the available land (Wu et al., 2019). Systems installed on rooftops or combined with crop production optimises land use even more. There are also opportunities to integrate renewable energy in the currently "empty land," which is relatively unproductive and economically non-viable, hence the prospects for employment and economic growth in the area (Hempleman, 2022). In the United States, the development of solar PV systems on marginal lands is identified as having a high prospect with enough space capable of catering to future solar energy requirements by the year 2050 (Kirui, 2019).

Renewable energy and green energy, subsets of the FEW nexus, are important. Attention must be paid to the FEW nexus as interlinked systems. What is represented by the Food-Energy-Water nexus are those key drivers of renewable and green energy solutions towards the sustainability, reliability, and performance improvement of many sectors in balance with sustainable development agendas.

3.9 VIRTUAL REALITY, AUGMENTED REALITY, AND FEW NEXUS 4.0 – ENHANCING HUMAN INTERACTION WITH TECHNOLOGY

VR and AR technologies, though a bit nouvelle, are gradually finding applications in a wide array of fields, spanning from science to service, education, medicine, and even the military (Shekar & Kachhi, 2024). Agriculture has not lagged in its benefits from computer systems, as Agriculture 4.0, Digital Agriculture, and PA have all in positive results of increased food production in a sustainable manner manifested into positive results of increased food production sustainably. The agricultural sector is, therefore, an attractive avenue to explore for professionals and researchers working with VR and AR technologies (De Oliveira & Correa, 2020)

3.9.1 Virtual Reality

VR is a disruptive technology that changes human experience like never before. Also known as immersive multimedia, VR involves a physical presence in reality or from imagination and takes the user to that destination. It uses sight and sound to trick one's perception of reality. Still, the most important aspect is the sensation of one's actual presence and immersion within the virtual environment, known as "presence" (Britannica, 2024).

VR is an advanced computer interface that enables real-time simulation and interaction through various sensory channels and including visual, auditory, tactile, smell, taste, and other natural phenomena perceptible to the senses (SraSontisirkit, 2017). The concept of VR traces back to Stanley G. Weinbaum's short story "Pygmalion's Spectacles" in 1935, which described a goggle based VR system featuring holographic demonstrations of fictional experiences, including sensory elements like aroma and touch (Anna, 2018). One crucial aspect of VR is the meticulously engineered environment in which it occurs, as even minor discrepancies can disrupt the entire experience. Thus, achieving immersion, where the user feels fully absorbed in the virtual environment and is a key objective (Britannica, 2024). VR environments are characterized by their simulated nature and computer generated graphics and three-dimensionality, interactivity, engagement of human senses, and diverse forms of existence (Balzerkiewitz & Stechert, 2020)

3.9.1.1 Applications of Virtual Reality

VR applications can be classified as follows:

1. Simulation of real environments: for example, the interior of a building or spacecraft with the aim of training or education.
2. Creating a fictional setting, usually for a game or educational adventure (Nasyrov & Excell, 2020).

Areas where VR applications are commonly used comprise a wide range of fields; some notable ones include:

1. Design Evaluation (Virtual Prototyping)
2. Architectural Walk-through
3. Development and Maintenance
4. Data Visualisation and Concept Visualisation
5. Operations in Dangerous or Remote Environments
6. Training and Simulation
7. Sales and Marketing
8. Entertainment and Leisure
9. Enhanced Realities
10. Human factors and Ergonomic Studies
11. Simulation of Assembly Sequences and Maintenance Tasks
12. Aid for the Handicapped
13. Study and Treatment of Phobias (e.g., fear of height, Beier, n.d.).

3.9.1.2 Types of Virtual Reality Environment

Present in today's technological world, each offering different levels of immersion and functionality. Some of them include semi-immersive VR, fully immersive VR, and collaborative virtual environments (Simsek et al., 2021).

Fully immersive VR environments offer a total sense of presence, and they are associated with collaborative virtual environment systems and a variety of research investigations aimed at exploring basic issues in the nature of VR. For some applications, such as training, semi-immersive environments are created; these are more cost-effective than fully immersive ones (Martirosov et al., 2022). Collaborative environments concern themselves with a joint presence within the virtual experience, which may be fully or semi-immersive. All these types of environments, of course, have implications for research questions, and choosing which to utilise for any particular project will depend on budget and project objectives (Balzerkiewitz & Stechert, 2020)

3.9.1.3 Immersion and Presence in Virtual Reality

Authors often define immersion as the capability of computer displays to present a virtual environment to users, while presence refers to the subjective sensation of "being there" experienced by users in that virtual environment (Szabó et al., 2020). Immersion mainly concerns the technical nature of the system, which involves the software and hardware. On the other hand, the feeling of presence is affected by factors like a user's life experiences, memories, skills, emotional situations, and past experiences. This means that various users may have various experiences of both (SraSontisirkit, 2017)

Interaction is an essential aspect in increasing immersion in VR. Interaction in a well-designed VR system should be capable of capturing user input seamlessly to allow for the effecting of dynamic changes within the virtual environment. For example, the visual display of a VR system should accurately respond to a user's physical movements, allowing them to navigate and manipulate virtual objects (Petersen et al., 2022). The haptic feedback mechanism provides a sense of realism by simulating touch feelings, such as resistance in touching virtual objects (Wee et al., 2021). In other words, interaction mechanisms must be able to support sound and interesting interaction paradigms with effective feedback for interesting and engaging VR applications (D. Wang et al., 2020). Figure 3.33 shows a block diagram of the structure of a user interface (UI) based on collaboration in immersive VR. It consists of a part concerned with direct interaction between the user and the virtual environment. The second consists of user interaction for collaboration, and the third involves the design of a communication interface that allows multiple users to exchange information for collaboration (Park et al., 2019).

3.9.1.4 Forms of Virtual Reality

Immersive VR systems involve a number of computer interface devices in an attempt to deeply involve a user in the environment. These might include a head-mounted display (HMD), position trackers, fibre-optic wired gloves, and audio systems that provide 3-D (binaural) sound. The use of these technologies, singly and especially

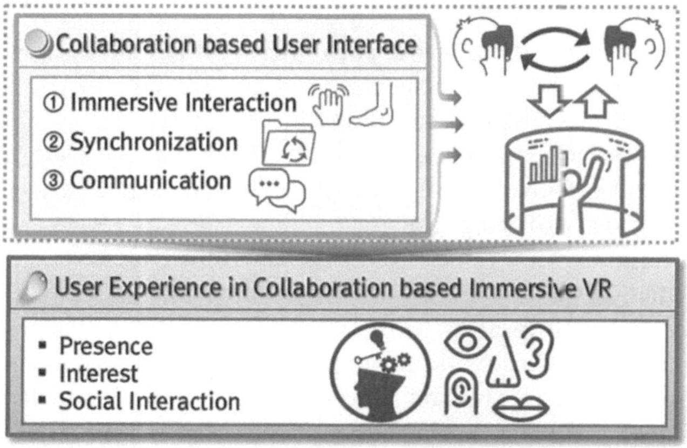

FIGURE 3.33 Structure of User Interface Based on Collaboration in Immersive Virtual Reality (W. Park et al., 2019).

in combination, creates a first-person view of a virtual environment for the user, engendering presence and realism (Al-Jundi & Tanbour, 2022; Blečić et al., 2021). In some applications of immersive VR, a treadmill interface is added to allow users to walk through virtual space. Other forms of HMDs, like the BOOM viewer from Fake Space Labs, hang in front of the user's face rather than attach to the user's head, providing a lighter and less cumbersome viewing experience (Cherni et al., 2021). In all cases, immersive VR immerses a person fully within a computer-generated image that appears and behaves in a realistic manner concerning perception, primarily visual and auditory, and sometimes even tactile (Dincelli & Yayla, 2022). Tactile perception in immersive VR is commonly referred to as the ability to feel or touch as well as to sensitiveness of bodily sensations in the course of interaction with virtual objects or surrounding space (S. Biswas & Visell, 2021). In an immersive VR environment, haptic interfaces include gloved or non-loved controllers that provide the haptic feedback to the user when interacting with virtual objects. These devices integrate components such as tactile feedback, for instance, vibration or pressure, which enhances immersion and realism within the virtual environment (Kim et al., 2020). Thus, the sensation of the presence experience is more excellent in immersive VR, so the user's senses are better stimulated when interacting with objects in the virtual environment.

3.9.2 Augmented Reality

AR is not the same as VR because the user can still perceive the physical environment. At the same time, graphics and audibles are laid over it to enhance the user's perception of reality or to point out some features (Y. Zhang & Zhao, 2024). For example, in aeronautics, AR can draw attention to controls or information, such as those required for landing an aircraft (Pradhan et al., 2023). In the same way, medicine is ripe for AR, with use cases as diverse as those of helping surgeons during surgery. Recently, a surgeon performed brain surgery assisted by an AR system where

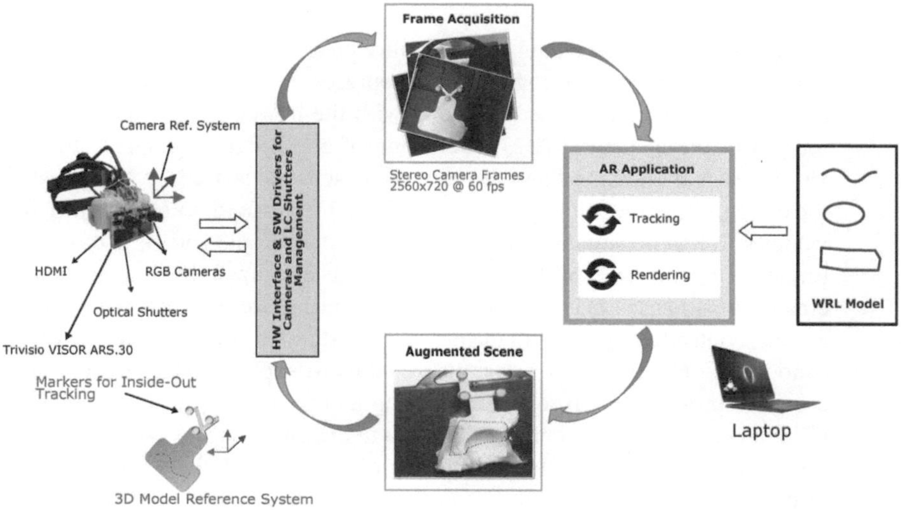

FIGURE 3.34 Typical Structure of a Wearable Augmented Reality (AR) platform (Covaciu et al., 2024).

a 3D image was inserted over the video image for a better view of the surgical area, thus increasing the efficiency of the surgery (Hey et al., 2023). Figure 3.34 illustrates the typical structure of a wearable AR platform. It consists of a laptop computer, SD model reference system, camera bot system, frame acquisition, and the AR application (Covaciu et al., 2024).

Types of VR are discussed in the following section.

i. Through the window:

   The most common form of VR is "through-the-window VR," which is heavily used in games and movie theatres (Barta et al., 2025). It lets participants see the virtual world but from within their real-world environment. The window into the virtual world may be a small computer display or a full-scale movie screen. Computer games provide the most significant degree of interactivity, the least by movie theatres (Moscoso et al., 2022). The video images displayed in the through-the-window virtual world are typically real-world scenes. In some cases, such as the rollercoaster ride simulation, the seats move and shake as an appropriate response to the images on the screen. (Parekh, 2025)

ii. Second person:

   Second-person VR records a participant's image and then inserts it into an immersive space where they can see the monitor and corresponding virtual interaction. Usually, this is done using Chroma-keying (Kim et al., 2020). The downside is that doing that lighting around the participant may be high, or the resolution of the image from the participant may not match the background. Most of the time, the users must get used to the body actions that they perform, which are shown on a screen (Parekh, 2025)

iii. Telepresence:

Telepresence is more than the capability of visualising totally synthetic, computer-generated environments. Telepresence marries the external physical environment sensed remotely with the power of a human operator (Pushpakumar et al., 2023). Detection of object can be carried out by various sensors mounted on the robot or by sensors carried by humans at the extremes of WALDO-like systems. Firefighters use remotely controlled vehicles to handle situations that may be too dangerous, and surgeons use tiny instruments transmitted by cables with accompanying video cameras to do surgeries that require very little cutting. Researchers have used remote-controlled robots with telepresence to study the ocean floor and volcanoes, and NASA hopes to use telerobotics to explore space. Designing using telepresence often requires building and testing physical prototypes (Klamt et al., 2020). Advantages include a reduction in design and manufacturing costs. The prototyping and testing of the physical phase are avoided or minimised, and efficient manufacturing processes can be implemented.

iv. Mixed reality:

By merging the telepresence and VR systems together, we can have mixed reality or seamless simulation systems. Here, the computer-generated inputs are fused with telepresence inputs and/or the user's view of the real world (Rauschnabel et al., 2022). For instance, a surgeon's view of a brain surgery could be superimposed with images from previous CAT scans and real-time ultrasound. Or, in the more traditional sense, a combatant pilot might see computer-generated maps, plans, and data displays within his helmet screen or on cockpit displays (Dixon et al., 2022).

3.9.3 Virtual Reality Display

According to Sherman and Craig, (2018), there are four primary displays in VR. These four primary displays in VR significantly enhance immersion. They are:

1. Visual display
2. Audio display
3. Haptic display
4. Vestibular display

They are briefly explained below:

3.9.3.1 Visual Display

End users of digital devices with visual output rely heavily on visual input to feel immersed in a VR environment. Hence, most VR systems prioritise visual display (Delgado et al., 2020). Visual immersion is influenced by several factors, some of which include:

Field of view (FOV): the extent of the visual field that can be seen at any given moment, measured in degrees of visual angle.

Field of regard (FOR): the total size of the visual field surrounding the user, also measured in degrees of visual angle.
Pixel per inch (PPI): the pixel density or resolution of the display.
Stereoscopy: presenting two slightly offset images to each eye to create a sense of depth.
Frame rate: the number of images rendered by the system per second.
Refresh rate: the number of images the visual display reconstructs per second. For example, if a TV has a refresh rate of 120Hz and displays content at 24 frames per second, each frame is repeated five times every 24th of a second.

This means that to create a realistic environment, a VR system also has to follow the position and rotation of the user's head to be able to render images based on the user's perspective (Sherman & Craig, 2018)

3.9.3.2 Audio Display

Sound is vital to inducing a sense of presence within the virtual environment. In this manner, good-quality sound would enhance the user experience even if there is an absence of good-quality visuals. For example, one clear advantage of sound over vision is that sound can give a sense of place and an understanding of activity within that place (Kern & Ellermeier, 2020).

This level of audio immersion in VR is based on several factors, which are:

3D localisation: The system should be able to track the position and rotation of the listener. As the listener moves around, the sounds should also change and increase in volume when the listener nears the sound source (Valzolgher et al., 2020). The proper position of the sound source should also be maintained even as the listener rotates their head.
Delivery method for sounds: The different spatial delivery methods of audio, such as 2.0, 2.1, 5.1, and 7.1 channels and headphone configurations, offer various levels of spatial audio fidelity and spatial audio immersion (Battah, 2019).
Variability: One of the things that may add to the dullness or the shallowness of a virtual environment is the reuse of sounds over and over. Replicating a large number of sounds that tend not to repeat at a noticeable rate increases immersion by giving the user a more natural and dynamic auditory experience (Valzolgher et al., 2020)

3.9.3.3 Haptic Displays

A haptic display is a UI device that stimulates users' sense of touch. Haptic devices are increasingly found in the gaming industry, such as driving wheel joysticks with force feedback, vibration-enabled gamepads, and mobile phones equipped with vibration feedback (Parisi, 2019). These devices generate tactile sensations for their users to enhance the gaming experience. The haptic display provides users with several information, such as surface properties of virtual objects, including texture, temperature, shape, viscosity, friction, deformation, inertia, and weight. For example, users can distinguish soft and hard tissues, an extremely important aspect in medical surgery applications when tactile feedback is critical for accurate manipulation and interaction with virtual anatomical models (Reddy et al., 2023).

3.9.3.4 Vestibular Display

In VR, vestibular display denotes any system designed to reproduce motion and spatial orientation in such a way that could enhance the sense of immersion by involving the user's sense of balance and movement (Sra et al., 2019). Unlike visual and auditory displays, which are mainly received through the eyes and ears, respectively, vestibular displays aim to recreate the feelings of motion and tilt through inner ear processing for the simulation of a sense of motion in virtual environments (Camilo Gil-Carvajal et al., 2024). Such displays are indispensable in applications where the realism desired is very high, such as flight simulators or training programmes based on VR, where truthful replicating physical sensations is essential for effective training and engaging the user. The integration challenges of vestibular displays include avoiding motion sickness and exact synchronisation with visual and audio cues to provide coherence and comfort (A. Javaid, Rasool, et al., 2024)

In the FEW Nexus 4.0, VR and AR have only been explored recently. As (Prabakaran et al., 2018) claimed in their article, there are also agricultural implications as new technologies are flowing in, and agriculture could reach a more efficient stage concerning farm output and land management. For instance, Huuskonen and Oksanen (2018) claimed that nowadays, "work in the field is done by machines controlled by a man." An indicator is that no practical experience is being developed in agriculture to identify the condition of the ground on which farmers work. Obviously, agriculture has moved far from old methods and into the digital age, even if many of these methods are pretty hard to understand by farmers themselves.

Workman (2023) believes that scientific visualisation is a useful technique used in many science and engineering disciplines. AR provides ways and approaches to develop scientific visualisation systems. Perifanis and Kitsios (2023) believe that to create a virtual environment and support natural interaction, it is important to have input and output devices that can recognise the variety of inputs by different users and produce instant answers to facilitate a meaningful dialogue between the user and the computer. All such agricultural systems can be used in the farm machinery assembly, agricultural education, and various applications.

AR techniques have been shown to enhance tasks' efficiency across many industries and provide exciting entertainment and educational experiences (van Lopik et al., 2020). The authors highlight that scant R&D have been undertaken in the agricultural sector regarding AR applications. Such applications include weed detection software, pest control, soil sample collection through drones, and autonomous vehicle guidance.

Embedding VR with FEW nexus improves the existing approaches and enables an easy understanding of human-oriented concerns in the FEW nexus. VR can be deployed in the FEW Nexus 4.0 in the following ways:

1. **Education and Training**: VR training programmes can create realistic platforms for experts and interested parties in food industries, energy

sectors, and water systems. Through realistic case scenarios, the users can familiarise themselves with the FEW nexus components, how they interrelate, and ways to promote effective resource conservation.

2. **Data Visualisation and Analysis**: Through the application of VR, it becomes possible to represent the structures and relations related to food, energy, and water. VR engages users and makes them interact with the world of data and analysis, making the system's dynamics, trends, and possible interventions easier to explain. It also enables stakeholders to distinguish the spatial-temporal patterns of occurrence and locate the hotspots to make the right choices in the right time and space to maximise benefits and minimise risks.
3. **Stakeholder Engagement and Decision-Making**: Through immersive virtual representations of key circumstances, this advanced interface empowers stakeholders to make well-informed decisions about the FEW nexus. In this regard, VR technology enables facilitating, managing, or designing talks and scenarios with 3D stakeholders. In particular, this will enable the joint discussion and agreement on strategies for distribution and policy initiatives that concern the use of resources.
4. **Planning and Resource Management**: Through VR, planners and policymakers will get the chance to see the functionalities of FEW nexus without physically having to build it out or apply scenarios that could potentially negatively impact on the environment. Simulation modelling has already been used to quantify the impact of diverse interventions on land use, water resource management, renewable energy, and the creation of food systems.
5. **Public Awareness and Outreach**: The education of the public about the complexities and potentials of the FEW nexus in VR can be an invaluable method. Immersive material such as games and VR shows, as well as field trips, may help to alter the behaviours of target groups and inform them about the protection of natural resources and the consequences of climate change and help communities prepare for it.
6. **Research and Innovation**: Thus, as an interfacing research domain, VR aids interdisciplinary FDI approaches in the FEW nexus. Technologists can incubate and experiment with ideas and subjects, prototype physical systems, and evaluate technologies that cannot be tested in the real world. Through software such as VR, one can practice various options and strategies and perfect them for addressing complex FEW problems.

Applying VR to the FEW nexus frameworks opens doors to practical and creative ways of educating, theorising, communicating, enhancing multidisciplinary collaboration, decision-making and resource management, public engagement, and innovation. This is an opportunity to analyse how the global community's challenges in food security, energy, water, and environmental issues may be solved using VR technology in close cooperation with stakeholders.

3.10 ADVANCEMENTS IN SENSORS AND MONITORING TECHNOLOGIES FOR FEW 4.0 SYSTEMS

Agri-food energy and water industries represent vital sectors for global economic development and for ensuring food safety (Pawlak & Kołodziejczak, 2020). However, this sector has numerous challenges, such as increasing demand for food, control of food security, climate changes, clean energy generation and environmental protection (Wakweya, 2023). The agri-food industries are developing novel strategies to stay productive, sustainable and competitive in this 21st century. The essence of adopting these new strategies is to benefit from the opportunities offered by Industry 4.0. Figure 3.35 shows a smart factory comprising wearables, UAVs, smart packaging, cobots, ML for data analytics, and industrial AR.

Technologically, the industry has been evolving with the manufacturing which has taken place in the industry. Systems' mechanisation during Agri-Food 1.0 and the utilisation of electricity during Agri-Food 2.0, as well as the emergence of robotics and automation during Agri-Food 3.0, have each experienced significant changes. The industry is being controlled by Industry 4.0, a cyber-physical system whereby multiple changes in this period are related to the use of technologies like intelligent GHGs, robots, drones, and intelligent sensors (Kunst et al., 2022). This new era is called Agri-Food 4.0 (Kunst et al., 2022). This era takes into account the aspect of digitalisation, modern machine tools, and sustainable practices. Strategies focus on minimising the process costs, resources used, and the promotion of renewable resources (Mondejar et al., 2021b). Generally, Agri-Food 4.0 is a total move taken and overcoming challenges to tackle and overcome them in the agro-industry while incorporating technological changes to increase efficiency and sustainability (Miranda et al., 2019). The rapid development of technology has led to massive

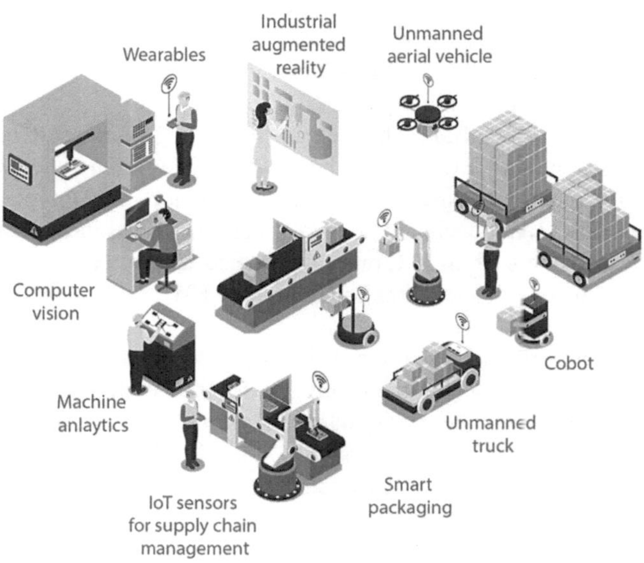

FIGURE 3.35 General Concept of a Smart Factory (Kalsoom et al., 2020).

growth of several technological concepts, eventually forming the basis of Industry 4.0. Some of these concepts have evolved over the years. In the case of sensing, the evolution started from Sensing and Smart Enterprise (S2 enterprise) model to the more advanced sensing, smart, and sustainable enterprise (S3 enterprise) framework. The core focus of these concepts is the adoption of current technologies for increased growth, productivity, competitiveness, and sustainability across sectors in the world's economy, of which the agri-food sector is not left out (Zhou et al., 2022).

Key features of the S3 enterprise are sensing, smart, and sustainable solutions, which are pivotal to the management of operations and the development of products. It is in this regard that current technological and social trends point towards products that will have a significant role in daily life and, in turn, improve the QoL in society (de Sousa et al., 2020).

Implementing sensing, smart, and sustainable technologies in the agri-food sector is proposed as a strategy to offer solutions that are primarily technology-based and aimed at attaining sustainability objectives (Lezoche et al., 2020). "Sensing" refers to the capacity of systems to detect events, acquire data, and measure changes in the physical environment (M. Javaid, Haleem, Rab, et al., 2021). "Smart" is the capacity of systems to respond to and control functions and display intelligence for adaptive or predictive decision-making based on available data. Smart products can also be connected to other devices. "Sustainability" is applied to products to balance or optimise their performance considering social, economic, and environmental objectives (Foresti et al., 2020). Besides the many benefits of the S3, one of the critical challenges that could arise is the high cost of adopting these technologies, especially for small and medium farmers in developing countries. So modern-day research is aimed at developing new technologies that can incorporate sensing, smart, and sustainable solutions that will give innovative and economically feasible impacts. The development of technologies to support the agri-food industry is discussed below. The discussions focused on the role of Agri-Food 4.0 technologies in curbing farming challenges through breakthroughs in smart farming, sensors, traceability, smart logistics, and smart food awareness (Miranda et al., 2017).

3.10.1 Sensing, Smart, and Sustainable Technologies

In the new age of technology, diverse branches of engineering have converged, leading to the development of new design methodologies such that one cannot define the interfaces between mechanical, electrical/electronic, and software anymore. These methods, which have been advanced through sensing and intelligent technologies, tend to make the mechanical, electrical/electronic, and software fields identical. This new world is increasingly embraced by designers and companies using new methods such as concurrent engineering to realise the whole product lifecycle (Rihar et al., 2021).

The concepts of sensing, smart, and sustainable are relevant in today's agro-sector and act as a key driver for the transformation of this sector. S3 technologies relate to those with sensing, intelligent decision-making functions and those that foster sustainability. These technologies improve continuously, and their functionality is robust because they incorporate mechanical, electrical/electronic, and software components (Miranda et al., 2017, Figure 3.36).

FIGURE 3.36 An S3 Farm (Kim et al., 2024).

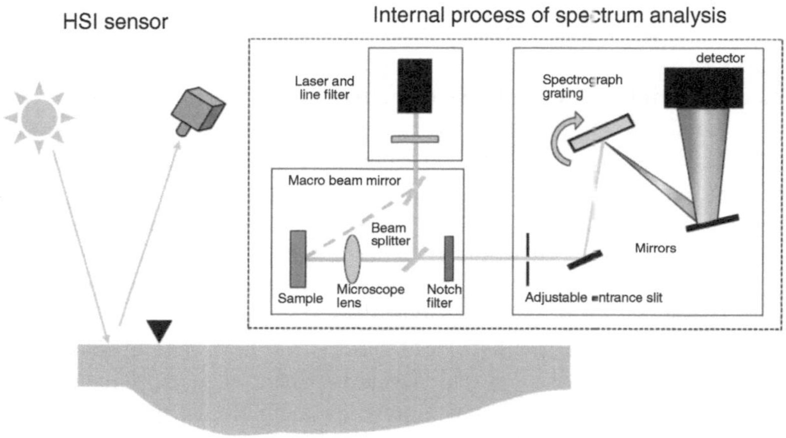

FIGURE 3.37 Use of Sensor Technology for Monitoring Water Quality (J. Park et al., 2020).

3.10.2 Sensing Technologies

Sensing technologies are the inputs or switches of any digital system that uses sensors. They form the basis for detecting events, data collection, and monitoring changes in the physical environment. Data collected from sensors are used for monitoring, control, and system optimisation. Integrating multiple sensors in a system enables the system to be smarter because data of more than one system is now fused together for better accuracy (Asachi & Alonso Camargo-Valero, 2023). Currently, there is ongoing research aimed at developing better-sensing systems. This research involves utilising various techniques and materials for improving the processing capabilities, size, power consumption, and prices of sensing systems. Advancements in virtual sensors and sensor-less techniques enable products to be developed that require sensing capabilities without the need for physical sensors (Verma et al., 2021). Figure 3.37 illustrates the use of sensors in water quality monitoring to manage smart water and waste treatment.

The concept of the smart sensor is an important development in the area of sensing technologies. Smart sensors are advanced sensors with higher sensing capabilities, faster processing speeds, higher and stronger memory, and better network controllers, which are all embedded in one unit.

Technological advancements such as sensor networks and wireless sensor networks bring us this new world of connected applications used to monitor various parameters, namely, temperature, sound, vibration, pressure, and mobility. These are the current and emerging trends in the field of sensing technologies. They foster efficient, networked, and general sensing technologies used in expert systems today (Arshi & Mondal, 2023).

3.10.3 Smart Technologies

Smart technologies, also known as "smart products," mediate between the real and virtual worlds and allow the infusion of digital information into physical objects (Katuk et al., 2023). The technologies have three key parts: physical parts, mechanical, and electrical/electronic parts. They include smart parts like processing elements, sensors and actuators, computer algorithms, and protocols. Some auxiliary parts also included in them are communication ports, antennae, and communication protocols, both wired and wireless (Wolf, 2023).

Smart products are designed based on CPSs. They are built by tightly joining computing and aspects of physical processes in order to achieve better intelligence and performance (Yao et al., 2019). They rely on AI control algorithms, such as fuzzy logic and evolutionary algorithms, to derive intelligent judgements and process information (Jia & Wang, 2024).

Smart products can be used in various application areas, such as automotive systems, medical devices, assisted living, traffic control and safety, process control, energy conservation, environmental control, robotics, instrumentation, critical infrastructure control, defence systems, and manufacturing. This is the again what smart technologies offer. They deeply impact the future with widespread application areas that bring about intelligent solutions (Islam et al., 2023). Figure 3.38 illustrates the application of smart technologies in IoT-based devices in agricultural production; the devices are based on IoT, which is a technology that complements most of these smart devices. From the diagram, smart devices are used in precision farming, agricultural drones, computer imaging, remote sensing, livestock monitoring, climate condition monitoring, and smart greenhouses (Ali et al., 2023)

3.10.4 Sustainable Technologies

Sustainable technologies are of products that integrate social, economic, and environmental objectives aimed at achieving minimal environmental load by using efficient resource deployment and manufacturing techniques (Diaz et al., 2021b). They aim to improve the QoL of people in the health, education, cultural, and improve people's QoL in the health, education, cultural, housing, manufacturing, creative and other sectors of the economy. Economically, sustainable technologies ensure productivity in the creation of low-priced goods. This, in turn, gives birth to new enterprises while generating employment and ensuring profitability in the long run (Shahid et al., 2023)

To develop technologies that are sustainable in the long run, various eco-design resources, namely eco-design, sustainable design, design for the environment or life

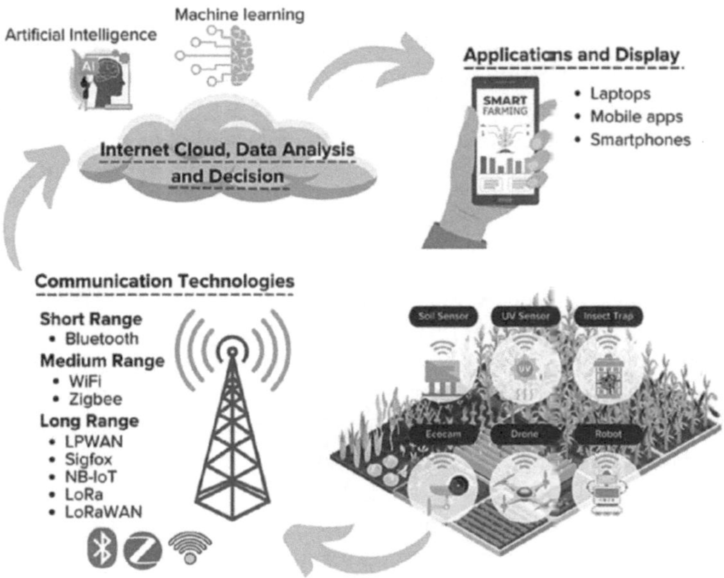

FIGURE 3.38 Application of Smart Technologies in Farming (A. Ali et al., 2023).

cycle assessment, are utilised (Hossain, 2019). The resources enable the designers to quantify and analyse different sustainability factors or indicators at each stage of product development and manufacturing, thus giving them the power to decide on the sustainability factor (Mengistu et al., 2024). Requirements for eco-design of sustainable products include selection of low-impact materials, reduction in mass or volume, clean manufacturing techniques, reduced environmental load due to packaging and distribution, optimisation of product life cycle or end life options by way of reuse or remanufacture or recycling or disposal (Miranda, Ponce, et al., 2019). The above principles will form the backbone of designing environmentally and socially sensitive technologies. Figure 3.39 shows the major research areas in environmental technology. They include clean and renewable energy, environmental policies and frameworks, solid waste and e-waste management, waste-to-energy conversion, sustainable materials for the cleanup of toxic contaminants, and recovery of metals and other beneficial products (Sahoo et al., 2024).

3.10.5 Integration of S3 Technologies for the Agri-Food Industry

The S3 Product Development Reference Framework realises innovative technologies that fuse sensing, smart, and sustainable solutions, streamlined for the agri-food 4.0 market (Miranda et al., 2017). This framework is a partial area of the Integrated Project Planning and Management Development reference model, but it guides the creation of goods, manufacturing processes, and systems (Molina et al., 2021). Innovation in technologies that considers the needs of the developing agri-food business will generate efficiency, intelligence, and environmental sustainability that protects all the resources in the food energy and water systems (Miranda et al., 2019).

FIGURE 3.39 Major Research Areas in Sustainable Environmental Technology (Sahoo et al., 2024).

In the design of innovative technology for the agri-food space and also for the components of the FEW nexus, we need to take into consideration the following components:

- What parameters that need to be sensed?

  In precision agri-food production and automated agricultural systems, the sensing system forms the system's core. It serves two purposes: to sample the environment and acquire information from different production system functions. Most of the environmental variables important for agri-food production can be measured by sensors, which represent indispensable input for control systems (Finger et al., 2019).

  Designing any form of technological sensing system in Industry 4.0 requires carefully selecting sensors for the specific S3 technology being developed. Designers must consider power, output signal orientation, communication protocols, and electronic conditioning. To achieve a quality sensing system, all these need to be considered alongside a good data transmission system and data acquisition modules. All these will be integrated into the structure of the S3 (Singh et al., 2022).

In a nutshell, the design of the sensing system is the strategic integration of sensors in a digital system to systematically collect data and monitor the environment. This is very important to ensure efficiency in the agricultural sector (Miranda et al., 2019).

- What aspects require smart control?

  Several subsystems and variables can be smartly controlled in agri-food production systems. For example, all variables necessary in controlling aeration, light, moisture, ventilation, and temperature, among others, can be controlled intelligently in production systems. Processes that run efficiently, with the right production quality, and with little or reduced wastage in energy and water use are controlled with sensors.

  These smart systems are usually designed with physical components, smart components, and connectivity systems. The implication, therefore, is that the design process will have to incorporate hardware design towards the development of the physical features of the technology, as well as mechanical and electrical/electronic design processes. The design processes will also have to incorporate software and firmware in implementing the solutions. It will focus on developing data and information processing algorithms, including the programming of the control system and interface designs (Miranda et al., 2017).

- What aspects require sustainability?

  When introducing a new technology, the designer must factor in the three important stages of the product life cycle: manufacturing, use, and end-of-life. From a sustainability point of view, the new technology must satisfy the following conditions: display better data on the manufacturing process, its use, and treatment of end-of-life compared with existing technologies (Taddese et al., 2020). Manufacturing involves all steps and strategies to convert raw materials into finished products. This stage makes use of energy, water and some consumable materials. The end-of-life stage includes aspects to do with the final disposition of the product's components and may consist of recycling, reuse, remanufacturing, disposal, and incineration (Ciliberto et al., 2021).

  These considerations are categorised under the triple bottom line: environment, economy, and society. Their categorisation may differ depending on the type of technology, its life cycle, or regulation. But understanding these influencing factors is very important in the design and development of sustainable technologies (Miranda et al., 2019).

3.10.6 Case Studies of S3 Technologies

Some cases where S3 technologies have been successfully deployed are presented in this section.

- Intelligent hydroponic greenhouse

  The Intelligent Hydroponic Greenhouse (IHG), developed using the S3 Product development approach, concerns tomato production. Its development is geared towards providing solutions to the problem of defining the optimal physical infrastructure and control systems for regulating climate conditions within the industry. Fitted with sensors for evaluating environmental factors such as rain, wind speed and direction, humidity, CO_2 levels, sun intensity, solar luminosity, and temperature, the IHG maintains optimal conditions to preserve crop integrity.

Data evaluated by the smart system of the IHG can be accessed wirelessly through a human-machine interface and a machine-machine system. Equally, the IHG incorporates a monitoring system that allows users to evaluate each subsystem's status. The IHG is housed at the Tecnológico de Monterrey in Mexico City Campus and is a fine embodiment of a high-tech solution for sustainable and efficient tomato production (Miranda et al., 2019).

- Sun-tracker for multi-purposes

  Solar panels have become widely used to harness solar energy, though their cost per watt remains high compared to other energy sources, such as wind. The proposed "Sun Tracker" by Kanwal et al. (2023) maximises the energy generation of solar cells to provide a competitive advantage in solar energy (Kanwal et al., 2023). The tracker is implemented based on two algorithms for optimisation positioning, they include: a genetic algorithm and a swarm optimisation algorithm. The genetic algorithm offers an offline solution that reduces computational cost since results are programmed into a lookup table. This method is suitable for regions with a constant climatic condition. However, in areas where climatic conditions are not continuous, an online solution is required (Miranda et al., 2019). The Sun Tracker is a reliable way to gather solar energy under different climatic conditions, and hence is a suitable device for powering different agri-food technologies (Kanwal et al., 2023). Automatic agricultural systems in crop fields are

FIGURE 3.40 A Physical Prototype of a Sun Tracker (a) front view (b) side view (Morón et al., 2017).

among those devices that use this technology. Figure 3.40a and b shows the design of a photovoltaic sun tracker.

- Quadrotor or drone (UAVs)

  An UAV quadrotor has a Type-2 Fuzzy Logic-based motion controller to maintain a horizontal position of its "x" and "y" axes, controlling Pitch and Roll without rotation along the "z" axis. A Parallax platform (Elev8) houses the redesigned quadrotor for applications. It has mechanic reinforcement in the structure to survive impacts and support extra hardware such as sensors, cameras, and processors (Al-Mahturi et al., 2021).

  The control embedded within the microprocessor maintains a reference within defined boundaries and can handle disturbances within a specific range. The UI allows real-time monitoring by showing the angle representations and numerical displays for each axis, as well as the derivatives of these angles. It also shows the Pulse Width Modulation (PWM) percentage values of each motor, the energy indicators, and the total torque generated by the motors, as well as charts for the PWM percentage that applies to the motors and the angle/angular velocity measures of each axis (Al-Mahturi et al., 2021)

  This technology of UAV finds application in agriculture where a UAV can be used for spraying (fertilisers, herbicides, pesticides), image capturing for 3D modelling, acquisition of data (using various cameras like infrared, thermal, etc., for field data capture), and monitoring crop growth (Miranda et al., 2017).

- Agricultural hexapod robot

  This hexapod robot with six legs acts as a mobile monitoring system that could acquire information about the condition of plants and soil for agricultural purposes. It roams around the field to monitor crop growth and detect disorders. Several sensor systems for various data collection are embedded in this robot. Neuro-fuzzy motion controlling was developed and implemented as a new concept for this hexapod, using ANNs and fuzzy clusters based on trigonometric series. All these units have been integrated to

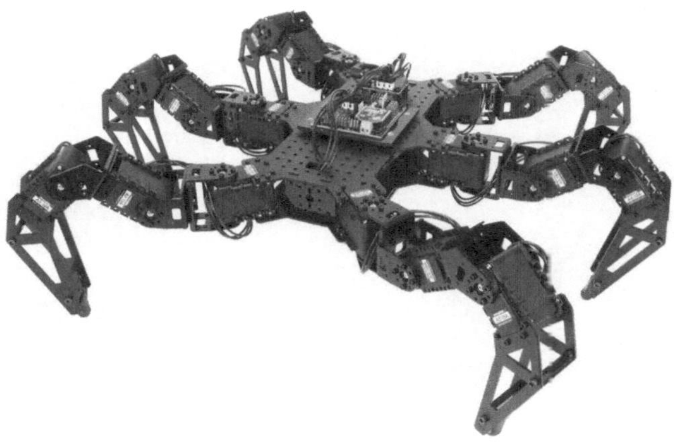

FIGURE 3.41 A picture of an Agricultural Hexapod (Arrigoni et al., 2024).

become the controller's core, offering better performance for the hexapod. The system was programmed using a high-level language and a Motorola microcontroller as the control unit. The system's architecture was made to be flexible and adaptable. Figure 3.41 shows a picture of a hexapod robot. This agricultural hexapod robot is more sensitive, intelligent, and durable to collect and process data adequately in agricultural fields i.e. (Miranda et al., 2017).

Advanced sensors and monitoring technologies have a huge potential to improve FEW systems growth in Industry 4.0. Technologies integrated within the FEW nexus will increase efficiency, sustainability, and resilience while managing this essential resource.

3.11 SUMMARY

This chapter discusses emerging technologies, explaining what they are and how they drive the FEW Nexus 4.0. It explains how IoT and AI are integrated into the FEW 4.0 systems for efficient and effective protection of the FEW resources. It discusses DL and ML, starting with the difference between explaining how it is used in the FEW Nexus 4.0. Apart from the earlier mentioned examples, it also explains in depth how blockchain technology, PA, SC, BDA, GIS, green energy and renewables, and AR/VR are all used in the FEW Nexus 4.0.

REFERENCES

A Tech Revolution will Help Farmers Harvest Sunshine with Their Crops | WIRED. (n.d.). Retrieved August 31, 2024, from https://www.wired.com/story/energy-agrivoltaic-farms/

Abdel-Basset, M., Gamal, A., Chakrabortty, R. K., & Ryan, M. (2021). Development of a hybrid multi-criteria decision-making approach for sustainability evaluation of bioenergy production technologies: A case study. *Journal of Cleaner Production*, *290*. https://doi.org/10.1016/J.JCLEPRO.2021.125805

Abiri, R., Rizan, N., Balasundram, S. K., Shahbazi, A. B., & Abdul-Hamid, H. (2023). Application of digital technologies for ensuring agricultural productivity. *Heliyon*, *9*(12), e22601. https://doi.org/10.1016/J.HELIYON.2023.E22601

Abraham, S., Shahbazian, A., Dao, K., Tran, H., & Thompson, P. (2017). *An Internet of Things (IoT)-Based Aquaponics Facility.* 1–1. https://doi.org/10.1109/GHTC.2017.8239339

Abu Hasan, H., Muhammad, M. H., & Ismail, N. I. (2020). A review of biological drinking water treatment technologies for contaminants removal from polluted water resources. *Journal of Water Process Engineering*, *33*, 101035. https://doi.org/10.1016/J.JWPE.2019.101035

Aćin, V., Mirosavljević, M., Živančev, D., Jocković, B., Brbaklić, L., & Jaćimović, G. (2023). Field management practices to produce nutritional and healthier main crops. *Developing Sustainable and Health-Promoting Cereals and Pseudocereals: Conventional and Molecular Breeding*, 137–173. https://doi.org/10.1016/B978-0-323-90566-4.00006-0

Adams, A., Jumpah, E. T., & Dramani, H. S. (2023). Dynamics of clean and sustainable households' energy technologies in developing countries: The case of improved cookstoves in Ghana. *Sustainable Futures*, *5*, 100108. https://doi.org/10.1016/J.SFTR.2023.100108

Aghimien, D., Ngcobo, N., Aigbavboa, C., Dixit, S., Vatin, N. I., Kampani, S., & Khera, G. S. (2022). Barriers to digital technology deployment in value management practice. *Buildings*, *12*(6), 731. https://doi.org/10.3390/BUILDINGS12060731

Agrivoltalics. (2023). *Agrivoltaics – Sheep grazing in solar farms; good for both – Energy smart farming | Energy smart farming*. https://extensionaus.com.au/energysmartfarming/agrivoltaics-sheep-grazing-in-solar-farms-good-for-both/

Ahmad, T., & Zhang, D. (2020). A critical review of comparative global historical energy consumption and future demand: The story told so far. *Energy Reports*, *6*, 1973–1991. https://doi.org/10.1016/J.EGYR.2020.07.020

Ajibola, O. A., & Ogbolumani, O. A. (2024). Smart battery storage integration in an IoT-based solar-powered waste management system. *Journal of Digital Food, Energy & Water Systems*, *5*(2), 17–42. https://doi.org/10.36615/HREWQ861

Akhter, M. N., Mekhilef, S., Mokhlis, H., & Shah, N. M. (2019). Review on forecasting of photovoltaic power generation based on machine learning and metaheuristic techniques. *IET Renewable Power Generation*, *13*(7), 1009–1023. https://doi.org/10.1049/IET-RPG.2018.5649

Albino, V., Berardi, U., & Dangelico, R. M. (2015). Smart cities: Definitions, dimensions, performance, and initiatives. *Journal of Urban Technology*, *22*(1), 3–21. https://doi.org/10.1080/10630732.2014.942092

Albrecht, T. R., Crootof, A., & Scott, C. A. (2018). The water-energy-food nexus: A systematic review of methods for nexus assessment. *Environmental Research Letters*, *13*(4), 043002. https://doi.org/10.1088/1748-9326/AAA9C6

Ale, S., Femeena, P. V., Mehan, S., & Cibin, R. (2019). Environmental impacts of bioenergy crop production and benefits of multifunctional bioenergy systems. *Bioenergy with Carbon Capture and Storage: Using Natural Resources for Sustainable Development*, 195–217. https://doi.org/10.1016/B978-0-12-816229-3.00010-7

Ali, A., & Alshmrany, S. (2020). Internet of Things (IoT) embedded smart sensors system for agriculture and farm management. *International Journal of Advanced and Applied Sciences*, *7*(10), 38–45. https://doi.org/10.21833/IJAAS.2020.10.005

Ali, A., Hussain, T., Tantashutikun, N., Hussain, N., & Cocetta, G. (2023). Application of smart techniques, Internet of Things and data mining for resource use efficient and sustainable crop production. *Agriculture*, *13*(2), 397. https://doi.org/10.3390/AGRICULTURE13020397

Ali, S., & Jang, C. M. (2020). Optimum design of hybrid renewable energy system for sustainable energy supply to a remote Island. *Sustainability*, *12*(3), 1280. https://doi.org/10.3390/SU12031280

Al-Jundi, H. A., & Tanbour, E. Y. (2022). A framework for fidelity evaluation of immersive virtual reality systems. *Virtual Reality*, *26*(3), 1103–1122. https://doi.org/10.1007/S10055-021-00618-Y

Al-Mahturi, A., Santoso, F., Garratt, M. A., & Anavatti, S. G. (2021). Modeling and control of a quadrotor unmanned aerial vehicle using type-2 fuzzy systems. In *Unmanned Aerial Systems: Theoretical Foundation and Applications: A Volume in Advances in Nonlinear Dynamics and Chaos (ANDC)* (pp. 25–46). https://doi.org/10.1016/B978-0-12-820276-0.00009-1

AL-Shammari, M. M. A., AL- Shamma'a, A. M., Al Maliki, A., Hussain, H. M., Yaseen, Z. M., & Armanuos, A. M. (2021). Integrated water harvesting and aquifer recharge evaluation methodology based on remote sensing and geographical information system: Case study in Iraq. *Natural Resources Research*, *30*(3), 2119–2143. https://doi.org/10.1007/S11053-021-09835-3

Androniceanu, A., & Sabie, O. M. (2022). Overview of green energy as a real strategic option for sustainable development. *Energies*, *15*(22), 8573. https://doi.org/10.3390/EN15228573

Ang, T. Z., Salem, M., Kamarol, M., Das, H. S., Nazari, M. A., & Prabaharan, N. (2022). A comprehensive study of renewable energy sources: Classifications, challenges and suggestions. *Energy Strategy Reviews*, *43*, 100939. https://doi.org/10.1016/J.ESR.2022.100939

Anna, E. K. (2018). *Adventures in Future Reality No. 2 — Pygmalion's Spectacles | by Anna Eva Kotyza | Medium.* https://medium.com/@aek452/pygmalions-spectacles-adventures-in-future-reality-no-2-d72465840a7a

Antar, M., Lyu, D., Nazari, M., Shah, A., Zhou, X., & Smith, D. L. (2021). Biomass for a sustainable bioeconomy: An overview of world biomass production and utilization. *Renewable and Sustainable Energy Reviews, 139*, 110691. https://doi.org/10.1016/J.RSER.2020.110691

Apeh, O. O., & Nwulu, N. I. (2025). Improving traceability and sustainability in the agri-food industry through blockchain technology: A bibliometric approach, benefits and challenges. *Energy Nexus, 17*, 100388. https://doi.org/10.1016/J.NEXUS.2025.100388

Arrigoni, S., Zangrandi, M., Bianchi, G., & Braghin, F. (2024). Control of a Hexapod Robot Considering Terrain Interaction. *Robotics, 13*(10), 142. https://doi.org/10.3390/ROBOTICS13100142

Arshi, O., & Mondal, S. (2023). Advancements in sensors and actuators technologies for smart cities: A comprehensive review. *Smart Construction and Sustainable Cities, 1*(1), 1–19. https://doi.org/10.1007/S44268-023-00022-2

Asachi, M., & Alonso Camargo-Valero, M. (2023). Multi-sensors data fusion for monitoring of powdered and granule products: Current status and future perspectives. *Advanced Powder Technology, 34*(7), 104055. https://doi.org/10.1016/J.APT.2023.104055

Aubert, B. A., Schroeder, A., & Grimaudo, J. (2012). IT as enabler of sustainable farming: An empirical analysis of farmers' adoption decision of precision agriculture technology. *Decision Support Systems, 54*(1), 510–520. https://doi.org/10.1016/J.DSS.2012.07.002

Azadi, H., Movahhed Moghaddam, S., Burkart, S., Mahmoudi, H., Van Passel, S., Kurban, A., & Lopez-Carr, D. (2021). Rethinking resilient agriculture: From climate-smart agriculture to vulnerable-smart agriculture. *Journal of Cleaner Production, 319*, 128602. https://doi.org/10.1016/J.JCLEPRO.2021.128602

Babatunde, O. M., Denwigwe, I. H., Adedoja, O. S., Babatunde, D. E., & Gbadamosi, S. L. (2019). Harnessing renewable energy for sustainable agricultural applications. *International Journal of Energy Economics and Policy, 9*(5), 308–315. https://doi.org/10.32479/IJEEP.7775

Babel, M. S., Shrestha, A., Anusart, K., & Shinde, V. (2021). Evaluating the potential for conserving water and energy in the water supply system of Bangkok. *Sustainable Cities and Society, 69*, 102857. https://doi.org/10.1016/J.SCS.2021.102857

Badzar, A., & Sternberg, H. (2016). *Blockchain for Securing Sustainable Transport Contracts and Supply Chain Transparency – An Explorative Study of Blockchain Technology in Logistics.* 2016–2021. https://lup.lub.lu.se/student-papers/record/8880383

Bakar, K. B. A., Zuhra, F. T., Isyaku, B., & Sulaiman, S. B. (2023). A review on the immediate advancement of the internet of things in wireless telecommunications. *IEEE Access, 11*, 21020–21048. https://doi.org/10.1109/ACCESS.2023.3250466

Balali, F., Nouri, J., Nasiri, A., & Zhao, T. (2020). Data analytics. *Data Intensive Industrial Asset Management*, 105–113. https://doi.org/10.1007/978-3-030-35930-0_7

Balzerkiewitz, H. P., & Stechert, C. (2020). The evolution of virtual reality towards the usage in early design phases. *Proceedings of the Design Society: DESIGN Conference, 1*, 91–100. https://doi.org/10.1017/DSD.2020.159

Barr, M. R., Volpe, R., & Kandiyoti, R. (2021). Liquid biofuels from food crops in transportation – A balance sheet of outcomes. *Chemical Engineering Science: X, 10*, 100090. https://doi.org/10.1016/J.CESX.2021.100090

Barron-Gafford, G. A., Pavao-Zuckerman, M. A., Minor, R. L., Sutter, L. F., Barnett-Moreno, I., Blackett, D. T., Thompson, M., Dimond, K., Gerlak, A. K., Nabhan, G. P., & Macknick, J. E. (2019). Agrivoltaics provide mutual benefits across the food–energy–water nexus in drylands. *Nature Sustainability, 2*(9), 848–855. https://doi.org/10.1038/S41893-019-0364-5

Barta, S., Gurrea, R., & Flavián, C. (2025). Augmented reality experiences: Consumer-centered augmented reality framework and research agenda. *Psychology & Marketing*, *42*(2), 634–650. https://doi.org/10.1002/MAR.22143

Bastidas, V., Reychav, I., Ofir, A., Bezbradica, M., & Helfert, M. (2022). Concepts for modeling smart cities: An archiMate extension. *Business and Information Systems Engineering*, *64*(3), 359–373. https://doi.org/10.1007/S12599-021-00724-W

Battah, A. (2019). *A User Study on User Experience of Spatial Audio in 360 Degree Music Videos*.

Batty, M., Axhausen, K. W., Giannotti, F., Pozdnoukhov, A., Bazzani, A., Wachowicz, M., Ouzounis, G., & Portugali, Y. (2012). Smart cities of the future. *The European Physical Journal Special Topics*, *214*(1), 481–518. https://doi.org/10.1140/EPJST/E2012-01703-3

Bayih, A. Z., Morales, J., Assabie, Y., & de By, R. A. (2022). Utilization of internet of things and wireless sensor networks for sustainable amallholder agriculture. *Sensors*, *22*(9), 3273. https://doi.org/10.3390/S22093273

Beatty, P. H., & Good, A. G. (2011). Plant science. Future prospects for cereals that fix nitrogen. *Science (New York)*, *333*(6041), 416–417. https://doi.org/10.1126/SCIENCE.1209467

Beaulieu, T. Y., Sarker, S., & Sarker, S. (2015). A conceptual framework for understanding crowdfunding. *Communications of the Association for Information Systems*, *37*, 1–31. https://doi.org/10.17705/1CAIS.03701

Beier. (n.d.). *UM-VRL: Virtual reality: A short introduction*. Retrieved August 28, 2024, from https://websites.umich.edu/~vrl/intro/

Berckmans, D. (2014). Precision livestock farming technologies for welfare management in intensive livestock systems. *Revue Scientifique et Technique (International Office of Epizootics)*, *33*(1), 189–196. https://doi.org/10.20506/RST.33.1.2273

Betz, U. A. K., Arora, L., Assal, R. A., Azevedo, H., Baldwin, J., Becker, M. S., Bostock, S., Cheng, V., Egle, T., Ferrari, N., Schneider-Futschik, E. K., Gerhardy, S., Hammes, A., Harzheim, A., Herget, T., Jauset, C., Kretschmer, S., Lammie, C., Kloss, N., … Zhao, G. (2023). Game changers in science and technology – Now and beyond. *Technological Forecasting and Social Change*, *193*, 122588. https://doi.org/10.1016/J.TECHFORE.2023.122588

Bharti, A., Paritosh, K., Mandla, V. R., Chawade, A., & Vivekanand, V. (2021). GIS application for the estimation of bioenergy potential from agriculture residues: An overview. *Energies*, *14*(4), 898. https://doi.org/10.3390/EN14040898

Bhat, S. A., & Huang, N. F. (2021). Big data and AI revolution in precision agriculture: Survey and challenges. *IEEE Access*, *9*, 110209–110222. https://doi.org/10.1109/ACCESS.2021.3102227

Bhatt, S. (2019). *(PDF) 172 AERC VVN SOLAR GUJ Report*. https://www.researchgate.net/publication/332383250_172_AERC_VVN_SOLAR_GUJ_Report

Bibri, S. E., & Krogstie, J. (2017). Smart sustainable cities of the future: An extensive interdisciplinary literature review. *Sustainable Cities and Society*, *31*, 183–212. https://doi.org/10.1016/J.SCS.2017.02.016

Biggs, E. M., Bruce, E., Boruff, B., Duncan, J. M. A., Horsley, J., Pauli, N., McNeill, K., Neef, A., Van Ogtrop, F., Curnow, J., Haworth, B., Duce, S., & Imanari, Y. (2015). Sustainable development and the water-energy-food nexus: A perspective on livelihoods. *Environmental Science & Policy*, *54*, 389–397. https://doi.org/10.1016/J.ENVSCI.2015.08.002

Biswas, K., Muthukkumarasamy, V., & Tan, W. L. (2017). *Blockchain Based Wine Supply Chain Traceability System*.

Biswas, S., & Visell, Y. (2021). Haptic perception, mechanics, and material technologies for virtual reality. *Advanced Functional Materials*, *31*(39). 2008186. https://doi.org/10.1002/ADFM.202008186

Blečić, I., Cuccu, S., Fanni, F. A., Frau, V., MacIs, R., Saiu, V., Senis, M., Spano, L. D., & Tola, A. (2021). First-person cinematographic videogames: Game model, authoring environment, and potential for creating affection for places. *Journal on Computing and Cultural Heritage*, *14*(2). https://doi.org/10.1145/3446977

Bódis, K., Kougias, I., Jäger-Waldau, A., Taylor, N., & Szabó, S. (2019). A high-resolution geospatial assessment of the rooftop solar photovoltaic potential in the European Union. *Renewable and Sustainable Energy Reviews*, *114*, 109309. https://doi.org/10.1016/J.RSER.2019.109309

Boehm, V. A. J., Kim, J., & Hong, J. W. K. (2018). Holistic tracking of products on the blockchain using NFC and verified users. *Lecture Notes in Computer Science (Including Subseries Lecture Notes in Artificial Intelligence and Lecture Notes in Bioinformatics)*, *10763 LNCS*, 184–195. https://doi.org/10.1007/978-3-319-93563-8_16

Bogdanov, D., Ram, M., Aghahosseini, A., Gulagi, A., Oyewo, A. S., Child, M., Caldera, U., Sadovskaia, K., Farfan, J., De Souza Noel Simas Barbosa, L., Fasihi, M., Khalili, S., Traber, T., & Breyer, C. (2021). Low-cost renewable electricity as the key driver of the global energy transition towards sustainability. *Energy*, *227*, 120467. https://doi.org/10.1016/J.ENERGY.2021.120467

Bohner, M., Jahangiri, J., Kiteto, M. K., & Mecha, C. A. (2024). Navigating the water–energy nexus: A mathematical approach. *Foundations*, *4*(4), 713–737. https://doi.org/10.3390/FOUNDATIONS4040045

Bolinger, M., Seel, J., & Robson, D. (2018). *Empirical Trends in Project Technology, Cost, Performance, and PPA Pricing in the United States-2018 Edition*. Lawrence Berkeley National Laboratory, Berkeley.

Boon, W., & Moors, E. (2008). Exploring emerging technologies using metaphors – A study of orphan drugs and pharmacogenomics. *Social Science & Medicine*, *66*(9), 1915–1927. https://doi.org/10.1016/J.SOCSCIMED.2008.01.012

Borge-Diez, D., García-Moya, F. J., & Rosales-Asensio, E. (2021). Comprehensive assessment of Gran Canaria water-energy-food nexus with GIS-based tool. *Journal of Cleaner Production*, *323*. https://doi.org/10.1016/J.JCLEPRO.2021.129197

Borkar, P. S., & Thakur, R. (2021). Smart environment monitoring models using cloud-based data analytics: A comprehensive study. *Machine Learning Approach for Cloud Data Analytics in IoT*, 227–271. https://doi.org/10.1002/9781119785873.CH10

Bramley, R. G. V. (2009). Lessons from nearly 20 years of precision agriculture research, development, and adoption as a guide to its appropriate application. *Crop and Pasture Science*, *60*(3), 197–217. https://doi.org/10.1071/CP08304

Britannica. (2024). *Virtual reality (VR) | Definition, Development, Technology, Examples, & Facts | Britannica*. https://www.britannica.com/technology/virtual-reality

Bronson, K., & Knezevic, I. (2016). Big Data in food and agriculture. *Big Data and Society*, *3*(1). https://doi.org/10.1177/2053951716648174

Brous, P., Janssen, M., & Herder, P. (2020). The dual effects of the Internet of Things (IoT): A systematic review of the benefits and risks of IoT adoption by organizations. *International Journal of Information Management*, *51*, 101952. https://doi.org/10.1016/J.IJINFOMGT.2019.05.008

Bukhary, S., Ahmad, S., & Batista, J. (2019). Analyzing land and water requirements for solar deployment in the Southwestern United States. *Renewable and Sustainable Energy Reviews*, *82*, 3288–3305. https://doi.org/10.1016/J.RSER.2017.10.016

Busse, M., Schwerdtner, W., Siebert, R., Doernberg, A., Kuntosch, A., König, B., & Bokelmann, W. (2015). Analysis of animal monitoring technologies in Germany from an innovation system perspective. *Agricultural Systems*, *138*, 55–65. https://doi.org/10.1016/j.agsy.2015.05.009

Calabi-Floody, M., Medina, J., Rumpel, C., Condron, L. M., Hernandez, M., Dumont, M., & Mora, M. L. (2018). Smart fertilizers as a strategy for sustainable agriculture. *Advances in Agronomy*, *147*, 119–157. https://doi.org/10.1016/BS.AGRON.2017.10.003

Camilo Gil-Carvajal, J., Soo Jo, E., Chul Park, D., Song, W., & Jeong, C. H. (2024). Driving simulator using virtual reality tools combining sound, vision, vibration, and motion. *Applied Acoustics*, *224*, 110137. https://doi.org/10.1016/J.APACOUST.2024.110137

Caprani, L. (2016). Five ways the sustainable development goals are better than the millennium development goals and why every educationalist should care. *Management in Education*, *30*(3), 102–104. https://doi.org/10.1177/0892020616653464

Carreño-Ortega, A., Galdeano-Gómez, E., Pérez-Mesa, J. C., & Galera-Quiles, M. D. C. (2017). Policy and environmental implications of photovoltaic systems in farming in Southeast Spain: Can greenhouses reduce the greenhouse effect? *Energies*, *10*(6), 761. https://doi.org/10.3390/EN10060761

Cassimon, D., Engelen, P.-J., & Van Cappellen, H. (2019). Globalization. *Encyclopedia of Law and Economics*, 972–980. https://doi.org/10.1007/978-1-4614-7753-2_194

Chamoso, P., González-Briones, A., Rodríguez, S., & Corchado, J. M. (2018). Tendencies of technologies and platforms in smart cities: A state-of-the-art review. *Wireless Communications and Mobile Computing*, *2018*(1), 3086854. https://doi.org/10.1155/2018/3086854

Chang, S., Saha, N., Castro-Lacouture, D., & Yang, P. P. J. (2019). Multivariate relationships between campus design parameters and energy performance using reinforcement learning and parametric modeling. *Applied Energy*, *249*, 253–264. https://doi.org/10.1016/j.apenergy.2019.04.109

Chen, L., Zheng, Y., Yu, J., Peng, Y., Li, R., & Han, S. (2024). A GIS-based approach for urban building energy modeling under climate change with high spatial and temporal resolution. *Energies*, *17*(17), 4313. https://doi.org/10.3390/EN17174313

Chen, X. H., Tee, K., Elnahass, M., & Ahmed, R. (2023). Assessing the environmental impacts of renewable energy sources: A case study on air pollution and carbon emissions in China. *Journal of Environmental Management*, *345*, 118525. https://doi.org/10.1016/J.JENVMAN.2023.118525

Cherni, H., Nicolas, S., & Métayer, N. (2021). Using virtual reality treadmill as a locomotion technique in a navigation task: Impact on user experience – Case of the KatWalk. *International Journal of Virtual Reality*, *21*(1), 1–14. https://doi.org/10.20870/IJVR.2021.21.1.3046

Chin, J., Callaghan, V., & Ben Allouch, S. (2019). The Internet-of-Things: Reflections on the past, present and future from a user-centered and smart environment perspective. *Journal of Ambient Intelligence and Smart Environments*, *11*(1), 45–69. https://doi.org/10.3233/AIS-180506

Cho, K., van Merriënboer, B., Bahdanau, D., & Bengio, Y. (2014). On the properties of neural machine translation: Encoder–decoder approaches. *Proceedings of SSST 2014 – 8th Workshop on Syntax, Semantics and Structure in Statistical Translation*, 103–111. https://doi.org/10.3115/V1/W14-4012

Choi, M., & Lee, I. (2017). Distributed energy resources monitoring and control based on IoT protocol. *International Conference on Information and Communication Technology Convergence: ICT Convergence Technologies Leading the Fourth Industrial Revolution, ICTC 2017*, *2017-December*, 1115–1117. https://doi.org/10.1109/ICTC.2017.8190871

Christidis, K., & Devetsikiotis, M. (2016). Blockchains and smart contracts for the Internet of Things. *IEEE Access*, *4*, 2292–2303. https://doi.org/10.1109/ACCESS.2016.2566339

Ciliberto, C., Szopik-Depczyńska, K., Tarczyńska-Łuniewska, M., Ruggieri, A., & Ioppolo, G. (2021). Enabling the circular economy transition: A sustainable lean manufacturing recipe for Industry 4.0. *Business Strategy and the Environment*, *30*(7), 3255–3272. https://doi.org/10.1002/BSE.2801

Clifton-Brown, J., Harfouche, A., Casler, M. D., Dylan Jones, H., Macalpine, W. J., Murphy-Bokern, D., Smart, L. B., Adler, A., Ashman, C., Awty-Carroll, D., Bastien, C., Bopper, S., Botnari, V., Brancourt-Hulmel, M., Chen, Z., Clark, L. V., Cosentino, S., Dalton, S., Davey, C., … Lewandowski, I. (2019). Breeding progress and preparedness for mass-scale deployment of perennial lignocellulosic biomass crops switchgrass, miscanthus, willow and poplar. *GCB Bioenergy*, *11*(1), 118–151. https://doi.org/10.1111/GCBB.12566

Climate Links. (2020). *Case Study – Lessons from Small-Scale Hydropower Development in eThekwini | Global Climate Change*. https://www.ctc-n.org/resources/powering-water-supply-south-africa-small-hydro-opportunities-existing-infrastructure#:~:text=BackgroundThe%20purpose%20of%20the%20project%20is%20to,using%20Durban%20City%20Council%20as%20a%20model.

Covaciu, F., Vaida, C., Gherman, B., Pisla, A., Tucan, P., & Pisla, D. (2024). Development of a virtual reality-based environment for telerehabilitation. *Applied Sciences*, *14*(24), 12022. https://doi.org/10.3390/APP142412022

Cowley, R., & Caprotti, F. (2019). Smart city as anti-planning in the UK. *Environment and Planning D: Society and Space*, *37*(3), 428–448. https://doi.org/10.1177/0263775818787506

Croman, K., Decker, C., Eyal, I., Gencer, A. E., Juels, A., Kosba, A., Miller, A., Saxena, P., Shi, E., Sirer, E. G., Song, D., & Wattenhofer, R. (2016). On scaling decentralized blockchains (A position paper). *Lecture Notes in Computer Science (Including Subseries Lecture Notes in Artificial Intelligence and Lecture Notes in Bioinformatics)*, *9604 LNCS*, 106–125. https://doi.org/10.1007/978-3-662-53357-4_8

Cui, X., & Gao, Z. (2020). A standard architecture of agricultural big data for deep learning. *Proceedings of 2020 IEEE International Conference on Advances in Electrical Engineering and Computer Applications, AEECA 2020*, 908–911. https://doi.org/10.1109/AEECA49918.2020.9213593

Cunningham, M. A., & Seidman, J. (2024). Competition for land: Equity and renewable energy in farmlands. *Land*, *13*(7), 939. https://doi.org/10.3390/LAND13070939

Dai, Y., Hasanefendic, S., & Bossink, B. (2024). A systematic literature review of the smart city transformation process: The role and interaction of stakeholders and technology. *Sustainable Cities and Society*, *101*, 105112. https://doi.org/10.1016/J.SCS.2023.105112

Dakshayini, M., & Balaji Prabhu, B. V. (2020). An effective big data and blockchain (BD-BC) based decision support model for sustainable agriculture system. *EAI/Springer Innovations in Communication and Computing*, 77–86. https://doi.org/10.1007/978-3-030-19562-5_8

D'amore, G., Di Vaio, A., Balsalobre-Lorente, D., & Boccia, F. (2022). Artificial intelligence in the water–energy–food model: A holistic approach towards sustainable development goals. *Sustainability*, *14*(2), 867. https://doi.org/10.3390/SU14020867

Dargan, S., Kumar, M., Ayyagari, M. R., & Kumar, G. (2020). A survey of deep learning and its applications: A new paradigm to machine learning. *Archives of Computational Methods in Engineering*, *27*(4), 1071–1092. https://doi.org/10.1007/S11831-019-09344-W

Davenport, T., Guha, A., Grewal, D., & Bressgott, T. (2020). How artificial intelligence will change the future of marketing. *Journal of the Academy of Marketing Science*, *48*(1), 24–42. https://doi.org/10.1007/S11747-019-00696-0

David, L. O., Aigbavboa, C., Nwulu, N., & Adepoju, O. O. (2024). Application of digital technologies of the fourth industrial revolution (4IR) in water, energy, and food (WEF) nexus. *Green Energy and Technology*, *Part F3772*, 287–317. https://doi.org/10.1007/978-3-031-72833-4_9

David, L. O., Nwulu, N. I., Aigbavboa, C. O., Adepoju, A., & Adepoju, O. O. (2025). Evaluating the use of blockchain technology and identifying critical success factors for the successful implementation of renewable energy projects in sub-Saharan Africa. *International Journal of Sustainable Energy*, *44*(1). https://doi.org/10.1080/14786451.2025.2449867

David, L. O., Nwulu, N. I., Aigbavboa, C. O., & Adepoju, O. C. (2022). Integrating fourth industrial revolution (4IR) technologies into the water, energy & food nexus for sustainable security: A bibliometric analysis. *Journal of Cleaner Production*, *363*, 132522. https://doi.org/10.1016/J.JCLEPRO.2022.132522

Davila Delgado, J. M., Oyedele, L., Demian, P., & Beach, T. (2020). A research agenda for augmented and virtual reality in architecture, engineering and construction. *Advanced Engineering Informatics*, *45*, 101122. https://doi.org/10.1016/J.AEI.2020.101122

De Oliveira, M. E., & Correa, C. G. (2020). Virtual reality and augmented reality applications in agriculture: A literature review. *Symposium on Virtual and Augmented Reality*, 1–9. https://doi.org/10.1109/SVR51698.2020.00017

de Sousa, T. B., Yamanari, J. S., & Guerrini, F. M. (2020). An enterprise model on sensing, smart, and sustainable (S^3) enterprises. *Gestao e Producao*, *27*(3). https://doi.org/10.1590/0104-530X5624-20

Debnath, D., Khanna, M., Rajagopal, D., & Zilberman, D. (2019). The future of biofuels in an electrifying global transportation sector: Imperative, prospects and challenges. *Applied Economic Perspectives and Policy*, *41*(4), 563–582. https://doi.org/10.1093/AEPP/PPZ023

Desai, H., Guruvayurappan, D., Merchant, M., Somaiya, S., & Mundra, H. (2017). IoT based grocery monitoring system. *IFIP International Conference on Wireless and Optical Communications Networks, WOCN*. https://doi.org/10.1109/WOCN.2017.8065839

Dey, N., Hassanien, A. E., Bhatt, C., Ashour, A. S., & Satapathy, S. C. (Eds.). (2018). *Internet of things and big data analytics toward next-generation intelligence. 30*. https://doi.org/10.1007/978-3-319-60435-0

Di Vaio, A., Hassan, R., & Alavoine, C. (2022). Data intelligence and analytics: A bibliometric analysis of human–Artificial intelligence in public sector decision-making effectiveness. *Technological Forecasting and Social Change*, *174*. https://doi.org/10.1016/j.techfore.2021.121201

Di Vaio, A., Palladino, R., Hassan, R., & Escobar, O. (2020). Artificial intelligence and business models in the sustainable development goals perspective: A systematic literature review. *Journal of Business Research*, *121*, 283–314. https://doi.org/10.1016/J.JBUSRES.2020.08.019

Diaz, A., Schöggl, J. P., Reyes, T., & Baumgartner, R. J. (2021). Sustainable product development in a circular economy: Implications for products, actors, decision-making support and lifecycle information management. *Sustainable Production and Consumption*, *26*, 1031–1045. https://doi.org/10.1016/J.SPC.2020.12.044

Dincelli, E., & Yayla, A. (2022). Immersive virtual reality in the age of the metaverse: A hybrid-narrative review based on the technology affordance perspective. *Journal of Strategic Information Systems*, *31*(2). https://doi.org/10.1016/j.jsis.2022.101717

Dixon, L., Lim, A., Grech-Sollars, M., Nandi, D., & Camp, S. (2022). Intraoperative ultrasound in brain tumor surgery: A review and implementation guide. *Neurosurgical Review*, *45*(4), 2503. https://doi.org/10.1007/S10143-022-01778-4

Dolci, R. (2017). IoT solutions for precision farming and food manufacturing: Artificial intelligence applications in digital food. *Proceedings – International Computer Software and Applications Conference*, *2*, 384–385. https://doi.org/10.1109/COMPSAC.2017.157

Duan, W., Chen, Y., Zou, S., & Nover, D. (2019). Managing the water-climate-food nexus for sustainable development in Turkmenistan. *Journal of Cleaner Production*, *220*, 212–224. https://doi.org/10.1016/J.JCLEPRO.2019.02.040

Duarah, P., Haldar, D., Patel, A. K., Dong, C. Di, Singhania, R. R., & Purkait, M. K. (2022). A review on global perspectives of sustainable development in bioenergy generation. *Bioresource Technology*, *348*. https://doi.org/10.1016/J.BIORTECH.2022.126791

Duarte, L., & Teodoro, A. C. (2021). GIS open-source plugins development: A 10-year bibliometric analysis on scientific literature. *Geomatics*, *1*(2), 206–245. https://doi.org/10.3390/GEOMATICS1020013

Dupraz, C., Marrou, H., Talbot, G., Dufour, L., Nogier, A., & Ferard, Y. (2011). Combining solar photovoltaic panels and food crops for optimising land use: Towards new agrivoltaic schemes. *Renewable Energy*, *36*(10), 2725–2732. https://doi.org/10.1016/J.RENENE.2011.03.005

Durga, N., Schmitter, P., Ringler, C., Mishra, S., Magombeyi, M. S., Ofosu, A., Pavelic, P., Hagos, F., Melaku, D., Verma, S., Minh, T., & Matambo, C. (2024). Barriers to the uptake of solar-powered irrigation by smallholder farmers in sub-saharan Africa: A review. *Energy Strategy Reviews*, *51*, 101294. https://doi.org/10.1016/J.ESR.2024.101294

Dwivedi, P., Wang, W., Hudiburg, T., Jaiswal, D., Parton, W., Long, S., Delucia, E., & Khanna, M. (2015). Cost of abating greenhouse gas emissions with cellulosic ethanol. *Environmental Science & Technology*, *49*(4), 2512–2522. https://doi.org/10.1021/ES5052588

Dwivedi, Y. K., Hughes, L., Ismagilova, E., Aarts, G., Coombs, C., Crick, T., Duan, Y., Dwivedi, R., Edwards, J., Eirug, A., Galanos, V., Ilavarasan, P. V., Janssen, M., Jones, P., Kar, A. K., Kizgin, H., Kronemann, B., Lal, B., Lucini, B., … Williams, M. D. (2021). Artificial intelligence (AI): Multidisciplinary perspectives on emerging challenges, opportunities, and agenda for research, practice and policy. *International Journal of Information Management*, *57*, 101994. https://doi.org/10.1016/J.IJINFOMGT.2019.08.002

Dwivedi, Y. K., Ismagilova, E., Hughes, D. L., Carlson, J., Filieri, R., Jacobson, J., Jain, V., Karjaluoto, H., Kefi, H., Krishen, A. S., Kumar, V., Rahman, M. M., Raman, R., Rauschnabel, P. A., Rowley, J., Salo, J., Tran, G. A., & Wang, Y. (2021). Setting the future of digital and social media marketing research: Perspectives and research propositions. *International Journal of Information Management*, *59*, 102168. https://doi.org/10.1016/J.IJINFOMGT.2020.102168

Prabakaran, G., Vaithiyanathan, D., & Ganesan, M. (2018). Fuzzy decision support system for improving the crop productivity and efficient use of fertilizers. *Computers and Electronics in Agriculture*, *150*, 88–97. https://doi.org/10.1016/j.compag.2018.03.030

EL-Mesery, H. S., EL-Seesy, A. I., Hu, Z., & Li, Y. (2022). Recent developments in solar drying technology of food and agricultural products: A review. *Renewable and Sustainable Energy Reviews*, *157*. https://doi.org/10.1016/J.RSER.2021.112070

Endo, A., Burnett, K., Orencio, P. M., Kumazawa, T., Wada, C. A., Ishii, A., Tsurita, I., & Taniguchi, M. (2015). Methods of the water-energy-food nexus. *Water*, *7*(10), 5806–5830. https://doi.org/10.3390/W7105806

Erkenntnis. (2021). *Digital SkillUp – What Are emerging Technologies and Why Are They Relevant?* https://www.digitalskillup.eu/de/articles/what-are-emerging-technologies/

Ersoy, P., Börühan, G., Kumar Mangla, S., Hormazabal, J. H., Kazancoglu, Y., & Lafcı, Ç. (2022). Impact of information technology and knowledge sharing on circular food supply chains for green business growth. *Business Strategy and the Environment*, *31*(5), 1875–1904. https://doi.org/10.1002/BSE.2988

Ertz, M., & Boily, É. (2019). The rise of the digital economy: Thoughts on blockchain technology and cryptocurrencies for the collaborative economy. *International Journal of Innovation Studies*, *3*(4), 84–93. https://doi.org/10.1016/J.IJIS.2019.12.002

Factors, W., Jian, M.-S., Cui, L., & Wang, W. (2023). Factors affecting the adoption of digital technology by farmers in China: A systematic literature review. *Sustainability*, *15*(20), 14824. https://doi.org/10.3390/SU152014824

Fakunle, S. O., Okunola, J. L., & Ajani, B. K. (2023). Technological advancement, globalization, and developing countries. *The Palgrave Handbook of Global Social Change*, 1–17. https://doi.org/10.1007/978-3-030-87624-1_160-1

Fan, X., Naz, M., Fan, X., Xuan, W., Miller, A. J., & Xu, G. (2017). Plant nitrate transporters: From gene function to application. *Journal of Experimental Botany*, *68*(10), 2463–2475. https://doi.org/10.1093/JXB/ERX011

Farrell, A. E., Plevin, R. J., Turner, B. T., Jones, A. D., O'Hare, M., & Kammen, D. M. (2006). Ethanol can contribute to energy and environmental goals. *Science (New York)*, *311*(5760), 506–508. https://doi.org/10.1126/SCIENCE.1121416

Faulkner, A., Mchenry, G., & Analyst, J. (2014). *Agriculture Gets Smart: The Rise of Data and Robotics.*

Fawzy, A. (2020). *(PDF) Desalination Using Renewable Energy*. https://www.researchgate.net/publication/341099488_Desalination_Using_Renewable_Energy

Ferreira, A., Sousa, V., Pinheiro, M., Meireles, I., Silva, C. M., Brito, J., & Mateus, R. (2023). Potential of rainwater harvesting in the retail sector A case study in Portugal. *Environmental Science and Pollution Research*, *30*(14), 42427–42442. https://doi.org/10.1007/S11356-023-25137-Y/FIGURES/10

Finger, R., Swinton, S. M., El Benni, N., & Walter, A. (2019). Precision farming at the nexus of agricultural production and the environment. *Annual Review of Resource Economics*, *11*, 313–335. https://doi.org/10.1146/annurev-resource-100518-093929

Foresti, R., Rossi, S., Magnani, M., Guarino Lo Bianco, C., & Delmonte, N. (2020). Smart society and artificial intelligence: Big data scheduling and the global standard method applied to smart maintenance. *Engineering*, *6*(7), 835–845. https://doi.org/10.1016/J.ENG.2019.11.014

Formolli, M., Kleiven, T., & Lobaccaro, G. (2023). Assessing solar energy accessibility at high latitudes: A systematic review of urban spatial domains, metrics, and parameters. *Renewable and Sustainable Energy Reviews*, *177*, 113231. https://doi.org/10.1016/J.RSER.2023.113231

Friedman, N., & Ormiston, J. (2022). Blockchain as a sustainability-oriented innovation?: Opportunities for and resistance to blockchain technology as a driver of sustainability in global food supply chains. *Technological Forecasting and Social Change*, *175*, 121403. https://doi.org/10.1016/J.TECHFORE.2021.121403

Gagnon, P., Margolis, R., Melius, J., Phillips, C., & Elmore, R. (2016). *Rooftop Solar Photovoltaic Technical Potential in the United States. A Detailed Assessment.* National Renewable Energy Laboratory. https://doi.org/10.2172/1236153

Gallardo, R. K., & Sauer, J. (2018). Adoption of labor-saving technologies in agriculture. *Annual Review of Resource Economics*, *10*, 185–206. https://doi.org/10.1146/ANNUREV-RESOURCE-100517-023018

García-Gil, D., Luengo, J., García, S., & Herrera, F. (2019). Enabling smart data: Noise filtering in big data classification. *Information Sciences*, *479*, 135–152. https://doi.org/10.1016/J.INS.2018.12.002

Gbashi, S., & Njobeh, P. B. (2024). Enhancing food integrity through artificial intelligence and machine learning: A comprehensive review. *Applied Sciences*, *14*(8), 3421. https://doi.org/10.3390/APP14083421

Geographic Information System (GIS): Definition, Development, Applications & Components. (2022). https://www.researchgate.net/publication/340182760_Geographic_Information_System_GIS_Definition_Development_Applications_Components

Gewirtz, D. (2018). *Volume, Velocity, and Variety: Understanding the Three V's of Big Data.* ZDNET. https://www.zdnet.com/article/volume-velocity-and-variety-understanding-the-three-vs-of-big-data/#google_vignette

Ghatrehsamani, S., Wade, T., & Ampatzidis, Y. (2020). The adoption of precision agriculture technologies by Florida growers: A comparison of 2005 and 2018 survey data. *Acta Horticulturae*, *1*(1279), 311–316. https://doi.org/10.17660/ACTAHORTIC.2020.1279.44

Gielen, D., Boshell, F., Saygin, D., Bazilian, M. D., Wagner, N., & Gorini, R. (2019). The role of renewable energy in the global energy transformation. *Energy Strategy Reviews*, *24*, 38–50. https://doi.org/10.1016/J.ESR.2019.01.006

Gil, D., Johnsson, M., Mora, H., & Szymański, J. (2019). Review of the complexity of managing big data of the Internet of Things. *Complexity*, *2019*. https://doi.org/10.1155/2019/4592902

Giller, K. E., Delaune, T., Silva, J. V., Descheemaeker, K., van de Ven, G., Schut, A. G. T., van Wijk, M., Hammond, J., Hochman, Z., Taulya, G., Chikowo, R., Narayanan, S., Kishore, A., Bresciani, F., Teixeira, H. M., Andersson, J. A., & van Ittersum, M. K. (2021). The future of farming: Who will produce our food? *Food Security*, *13*(5), 1073–1099. https://doi.org/10.1007/S12571-021-01184-6

Giusti, E., & Marsili-Libelli, S. (2015). A fuzzy decision support system for irrigation and water conservation in agriculture. *Environmental Modelling & Software*, *63*, 73–86. https://doi.org/10.1016/J.ENVSOFT.2014.09.020

Goldberg, Z. A. (2023). Solar energy development on farmland: Three prevalent perspectives of conflict, synergy and compromise in the United States. *Energy Research & Social Science*, *101*, 103145. https://doi.org/10.1016/j.erss.2023.103145

Gomiero, T. (2018). Large-scale biofuels production: A possible threat to soil conservation and environmental services. *Applied Soil Ecology*, *123*, 729–736. https://doi.org/10.1016/J.APSOIL.2017.09.028

Gonzalez Sanchez, R., Seliger, R., Fahl, F., De Felice, L., Ouarda, T. B. M. J., & Farinosi, F. (2020). Freshwater use of the energy sector in Africa. *Applied Energy*, *270*, 115171. https://doi.org/10.1016/J.APENERGY.2020.115171

Gracias, J. S., Parnell, G. S., Specking, E., Pohl, E. A., & Buchanan, R. (2023). Smart cities—A structured literature review. *Smart Cities*, *6*(4), 1719–1743. https://doi.org/10.3390/SMARTCITIES6040080

Guerin, T. F. (2019). Impacts and opportunities from large-scale solar photovoltaic (PV) electricity generation on agricultural production. *Environmental Quality Management*. https://doi.org/10.1002/TQEM.21629

Guo, W., Che, L., Shahidehpour, M., & Wan, X. (2021). Machine-learning based methods in short-term load forecasting. *The Electricity Journal*, *34*(1), 106884. https://doi.org/10.1016/J.TEJ.2020.106884

Gupta, V. (2017). *A Brief History of Blockchain*. https://hbr.org/2017/02/a-brief-history-of-blockchain

Haenlein, M., Kaplan, A., Tan, C. W., & Zhang, P. (2019). Artificial intelligence (AI) and management analytics. *Journal of Management Analytics*, *6*(4), 341–343. https://doi.org/10.1080/23270012.2019.1699876

Haessler, P., Giones, F., & Brem, A. (2023). The who and how of commercializing emerging technologies: A technology-focused review. *Technovation*, *121*, 102637. https://doi.org/10.1016/J.TECHNOVATION.2022.102637

Hammi, B., Khatoun, R., Zeadally, S., Fayad, A., & Khoukhi, L. (2018). IoT technologies for smart cities. *IET Networks*, *7*(1), 1–13. https://doi.org/10.1049/IET-NET.2017.0163

Han, C. S., Lotfi Ghahroud, M., Park, M. J., & Ghassemi Nejad, J. (2023). A new blockchain investment and energy certificate platform. *Cogent Engineering*, *10*(2). https://doi.org/10.1080/23311916.2023.2260226

Han, T., Jiang, D., Zhao, Q., Wang, L., & Yin, K. (2018). Comparison of random forest, artificial neural networks and support vector machine for intelligent diagnosis of rotating machinery. *Transactions of the Institute of Measurement and Control*, *40*(8), 2681–2693. https://doi.org/10.1177/0142331217708242

Hannan, M. A., Begum, R. A., Abdolrasol, M. G., Hossain Lipu, M. S., Mohamed, A., & Rashid, M. M. (2018). Review of baseline studies on energy policies and indicators in Malaysia for future sustainable energy development. *Renewable and Sustainable Energy Reviews*, *94*, 551–564. https://doi.org/10.1016/J.RSER.2018.06.041

Hashem, I. A. T., Anuar, N. B., Gani, A., Yaqoob, I., Xia, F., & Khan, S. U. (2016). MapReduce: Review and open challenges. *Scientometrics*, *109*(1), 389–422. https://doi.org/10.1007/S11192-016-1945-Y

Hemeida, M. G., Hemeida, A. M., Senjyu, T., & Osheba, D. (2022). Renewable energy resources technologies and life cycle assessment: Review. *Energies*, *15*(24), 9417. https://doi.org/10.3390/EN15249417

Hempleman, J. M. (2022). *Digital Commons @ DU Digital Commons @ DU Sustainable Rural Development: Is It Possible to Boost Rural Sustainable Rural Development: Is It Possible to Boost Rural Economies While Protecting the Environment? Economies While Protecting the Environment?* https://digitalcommons.du.edu/

Herrando, M., Pantaleo, A. M., Wang, K., & Markides, C. N. (2019). Solar combined cooling, heating and power systems based on hybrid PVT, PV or solar-thermal collectors for building applications. *Renewable Energy*, *143*, 637–647. https://doi.org/10.1016/J.RENENE.2019.05.004

Hey, G., Guyot, M., Carter, A., & Lucke-Wold, B. (2023). Augmented reality in neurosurgery: A new paradigm for training. *Medicina, 59*(10). https://doi.org/10.3390/MEDICINA59101721

Hochreiter, S., & Schmidhuber, J. (1997). Long short-term memory. *Neural Computation*, *9*(8), 1735–1780. https://doi.org/10.1162/NECO.1997.9.8.1735

Hopkins, J. L. (2021). An investigation into emerging industry 4.0 technologies as drivers of supply chain innovation in Australia. *Computers in Industry*, *125*, 103323. https://doi.org/10.1016/J.COMPIND.2020.103323

Hossain, Md. F. (2019). Introduction. *Sustainable Design and Build*, 1–10. https://doi.org/10.1016/B978-0-12-816722-9.00001-X

Hudiburg, T. W., Wang, W., Khanna, M., Long, S. P., Dwivedi, P., Parton, W. J., Hartman, M., & Delucia, E. H. (2016). Impacts of a 32-billion-gallon bioenergy landscape on land and fossil fuel use in the US. *Nature Energy*, *1*(1), 15005. https://doi.org/10.1038/NENERGY.2015.5

Huuskonen, J., & Oksanen, T. (2018). Soil sampling with drones and augmented reality in precision agriculture. *Computers and Electronics in Agriculture*, *154*, 25–35. https://doi.org/10.1016/J.COMPAG.2018.08.039

icct. (2023). *Home – International Council on Clean Transportation*. Icct. https://theicct.org/

Iheanachor, N., Umukoro, I. O., & David-West, O. (2021). The role of product development practices on new product performance: Evidence from Nigeria's financial services providers. *Technological Forecasting and Social Change*, *164*, 120470. https://doi.org/10.1016/J.TECHFORE.2020.120470

Ikegwu, A. C., Nweke, H. F., Anikwe, C. V., Alo, U. R., & Okonkwo, O. R. (2022). Big data analytics for data-driven industry: A review of data sources, tools, challenges, solutions, and research directions. *Cluster Computing*, *25*(5), 3343–3387. https://doi.org/10.1007/S10586-022-03568-5

Iqbal, R., Doctor, F., More, B., Mahmud, S., & Yousuf, U. (2020). Big data analytics: Computational intelligence techniques and application areas. *Technological Forecasting and Social Change*, *153*, 119253. https://doi.org/10.1016/J.TECHFORE.2018.03.024

IRENA. (2014). *(PDF) Setting Renewable Energy Targets*. https://www.researchgate.net/publication/322420240_Setting_Renewable_Energy_Targets

Islam, M. M., Nooruddin, S., Karray, F., & Muhammad, G. (2023). Internet of Things: Device capabilities, architectures, protocols, and smart applications in healthcare domain. *IEEE Internet of Things Journal*, *10*(4), 3611–3641. https://doi.org/10.1109/JIOT.2022.3228795

Issa Zadeh, S. B., & Garay-Rondero, C. L. (2023). Enhancing urban sustainability: Unravelling carbon footprint reduction in smart cities through modern supply-chain measures. *Smart Cities*, *6*(6), 3225–3250. https://doi.org/10.3390/SMARTCITIES6060143

Jaeger, J. (2021). *Explaining the Exponential Growth of Renewable Energy*. https://www.wri.org/insights/growth-renewable-energy-sector-explained

Jaganathan, D., Ramasamy, K., Sellamuthu, G., Jayabalan, S., & Venkataraman, G. (2018). CRISPR for crop improvement: An update review. *Frontiers in Plant Science*, *9*, 364675. https://doi.org/10.3389/FPLS.2018.00985

Jaiswal, K. K., Chowdhury, C. R., Yadav, D., Verma, R., Dutta, S., Jaiswal, K. S., Sangmesh B, & Karuppasamy, K. S. K. (2022). Renewable and sustainable clean energy development and impact on social, economic, and environmental health. *Energy Nexus*, *7*. https://doi.org/10.1016/J.NEXUS.2022.100118

Jamwal, A., Agrawal, R., & Sharma, M. (2022). Deep learning for manufacturing sustainability: Models, applications in Industry 4.0 and implications. *International Journal of Information Management Data Insights*, *2*(2), 100107. https://doi.org/10.1016/J.JJIMEI.2022.100107

Javaid, A., Rasool, S., & Maqsood, A. (2024). Analysis of visual and vestibular information on motion sickness in flight simulation. *Aerospace*, *11*(2), 139. https://doi.org/10.3390/AEROSPACE11020139

Javaid, M., Haleem, A., Rab, S., Pratap Singh, R., & Suman, R. (2021). Sensors for daily life: A review. *Sensors International*, *2*, 100121. https://doi.org/10.1016/J.SINTL.2021.100121

Javan, K., Altaee, A., BaniHashemi, S., Darestani, M., Zhou, J., & Pignatta, G. (2024). A review of interconnected challenges in the water–energy–food nexus: Urban pollution perspective towards sustainable development. *Science of The Total Environment*, *912*, 169319. https://doi.org/10.1016/J.SCITOTENV.2023.169319

Jia, Y., & Wang, Z. (2024). Application of artificial intelligence based on the fuzzy control algorithm in enterprise innovation. *Heliyon*, *10*(6), e28116. https://doi.org/10.1016/J.HELIYON.2024.E28116

Joardder, M. U. H., & Islam, M. K. (2022). Thermophysical aspects of cooking processes. In *High-Temperature Processing of Food Products: Unit Operations and Processing Equipment in the Food Industry* (pp. 47–57). https://doi.org/10.1016/B978-0-12-818618-3.00009-4

Johansson, I., Stenqvist, C., & Thollander, P. (2017). *Energy Efficiency Networks for SMEs – Program Theory and Ongoing Evaluation*. https://urn.kb.se/resolve?urn=urn:nbn:se:hig:diva–26503

Jordan, C. F. (2013). *An Ecosystem Approach to Sustainable Agriculture*. *1*. https://doi.org/10.1007/978-94-007-6790-4

Julieta Cedillo-Elias, E., Antonio Orizaga-Trejo, J., Larios, V. M., & Maciel Arellano, L. A. (2018). Smart Government infrastructure based in SDN Networks: The case of Guadalajara Metropolitan Area. *2018 IEEE International Smart Cities Conference, ISC2 2018*. https://doi.org/10.1109/ISC2.2018.8656801

Juszczyk, O., & Shahzad, K. (2022). Blockchain technology for renewable energy: Principles, applications and prospects. *Energies*, *15*(13), 4603. https://doi.org/10.3390/EN15134603

Kadhim, N., Mourshed, M., & Bray, M. (2016). Advances in remote sensing applications for urban sustainability. *Euro-Mediterranean Journal for Environmental Integration*, *1*(1), 1–22. https://doi.org/10.1007/S41207-016-0007-4

Kalsoom, T., Ramzan, N., Ahmed, S., & Ur-Rehman, M. (2020). Advances in sensor technologies in the era of smart factory and industry 4.0. *Sensors*, *20*(23), 6783. https://doi.org/10.3390/S20236783

Kamath, R. (2018). Food traceability on blockchain: Walmart's pork and mango pilots with IBM. *The Journal of The British Blockchain Association*, *1*(1), 1–12. https://doi.org/10.31585/JBBA-1-1-(10)2018

Kamilaris, A., Kartakoullis, A., & Prenafeta-Boldú, F. X. (2017). A review on the practice of big data analysis in agriculture. *Computers and Electronics in Agriculture*, *143*, 23–37. https://doi.org/10.1016/J.COMPAG.2017.09.037

Kanwal, T., Rehman, S. U., Ali, T., Mahmood, K., Villar, S. G., Lopez, L. A. D., & Ashraf, I. (2023). An intelligent dual-axis solar tracking system for remote weather monitoring in the agricultural field. *Agriculture*, *13*(8), 1600. https://doi.org/10.3390/AGRICULTURE13081600

Karavolias, N. G., Horner, W., Abugu, M. N., & Evanega, S. N. (2021). Application of gene editing for climate change in agriculture. *Frontiers in Sustainable Food Systems*, *5*, 685801. https://doi.org/10.3389/FSUFS.2021.685801

Kashiwao, T., Nakayama, K., Ando, S., Ikeda, K., Lee, M., & Bahadori, A. (2017). A neural network-based local rainfall prediction system using meteorological data on the internet: A case study using data from the Japan Meteorological Agency. *Applied Soft Computing*, *56*, 317–330. https://doi.org/10.1016/J.ASOC.2017.03.015

Katuk, N., Wan Abdullah, W. A. N., Sugiharto, T., & Ahmad, I. (2023). Smart technology: Ecosystem, impacts, challenges and the path forward. *Information System and Smart City*, *1*(1). https://doi.org/10.59400/ISSC.V1I1.63

Kern, A. C., & Ellermeier, W. (2020). Audio in VR: Effects of a soundscape and movement-triggered step sounds on presence. *Frontiers in Robotics and AI*, *7*, 494828. https://doi.org/10.3389/FROBT.2020.00020

Khanna, M. (2021). Digital transformation of the agricultural sector: Pathways, drivers and policy implications. *Applied Economic Perspectives and Policy*, *43*(4), 1221–1242. https://doi.org/10.1002/AEPP.13103

Khanna, M., & Crago, C. L. (2012). Measuring indirect land use change with biofuels: Implications for policy. *Annual Review of Resource Economics*, *4*, 161–184. https://doi.org/10.1146/annurev-resource-110811-114523

Khatoon, A., Verma, P., Southernwood, J., Massey, B., & Corcoran, P. (2019). Blockchain in energy efficiency: Potential applications and benefits. *Energies*, *12*(17), 3317. https://doi.org/10.3390/EN12173317

Kibria, M. G., Nguyen, K., Villardi, G. P., Zhao, O., Ishizu, K., & Kojima, F. (2018). Big data analytics, machine learning, and artificial intelligence in next-generation wireless networks. *IEEE Access*, *6*, 32328–32338. https://doi.org/10.1109/ACCESS.2018.2837692

Kim, D., Yoon, Y., Lee, J., Mago, P. J., Lee, K., & Cho, H. (2022). Design and implementation of smart buildings: A review of current research trend. *Energies*, *15*(12), 4278. https://doi.org/10.3390/EN15124278

Kim, J., Kim, S., Ju, C., & Son, H. I. (2019). Unmanned aerial vehicles in agriculture: A review of perspective of platform, control, and applications. *IEEE Access*, *7*, 105100–105115. https://doi.org/10.1109/ACCESS.2019.2932119

Kim, J., Park, H., Seo, C., Kim, H., Choi, G., Kim, M., Kim, B., & Lee, W. (2024). Sustainable and inflatable aeroponics smart farm system for water efficiency and high-value crop production. *Applied Sciences (Switzerland)*, *14*(11), 4931. https://doi.org/10.3390/APP14114931

Kim, M., Kim, J., Jeong, K., & Kim, C. (2020). Grasping VR: Presence of pseudo-haptic interface based portable hand grip system in immersive virtual reality. *International Journal of Human-Computer Interaction*, *36*(7), 685–698. https://doi.org/10.1080/10447318.2019.1680920

Kirli, D., Couraud, B., Robu, V., Salgado-Bravo, M., Norbu, S., Andoni, M., Antonopoulos, I., Negrete-Pincetic, M., Flynn, D., & Kiprakis, A. (2022). Smart contracts in energy systems: A systematic review of fundamental approaches and implementations. *Renewable and Sustainable Energy Reviews*, *158*, 112013. https://doi.org/10.1016/J.RSER.2021.112013

Kirui, O. (2019). The agricultural mechanization in Africa: Micro-level analysis of state drivers and effects. *SSRN Electronic Journal*. https://doi.org/10.2139/SSRN.3368103

Klamt, T., Schwarz, M., Lenz, C., Baccelliere, L., Buongiorno, D. Cichon, T., DiGuardo, A., Droeschel, D., Gabardi, M., Kamedula, M., Kashiri, N., Laurenzi, A., Leonardis, D., Muratore, L., Pavlichenko, D., Periyasamy, A. S., Rodriguez, D., Solazzi, M.,

Frisoli, A., … Behnke, S. (2020). Remote mobile manipulation with the centauro robot: Full-body telepresence and autonomous operator assistance. *Journal of Field Robotics*, *37*(5), 889–919. https://doi.org/10.1002/ROB.21895

Kolawole, I. D., Kolawole, G. O., Sanni-manuel, B. A., Kolawole, S. K., Ewansiha, J. U., Kolawole, V. A., & Kolawole, F. O. (2024). Economic impact of waste from food, water, and agriculture in Nigeria: Challenges, implications, and applications—a review. *Discover Environment*, *2*(1), 1–28. https://doi.org/10.1007/S44274-024-00086-6

Krishna, R. R., Priyadarshini, A., Jha, A. V., Appasani, B., Srinivasulu, A., & Bizon, N. (2021). State-of-the-art review on IoT threats and attacks: Taxonomy, challenges and solutions. *Sustainability*, *13*(16), 9463. https://doi.org/10.3390/SU13169463

Krizhevsky, A., Sutskever, I., & Hinton, G. E. (2012). ImageNet classification with deep convolutional neural networks. *Advances in Neural Information Processing Systems*, *25*. https://code.google.com/p/cuda-convnet/

Kshetri, N. (2018). 1 Blockchain's roles in meeting key supply chain management objectives. *International Journal of Information Management*, *39*, 80–89. https://doi.org/10.1016/J.IJINFOMGT.2017.12.005

Kufel, J., Bargieł-Łączek, K., Kocot, S., Koźlik, M., Bartnikowska, W., Janik, M., Czogalik, Ł., Dudek, P., Magiera, M., Lis, A., Paszkiewicz, I., Nawrat, Z., Cebula, M., & Gruszczyńska, K. (2023). What is machine learning, artificial neural networks and deep learning?—Examples of practical applications in medicine. *Diagnostics*, *13*(15). https://doi.org/10.3390/DIAGNOSTICS13152582

Kul, B., & Şen, M. (2017). Energy saving IoT-based advanced load limiter. *2017 26th International Scientific Conference Electronics, ET 2017 – Proceedings*, *2017-January*, 1–5. https://doi.org/10.1109/ET.2017.8124362

Kumar, N. M., Chand, A. A., Malvoni, M., Prasad, K. A., Mamun, K. A., Islam, F. R., & Chopra, S. S. (2020). Distributed energy resources and the application of AI, IoT, and blockchain in smart grids. *Energies*, *13*(21), 5739. https://doi.org/10.3390/EN13215739

Kumar, N. M., Dash, A., & Singh, N. K. (2018). Internet of Things (IoT): An opportunity for energy-food-water nexus. *2018 International Conference on Power Energy, Environment and Intelligent Control, PEEIC 2018*, 68–72. https://doi.org/10.1109/PEEIC.2018.8665632

Kumar, N. M., & Mallick, P. K. (2018). Blockchain technology for security issues and challenges in IoT. *Procedia Computer Science*, *132*, 1815–1823. https://doi.org/10.1016/J.PROCS.2018.05.140

Kumar, V., Ashraf, A. R., & Nadeem, W. (2024). AI-powered marketing: What, where, and how? *International Journal of Information Management*, *77*, 102783. https://doi.org/10.1016/J.IJINFOMGT.2024.102783

Kunst, R., Ramos, G., Righ, R., André da Costa, C., Pignaton, E., Binotto, A., Favilla, J., Ohta, R., & High, R. (2022). Industry 4.0: Applications and future perspectives. *Handbook of Smart Materials, Technologies, and Devices: Applications of Industry 4.0: Volume 1–3*, *2*, 1277–1306. https://doi.org/10.1007/978-3-030-84205-5_33

Kussul, N., Lavreniuk, M., Skakun, S., & Shelestov, A. (2017). Deep learning classification of land cover and crop types using remote sensing data. *IEEE Geoscience and Remote Sensing Letters*, *14*(5), 778–782. https://doi.org/10.1109/LGRS.2017.2681128

La Rosa, D., Barbarossa, L., Privitera, R., & Martinico, F. (2014). Agriculture and the city: A method for sustainable planning of new forms of agriculture in urban contexts. *Land Use Policy*, *41*, 290–303. https://doi.org/10.1016/J.LANDUSEPOL.2014.06.014

Langholtz, M. H., Stokes, B. J., & Eaton, L. M. (2016). 2016 Billion-ton report: Advancing domestic resources for a thriving bioeconomy, Volume 1: Economic availability of feedstock. *Oak Ridge National Laboratory, Oak Ridge, Tennessee, Managed by UT-Battelle, LLC for the U.S. Department Of Energy*, *2016*, 1–411. https://research.fs.usda.gov/treesearch/53253

Lazaro, L. L. B., Bellezoni, R. A., Puppim de Oliveira, J. A., Jacobi, P. R., & Giatti, L. L. (2022). Ten years of research on the water-energy-food nexus: An analysis of topics evolution. *Frontiers in Water*, *4*, 859891. https://doi.org/10.3389/FRWA.2022.859891

Leck, H., Conway, D., Bradshaw, M., & Rees, J. (2015). Tracing the water–energy–food nexus: Description, theory and practice. *Geography Compass*, *9*(8), 445–460. https://doi.org/10.1111/GEC3.12222

Lee, G. (2019). What roles should the government play in fostering the advancement of the Internet of Things? *Telecommunications Policy*, *43*(5), 434–444. https://doi.org/10.1016/j.telpol.2018.12.002

Lee, J., Babcock, J., Pham, T. S., Bui, T. H., & Kang, M. (2023). Smart city as a social transition towards inclusive development through technology: A tale of four smart cities. *International Journal of Urban Sciences*, *27*(S1), 75–100. https://doi.org/10.1080/12265934.2022.2074076

Leng, K., Bi, Y., Jing, L., Fu, H. C., & Van Nieuwenhuyse, I. (2018). Research on agricultural supply chain system with double chain architecture based on blockchain technology. *Future Generation Computer Systems*, *86*, 641–649. https://doi.org/10.1016/J.FUTURE.2018.04.061

Lezoche, M., Panetto, H., Kacprzyk, J., Hernandez, J. E., & Alemany Díaz, M. M. E. (2020). Agri-food 4.0: A survey of the supply chains and technologies for the future agriculture. *Computers in Industry*, *117*. https://doi.org/10.1016/J.COMPIND.2020.103187

Li, M., Li, X., Zhou, Z., Wu, P., Fang, M., Pan, X., Lin, Q., Luo, W., Wu, G., & Li, H. (2016). Reassessment of the four yield-related genes Gn1a, DEP1, GS3, and IPA1 in rice using a CRISPR/Cas9 system. *Frontiers in Plant Science*, *7*(MAR2016), 186370. https://doi.org/10.3389/FPLS.2016.00377

Li, S., Zhang, R., Lei, D., Huang, Y., Cheng, S., Zhu, Z., Wu, Z., & Cravotto, G. (2021). Impact of ultrasound, microwaves and high-pressure processing on food components and their interactions. *Trends in Food Science & Technology*, *109*, 1–15. https://doi.org/10.1016/j.tifs.2021.01.017

Li, Z., Wang, W. M., Liu, G., Liu, L., He, J., & Huang, G. Q. (2018). Toward open manufacturing a cross-enterprises knowledge and services exchange framework based on blockchain and edge computing. *Industrial Management and Data Systems*, *118*(1), 303–320. https://doi.org/10.1108/IMDS-04-2017-0142

Liaghat, S., & Balasundram, S. K. (2010). A review: The role of remote sensing in precision agriculture. *American Journal of Agricultural and Biological Science*, *5*(1), 50–55. https://doi.org/10.3844/AJABSSP.2010.50.55

Lin, J., Yu, W., Zhang, N., Yang, X., Zhang, H., & Zhao, W. (2017). A survey on internet of things: Architecture, enabling technologies, security and privacy, and applications. *IEEE Internet of Things Journal*, *4*(5), 1125–1142. https://doi.org/10.1109/JIOT.2017.2683200

Litvinski, O. (2018). Emerging technology: Toward a conceptual definition. *International Journal of Trade, Economics and Finance*, *9*(6), 257–263. https://doi.org/10.18178/ijtef.2018.9.6.625

Liu, B., & Rajagopal, D. (2019). Life-cycle energy and climate benefits of energy recovery from wastes and biomass residues in the United States. *Nature Energy*, *4*(8), 700–708. https://doi.org/10.1038/S41560-019-0430-2

Liu, S., Zhu, Y., Yue, E., & Guo, X. (2019). Key technology and application platform of agricultural economic spatial information service-chinese agricultural economic electronic map. *China's e-Science Blue Book* 2018, 269–289. https://doi.org/10.1007/978-981-13-9390-7_16

Liu, Y., Gu, J., & Zhang, M. (2020). Energy use and challenges in current wastewater treatment plants. *A-B Processes: Towards Energy Self-Sufficient Municipal Wastewater Treatment*, 1–28. https://doi.org/10.2166/9781789060089_0001

Logan, B. E., Rossi, R., Baek, G., Shi, L., Oconnor, J., & Peng, W. (2020). Energy use for electricity generation requires an assessment more directly relevant to climate change. *ACS Energy Letters*, *5*(11), 3514–3517. https://doi.org/10.1021/ACSENERGYLETT.0C02093

Lokers, R., Knapen, R., Janssen, S., Van Randen, Y., & Jansen, J. (2016). Analysis of Big Data technologies for use in agro-environmental science. *Environmental Modelling & Software*, *84*, 494–504. https://doi.org/10.1016/j.envsoft.2016.07.017

Lombardi, M., Pascale, F., & Santaniello, D. (2021). Internet of Things: A General Overview between Architectures, Protocols and Applications. *Information*, *12*(2), 87. https://doi.org/10.3390/INFO12020087

Lozano, M. G., Brynielsson, J., Franke, U., Rosell, M., Tjörnhammar, E., Varga, S., & Vlassov, V. (2020). Veracity assessment of online data. *Decision Support Systems*, *129*, 113132. https://doi.org/10.1016/j.dss.2019.113132

Lowenberg-DeBoer, J. (2015a). *The Precision Agriculture Revolution | Foreign Affairs*. https://www.foreignaffairs.com/articles/united-states/2015-04-20/precision-agriculture-revolution

Lowenberg-DeBoer, J. (2015b). *The Precision Agriculture Revolution Making the Modern Farmer | Request PDF*. The Precision Agriculture Revolution Making the Modern Farmer I Request PDF. https://www.researchgate.net/publication/288005928_The_Precision_Agriculture_Revolution_Making_the_Modern_Farmer

Lowenberg-DeBoer, J., Huang, I. Y., Grigoriadis, V., & Blackmore, S. (2020). Economics of robots and automation in field crop production. *Precision Agriculture*, *21*(2), 278–299. https://doi.org/10.1007/S11119-019-09667-5

Lu, X., Zhou, K., Zhang, X., & Yang, S. (2018). A systematic review of supply and demand side optimal load scheduling in a smart grid environment. *Journal of Cleaner Production*, *203*, 757–768. https://doi.org/10.1016/J.JCLEPRO.2018.08.301

Lucena, P., Binotto, A. P. D., Momo, F. da S., & Kim, H. (2018). *A Case Study for Grain Quality Assurance Tracking based on a Blockchain Business Network*. https://arxiv.org/abs/1803.07877v1

Lúcio, C., Silva, C. M., & Sousa, V. (2020). A scale-adaptive method for urban rainwater harvesting simulation. *Environmental Science and Pollution Research*, *27*(5), 4557–4570. https://doi.org/10.1007/S11356-019-04889-6

Madumidha, S., Ranjani, P. S., Varsinee, S. S., & Sundari, P. S. (2019). Transparency and traceability: In food supply chain system using blockchain technology with Internet of Things. *Proceedings of the International Conference on Trends in Electronics and Informatics, ICOEI 2019*, 983–987. https://doi.org/10.1109/ICOEI.2019.8862726

Mahlknecht, J., González-Bravo, R., & Loge, F. J. (2020). Water-energy-food security: A Nexus perspective of the current situation in Latin America and the Caribbean. *Energy*, *194*, 116824. https://doi.org/10.1016/J.ENERGY.2019.116824

Majeed, Y., Khan, M. U., Waseem, M., Zahid, U., Mahmood, F., Majeed, F., Sultan, M., & Raza, A. (2023). Renewable energy as an alternative source for energy management in agriculture. *Energy Reports*, *10*, 344–359. https://doi.org/10.1016/J.EGYR.2023.06.032

Majumdar, J., Naraseeyappa, S., & Ankalaki, S. (2017). Analysis of agriculture data using data mining techniques: Application of big data. *Journal of Big Data*, *4*(1). https://doi.org/10.1186/S40537-017-0077-4

Makinde, P., & Obikoya, E. (2024). Implementation of solar system for electricity generation for rural farmers: A review. *World Journal of Advanced Research and Reviews*, *22*(3), 458–471. https://doi.org/10.30574/WJARR.2024.22.3.1705

Maksimović, M., Vujović, V., & Omanović-Mikličanin, E. (2015). Application of Internet of Things in food packaging and transportation. *International Journal of Sustainable Agricultural Management and Informatics*, *1*(4), 333–350. https://doi.org/10.1504/IJSAMI.2015.075053

Manglani, H., Hodge, G. L., & Oxenham, W. (2019). Application of the Internet of Things in the textile industry. *Textile Progress*, *51*(3), 225–297. https://doi.org/10.1080/00405167.2020.1763701

Maouchi, M. E., Ersoy, O., & Erkin, Z. (2018). *TRADE: A Transparent, Decentralized Traceability System for the Supply Chain*. 1–6. https://doi.org/10.18420/BLOCKCHAIN2018_01

Marchant, N. (2021). *What is the Internet of Things – An explainer | World Economic Forum*. https://www.weforum.org/agenda/2021/03/what-is-the-internet-of-things/

Martín-Retortillo, M., & Pinilla, V. (2015). On the causes of economic growth in Europe: Why did agricultural labour productivity not converge between 1950 and 2005? *Cliometrica, Journal of Historical Economics and Econometric History*, *9*(3), 359–396. https://doi.org/10.1007/S11698-014-0119-5

Martirosov, S., Bureš, M., & Zítka, T. (2022). Cyber sickness in low-immersive, semi-immersive, and fully immersive virtual reality. *Virtual Reality*, *26*(1), 15–32. https://doi.org/10.1007/S10055-021-00507-4

Matsui, T., Suzuki, K., Ando, K., Kitai, Y., Haga, C., Masuhara, N., & Kawakubo, S. (2022). A natural language processing model for supporting sustainable development goals: Translating semantics, visualizing nexus, and connecting stakeholders. *Sustainability Science*, *17*(3), 969–985. https://doi.org/10.1007/S11625-022-01093-3

Mawusi, S. K., Shrestha, P., Xue, C., & Liu, G. (2023). A comprehensive review of the production, adoption and sustained use of biomass pellets in Ghana. *Heliyon*, *9*(6), e16416. https://doi.org/10.1016/J.HELIYON.2023.E16416

Mengistu, A. T., Dieste, M., Panizzolo, R., & Biazzo, S. (2024). Sustainable product design factors: A comprehensive analysis. *Journal of Cleaner Production*, *463*, 142260. https://doi.org/10.1016/J.JCLEPRO.2024.142260

Meran, G., Siehlow, M., & von Hirschhausen, C. (2021). Integrated water resource management: Principles and applications. *Springer Water*, 23–121. https://doi.org/10.1007/978-3-030-48485-9_3

Miao, R., & Khanna, M. (2017). Effectiveness of the biomass crop assistance program: Roles of behavioral factors, credit constraint, and program design. *Applied Economic Perspectives and Policy*, *39*(4), 584–608. *https://doi.org/10.1093/aepp/ppx031*

Miao, R., & Khanna, M. (2020). Harnessing advances in agricultural technologies to optimize resource utilization in the food-energy-water nexus. *Annual Review of Resource Economics*, *12*, 65–85. http://dx.doi.org/10.1146/annurev-resource-110319-115428

Miranda, J., Pérez-Rodríguez, R., Borja, V., Wright, P. K., & Molina, A. (2017). Sensing, smart and sustainable product development (S3 product) reference framework. *International Journal of Production Research*, *57*(14), 4391–4412. https://doi.org/10.1080/00207543.2017.1401237

Miranda, J., Ponce, P., Molina, A., & Wright, P. (2019). Sensing, smart and sustainable technologies for agri-food 4.0. *Computers in Industry*, *108*, 21–36. https://doi.org/10.1016/J.COMPIND.2019.02.002

Mishra, C., & Gupta, D. L. (2017). Deep machine learning and neural networks: An overview. *IAES International Journal of Artificial Intelligence (IJ-AI*, *6*(2), 66–73.

Mishra, S., & Tyagi, A. K. (2022). The role of machine learning techniques in internet of things-based cloud applications. *Internet of Things*, 105–135. https://doi.org/10.1007/978-3-030-87059-1_4

Mofidi, F., & Akbari, H. (2020). Intelligent buildings: An overview. *Energy and Buildings*, *223*, 110192. https://doi.org/10.1016/J.ENBUILD.2020.110192

Mohamed, A., Najafabadi, M. K., Wah, Y. B., Zaman, E. A. K., & Maskat, R. (2020). The state of the art and taxonomy of big data analytics: View from new big data framework. *Artificial Intelligence Review*, *53*(2), 989–1037. https://doi.org/10.1007/S10462-019-09685-9

Mohammadpoor, M., & Torabi, F. (2020). Big data analytics in oil and gas industry: An emerging trend. *Petroleum*, *6*(4), 321–328. https://doi.org/10.1016/j.petlm.2018.11.001

Mohiuddin, M., Reza, M. N. H., Jayashree, S., Al-Azad, M. S., & Ed-Dafali, S. (2023). The role of governments in driving industry 4.0 adoption in emerging countries: Mediating effect of organizational structure. *Journal of Global Information Management*, *31*(1), 1–31. https://doi.org/10.4018/jgim.323439

Mohsin, M., Hanif, I., Taghizadeh-Hesary, F., Abbas, Q., & Iqbal, W. (2021). Nexus between energy efficiency and electricity reforms: A DEA-Based way forward for clean power development. *Energy Policy*, *149*. https://doi.org/10.1016/j.enpol.2020.112052

Molina, A., Ponce, P., Miranda, J., & Cortés, D. (2021). Integrated product, process and manufacturing system development reference model framework. *Enabling Systems for Intelligent Manufacturing in Industry 4.0*, 29–69. https://doi.org/10.1007/978-3-030-65547-1_2

Mondejar, M. E., Avtar, R., Diaz, H. L. B., Dubey, R. K., Esteban, J., Gómez-Morales, A., Hallam, B., Mbungu, N. T., Okolo, C. C., Prasad, K. A., She, Q., & Garcia-Segura, S. (2021). Digitalization to achieve sustainable development goals: Steps towards a smart green planet. *Science of the Total Environment*, *794*, 148539. https://doi.org/10.1016/J.SCITOTENV.2021.148539

Mondillo, G., Colosimo, S., Perrotta, A., Frattolillo, V., & Gicchino, M. F. (2024). Unveiling artificial intelligence's power: Precision, personalization, and progress in rheumatology. *Journal of Clinical Medicine*, *13*(21), 6559. https://doi.org/10.3390/JCM13216559

Montealegre, A. L., García-Pérez, S., Guillén-Lambea, S., Monzón-Chavarrías, M., & Sierra-Pérez, J. (2022). GIS-based assessment for the potential of implementation of food-energy-water systems on building rooftops at the urban level. *Science of The Total Environment*, *803*, 149963. https://doi.org/10.1016/J.SCITOTENV.2021.149963

Moore, S., Graff, H., Ouellet, C., Leslie, S., & Olweean, D. (2022). Can we have clean energy and grow our crops too? Solar siting on agricultural land in the United States. *Energy Research & Social Science*, *91*, 102731. https://doi.org/10.1016/J.ERSS.2022.102731

Morales, M., Quintero, J., Conejeros, R., & Aroca, G. (2015). Life cycle assessment of lignocellulosic bioethanol: Environmental impacts and energy balance. *Renewable and Sustainable Energy Reviews*, *42*, 1349–1361. https://doi.org/10.1016/J.RSER.2014.10.097

Morley, J., Widdicks, K., & Hazas, M. (2018). Digitalisation, energy and data demand: The impact of Internet traffic on overall and peak electricity consumption. *Energy Research & Social Science*, *38*, 128–137. https://doi.org/10.1016/J.ERSS.2018.01.018

Morón, C., Ferrández, D., Saiz, P., Vega, G., & Díaz, J. P. (2017). New prototype of photovoltaic solar tracker based on arduino. *Energies*, *10*(9), 1298. https://doi.org/10.3390/EN10091298

Moscoso, C., Chamilothori, K., Wienold, J., Andersen, M., & Matusiak, B. (2022). Regional differences in the perception of daylit scenes across Europe using virtual reality. Part I: Effects of window size. *LEUKOS: The Journal of the Illuminating Engineering Society of North America*, *18*(3), 294–315. https://doi.org/10.1080/15502724.2020.1854779

Mosher, J. N., & Corscadden, K. W. (2019). Agriculture's contribution to the renewable energy sector: Policy and economics – Do they add up? *Renewable and Sustainable Energy Reviews*, *16*(6), 4157–4164. https://doi.org/10.1016/J.RSER.2012.03.027

Mostajabi, F., Safaei, A. A., & Sahafi, A. (2021). A systematic review of data models for the big data problem. *IEEE Access*, *9*, 128889–128904. https://doi.org/10.1109/ACCESS.2021.3112880

Movellan, J. (2013). *Japan Next-Generation Farmers Cultivate Crops and Solar Energy*. Japan Next-Generation Farmers Cultivate Crops and Solar Energy. https://www.renewableenergyworld.com/solar/japan-next-generation-farmers-cultivate-agriculture-and-solar-energy/#gref

Mukri, B. (2008). Blockchain technology in supply chain management: A review. *International Research Journal of Engineering and Technology*. www.irjet.net

Nakamoto, S. (n.d.). *Bitcoin: A Peer-to-Peer Electronic Cash System*. www.bitcoin.org
Narendran, S., Pradeep, P., & Ramesh, M. V. (2017). An Internet of Things (IoT) based sustainable water management. *GHTC 2017 – IEEE Global Humanitarian Technology Conference, Proceedings, 2017-January*, 1–6. https://doi.org/10.1109/GHTC.2017.8239320
Nasyrov, R. R., & Excell, P. S. (2020). Creation of interactive virtual reality scenarios as a training and education tool. *Springer Series on Cultural Computing*, 353–369. https://doi.org/10.1007/978-3-030-42097-0_19
Nath, F., Chowdhury, M. O. S., & Rhaman, M. M. (2023). Navigating produced water sustainability in the oil and gas sector: A critical review of reuse challenges, treatment technologies, and prospects ahead. *Water*, *15*(23), 4088. https://doi.org/10.3390/W15234088
Navarro-Hellín, H., Martínez-del-Rincon, J., Domingo-Miguel, R., Soto-Valles, F., & Torres-Sánchez, R. (2016). A decision support system for managing irrigation in agriculture. *Computers and Electronics in Agriculture*, *124*, 121–131. https://doi.org/10.1016/J.COMPAG.2016.04.003
Nawar, S., Corstanje, R., Halcro, G., Mulla, D., & Mouazen, A. M (2017). Delineation of soil management zones for variable-rate fertilization: A review. *Advances in Agronomy*, *143*, 175–245. https://doi.org/10.1016/BS.AGRON.2017.01.003
Nawaz, N., Arunachalam, H., Pathi, B. K., & Gajenderan, V. (2024). The adoption of artificial intelligence in human resources management practices. *International Journal of Information Management Data Insights*, *4*(1), 100208. https://doi.org/10.1016/J.JJIMEI.2023.100208
Neisse, R., Steri, G., & Nai-Fovino, I. (2017). A blockchain-based approach for data accountability and provenance tracking. *ACM International Conference Proceeding Series, Part F130521*. https://doi.org/10.1145/3098954.3098958
Nerurkar, P., Patel, D., Busnel, Y., Ludinard, R., Kumari, S., & Khan, M. K. (2021). Dissecting bitcoin blockchain: Empirical analysis of bitcoin network (2009–2020). *Journal of Network and Computer Applications*, *177*. https://doi.org/10 1016/j.jnca.2020.102940
Nishant, R., Kennedy, M., & Corbett, J. (2020). Artificial intelligence for sustainability: Challenges, opportunities, and a research agenda. *International Journal of Information Management*, *53*, 102104. https://doi.org/10.1016/J.IJINFOMGT.2020.102104
Noureldeen Mohamed, N. (2020). *Energy in Agriculture under Climate Change*. https://doi.org/10.1007/978-3-030-38010-6
Nwulu, N., & Damisa, U. (2023). Blockchain and smart contracts *Energy 4.0*, 1–20. https://doi.org/10.1063/9780735425163_006
Obaideen, K., Shehata, N., Sayed, E. T., Abdelkareem, M. A., Mahmoud, M. S., & Olabi, A. G. (2022). The role of wastewater treatment in achieving sustainable development goals (SDGs) and sustainability guideline. *Energy Nexus*, *7*, 100112. https://doi.org/10.1016/J.NEXUS.2022.100112
Ogbolumani, O. A., & Adekoya, M. (2025). Intelligent waste management optimization through machine learning analytics. *Journal of Science Research and Reviews*, *2*(1), 7–26. https://doi.org/10.70882/JOSRAR.2025.V2I1.25
Ogbolumani, O. A., & Mabaso, B. (2023). An IoT-based hydroponic monitoring and control system for sustainable food production. *Journal of Digital Food, Energy & Water Systems*, *4*(2), 106–140.
Ogbolumani, O. A., & Nwulu, N. (2020). Integrated appliance scheduling and optimal sizing of an autonomous hybrid renewable energy system for agricultural food production. *Lecture Notes in Mechanical Engineering*, 651–660. https://doi.org/10.1007/978-981-15-5753-8_60
Ogbolumani, O. A., & Nwulu, N. I. (2021). Multi-objective optimisation of constrained food-energy-water-nexus systems for sustainable resource allocation. *Sustainable Energy Technologies and Assessments*, *44*, 100967. https://doi.org/10.1016/j.seta.2020.100967

Ogbolumani, O. A., & Nwulu, N. I. (2022). A food-energy-water nexus meta-model for food and energy security. *Sustainable Production and Consumption*, *30*, 438–453. https://doi.org/10.1016/J.SPC.2021.12.019

Ogedengbe, D. E., James, O. O., Afolabi, J. O. A., Olatoye, F. O., & Eboigbe, E. O. (2023). Human resources in the era of the Fourth Industrial Revolution (4IR): Strategies and innovations in the global south. *Engineering Science & Technology Journal*, *4*(5), 308–322. https://doi.org/10.51594/ESTJ.V4I5.617

Okonta, D. E., & Vukovic, V. (2024). Smart cities software applications for sustainability and resilience. *Heliyon*, *10*(12), e32654. https://doi.org/10.1016/J.HELIYON.2024.E32654

Olawuyi, D. (2020). Sustainable development and the water-energy-food nexus: Legal challenges and emerging solutions. *Environmental Science & Policy*, *103*, 1–9. https://doi.org/10.1016/J.ENVSCI.2019.10.009

Oliver, M. A., Bishop, T. F. A., & Marchant, B. P. (2013). Precision agriculture for sustainability and environmental protection. *Precision Agriculture for Sustainability and Environmental Protection*, *9780203128329*, 1–283. https://doi.org/10.4324/9780203128329

Omar, A., Nashed, A., Li, Q., Leslie, G., & Taylor, R. A. (2020). Pathways for integrated concentrated solar power – Desalination: A critical review. *Renewable and Sustainable Energy Reviews*, *119*, 109609. https://doi.org/10.1016/J.RSER.2019.109609

Orsi, E., & Nesmachnow, S. (2017). Smart home energy planning using IoT and the cloud. *2017 IEEE URUCON, URUCON 2017, 2017-December*, 1–4. https://doi.org/10.1109/URUCON.2017.8171843

Osakabe, Y., Watanabe, T., Sugano, S. S., Ueta, R., Ishihara, R., Shinozaki, K., & Osakabe, K. (2016). Optimization of CRISPR/Cas9 genome editing to modify abiotic stress responses in plants. *Scientific Reports*, *6*(1), 1–10. https://doi.org/10.1038/srep26685

Pal, D., Funilkul, S., Charoenkitkarn, N., & Kanthamanon, P. (2018). Internet-of-things and smart homes for elderly healthcare: An end user perspective. *IEEE Access*, *6*, 10483–10496. https://doi.org/10.1109/ACCESS.2018.2808472

Panagopoulos, A. (2021). Water-energy nexus: Desalination technologies and renewable energy sources. *Environmental Science and Pollution Research International*, *28*(17), 21009–21022. https://doi.org/10.1007/S11356-021-13332-8

Papa, S. F. (2017). *Use of Blockchain Technology in Agribusiness: Transparency and Monitoring in Agricultural Trade*. 38–40. https://doi.org/10.2991/MSMI-17.2017.9

Parekh, R. (2025). *Principles of Multimedia* (3rd ed.). 101–103. CRC Press. https://doi.org/10.1201/9781003588030

Parisi, D. (2019). Rumble/control: Toward a critical history of touch feedback in video games. *ROMchip*, *1*(2). https://romchip.org/index.php/romchip-journal/article/view/86

Park, J., Kim, K. T., & Lee, W. H. (2020). Recent advances in Information and Communications Technology (ICT) and sensor technology for monitoring water quality. *Water*, *12*(2), 510. https://doi.org/10.3390/W12020510

Park, W., Heo, H., Park, S., & Kim, J. (2019). A study on the presence of immersive user interface in collaborative virtual environments application. *Symmetry*, *11*(4), 476. https://doi.org/10.3390/SYM11040476

Parry, E., & Battista, V. (2019). The impact of emerging technologies on work: A review of the evidence and implications for the human resource function. *Emerald Open Research*, *1*(4). https://doi.org/10.1108/EOR-04-2023-0001

Pawlak, K., & Kołodziejczak, M. (2020). The role of agriculture in ensuring food security in developing countries: Considerations in the context of the problem of sustainable food production. *Sustainability*, *12*(13), 5488. https://doi.org/10.3390/SU12135488

Perifanis, N. A., & Kitsios, F. (2023). Investigating the influence of artificial intelligence on business value in the digital era of strategy: A literature review. *Information*, *14*(2), 85. https://doi.org/10.3390/INFO14020085

Perwej, Y. (2017). An experiential study of the big data. *International Transaction of Electrical and Computer Engineers System*, *4*(1), 14–25. doi: 10.12691/iteces-4-1-3

Petersen, G. B., Petkakis, G., & Makransky, G. (2022). A study of how immersion and interactivity drive VR learning. *Computers & Education*, *179*, 104429. https://doi.org/10.1016/J.COMPEDU.2021.104429

Pham, X., & Stack, M. (2018). How data analytics is transforming agriculture. *Business Horizons*, *61*(1), 125–133. https://doi.org/10.1016/J.BUSHOR.2017.09.011

Phiri, E., Rowley, P. N., & Blanchard, R. E. (2023). Addressing challenges in delivering sustainable rural water provision using solar water pumping in Malawi: A stakeholder analysis. *Energies*, *16*(23), 7758. https://doi.org/10.3390/EN16237758

Pradhan, P., Rostami, M., Kamoonpuri, J., & Chung, J. (2023). The state of augmented reality in aerospace navigation and engineering. *Applications of Augmented Reality – Current State of the Art*. https://doi.org/10.5772/INTECHOPEN.1002358

Predescu, A., Mocanu, M., & Lupu, C. (2017). Real time implementation of IoT structure for pumping stations in a water distribution system. *2017 21st International Conference on System Theory, Control and Computing, ICSTCC 2017*, 529–534. https://doi.org/10.1109/ICSTCC.2017.8107089

Pueppke, S. G. (2021). Ancient WEF: Water–energy–food nexus in the distant past. *Water*, *13*(7), 925. https://doi.org/10.3390/W13070925

Pundir, A., Singh, S., Kumar, M., Bafila, A., & Saxena, G. J. (2022). Cyber-physical systems enabled transport networks in smart cities: Challenges and enabling technologies of the new mobility era. *IEEE Access*, *10*, 16350–16364. https://doi.org/10.1109/ACCESS.2022.3147323

Purvis, L., & Dinar, A. (2020). Are intra- and inter-basin water transfers a sustainable policy intervention for addressing water scarcity? *Water Security*, *9*, 100058. https://doi.org/10.1016/J.WASEC.2019.100058

Pushpakumar, R., Sanjaya, K., Rathika, S., Alawadi, A. H., Makhzuna, K., Venkatesh, S., & Rajalakshmi, B. (2023). Human-computer interaction: Enhancing user experience in interactive systems. *E3S Web of Conferences*, *399*. https://doi.org/10.1051/E3SCONF/202339904037

Rajagukguk, R. A., Ramadhan, R. A. A., & Lee, H. J. (2020). A review on deep learning models for forecasting time series data of solar irradiance and photovoltaic power. *Energies*, *13*(24), 6623. https://doi.org/10.3390/en13246623

Ramesh, M. V., Nibi, K. V., Kurup, A., Mohan, R., Aiswarya, A., Arsha, A., & Sarang, P. R. (2017). Water quality monitoring and waste management using IoT. *GHTC 2017 – IEEE Global Humanitarian Technology Conference, Proceedings*, *2017-January*, 1–7. https://doi.org/10.1109/GHTC.2017.8239311

Ramírez-Moreno, M. A., Keshtkar, S., Padilla-Reyes, D. A., Ramos-López, E., García-Martínez, M., Hernández-Luna, M. C., Mogro, A. E., Mahlknecht, J., Huertas, J. I., Peimbert-García, R. E., Ramírez-Mendoza, R. A., Mangini, A. M., Roccotelli, M., Pérez-Henríquez, B. L., Mukhopadhyay, S. C., & Lozoya-Santos, J. J. (2021). Sensors for sustainable smart cities: A review. *Applied Sciences*, *11*(17), 8198. https://doi.org/10.3390/APP11178198

Ramundo, L., Taisch, M., & Terzi, S. (2016). State of the art of technology in the food sector value chain towards the IoT. *2016 IEEE 2nd International Forum on Research and Technologies for Society and Industry Leveraging a Better Tomorrow, RTSI 2016*. https://doi.org/10.1109/RTSI.2016.7740612

Rao, T. R., Mitra, P., Bhatt, R., & Goswami, A. (2019). The big data system, components, tools, and technologies: A survey. *Knowledge and Information Systems*, *60*(3), 1165–1245. https://doi.org/10.1007/s10115-018-1248-0

Rashidi, M. M., Mahariq, I., Murshid, N., Wongwises, S., Mahian, O., & Alhuyi Nazari, M. (2022). Applying wind energy as a clean source for reverse osmosis desalination: A comprehensive review. *Alexandria Engineering Journal*, *61*(12), 12977–12989. https://doi.org/10.1016/j.aej.2022.06.056

Rasul, G., & Sharma, B. (2016). The nexus approach to water–energy–food security: An option for adaptation to climate change. *Climate Policy*, *16*(6), 682–702. https://doi.org/10.1080/14693062.2015.1029865

Rauschnabel, P. A., Felix, R., Hinsch, C., Shahab, H., & Alt, F. (2022). What is XR? Towards a framework for augmented and virtual reality. *Computers in Human Behavior*, *133*, 107289. https://doi.org/10.1016/J.CHB.2022.107289

Ray, P. P. (2018). A survey on internet of things architectures. *Journal of King Saud University – Computer and Information Sciences*, *30*(3), 291–319. https://doi.org/10.1016/J.JKSUCI.2016.10.003

Rayes, A., & Salam, S. (2016). Internet of Things-from hype to reality: The road to digitization. In *Internet of Things from Hype to Reality: The Road to Digitization* (pp. 1–328). https://doi.org/10.1007/978-3-319-44860-2

Reddy, K., Gharde, P., Tayade, H., Patil, M., Reddy, L. S., & Surya, D. (2023). Advancements in robotic surgery: A comprehensive overview of current utilizations and upcoming frontiers. *Cureus*, *15*(12). https://doi.org/10.7759/CUREUS.50415

Rejeb, A., Rejeb, K., Zrelli, I., Süle, E., & Iranmanesh, M. (2024). Blockchain technology in the renewable energy sector: A co-word analysis of academic discourse. *Heliyon*, *10*(8), e29600. https://doi.org/10.1016/J.HELIYON.2024.E29600

Renewable Energy Agency. (2015). *Renewable Energy in the Water, Energy and Food Nexus*. www.irena.org

Reyna, A., Martín, C., Chen, J., Soler, E., & Díaz, M. (2018). On blockchain and its integration with IoT. Challenges and opportunities. *Future Generation Computer Systems*, *88*, 173–190. https://doi.org/10.1016/J.FUTURE.2018.05.046

Riahi, Y., & Riahi, S. (2018a). Big data and big data analytics: Concepts, types and technologies. *International Journal of Research and Engineering*, *5*(9), 524–528. https://doi.org/10.21276/IJRE.2018.5.9.5

Rihar, L., Žužek, T., & Kušar, J. (2021). How to successfully introduce concurrent engineering into new product development? *Concurrent Engineering Research and Applications*, *29*(2), 87–101. https://doi.org/10.1177/1063293X20967929

Robertson, G. P., Hamilton, S. K., Barham, B. L., Dale, B. E., Izaurralde, R. C., Jackson, R. D., Landis, D. A., Swinton, S. M., Thelen, K. D., & Tiedje, J. M. (2017). Cellulosic biofuel contributions to a sustainable energy future: Choices and outcomes. *Science (New York)*, *356*(6345). https://doi.org/10.1126/SCIENCE.AAL2324

Rosi-Marshall, E. J., Tank, J. L., Royer, T. V., Whiles, M. R., Evans-White, M., Chambers, C., Griffiths, N. A., Pokelsek, J., & Stephen, M. L. (2007). Toxins in transgenic crop byproducts may affect headwater stream ecosystems. *Proceedings of the National Academy of Sciences of the United States of America*, *104*(41), 16204–16208. https://doi.org/10.1073/PNAS.0707177104

Rotolo, D., Hicks, D., & Martin, B. R. (2015). What is an emerging technology? *Research Policy*, *44*(10), 1827–1843. https://doi.org/10.1016/j.respol.2015.06.006

Rufí-Salís, M., Petit-Boix, A., Villalba, G., Sanjuan-Delmás, D., Parada, F., Ercilla-Montserrat, M., Arcas-Pilz, V., Muñoz-Liesa, J., Rieradevall, J., & Gabarrell, X. (2020). Recirculating water and nutrients in urban agriculture: An opportunity towards environmental sustainability and water use efficiency? *Journal of Cleaner Production*, *261*. https://doi.org/10.1016/J.JCLEPRO.2020.121213

Saha, M., & Eckelman, M. J. (2017). Growing fresh fruits and vegetables in an urban landscape: A geospatial assessment of ground level and rooftop urban agriculture potential in Boston, USA. *Landscape and Urban Planning*, *165*, 130–141. https://doi.org/10.1016/J.LANDURBPLAN.2017.04.015

Şahin, U., Ballı, S., & Chen, Y. (2021). Forecasting seasonal electricity generation in European countries under Covid-19-induced lockdown using fractional grey prediction models and machine learning methods. *Applied Energy*, *302*, 117540. https://doi.org/10.1016/J.APENERGY.2021.117540

Sahoo, P. K., Datta, R., Rahman, M. M., & Sarkar, D. (2024). Sustainable environmental technologies: Recent development, opportunities, and key challenges. *Applied Sciences*, *14*(23), 10956. https://doi.org/10.3390/APP142310956

Said Mohamed, E., Belal, A. A., Kotb Abd-Elmabod, S., El-Shirbeny, M. A., Gad, A., & Zahran, M. B. (2021). Smart farming for improving agricultural management. *Egyptian Journal of Remote Sensing and Space Science*, *24*(3), 971–981. https://doi.org/10.1016/J.EJRS.2021.08.007

Saiz-Rubio, V., & Rovira-Más, F. (2020). From smart farming towards Agriculture 5.0: A review on crop data management. *Agronomy*, *10*(2), 207. https://doi.org/10.3390/AGRONOMY10020207

Salloom, T., Kaynak, O., & He, W. (2021). A novel deep neural network architecture for real-time water demand forecasting. *Journal of Hydrology*, *599*, 126353. https://doi.org/10.1016/J.JHYDROL.2021.126353

Salvador, D. S., Toboso-Chavero, S., Nadal, A., Gabarrell, X., Rieradevall, J., & da Silva, R. S. (2019). Potential of technology parks to implement Roof Mosaic in Brazil. *Journal of Cleaner Production*, *235*, 166–177. https://doi.org/10.1016/J.JCLEPRO.2019.06.214

Sánchez-Corcuera, R., Nuñez-Marcos, A., Sesma-Solance, J., Bilbao-Jayo, A., Mulero, R., Zulaika, U., Azkune, G., & Almeida, A. (2019). Smart cities survey: Technologies, application domains and challenges for the cities of the future. *International Journal of Distributed Sensor Networks*, *15*(6). https://doi.org/10.1177/1550147719853984

Sanjuan-Delmás, D., Llorach-Massana, P., Nadal, A., Ercilla-Montserrat, M., Muñoz, P., Montero, J. I., Josa, A., Gabarrell, X., & Rieradevall, J. (2018). Environmental assessment of an integrated rooftop greenhouse for food production in cities. *Journal of Cleaner Production*, *177*, 326–337. https://doi.org/10.1016/J.JCLEPRO.2017.12.147

Sarkar, N., Gupta, R., Keserwani, P. K., & Govil, M. C. (2022). Air quality index prediction using an effective hybrid deep learning model. *Environmental Pollution (Barking, Essex: 1987)*, *315*. https://doi.org/10.1016/J.ENVPOL.2022.120404

Sarkar, R. (2012). *Decision Support Systems for Agrotechnology Transfer*. 263–299. https://doi.org/10.1007/978-94-007-4113-3_10

Sarker, I. H. (2021a). Deep learning: A comprehensive overview on techniques, taxonomy, applications and research directions. *SN Computer Science*, *2*(6), 1–20. https://doi.org/10.1007/S42979-021-00815-1/FIGURES/11

Sarker, I. H. (2021b). Machine learning: Algorithms, real-world applications and research directions. *SN Computer Science*, *2*(3), 1–21. https://doi.org/10.1007/S42979-021-00592-X/FIGURES/11

Sedeek, K. E. M., Mahas, A., & Mahfouz, M. (2019). Plant genome engineering for targeted improvement of crop traits. *Frontiers in Plant Science*, *10*, 429493. https://doi.org/10.3389/fpls.2019.00114

Sekiyama, T., & Nagashima, A. (2019). Solar sharing for both food and clean energy production: Performance of agrivoltaic systems for corn, a typical shade-intolerant crop. *Environments*, *6*(6), 65. https://doi.org/10.3390/ENVIRONMENTS6060065

Sethi, P., & Sarangi, S. R. (2017). Internet of things: Architectures, protocols, and applications. *Journal of Electrical and Computer Engineering*, *2017*(1), 9324035. https://doi.org/10.1155/2017/9324035

Shafiq, A., Naz, S., Butt, S. A., & Piñeres-Espitia, G. (2024). Enhancing learning environments with IoT: A novel decision-making approach using probabilistic linguistic T-spherical fuzzy set. *Journal of Supercomputing*, *80*(12), 17524–17574. https://doi.org/10.1007/s11227-024-06129-2

Shahid, M. S., Hossain, M., Shahid, S., & Anwar, T. (2023). Frugal innovation as a source of sustainable entrepreneurship to tackle social and environmental challenges. *Journal of Cleaner Production*, *406*, 137050. https://doi.org/10.1016/J.JCLEPRO.2023.137050

Shekar, V., & Kachhi, Z. (2024). Technology applications in virtual and augmented reality for human welfare: the new future. In *Entrepreneurship and Creativity in the Metaverse* (pp. 179–197). IGI Global Scientific Publishing.

Sherman, W. R., & Craig, A. B. (2018). Bringing the virtual world to life. *Understanding Virtual Reality*, 658–723. https://doi.org/10.1016/B978-0-12-800965-9.00008-8

Shokri, A., & Sanavi Fard, M. (2022). A sustainable approach in water desalination with the integration of renewable energy sources: Environmental engineering challenges and perspectives. *Environmental Advances*, *9*, 100281. https://doi.org/10.1016/J.ENVADV.2022.100281

Silva, M., Oliveira, V., & Leal, V. (2017). Urban form and energy demand: A review of energy-relevant urban attributes. *Journal of Planning Literature*, *32*(4). 346–365. https://doi.org/10.1177/0885412217706900

Silvestre, B. S., & Ţîrcă, D. M. (2019). Innovations for sustainable development: Moving toward a sustainable future. *Journal of Cleaner Production*, *208*, 325–332. https://doi.org/10.1016/J.JCLEPRO.2018.09.244

Simsek, M., Oteafy, S., Dawy, Z., Eid, M., Holland, O., & Steinbach, E. (2021). Tactile internet standards of the IEEE P1918.1 working group. *Tactile Internet: With Human-in-the-Loop*, 351–374</genre>. https://doi.org/10.1016/B978-0-12-821343-8.00028-9

Singh, R. R., Bhatti, G., & Saravanan, D. (2022). New-age condition monitoring of on-load tap changing transformers in distributed energy systems for Industry 4.0. *E-Prime – Advances in Electrical Engineering, Electronics and Energy*, *2*, 100087. https://doi.org/10.1016/J.PRIME.2022.100087

Siow, E., Tiropanis, T., & Hall, W. (2018). Analytics for the internet of things. *ACM Computing Surveys (CSUR)*, *51*(4). https://doi.org/10.1145/3204947

Sishodia, R. P., Ray, R. L., & Singh, S. K. (2020). Applications of remote sensing in precision agriculture: A review. *Remote Sensing*, *12*(19), 1–31. https://doi.org/10.3390/RS12193136

Sivarajah, U., Kamal, M. M., Irani, Z., & Weerakkody, V. (2017). Critical analysis of big data challenges and analytical methods. *Journal of Business Research*, *70*, 263–286. https://doi.org/10.1016/J.JBUSRES.2016.08.001

Small, H., Boyack, K. W., & Klavans, R. (2014). Identifying emerging topics in science and technology. *Research Policy*, *43*(8), 1450–1467. https://doi.org/10.1016/j.respol.2014.02.005

Solar – Idyllwild Water District. (n.d.). Retrieved August 5, 2024, from https://www.idyllwildwater.com/solar

Sonar, D. (2020). Renewable energy based trigeneration systems—Technologies, challenges and opportunities. In *Renewable-Energy-Driven Future: Technologies, Modelling, Applications, Sustainability and Policies* (pp. 125–168). https://doi.org/10.1016/B978-0-12-820539-6.00004-2

Soori, M., Arezoo, B., & Dastres, R. (2023). Artificial intelligence, machine learning and deep learning in advanced robotics, a review. *Cognitive Robotics*, *3*, 54–70. https://doi.org/10.1016/J.COGR.2023.04.001

Sra, M., Jain, A., & Maes, P. (2019). Adding proprioceptive feedback to virtual reality experiences using galvanic vestibular stimulation. *Conference on Human Factors in Computing Systems – Proceedings*. https://doi.org/10.1145/3290605.3300905

SraSontisirkit. (2017). *A Study of Virtual Reality*. https://www.researchgate.net/publication/320517587_A_Study_of_Virtual_Reality

Stubbs, M. (2016). *Big Data in U.S. Agriculture*.

Sukhwani, V., Shaw, R., Deshkar, S., Mitra, B. K., & Yan, W. (2020). Role of smart cities in optimizing water-energy-food nexus: Opportunities in Nagpur, India. *Smart Cities*, *3*(4), 1266–1292. https://doi.org/10.3390/SMARTCITIES3040062

Swamy, S. N., & Kota, S. R. (2020). An empirical study on system level aspects of Internet of Things (IoT). *IEEE Access*, *8*, 188082–188134. https://doi.org/10.1109/ACCESS.2020.3029847

Swinton, S. M., & Lowenberg-DeBoer, J. (1998). Evaluating the profitability of site-specific farming. *Journal of Production Agriculture*, *11*(4), 439–446. https://doi.org/10.2134/JPA1998.0439

Szabó, B. K., Gilányi, A., Szabó, B. K., & Gilányi, A. (2020). *The Notion of Immersion in Virtual Reality Literature and Related Sources.* https://doi.org/10.1109/CogInfoCom50765.2020.9237875

Szabo, N. (2018). *Smart Contracts: Building Blocks for Digital Markets.*

Taddese, G., Durieux, S., & Duc, E. (2020). Sustainability performance indicators for additive manufacturing: A literature review based on product life cycle studies. *International Journal of Advanced Manufacturing Technology, 107*(7–8), 3109–3134. https://doi.org/10.1007/S00170-020-05249-2

Talebkhah, M., Sali, A., Gordan, M., Hashim, S. J., & Rokhani, F. Z. (2023). Comprehensive review on development of smart cities using Industry 4.0 technologies. *IEEE Access, 11*, 91981–92030. https://doi.org/10.1109/ACCESS.2023.3302262

Tantalaki, N., Souravlas, S., & Roumeliotis, M. (2019). Data-driven decision making in precision agriculture: The rise of big data in agricultural systems. *Journal of Agricultural and Food Information, 20*(4), 344–380. https://doi.org/10.1080/10496505.2019.1638264

Tarik, H., & Mohammed, O. J. (2020). Big data analytics and artificial intelligence serving agriculture. *Advances in Intelligent Systems and Computing, 1103 AISC*, 57–65. https://doi.org/10.1007/978-3-030-36664-3_7

Taye, M. M., Ali, Dewan, A., Sahu, A. K., & Taye, M. M. (2023). Understanding of machine learning with deep learning: Architectures, workflow, applications and future directions. *Computers, 12*(5), 91. https://doi.org/10.3390/COMPUTERS12050091

Teixeira, H., Magalhães, A., Ramalho, A., Pina, M. de F., & Gonçalves, H. (2021). Indoor environments and geographical information systems: A systematic literature review. *Sage Open, 11*(4). https://doi.org/10.1177/21582440211050379

Tey, Y. S., & Brindal, M. (2012). Factors influencing the adoption of precision agricultural technologies: A review for policy implications. *Precision Agriculture, 13*(6), 713–730. https://doi.org/10.1007/S11119-012-9273-6

Tian, F. (2017). A supply chain traceability system for food safety based on HACCP, blockchain & Internet of Things. *14th International Conference on Services Systems and Services Management, ICSSSM 2017 – Proceedings.* https://doi.org/10.1109/ICSSSM.2017.7996119

Timilsina, G. R., Kurdgelashvili, L., & Narbel, P. A. (2012). Solar energy: Markets, economics and policies. *Renewable and Sustainable Energy Reviews, 16*(1), 449–465. https://doi.org/10.1016/J.RSER.2011.08.009

Todeschi, V., Mutani, G., Baima, L., Nigra, M., & Robiglio, M. (2020). Smart solutions for sustainable cities—The re-coding experience for harnessing the potential of urban rooftops. *Applied Sciences, 10*(20), 7112. https://doi.org/10.3390/APP10207112

Torres, J. F., Martínez-Álvarez, F., & Troncoso, A. (2022). A deep LSTM network for the Spanish electricity consumption forecasting. *Neural Computing and Applications, 34*(13), 10533–10545. https://doi.org/10.1007/S00521-021-06773-2

Toufaily, E., Zalan, T., & Dhaou, S. B. (2021). A framework of blockchain technology adoption: An investigation of challenges and expected value. *Information and Management, 58*(3). https://doi.org/10.1016/J.IM.2021.103444

Tow, E. W., Hartman, A. L., Jaworowski, A., Zucker, I., Kum, S., AzadiAghdam, M., Blatchley, E. R., Achilli, A., Gu, H., Urper, G. M., & Warsinger, D. M. (2021). Modeling the energy consumption of potable water reuse schemes. *Water Research X, 13*, 100126. https://doi.org/10.1016/J.WROA.2021.100126

Tricomi, G., Giacobbe, M., Ficili, I., Peditto, N., & Puliafito, A. (2024). Smart city as cooperating smart areas: On the way of symbiotic cyber–physical systems environment. *Sensors, 24*(10), 3108. https://doi.org/10.3390/s24103108

Trindade, E. P., Hinnig, M. P. F., da Costa, E. M., Marques, J. S., Bastos, R. C., & Yigitcanlar, T. (2017). Sustainable development of smart cities: A systematic review of

the literature. *Journal of Open Innovation: Technology, Market, and Complexity*, *3*(3), 1–14. https://doi.org/10.1186/S40852-017-0063-2

Tripathi, G., Ahad, M. A., & Casalino, G. (2023). A comprehensive review of blockchain technology: Underlying principles and historical background with future challenges. *Decision Analytics Journal*, *9*, 100344. https://doi.org/10.1016/J.DAJOUR.2023.100344

Trommsdorff, M., Kang, J., Reise, C., Schindele, S., Bopp, G., Ehmann, A., Weselek, A., Högy, P., & Obergfell, T. (2021). Combining food and energy production: Design of an agrivoltaic system applied in arable and vegetable farming in Germany. *Renewable and Sustainable Energy Reviews*, *140*, 110694. https://doi.org/10.1016/J.RSER.2020.110694

Truman, B. (2019). *Fostering Collaborative Open Simulation for Next-Gen Enterprise Learning Ecosystems* (pp. 255–280). https://Services.Igi-Global.Com/Resolvedoi/Resolve.Aspx?Doi=10.4018/978-1-5225-9679-0.Ch013; https://doi.org/10.4018/978-1-5225-9679-0.CH013

Tse, D., Zhang, B., Yang, Y., Cheng, C., & Mu, H. (2017). Blockchain application in food supply information security. *IEEE International Conference on Industrial Engineering and Engineering Management*, *2017-December*, 1357–1361. https://doi.org/10.1109/IEEM.2017.8290114

Tummers, J., Kassahun, A., & Tekinerdogan, B. (2021). Reference architecture design for farm management information systems: A multi-case study approach. *Precision Agriculture*, *22*(1), 22–50. https://doi.org/10.1007/S11119-020-09728-0

Tunisia – Electrical Power Systems and Renewable Energy. (2024). https://www.trade.gov/country-commercial-guides/tunisia-electrical-power-systems-and-renewable-energy

Tuomi, I. (2018). The impact of artificial intelligence on learning, teaching, and education. *Science for Policy, December*, 147–155. https://doi.org/10.2760/12297

Ukoba, K., Yoro, K. O., Eterigho-Ikelegbe, O., Ibegbulam, C., & Jen, T. C. (2024). Adaptation of solar energy in the Global South: Prospects, challenges and opportunities. *Heliyon*, *10*(7), e28009. https://doi.org/10.1016/J.HELIYON.2024.E28009

UNDESA. (n.d.). *International Decade for Action "Water for Life" 2005–2015. Focus Areas: Water scarcity.*

Vahabzadeh, M., Afshar, A., & Molajou, A. (2023). Framing a novel holistic energy subsystem structure for water-energy-food nexus based on existing literature (basic concepts). *Scientific Reports*, *13*(1). https://doi.org/10.1038/S41598-023-33385-8

Valzolgher, C., Alzhaler, M., Gessa, E., Todeschini, M., Nieto, P., Verdelet, G., Salemme, R., Gaveau, V., Marx, M., Truy, E., Barone, P., Farnè, A., & Pavani, F. (2020). The impact of a visual spatial frame on real sound-source localization in virtual reality. *Current Research in Behavioral Sciences*, *1*, 100003. https://doi.org/10.1016/J.CRBEHA.2020.100003

van Lopik, K., Sinclair, M., Sharpe, R., Conway, P., & West, A. (2020). Developing augmented reality capabilities for industry 4.0 small enterprises: Lessons learnt from a content authoring case study. *Computers in Industry*, *117*, 103208. https://doi.org/10.1016/j.compind.2020.103208

Vera, I., Wicke, B., Lamers, P., Cowie, A., Repo, A., Heukels, B., Zumpf, C., Styles, D., Parish, E., Cherubini, F., Berndes, G., Jager, H., Schiesari, L., Junginger, M., Brandão, M., Bentsen, N. S., Daioglou, V., Harris, Z., & van der Hilst, F. (2022). Land use for bioenergy: Synergies and trade-offs between sustainable development goals. *Renewable and Sustainable Energy Reviews*, *161*, 112409. https://doi.org/10.1016/J.RSER.2022.112409

Verma, M., Dehaeze, T., Zhao, G., Watchi, J., & Collette, C. (2021). Virtual sensor fusion for high precision control. *Mechanical Systems and Signal Processing*, *150*, 107241. https://doi.org/10.1016/J.YMSSP.2020.107241

Wadhwa, A. (2016). Research and development (R&D) investment. *The Palgrave Encyclopedia of Strategic Management*, 1–4. https://doi.org/10.1057/978-1-349-94848-2_795-1

Wakweya, R. B. (2023). Challenges and prospects of adopting climate-smart agricultural practices and technologies: Implications for food security. *Journal of Agriculture and Food Research, 14*, 100698. https://doi.org/10.1016/J.JAFR.2023.100698

Walter, A., Finger, R., Huber, R., & Buchmann, N. (2017). Smart farming is key to developing sustainable agriculture. *Proceedings of the National Academy of Sciences of the United States of America, 114*(24), 6148–6150. https://doi.org/10.1073/PNAS.1707462114

Wang, D., Wang, X., Zheng, Q., Tao, B., & Zheng, G. (2020). How interaction paradigms affect user experience and perceived interactivity in virtual reality environment. *Lecture Notes in Computer Science (Including Subseries Lecture Notes in Artificial Intelligence and Lecture Notes in Bioinformatics), 12190 LNCS*, 223–234. https://doi.org/10.1007/978-3-030-49695-1_15

Wang, K., Yin, H., Quan, W., & Min, G. (2018). Enabling collaborative edge computing for software defined vehicular networks. *IEEE Network, 32*(5), 112–117. https://doi.org/10.1109/MNET.2018.1700364

Wang, X., Yao, F., Wen, F., Wang, X., Yao, F., & Wen, F. (2022). Applications of blockchain technology in modern power systems: A brief survey. *Energies, 15*(13), 4516. https://doi.org/10.3390/EN15134516

Wee, C., Yap, K. M., & Lim, W. N. (2021). Haptic interfaces for virtual reality: Challenges and research directions. *IEEE Access, 9*, 112145–112162. https://doi.org/10.1109/ACCESS.2021.3103598

Weersink, A., Fraser, E., Pannell, D., Duncan, E., & Rotz, S. (2018). Opportunities and challenges for big data in agricultural and environmental analysis. *Annual Review of Resource Economics, 10*, 19–37. https://doi.org/10.1146/annurev-resource-100516-053654

Weselek, A., Ehmann, A., Zikeli, S., Lewandowski, I., Schindele, S., & Högy, P. (2019). Agrophotovoltaic systems: Applications, challenges, and opportunities. A review. *Agronomy for Sustainable Development, 39*(4), 1–20. https://doi.org/10.1007/S13593-019-0581-3

What Is Deep Learning? | IBM. (n.d.). Retrieved August 30, 2024, from https://www.ibm.com/topics/deep-learning

Williams, R. J., & Zipser, D. (1989). A learning algorithm for continually running fully recurrent neural networks. *Neural Computation, 1*(2), 270–280. https://doi.org/10.1162/NECO.1989.1.2.270

Winston & Strawn LLP. (2023). *What Is Emerging Technology?* https://www.winston.com/en/legal-glossary/emerging-technology

Woetzel, L., Remes, J., Boland, B., Lv, K., Sinha, S., Strube, G., Means, J., Law, J., Cadena, A. & von der Tann, V. (2018). *Smart City Technology for a More Liveable Future | McKinsey*. https://www.mckinsey.com/capabilities/operations/our-insights/smart-cities-digital-solutions-for-a-more-livable-future

Wolf, J., Hubbard, S., Brauer, M., Ambelu, A., Arnold, B. F., Bain, R., Bauza, V., Brown, J., Caruso, B. A., Clasen, T., Colford, J. M., Freeman, M. C., Gordon, B., Johnston, R. B., Mertens, A., Prüss-Ustün, A., Ross, I., Stanaway, J., Zhao, J. T., … Boisson, S. (2022). Effectiveness of interventions to improve drinking water, sanitation, and handwashing with soap on risk of diarrhoeal disease in children in low-income and middle-income settings: A systematic review and meta-analysis. *Lancet (London, England), 400*(10345), 27. https://doi.org/10.1016/S0140-6736(22)00937-0

Wolf, M. (2023). Automotive and aerospace systems. *Computers as Components*, 437–452. https://doi.org/10.1016/B978-0-323-85128-2.00009-8

Wolfert, S., Ge, L., Verdouw, C., & Bogaardt, M. J. (2017). Big data in smart farming – A review. *Agricultural Systems, 153*, 69–80. https://doi.org/10.1016/J.AGSY.2017.01.023

Wolfewicz, A. (2023). *Deep Learning vs. Machine Learning – What's The Difference?* https://levity.ai/blog/difference-machine-learning-deep-learning

Wolt, J. D., Wang, K., & Yang, B. (2016). The regulatory status of genome-edited crops. *Plant Biotechnology Journal*, *14*(2), 510–518. https://doi.org/10.1111/PBI.12444

Workman, A. (2023). *Scientific Visualization | Overview, Purpose & Examples | Study.com.* https://study.com/academy/lesson/scientific-visualization-definition-examples.html

Wu, X., Hu, W., Huang, Q., Chen, C., Chen, Z., & Blaabjerg, F. (2019). Optimized placement of onshore wind farms considering topography. *Energies*, *12*(15), 2944. https://doi.org/10.3390/EN12152944

Wynn, M., & Bechkoum, K. (2024). Emerging technologies, sustainable engineering and cybersecurity in the digital age. *Sustainability*, *16*(17), 7364. https://doi.org/10.3390/SU16177364

Xu, L. Da, Xu, E. L., & Li, L. (2018). Industry 4.0: State of the art and future trends. *International Journal of Production Research*, *56*(8), 2941–2962. https://doi.org/10.1080/00207543.2018.1444806

Xu, Q., Zhong, M., & Li, X. (2022). How does digitalization affect energy? International evidence. *Energy Economics*, *107*, 105879. https://doi.org/10.1016/J.ENECO.2022.105879

Xu, R., Yang, Y., Qin, R., Li, H., Qiu, C., Li, L., Wei, P., & Yang, J. (2016). Rapid improvement of grain weight via highly efficient CRISPR/Cas9-mediated multiplex genome editing in rice. *Journal of Genetics and Genomics = Yi Chuan Xue Bao*, *43*(8), 529–532. https://doi.org/10.1016/J.JGG.2016.07.003

Yao, X., Zhou, J., Lin, Y., Li, Y., Yu, H., & Liu, Y. (2019). Smart manufacturing based on cyber-physical systems and beyond. *Journal of Intelligent Manufacturing*, *30*(8), 2805–2817. https://doi.org/10.1007/S10845-017-1384-5

Yassine, A., Singh, S., Hossain, M. S., & Muhammad, G. (2019). IoT big data analytics for smart homes with fog and cloud computing. *Future Generation Computer Systems*, *91*, 563–573. https://doi.org/10.1016/J.FUTURE.2018.08.040

Yiannas, F. (2018). A new era of food transparency powered by blockchain. *Innovations: Technology, Governance, Globalization*, *12*(1–2), 46–56. https://doi.org/10.1162/INOV_A_00266

Yli-Huumo, J., Ko, D., Choi, S., Park, S., & Smolander, K. (2016). Where is current research on blockchain technology?—A systematic review. *PLoS One*, *11*(10), e0163477. https://doi.org/10.1371/JOURNAL.PONE.0163477

Yu, Y., Li, Y., Tian, J., & Liu, J. (2018). Blockchain-based solutions to security and privacy issues in the Internet of Things. *IEEE Wireless Communications*, *25*(6), 12–18. https://doi.org/10.1109/MWC.2017.1800116

Zaidi, S. M. A., Chandola, V., Allen, M. R., Sanyal, J., Stewart, R. N., Bhaduri, B. L., & McManamay, R. A. (2018). Machine learning for energy-water nexus: Challenges and opportunities. *Big Earth Data*, *2*(3), 228–267. https://doi.org/10.1080/20964471.2018.1526057

Zakariazadeh, A., Ahshan, R., Al Abri, R., & Al-Abri, M. (2024). Renewable energy integration in sustainable water systems: A review. *Cleaner Engineering and Technology*, *18*, 100722. https://doi.org/10.1016/J.CLET.2024.100722

Zhang, N., Wang, M., & Wang, N. (2002). Precision agriculture – A worldwide overview. *Computers and Electronics in Agriculture*, *36*(2–3), 113–132. https://doi.org/10.1016/S0168-1699(02)00096-0

Zhang, X., Zhao, W., Zhang, Y., & Jegatheesan, V. (2021). A review of resource recovery from seawater desalination brine. *Reviews in Environmental Science and Biotechnology*, *20*(2), 333–361. https://doi.org/10.1007/S11157-021-09570-4

Zhang, Y., Liang, Z., Zong, Y., Wang, Y., Liu, J., Chen, K., Qiu, J. L., & Gao, C. (2016). Efficient and transgene-free genome editing in wheat through transient expression of CRISPR/Cas9 DNA or RNA. *Nature Communications*, *7*(1), 1–8. https://doi.org/10.1038/ncomms12617

Zhang, Y., & Zhao, Q. (2024). Interactive augmented reality to support education. *Encyclopedia of Computer Graphics and Games*, 993–1000. https://doi.org/10.1007/978-3-031-23161-2_82

Zhao, G., Liu, S., Lopez, C., Lu, H., Elgueta, S., Chen, H., & Boshkoska, B. M. (2019). Blockchain technology in agri-food value chain management: A synthesis of applications, challenges and future research directions. *Computers in Industry*, *109*, 83–99. https://doi.org/10.1016/J.COMPIND.2019.04.002

Zhou, Z., Liu, W., Cheng, P., Li, Z., Aldieri, L., Zhou, Z., Liu, W., Cheng, P., & Li, Z. (2022). The impact of the digital economy on enterprise sustainable development and its spatial-temporal evolution: An empirical analysis based on urban panel data in China. *Sustainability*, *14*(19), 11948. https://doi.org/10.3390/SU141911948

4 Application of FEW Nexus 4.0 in Food

4.1 FOOD SECURITY AND RESILIENCE

Based on the 1999 World Food Summit, the Food and Agriculture Organization of the United Nations (FAO) gave a concise definition of food security as it relates to physical, social, and economic access by various people in existence; food security is the provision of safe and nutritious food (both liquid and solid) that satisfies the dietary needs for an active and healthy lifestyle (*Policy Brief Changing Policy Concepts of Food Security*, 2006; *The State of Food Security and Nutrition in the World 2024*, 2024). The basis for understanding and formulating policies and designing solutions to tackle food insecurity was made on this foundation (Barrett, 2010; Clapp et al., 2022). Food security measurement is essential for diagnosis and response. However, the concept of food security is not recent; it has evolved over the years. Historically, it is worth noting that from the 1950s through the 1970s, food security was assessed from the national supply side, including food availability, factoring in population growth, and shocks in production. The early days of the 1980s saw food insecurity measurement systems shift attention to the demand side, which emphasised access to food by individual persons (Maxwell Simon et al., 1992). Recently, there has been a focus on food utilisation, particularly on proper nutrition and feeding practices and the stability of these conditions over time (FAO, 1996; FAO et al., 2024).

Contemporary research suggests that the four pillars of food security are widely recognised. They are availability, access, utilisation, and stability. Even though food security is an enormously recognised issue, it is difficult to measure since many of its dimensions are not visible, and the aggregations for large areas are tricky. In this respect, the definitions and measures adopted by the FAO are not easy to apply consistently (Barrett, 2002, 2010; Fanzo, 2015; Headey & Ecker, 2013; Qazi & Al-Mhdawi, 2024; Ume et al., 2022)

While efforts have been made to measure food security, many fall below standard since measuring changes and aggregating data involves weaknesses. Assuming aggregate results do not accurately capture the utilisation of foods or changes over time because much of this literature depends on indices – whether at the household or individual levels. Household indices will show aspects of consumption and access while not showing health or nutrition dimensions, and they are usually static (Sanga et al., 2021).

Some aspects, such as political factors, make measuring food security difficult. Among these is national sovereignty over data collection, further complemented by the need for nationally representative samples. The technology and investment by

DOI: 10.1201/9781003633853-4

the World Bank in funding good, nationally representative longitudinal household surveys create an opportunity for accurate measurement (*Food Insecurity | Food Systems and Hunger Studies*, 2024).

This research domain on resilience, food security, and food systems is growing daily. It accommodates essential issues related to these topics, such as nutrition, biodiversity, political economy, and comparative analyses (Lindgren & Lang, 2023). However, with the ever-increasing literature comes a host of questions about these concepts and their interplay that continue to be discussed despite being resolved. One such question is establishing the adequate practical level at which a policy associated with resilience needs to be grounded. Is it preferable to focus on resilience at the local scale within the food system? If yes, should our focus be on the local food system, with locally driven food system management in times of food system crisis, or towards the more open global systems with food system conduits that connect to the globe? (Fanzo et al., 2021). Some notable scholars, such as Losch and May, tend to think that a "place-based" approach would essentially optimise food system resilience (Béné & Devereux, 2023). Some other notable scholars, such as Hoddinott, pulled out another argument for the respective vulnerability of in-looking systems and instead proposed that systems are integrated based on rich, robust trade markets (Béné & Devereux, 2023).

However, the COVID-19 pandemic brought up issues of many connections in global food systems, leading to the potential for "concatenated shocks" (Swinnen & Vos, 2021). The jury, therefore, had mixed reactions to the question concerning the idea of a local versus a global solution in food security. Some researchers, including Caron, Daguet, and Dury, believe that a possible mixed approach should accommodate local, regional, national, and international processes that allow for fostering diversity and cohabitation of context-specific pathways (Béné & Devereux, 2023)

This following section will first provide a brief overview of food security and resilience before focusing on the economic future evolution of these concepts.

4.1.1 Food Security

The history of food security dates back to the works of the famous British economist Robert Malthus and Italian political economist Giovanni Botero. Both of them focused on balancing the potential capacity for food production and the population size in cities. For centuries, Malthusian interpretations dominated thinking around food security until Amartya Sen's groundbreaking work in 1981. Sen's work detected problems that revolutionised the view of food security by pointing out that famines are caused by the inability of a particular population to access or purchase food rather than mere food shortages. This redefinition expanded the meaning of food security beyond its earlier origins. This new definition eventually gave birth to the four pillars of food security (World Bank Group, 2022).

During the early days of the 1970s, the world food crisis was growing, and the scope of food security then was focused on food availability. Severe fluctuations in agricultural commodity prices and disturbances in currency and energy markets triggered concerns about the price stability of major foodstuffs. Therefore, a definition was needed that would consider the critical needs and behaviours of potentially vulnerable groups affected by the urgency of famine, hunger, and food crises (A Global Food Crisis | World Food Programme, 2024). The World Food Conference held in 1974 provided the following definition of food security: “the availability at all times of adequate food supplies of basic foodstuffs to sustain a steady expansion of food consumption and to offset fluctuations in production and prices to the world”. This emphasises the need for enhanced production to meet the increasing population’s needs to offset protein-energy malnutrition that has affected more than 25% of the global population (Erenstein et al., 2022).

A deep analysis of food security crises raised the need for a shift of focus from the principle of food availability to a much broader approach, instead taking knowledge about the functioning of agricultural markets under stress conditions and the inability of at-risk population groups to access food. As a result, the definition of FAO has widened in scope to include securing access to available supplies by vulnerable groups of people and economic access to food. This prompted an updated definition that emphasises the necessity of ensuring physical and economic access to basic food by all people at all times (Pawlak & Kołodziejczak, 2020).

In the mid-1990s, the terminologies of nutrition security and food and nutrition security began to appear, and finally, food security was adopted under another name that defined the whole concept without separating its parts. Another defining moment occurred in *The State of Food Insecurity in the World* in 2001, where the definition was expanded to add a social dimension to the problem, and it was admitted that poverty alone was not a sufficient condition for inadequate food security (Food Security and Nutrition in the World, 2019).

At the 2009 World Summit on Food Security, food security was last conceptually updated, and a fourth dimension, stability, was added to the concept. Sometime in the year 2020, the Committee on World Food Security proposed the integration of two additional pillars into its definition of food security analytic framework: agency and sustainability (*HLPE Report #15 – Food Security and Nutrition: Building a Global Narrative towards 2030*, 2020). This acceptance of agency represents the ability of persons or groups to negotiate conditions at different levels of food systems associated with decision-making factors in the space of food production processes or activities related to food distribution and policy shaping. This addition underlines the importance of the role of empowerment in food security (Mewes, 2024).

Sustainability is another new pillar that describes the sustainability of food system approaches that protect ecosystems over the long term. Both agency and sustainability shape complex and dynamic interactions among economic, social and environmental systems, without which food security cannot be attained (HLPE Report #15 – Food Security and Nutrition: Building a Global Narrative towards 2030, 2020). The addition of these set of pillars complements that of 2006 and emphasises that it is calling for consideration to be offered both to long-term effects and the needs of future generations (Stephanie Safdie, 2024).

Putting all the earlier theories together, we can define food security in simple terms to mean the availability of adequate, affordable, and nutritious food that does not compromise an individual’s ability to meet other essential basic requirements of life while being food secure, regardless of their religious or social status, gender, economic class, or geographical location (*Brundtland Report | Sustainable Development & Global*

Environmental Issues | Britannica, 2019). It implies access to food for all, a prerequisite to staying active and healthy (Odey et al., 2022). Another component of food security is household food security. Household food security, on the other hand, implies that all family unit members have access to enough food at all times. This implies the state of being free of hunger and having no anxiety of going without food. Food security also means the capacity of a population to absorb and stop anticipating the effects of food supply shocks like floods and droughts, economic and political instability, and other man-made or natural catastrophes (Tchonkouang et al., 2024).

Food insecurity, on the other hand, denotes limitations and uncertainty in the acquisition or consumption of appropriate food in a manner that compromises the ability or readiness to live healthily. Chronic food insecurity, also called structural food insecurity, is a condition of the constant and limited ability of access to food in other access to food to meet household needs due to lack of income. This is very closely related to transitory food insecurity, where annual cycles of transitory food insecurity episodically determine the vulnerability of households to long-term food insecurity (USDA, 2023).

4.1.1.1 Understanding Food Security: A View from Core Concepts to Practical Approaches

Food security is best understood as a dynamic pathway that links production to consumption, covering integrated areas rather than being viewed as four independent pillars (Clapp et al., 2022). While the 2009 World Summit on Food Security did attempt to define its four pillars, namely availability, accessibility, utilisation, and stability, in order to encapsulate what is now considered food security in actuality, this visualisation can be deceptive because it does not capture the interconnected and interdependent nature of those dimensions. Viewing food security as pillars suggests static and separate elements, whereas they are intricately linked in reality.

The analogy of pillars further suggests equal weight, just as if each of the four dimensions carries the same value at 25%, but this may not be true. In an actual sense, the importance of each dimension varies across different contexts, and it is country-specific (Clapp et al., 2022).For instance, in most developing countries, the accessibility to food may be limited physically due to a lack of transport infrastructure. Major hindrances in developed countries may be economic access. Similarly, following a natural disaster, such as an earthquake, all dimensions comprising availability, accessibility, utilisation, and stability have become equally important. Such situations, therefore, call for differentiated responses. Consequently, the emphasis on dimensions in issues relating to food security does not carry equitable weight in either context or between climes (Peng & Berry, 2018). A representation of the interconnection between sustainable food security, food availability, nutrition, and health outcomes is shown in Figure 4.1.

The relationships among the four dimensions of food security can also be represented as follows: food production (i.e., availability), household access (i.e., accessibility), and individual utilisation. Accessibility includes physical (transport and infrastructure) and economic means (food purchasing power) as well as socio-cultural access and preferences, hence considering social protection (Reyes et al., 2021); stability gives a time dimension, not a long term, to food security. Besides the one-way pathway, food security can also be seen as circular, with a feedback loop from utilisation to availability, because human capital depends on productivity and optimal nutrition status for the workforce of all production sectors (Clapp et al., 2022). These concepts are summarised in Figure 4.1. An important insight from this figure is the significant role of food losses (in agriculture, post-harvest, and distribution) and food waste (from processing, and households, and communities), which may amount globally to one-third of the food available. It is extremely important that these issues are addressed to improve food security (Chauhan et al., 2021); however, it is a great challenge to secure world food availability in the future. From a systemic perspective, obesity can be considered a form of food waste. These pillars are summarised below:

Availability: Food access, which is one of the dimensions of the availability piller of food security, comprises the process of getting food through purchase, production, distribution or barter (Ashley, 2018). Availability of productive land, management of the existing land, crop and animal husbandry methods and climatic conditions play an essential role in food production (Jónsdóttir & Gísladóttir, 2023). However, factors such as competition for land for farming activities urban development, adverse effects on the natural environment, and poor and reckless approaches to farming may pull down production (FAO, 2023). For instance, while nations such as Japan

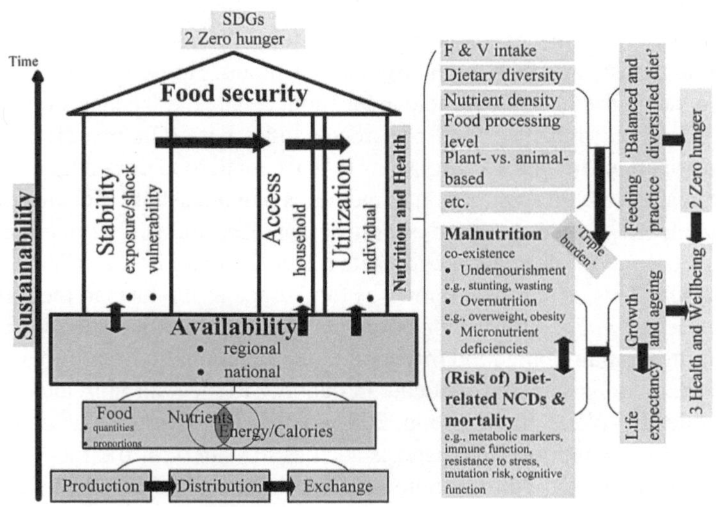

FIGURE 4.1 The Dimensions of Food Security (Moda et al., 2025).

In Figure 4.1, F & V = fruit and vegetable; NCD = non-communicable disease; SDGs = Sustainable Development Goals. Apricot text boxes refer to SDGs. Blue text boxes represent components of the "utilization" pillar to which sustainability indicators are applied. Green text boxes represent components of the "availability" pillar, the basis of food security. Arrows indicate the direction of contribution when variation in sustainability indicators occurs. The time axis on the left represents the time dimension of food security (Moda et al., 2025).

and Singapore embrace food security, they lack ample arable land, which is good enough for crop farming (FAO, 2021).

Another aspect of food production is the task of food distribution facilities. Food distribution includes activities such as storing, preliminary processing, transporting, packing, and promoting food products, which are essential for supplying food to different areas or countries. Physical facilities and structures of storage affect the magnitude of wastage during distribution and transportation, while unfavourable transport networks affect the costs of supplies. Also, because very few persons or households can be complete producers of food crops, effective trading mechanisms and market structures are a necessity in the exchange of food (García-Flores et al., 2015).

However, despite the increased global per capita food availability, people still face food insecurity due to variability in accessibility. Having prosperous economies is vital in obtaining food, and this shows that trading and market structures are essential in food security (World Bank Group, 2023).

Access: Food access is the affordability of food, the distribution of food, and an individual or family's choice of food (USDA, 2024). Poverty remains the major argument for the lead to high vulnerability if the issue of limited access to food and most probably leads to high vulnerability if price spikes in food continue to hold (Headey, 2016). Access depends on whether a household has enough income to purchase food or has sufficient resources for growing its food (Ogunniyi et al., 2024). There are basically two types of access: direct access, in which the household produces its food, and economic access, in which the household buys food. Household assets, household demographics, household education level, and gender of the head of household influence food preference and access (Olumba et al., 2023). Investigations in food security show that access to food does not necessarily mean adequate intakes for all the members of a household. Socially acceptable means of access to food are important. Research shows that the value of global food exports has skyrocketed from 2000 until 2021 (FAO, 2023).

Utilisation: This term is also known as food usage as it relates to the pillars of food security. Food utilisation covers a large variety of metabolic functions ranging from in-home food ingestion to the functionality required to ensure that the food is safe and adequate in quantity to meet an individual's nutrition needs (World Bank Group, 2024). A food product can be considered safe for consumption from aspects such as the safety of the food, preparation methods, and culture, among other things (Nordhagen et al., 2022). Also, healthcare services and sanitation facilities are largely responsible for improving food consumption, as they prevent infections that otherwise might hamper the nutritional acquisition process (Herrero et al., 2023). Nutrition education and knowledge transfer in food processing and preparation support this pillar of food security (Herrero et al., 2023)

Stability: Food stability is the availability of food at all times and seasons. This may be threatened by several factors that result in food insecurity (Mirzabaev et al., 2023). Transitory food insecurity is the temporary unavailability of food, which is often caused by disasters such as acts of nature, droughts, civil strife, and instability in the markets that cause an increase in prices. The other form of transitory insecurity is seasonal food insecurity, which arises from the predictable changes in food availability because of different production seasons. Moreover, temporary food insecurity could be caused by reasons such as loss of employment or productivity due to illness (World Bank Group, 2022).

Sustainability: Sustainability refers to the long-term capacity of food systems to provide both food security and nutrition and food sufficiency in ways that do not jeopardise or distort, either at local, national, regional, or global levels, the economic, social, and environmental bases that generate food security and nutrition for present and future generations (Thomas et al., 2022).

4.2 INTERCONNECTION BETWEEN FOOD SECURITY AND SUSTAINABILITY

Sustainable diets create a relationship between food security and sustainable food systems. The definition of sustainable diets shows this relation. Typical examples of sustainable diets are considered those that are "protective and respectful of biodiversity and ecosystems, culturally acceptable, accessible, economically fair and affordable; nutritionally adequate, safe and healthy; while optimising natural and human resources." (Brouwer et al., 2020). Similarly, a sustainable food system means one that ensures Food Security and Nutrition (FSN) for all without compromising the economic, social, and environmental bases upon which the provision for FSN rests for future generations (Brouwer et al., 2020).

It is widely recognised globally that climate change poses a severe threat to the sustainability of food security. However, actions taken in food systems generate about 20%–30% of all human-associated GHG emissions that contribute to climate change (Mirzabaev et al., 2023). This potential trade-off relationship exists between lowering human-associated GHG emissions and ensuring food security under the existing food system structure. A systematic and integrated approach is needed to ensure both the FSN requirements of the short and longer term and simultaneously lower its negative environmental impact due to GHG emissions from food system function activities. Although the precise features of sustainable food systems are uncertain, our understanding of them will be necessary (Fanzo et al., 2022).

4.3 ACHIEVING FOOD SECURITY THROUGH A FOOD SYSTEMS APPROACH

The concept of food security, as defined by the 1996 World Food Summit, is much broader than it is in today's context. The definition covers not only the physical and economic availability of food but also several other societal elements, including the built environment and economic conditions to the state of being free from fear related to food insecurity (Clapp et al., 2022). The COVID-19 pandemic has further exposed how food systems are deeply interconnected with other societal systems, underlining how income loss and reduced economic activity have increased food poverty (Davila et al., 2021).

The original idea behind the modern food security efforts dates as far back as the early years of the UN, back in 1943, which had the goal of zero hunger, through to some of the more recent initiatives, such as those of SDG 2 and the Food Systems Summit in 2021 (Fanzo, 2023). Despite these efforts, the problem of food security continues. Undernourishment and malnutrition continue to rise and ravage millions around the globe (Adeyeye et al., 2023). Initial strategies focused on increasing food supply have been broadened to address malnutrition through the availability, access, utilisation, and stability pillars (Clapp et al., 2022). Recently, these strategies have encountered environmental sustainability-focused interest in agricultural productivity (Muhie, 2022). More recently, what constitutes food security is pinned under a food system approach, constituting complex interacting factors like poverty, war, and climatic change (Britwum & Demont, 2022). Sector-wise strategies for food security, such as the Green Revolution, involve technological solutions like genetically modified organisms and short-term measures of food aid. Such challenges need a proper balance between innovation and sustainability (Muhie, 2022). Innovation in food security has led to sustainable solutions by sourcing comprehensive solutions linking sectorial and community needs (Adenle et al., 2019).

The concept of food systems has been instrumental in addressing food security by providing a holistic approach beyond the conventional fragmented methods known to us. Food systems thinking is drawn from a holistic perspective, whereby sectoral and technical strategies deal only with part of the system and overlook the interlinked actors and governance structures. This approach considers complex influences and relationships within the system and analyses the positive and negative impacts (Leach et al., 2020). Figure 4.2 shows a diagrammatic representation of the food systems framework.

Components of a food system include food supply chains, the environment, individual behaviours, and dietary patterns. It also includes all the various outcomes of a food system represented by nutrition and health, environmental and economic effects on livelihoods, wages, and social equity considerations. It also puts forward drivers, many of which are exogenous to the food system; however, these drivers push or pull the system in other directions with great strength (Tefft et al., 2020). Therefore, this becomes a more comprehensive approach toward understanding what is happening within the food systems, thus making it possible to devise strategies to meaningfully improve food security and sustainability.

4.3.1 Food Systems Interactions with Food Chains

Food production, processing, storage, distribution, retailing, and consumption form a complex chain-like structure known as the food chain. The food supply chains are complex networks covering all stages from production to storage, processing, distribution, and retailing. These stages are shaped by decisions at multiple geographical levels ranging from local to global (Carbone, 2017). These chains connect the process of producing food with that of consuming it and, in so doing, affect the availability, accessibility, and diversity of food that is made available. The food environment is

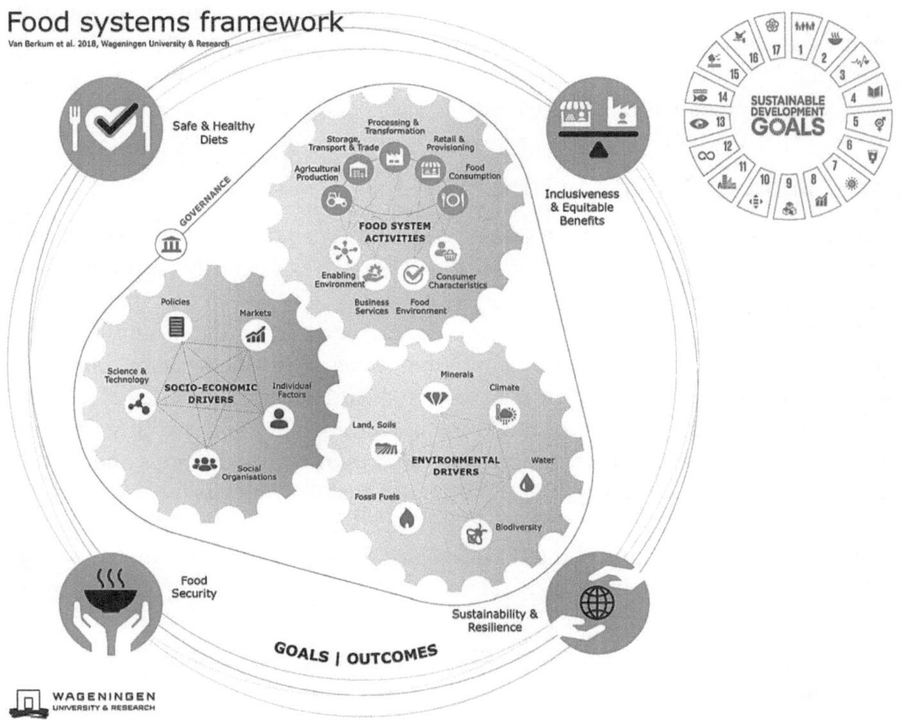

FIGURE 4.2 A Diagrammatic Representation of the Food System Framework (Brouwer et al., 2020).

strongly related to informal and formal market chains because it affects consumption patterns, choice of food, diet, and nutritional health outcomes (Ambikapathi et al., 2021). Furthermore, links across food systems bear key health, nutritional, environmental, and social equity implications (Ericksen, 2008; Fanzo et al., 2021; Martin et al., 2024; Webb et al., 2006). These links have been positively and negatively affected by drivers like climate variability, urbanisation, population growth, and policy changes (Béné, 2020).

The food systems approach brings out the interdependence between food production, distribution, and consumption within and across such multiple sectors as agriculture, trade, policy, health, environment, education, transport, and infrastructure. According to Fanzo et al. (2021), it provides a holistic and interlinked strategy for dealing effectively with food security concerns.

4.3.2 Connections between Food Security and Food Insecurity

Concepts of food security and insecurity take the shape of dynamic, reciprocal, and time-dependent relationships often characterised by the interaction of challenges brought forth by food insecurity and the strategies applied to address it. At any stage of food security, challenges may manifest. The resultant coping mechanisms are then activated at the level of the challenge, which may be at the national, household, or individual levels. These twin processes are linked in a linear order and contain a feedback loop that demands adjustments in the coping mechanism to arrive at a food-secure situation (Peng et al., 2018).

4.3.3 Resilience Thinking in Food Systems

Resilience thinking in social-ecological systems has been derived from the concept of ecological resilience introduced by Holling (1973). It is variously described as an approach rather than a set of hypotheses.

There are numerous definitions of resilience depending on the disciplinary setting; the unifying feature in all of them is that it comprises the systems' ability to cope and continue functioning in response to disturbances (Boston et al., 2024). Resilience can generally be defined in terms of sustainability as the ability to retain performance and developmental progress in the presence of disruptions and perturbations (Larrea-Gallegos et al., 2022). While sustainability implies the ability of the system to function in the future (Brown et al., 1987; Darvishi et al., 2023). Therefore, resilience and sustainability are intertwined since resilience helps sustain the system after a disturbance (Carissimi et al., 2023).

4.4 APPLICATIONS OF FOOD SECURITY AND RESILIENCE TO THE FEW 4.0 NEXUS

A new range of approaches applicable to the governance of natural resources so that basic human needs for water, energy, and food are satisfied in an environmentally sustainable, economically viable, and socially inclusive way would be an excellent idea for the 21st century. Our natural resources would have to be resilient to withstand and recover from a shock or disaster (Hogeboom et al., 2021; Platt et al., 2016).

Resilience thinking takes centre stage as a response to tackle the challenges of the 21st century, such as ensuring water, energy, and food for all within the boundaries of an increasingly changing Earth system and under the expanding pressure of climate change (Hogeboom et al., 2021)

The significant economic progress of the 20th century, noted by United Nations Development Programme in 2016, has pulled millions of people out of poverty while improving access to water, energy, and food for many. However, this progress has been achieved at the expense of massive losses to natural resources. Examples are the degradation of aquatic and terrestrial ecosystems is now widespread; most resources are depleted; extinction rates of species have attained alarming proportions, and many regions are increasingly vulnerable to shocks (*Sustainable Development Goals | United Nations Development Programme*, 2021).

The global development boom has, therefore, left millions of people behind. Whereas three in every ten people lack access to safe drinking water, equivalent to 2.1 billion people worldwide, six in every ten lack safely managed facilities for sanitation purposes (UNICEF, 2021). Furthermore, almost 1 billion people still lack access to electricity, more than 820 million lack adequate food supply, and many more lack healthy diets, which causes premature death and morbidity.

The insecurities of water and food supply and their associated negative impacts are likely to become even more severe due to population growth, resource-intensive lifestyles, and increased vulnerability to disruptive shocks such as climate change (Abbass et al., 2022). The achievement of numerous UN SDGs, particularly the food, water, and energy-related SDGs, that is, SDG 2, SDG 6, and SDG 7, respectively, calls for transformational change at all levels: environmental, social, and economic, as duly noted by United Nations (2023). One of the significant challenges of the Anthropogenic epoch is to provide the basic human necessities of water, energy, and food in an environmentally sustainable, economically viable, and socially inclusive way that has the resilience to cope with shocks and disasters (Roka, 2020).

The term "resilience thinking" was developed about scientific discourse, especially when Holling put it forward in 1973. The word originally came from the subject of ecology and resilience thinking, which is based on sustainability science and research for global changes (Scordato & Gulbrandsen, 2024). In complex and unpredictable environments, there are high levels of uncertainty, which could sometimes mean the availability of shocks and disasters and their capability to spread across various scales and systems; this negatively affects the performance of environmental systems (Kourtit et al., 2023). Hence, the idea of resilience thinking can proactively achieve the development of robust and durable structures that operate reliably amid inevitable disturbances, whether acute, such as a pandemic, or chronic, such as climate change (Folke et al., 2020).

Resilience and nexus frameworks are old. Both enjoy the active support of different scholarly communities. As (Al-Saidi & Elagib, 2017) said, "Nexus thinking traces back to environmental policy studies-debates and public discourse on natural resources development-planning and management." Similarly, resilience also had its genesis in related scientific discourse on questions of sustainability and system thinking. The word nexus was first presented at the World Economic Forum in 2011, and many authors marked the start of resilience thinking from one of the most cited works by Holling (1973) especially while discussing the FEW nexus. While the term nexus already conveys inter-linkages between the water, energy, and food systems, resilience is more about the capacity of a system to react to threats and preserve its functionality to deliver benefits (Nhamo et al., 2023)

This relationship between food security, resilience, and the FEW nexus is one of closeness. Their proximity makes these topics critical areas of attention for sustainable development, environmental management, and global resilience to climate change and systemic challenges. These topics find important applications in the FEW nexus.

Some of the applications of food security and resilience in the context of food energy and water nexus 4.0 include:

A. **Integrated Resource Management**: Building food security and resilience in the food-energy and water nexus allows for implementing integrated water, energy, and agricultural planning systems to optimise resource use and minimise conflicts within the context of the FEW Nexus 4.0. In addition, deploying precision agriculture techniques and the prospect of integrated resource management will be an added advantage in developing systems that can optimise water and energy inputs based on real-time data and environmental conditions (Kushwaha et al., 2024; Ogbolumani & Nwulu, 2024).

B. **Climate Resilience**: Climate change needs to be managed flexibly where all climatic variations, in general, and those of extreme climate fluctuation, in particular, are managed. Another area where new technologies for FEW will come in handy is in conjunction with better planning of land use and more efficient water resources to improve the ecological aspects within the systems. For instance, the water and fishery resources of the continent may be affected by irrigation development, leading to an increase in the level of pollutants in water due to competing water needs. Therefore, many resilience strategies of FEW have social implications since some options can cause huge impacts on sectors, and many people dependent on FEW resources are selected out of several options (Vörösmarty et al., 2023). For the concepts of FEW and food resilience to be effective when integrated with climate change, we need to design systems for climate-resilient agricultural technologies and practices, for instance, drought-resistant crops, efficient irrigation systems, and agroforestry to enhance yields under unpredictable climatic conditions (Apeh & Nwulu, 2024). Serious investment needs to be put into renewable energy sources and efficient energy technologies to reduce dependence on fossil fuels and minimise hazards relating to climate impacts.

C. **Policy and Governance**: Water, energy, food, and other related sectors like these offer a lot of opportunities for investment in creating jobs and innovations for better competitiveness (Orimoloye, 2022). It is, however, notable that not all the listed techniques may be suitable for use in all countries or will be adopted. Some studies underscore the need for countries to take integrated approaches seriously. Policies will be formulated to sustain resource management, foster innovation in food production and energy efficiency, assure equality in resource access, strengthen governance structures, and coordinate actions across sectors and ensure multiple stakeholders' involvement. Resilience and food security issues must also be embedded in national and regional planning procedures (Ogbolumani & Nwulu, 2021, 2024; Rasul, 2016).

D. **Community Engagement and Capacity Building**: The deployment of Public Participation in Scientific Research (PPSR) extends beyond the food-energy-water research field in efforts toward enhancing local, social, technical, political, and scientific capacity in communities (Hinojosa et al., 2021). Most approaches utilised in PPSR consider multiple levels of engagement lined up to achieve various social, political, or economic community-based goals (Hibbett et al., 2020). Approaches of this nature are widely employed in environmental science and conservation contexts to gather localised data on socio-environmental systems, including marine conservation and sustainable water management (Picone et al., 2020). It builds community scientific literacy, analytical skills, and an understanding of the uncertainties and complexities inherent to local systems, thereby improving community policy-making capacities beyond just collecting "last mile" data (Betz et al., 2023). Data-centred dialogues enhance social networks and build citizens (Tye et al., 2022) awareness about nexus thinking concepts that are otherwise hard to make into policy. It builds stakeholders' participation at all levels, which is necessary for adaptive and flexible governance and, hence, for effective socio-environmental management. Another important outcome of PPSR is empowerment, ensuring that the vulnerable population affected by threats to socio-environmental systems gets space in the decision-making processes. Empowerment is achieved by equipping citizens with analytical tools and data to underpin perspectives (Bonney et al., 2009). Whereas many citizen-led approaches characterise environmental management and conservation, their application in nexus management, represents untapped potential for resolving FEW nexus challenges through building community resilience (Hibbett et al., 2020).

4.5 SUMMARY

This chapter explains how FEW 4.0 is applied to food security and resilience. It defined the four pillars of food security as availability, access, utilisation, and stability before explaining each pillar in depth in later chapters. This chapter shows that despite the awareness of the importance of food security, ensuring food meets its standard quality has proven difficult due to the flaws in measuring food security. It explains how to achieve food security through a food systems approach and how these systems interact with food chains. Then, it describes how FEW Nexus 4.0 fits into the vision of food security.

REFERENCES

A Global Food Crisis | World Food Programme. (2024). https://www.wfp.org/global-hunger-crisis

Abbass, K., Qasim, M. Z., Song, H., Murshed, M., Mahmood, H., & Younis, I. (2022). A review of the global climate change impacts, adaptation, and sustainable mitigation measures. *Environmental Science and Pollution Research*, *29*(28), 42539–42559. https://doi.org/10.1007/S11356-022-19718-6

Adenle, A. A., Wedig, K., & Azadi, H. (2019). Sustainable agriculture and food security in Africa: The role of innovative technologies and international organizations. *Technology in Society*, *58*, 101143. https://doi.org/10.1016/J.TECHSOC.2019.05.007

Adeyeye, S. A. O., Ashaolu, T. J., Bolaji, O. T., Abegunde, T. A., & Omoyajowo, A. O. (2023). Africa and the nexus of poverty, malnutrition and diseases. *Critical Reviews in Food Science and Nutrition*, *63*(5), 641–656. https://doi.org/10.1080/10408398.2021.1952160

Al-Saidi, M., & Elagib, N. A. (2017). Towards understanding the integrative approach of the water, energy and food nexus. *Science of The Total Environment*, *574*, 1131–1139. https://doi.org/10.1016/J.SCITOTENV.2016.09.046

Ambikapathi, R., Shively, G., Leyna, G., Mosha, D., Mangara, A., Patil, C. L., Boncyk, M., Froese, S. L., Verissimo, C. K., Kazonda, P., Mwanyika-Sando, M., Killewo, J., & Gunaratna, N. S. (2021). Informal food environment is associated with household vegetable purchase patterns and dietary intake in the DECIDE study: Empirical evidence from food vendor mapping in peri-urban Dar es Salaam, Tanzania. *Global Food Security*, *28*, 100474. https://doi.org/10.1016/J.GFS.2020.100474

Apeh, O. O., & Nwulu, N. I. (2024). The water-energy-food-ecosystem nexus scenario in Africa: Perspective and policy implementations. *Energy Reports*, *11*, 5947–5962. https://doi.org/10.1016/J.EGYR.2024.05.060

Ashley, J. M. (2018). Importance of local knowledge in building resilience. *Human Resilience Against Food Insecurity*, 125–145. https://doi.org/10.1016/B978-0-12-811052-2.00008-1

Barrett, C. B. (2002). Food security and food assistance programs. *Handbook of Agricultural Economics*, *2, Part* 2, 2103–2190. https://ideas.repec.org/h/eee/hagchp/4-40.html

Barrett, C. B. (2010). Measuring food insecurity. *Science*, *327*(5967), 825–828. https://doi.org/10.1126/SCIENCE.1182768

Béné, C. (2020). Resilience of local food systems and links to food security – A review of some important concepts in the context of COVID-19 and other shocks. *Food Security*, *12*(4), 805–822. https://doi.org/10.1007/S12571-020-01076-1

Béné, C., & Devereux, S. (Eds.). (2023). *Resilience and Food Security in a Food Systems Context*. https://doi.org/10.1007/978-3-031-23535-1

Betz, U. A. K., Arora, L., Assal, R. A., Azevedo, H., Baldwin, J., Becker, M. S., Bostock, S., Cheng, V., Egle, T., Ferrari, N., Schneider-Futschik, E. K., Gerhardy, S., Hammes, A., Harzheim, A., Herget, T., Jauset, C., Kretschmer, S., Lammie, C., Kloss, N., … Zhao, G. (2023). Game changers in science and technology – Now and beyond. *Technological Forecasting and Social Change*, *193*, 122588. https://doi.org/10.1016/J.TECHFORE.2023.122588

Bonney, R., Cooper, C. B., Dickinson, J., Kelling, S., Phillips, T., Rosenberg, K. V., & Shirk, J. (2009). Citizen science: A developing tool for expanding science knowledge and scientific literacy. *BioScience*, *59*(11), 977–984. https://doi.org/10.1525/bio.2009.59.11.9

Boston, M., Bernie, D., Brogden, L., Forster, A., Galbrun, L., Hepburn, L. A., Lawanson, T., & Morkel, J. (2024). Community resilience: A multidisciplinary exploration for inclusive strategies and scalable solutions. *Resilient Cities and Structures*, *3*(1), 114–130. https://doi.org/10.1016/J.RCNS.2024.03.005

Britwum, K., & Demont, M. (2022). Food security and the cultural heritage missing link. *Global Food Security*, *35*, 100660. https://doi.org/10.1016/J.GFS.2022.100660

Brouwer, I. D., McDermott, J., & Ruben, R. (2020). Food systems everywhere: Improving relevance in practice. *Global Food Security*, *26*, 100398. https://doi.org/10.1016/J.GFS.2020.100398

Brown, B. J., Hanson, M. E., Liverman, D. M., & Merideth, R. W. (1987). Global sustainability: Toward definition. *Environmental Management*, *11*(6), 713–719. https://doi.org/10.1007/BF01867238

Brundtland Report | Sustainable Development & Global Environmental Issues | Britannica. (2019). https://www.britannica.com/topic/Brundtland-Report

Carbone, A. (2017). Food supply chains: Coordination governance and other shaping forces. *Agricultural and Food Economics*, *5*(1), 1–23. https://doi.org/10.1186/S40100-017-0071-3

Carissimi, M. C., Creazza, A., & Colicchia, C. (2023). Crossing the chasm: Investigating the relationship between sustainability and resilience in supply chain management. *Cleaner Logistics and Supply Chain*, *7*, 100098. https://doi.org/10.1016/J.CLSCN.2023.100098

Chauhan, C., Dhir, A., Akram, M. U., & Salo, J. (2021). Food loss and waste in food supply chains. A systematic literature review and framework development approach. *Journal of Cleaner Production*, *295*, 126438. https://doi.org/10.1016/J.JCLEPRO.2021.126438

Clapp, J., Moseley, W. G., Burlingame, B., & Termine, P. (2022). Viewpoint: The case for a six-dimensional food security framework. *Food Policy*, *106*, 102164. https://doi.org/10.1016/J.FOODPOL.2021.102164

Darvishi, Y., Karami, H., & Goodarzian, F. (2023). Sustainable development in oxygenated fuels. *Advancement in Oxygenated Fuels for Sustainable Development*, 315–330. https://doi.org/10.1016/B978-0-323-90875-7.00013-7

Davila, F., Bourke, R. M., McWilliam, A., Crimp, S., Robins, L., van Wensveen, M., Alders, R. G., & Butler, J. R. A. (2021). COVID-19 and food systems in Pacific Island Countries, Papua New Guinea, and Timor-Leste: Opportunities for actions towards the sustainable development goals. *Agricultural Systems*, *191*, 103137. https://doi.org/10.1016/J.AGSY.2021.103137

Erenstein, O., Jaleta, M., Sonder, K., Mottaleb, K., & Prasanna, B. M. (2022). Global maize production, consumption and trade: Trends and R&D implications. *Food Security*, *14*(5), 1295–1319. https://doi.org/10.1007/S12571-022-01288-7

Ericksen, P. J. (2008). Conceptualizing food systems for global environmental change research. *Global Environmental Change*, *18*(1), 234–245. https://doi.org/10.1016/J.GLOENVCHA.2007.09.002

Fanzo, J. (2015). Ethical issues for human nutrition in the context of global food security and sustainable development. *Global Food Security*, *7*, 15–23. https://doi.org/10.1016/J.GFS.2015.11.001

Fanzo, J. (2023). *Achieving Food Security through a Food Systems Lens*. 31–52. https://doi.org/10.1007/978-3-031-23535-1_2

Fanzo, J., Haddad, L., Schneider, K. R., Béné, C., Covic, N. M., Guarin, A., Herforth, A. W., Herrero, M., Sumaila, U. R., Aburto, N. J., Amuyunzu-Nyamongo, M., Barquera, S., Battersby, J., Beal, T., Bizzotto Molina, P., Brusset, E., Cafiero, C., Campeau, C., Caron, P., … Rosero Moncayo, J. (2021). Viewpoint: Rigorous monitoring is necessary to guide food system transformation in the countdown to the 2030 global goals. *Food Policy*, *104*, 102163. https://doi.org/10.1016/J.FOODPOL.2021.102163

Fanzo, J., Rudie, C., Sigman, I., Grinspoon, S., Benton, T. G., Brown, M. E., Covic, N., Fitch, K., Golden, C. D., Grace, D., Hivert, M. F., Huybers, P., Jaacks, L. M., Masters, W. A., Nisbett, N., Richardson, R. A., Singleton, C. R., Webb, P., & Willett, W. C. (2022). Sustainable food systems and nutrition in the 21st century: A report from the 22nd annual Harvard Nutrition Obesity Symposium. *The American Journal of Clinical Nutrition*, *115*(1), 18. https://doi.org/10.1093/AJCN/NQAB315

FAO. (1996, November 13). *World Food Summit – Final Report – Part 1*. FAO Rome. https://www.fao.org/4/w3548e/w3548e00.htm

FAO. (2021). *Food security and nutrition in the world the state of transforming food systems for food security, improved nutrition and affordable healthy diets for all*. https://doi.org/10.4060/cb4474en

FAO. (2023). The State of food security and nutrition in the world 2023. In *The State of Food Security and Nutrition in the World 2023*. https://doi.org/10.4060/CC3017EN

FAO, IFAD, UNICEF, WFP, & WHO. (2024). *Financing to End Hunger, Food Insecurity, and Malnutrition in All Its Forms*. https://doi.org/10.4060/cd1276en

Folke, C., Carpenter, S. R., Chapin, F., Gaffney, O., Galaz, V., Hoffmann, H., Lamont, M., Polasky, S., Rockstrom, J., Scheffer, M., Westley, F., & Österblom, H. (2020). Our future in the Anthropocene Biosphere: Global sustainability and resilient societies. *SSRN Electronic Journal*. https://doi.org/10.2139/SSRN.3671766

Food Insecurity | Food Systems and Hunger Studies. (2024, August 23). https://hunger.ca.uky.edu/food-insecurity

Food Security and Nutrition in the World the State of. (2019). www.fao.org/publications

García-Flores, R., de Souza Filho, O. V., Martins, R. S., Martins, C. V. B., & Juliano, P. (2015). Using logistic models to optimize the food supply chain. *Modeling Food Processing Operations*, 307–330. https://doi.org/10.1016/B978-1-78242-284-6.00011-8

Headey, D. (2016). Food Prices and Poverty. Policy Research Working Paper; No. 7898. © World Bank. http://hdl.handle.net/10986/25701

Headey, D., & Ecker, O. (2013). Rethinking the measurement of food security: From first principles to best practice. *Food Security*, *5*(3), 327–343. https://doi.org/10.1007/s12571-013-0253-0

Herrero, M., Hugas, M., Lele, U., Wirakartakusumah, A., & Torero, M. (2023). A shift to healthy and sustainable consumption patterns. *Science and Innovations for Food Systems Transformation*, 59–85. https://doi.org/10.1007/978-3-031-15703-5_5

Hibbett, E., Rushforth, R. R., Roberts, E., Ryan, S. M., Pfeiffer, K., Bloom, N. E., & Ruddell, B. L. (2020). Citizen-led community innovation for food energy water nexus resilience. *Frontiers in Environmental Science*, *8*, 571614. https://doi.org/10.3389/fenvs.2020.571614

Hinojosa, L., Riedy, R., Polman, J., Swanson, R., Nuessle, T., & Garneau, N. (2021). Expanding public participation in science practices beyond data collection. *Citizen Science: Theory and Practice*, *6*(1), 1–15. https://doi.org/10.5334/CSTP.292

HLPE Report #15 – Food Security and Nutrition: Building a Global Narrative towards 2030. (2020). https://www.unscn.org/en/news-events/recent-news?idnews=2091

Hogeboom, R. J., Borsje, B. W., Deribe, M. M., van der Meer, F. D., Mehvar, S., Meyer, M. A., Özerol, G., Hoekstra, A. Y., & Nelson, A. D. (2021). Resilience meets the water–energy–food nexus: Mapping the research landscape. In *Frontiers in Environmental Science* (Vol. 9). Frontiers Media S.A. https://doi.org/10.3389/fenvs.2021.630395

Holling, C. S. (1973). Resilience and stability of ecological systems. *Annual Review of Ecology and Systematics*, *4*(1), 1–23. https://doi.org/10.1146/ANNUREV.ES.04.110173.000245

Jónsdóttir, S., & Gísladóttir, G. (2023). Land use planning, sustainable food production and rural development: A literature analysis. *Geography and Sustainability*, *4*(4), 391–403. https://doi.org/10.1016/J.GEOSUS.2023.09.004

Kourtit, K., Nijkamp, P., & Banica, A. (2023). An analysis of natural disasters' effects – A global comparative study of 'Blessing in Disguise.' *Socio-Economic Planning Sciences*, *88*, 101599. https://doi.org/10.1016/J.SEPS.2023.101599

Kushwaha, M., Singh, S., Singh, V., & Dwivedi, S. (2024). Precision farming: A review of methods, technologies, and future prospects. *International Journal of Environment, Agriculture and Biotechnology*, *9*(2), 242–253. https://doi.org/10.22161/IJEAB.92.27

Larrea-Gallegos, G., Benetto, E., Marvuglia, A., & Gutiérrez, T. N. (2022). Sustainability, resilience and complexity in supply networks: A literature review and a proposal for an integrated agent-based approach. *Sustainable Production and Consumption*, *30*, 946–961. https://doi.org/10.1016/J.SPC.2022.01.009

Leach, M., Nisbett, N., Cabral, L., Harris, J., Hossain, N., & Thompson, J. (2020). Food politics and development. *World Development*, *134*, 105024. https://doi.org/10.1016/J.WORLDDEV.2020.105024

Lindgren, K.-A., & Lang, T. (2023). *Food Security and the Fractured Consensus on Food Resilience: An Analysis of Development Agency Narratives* (pp. 81–146). https://doi.org/10.1007/978-3-031-23535-1_4

Martin, A. E., Haines, J., & Fraser, E. D. G. (2024). Development of the food systems literacy competencies framework for youth: A modified Delphi study with experts. *Food Policy*, *128*. https://doi.org/10.1016/j.foodpol.2024.102702

Simon, M., Marisol, S., & Pointing, J. (1992). IFAD-HH-food-security-definitions. *Household Food Security: Concepts, Indicators, Measurements*, *1*(72), 1–72.

Mewes, A. (2024). *Agency in Food Systems – A Metric Approach and Its Implications for Food Security —Research@WUR*. https://research.wur.nl/en/projects/agency-in-food-systems-a-metric-approach-and-its-implications-for

Mirzabaev, A., Bezner Kerr, R., Hasegawa, T., Pradhan, P., Wreford, A., Cristina Tirado von der Pahlen, M., & Gurney-Smith, H. (2023). Severe climate change risks to food security and nutrition. *Climate Risk Management*, *39*, 100473. https://doi.org/10.1016/J.CRM.2022.100473

Moda, M., Anang, D. M., Sobukola, O., Hussain, M. A., Li, L., Kalu, A., Wu, X., & Naumovski, N. (2025). Sustainable food security and nutritional challenges. *Sustainability, 17*(3), 874. https://doi.org/10.3390/SU17030874

Muhie, S. H. (2022). Novel approaches and practices to sustainable agriculture. *Journal of Agriculture and Food Research, 10*, 100446. https://doi.org/10.1016/J.JAFR.2022.100446

Nhamo, L., Tazvinga, H., Mpandeli, S., Liphadzi, S., Botai, J., & Mabhaudhi, T. (2023). Measuring the sustainability of water, energy and food resources in the context of the WEF nexus. In *Nourishing Tomorrow: Clean Engineering and Nature-Friendly Living* (pp. 131–157). https://doi.org/10.1142/9789811264375_0003

Nordhagen, S., Lambertini, E., DeWaal, C. S., McClafferty, B., & Neufeld, L. M. (2022). Integrating nutrition and food safety in food systems policy and programming. *Global Food Security, 32*, 100593. https://doi.org/10.1016/J.GFS.2021.100593

Odey, G. O., Adegbite, M. A., Denkyira, S. A., Alhaj, S. M., & Lucero-Prisno, D. E. (2022). Women and food security in Africa: The double burden in addressing gender equality and environmental sustainability. *Advances in Food Security and Sustainability, 7*, 35–59. https://doi.org/10.1016/bs.af2s.2022.07.001

Ogbolumani, O. A., & Nwulu, N. I. (2021). Multi-objective optimisation of constrained food-energy-water-nexus systems for sustainable resource allocation. *Sustainable Energy Technologies and Assessments, 44*, 100967. https://doi.org/10.1016/j.seta.2020.100967

Ogbolumani, O. A., & Nwulu, N. I. (2024). Environmental impact assessment for a meta-model-based food-energy-water-nexus system. *Energy Reports, 11*, 218–232. https://doi.org/10.1016/j.egyr.2023.11.033

Ogunniyi, A. I., Omotayo, A. O., Olagunju, K. O., Rufai, M. A., Salman, K. K., Omotayo, O. P., & Aremu, A. O. (2024). Evaluating the role of households' food security status and socioeconomic determinants on child mortality in Nigeria. *Child Indicators Research, 17*(4), 1687–1714. https://doi.org/10.1007/s12187-024-10134-5

Olumba, C. C., Olumba, C. N., & Okpara, U. (2023). Household headship, resource ownership and food security: New evidence from Southeast Nigerian cities. *Scientific African, 22*, e01974. https://doi.org/10.1016/J.SCIAF.2023.E01974

Orimoloye, I. R. (2022). Water, energy and food nexus: Policy relevance and challenges. *Frontiers in Sustainable Food Systems, 5*, 824322. https://doi.org/10.3389/fsufs.2021.824322

Pawlak, K., & Kołodziejczak, M. (2020). The role of agriculture in ensuring food security in developing countries: Considerations in the context of the problem of sustainable food production. *Sustainability, 12*(13), 5488. https://doi.org/10.3390/SU12135488

Peng, W., & Berry, E. M. (2018). The concept of food security. *Encyclopedia of Food Security and Sustainability*, 1–7. https://doi.org/10.1016/B978-0-08-100596-5.22314-7

Peng, W., Dernini, S., & Berry, E. M. (2018). Coping with food insecurity using the sociotype ecological framework. *Frontiers in Nutrition, 5*, 107. https://doi.org/10.3389/FNUT.2018.00107

Picone, F., Buonocore, E., Claudet, J., Chemello, R., Russo, G. F., & Franzese, P. P. (2020). Marine protected areas overall success evaluation (MOSE): A novel integrated framework for assessing management performance and social-ecological benefits of MPAs. *Ocean & Coastal Management, 198*, 105370. https://doi.org/10.1016/J.OCECOAMAN.2020.105370

Platt, S., Brown, D., & Hughes, M. (2016). Measuring resilience and recovery. *International Journal of Disaster Risk Reduction, 19*, 447–460. https://doi.org/10.1016/J.IJDRR.2016.05.006

Policy Brief Changing Policy Concepts of Food Security. (2006). https://www.foodsecinfoaction.org/

Qazi, A., & Al-Mhdawi, M. K. S. (2024). Sustainability and adaptation dynamics in global food security: A bayesian belief network approach. *Journal of Cleaner Production*, *467*. https://doi.org/10.1016/j.jclepro.2024.142931

Rasul, G. (2016). Managing the food, water, and energy nexus for achieving the sustainable development goals in South Asia. *Environmental Development*, *18*, 14–25. https://doi.org/10.1016/J.ENVDEV.2015.12.001

Reyes, L. I., Constantinides, S. V., Bhandari, S., Frongillo, E. A., Schreinemachers, P., Wertheim-Heck, S., Walls, H., Holdsworth, M., Laar, A., Nguyen, T., Turner, C., Wellard, K., & Blake, C. E. (2021). Actions in global nutrition initiatives to promote sustainable healthy diets. *Global Food Security*, *31*, 100585. https://doi.org/10.1016/J.GFS.2021.100585

Roka, K. (2020). *Anthropocene and Climate Change* (pp. 20–32). https://doi.org/10.1007/978-3-319-95885-9_26

Sanga, U., Sidibé, A., & Olabisi, L. S. (2021). Dynamic pathways of barriers and opportunities for food security and climate adaptation in Southern Mali. *World Development*, *148*, 105663. https://doi.org/10.1016/J.WORLDDEV.2021.105663

Scordato, L., & Gulbrandsen, M. (2024). Resilience perspectives in sustainability transitions research: A systematic literature review. *Environmental Innovation and Societal Transitions*, *52*, 100887. https://doi.org/10.1016/J.EIST.2024.100887

Stephanie Safdie. (2024). *What are the Three Pillars of Sustainable Development?* https://greenly.earth/en-us/blog/company-guide/3-pillars-of-sustainable-development

Sustainable Development Goals | United Nations Development Programme. (2021). https://www.undp.org/sustainable-development-goals

Swinnen, J., & Vos, R. (2021). COVID-19 and impacts on global food systems and household welfare: Introduction to a special issue. *Agricultural Economics (Amsterdam, Netherlands)*, *52*(3), 365. https://doi.org/10.1111/AGEC.12623

Tchonkouang, R. D., Onyeaka, H., & Nkoutchou, H. (2024). Assessing the vulnerability of food supply chains to climate change-induced disruptions. *Science of The Total Environment*, *920*, 171047. https://doi.org/10.1016/J.SCITOTENV.2024.171047

Tefft, J., Jonasova, M., Tefft, J., & Jonasova, M. (2020). *Food Systems Transformation in An Urbanizing World*, 34–61. https://EconPapers.repec.org/RePEc:elg:eechap:17352_3

The State of Food Security and Nutrition in the World 2024. (2024). https://doi.org/10.4060/CD1254EN

Thomas, A., Alpha, A., Barczak, A., Zakhia-Rozis, N., & Zakhia, N. (2022). *Sustainable Food Systems for Food Security. Need for Combination of Local and Global Approaches*. https://hal.science/hal-03699725

Tye, M. R., Wilhelmi, O. V., Pierce, A. L., Sharma, S., Nichersu, I., Wróblewski, M., Goszczyński, W., Wendel, J., Laborgne, P., Heyder, M., & Nichersu, I. (2022). The food water energy nexus in an urban context: Connecting theory and practice for nexus governance. *Earth System Governance*, *12*, 100143. https://doi.org/10.1016/J.ESG.2022.100143

Ume, C., Nuppenau, E. A., & Domptail, S. E. (2022). A feminist economics perspective on the agroecology-food and nutrition security nexus. *Environmental and Sustainability Indicators*, *16*. https://doi.org/10.1016/j.indic.2022.100212

UNICEF. (2021). *Billions of People will Lack Access to Safe Water, Sanitation and Hygiene in 2030 unless Progress Quadruples – Warn WHO, UNICEF*. UNICEF. https://www.unicef.org/press-releases/billions-people-will-lack-access-safe-water-sanitation-and-hygiene-2030-unless

United Nations. (2023). *THE 17 GOALS | Sustainable Development*. https://sdgs.un.org/goals

USDA. (2023). *USDA ERS – Frequency of Food Insecurity*. https://www.ers.usda.gov/topics/food-nutrition-assistance/food-security-in-the-u-s/frequency-of-food-insecurity/

USDA. (2024). *USDA ERS – Food Prices and Spending*. https://www.ers.usda.gov/data-products/ag-and-food-statistics-charting-the-essentials/food-prices-and-spending/

Vörösmarty, C. J., Campana, P. E., Jewitt, G., Lawford, R., & Wuebbles, D. (2023). Editorial: Food-energy-water systems: Achieving climate resilience and sustainable development in the 21st century. *Frontiers in Environmental Science*, *11*, 1334892. https://doi.org/10.3389/fenvs.2023.1334892

Webb, P., Coates, J., Frongillo, E. A., Rogers, B. L., Swindale, A., & Bilinsky, P. (2006). Measuring household food insecurity: Why it's so important and yet so difficult to do. *The Journal of Nutrition*, *136*(5). https://doi.org/10.1093/JN/136.5.1404S

Wikipdia. (2021). The state of food security and nutrition in the world 2021: Transforming food systems for food security, improved nutrition and affordable healthy diets for all. In *In Brief to The State of Food Security and Nutrition in the World 2021* (2021st ed.). FAO. https://doi.org/10.4060/cb5409en

World Bank Group. (2022, July 23). *What Is Food Security? There Are Four Dimensions*. https://www.worldbank.org/en/topic/agriculture/brief/food-security-update/what-is-food-security

World Bank Group. (2023). *Food Security | Rising Food Insecurity*. https://www.worldbank.org/en/topic/agriculture/brief/food-security-update

World Bank Group. (2024). *What is Food Security? There Are Four Dimensions*. https://www.worldbank.org/en/topic/agriculture/brief/food-security-update/what-is-food-security

5 Application of FEW Nexus 4.0 in Energy

5.1 ENERGY EFFICIENCY AND OPTIMISATION

Due to the rise in energy requirements and the need to mitigate the environmental impacts of fossil fuel consumption, energy efficiency has become a critical topic. It has to do with using less energy to get the same task done to avoid wasting energy and is essential for developed and developing nations alike. The significance of energy efficiency can't be overemphasised due to its essential part in reducing greenhouse gas (GHG) emissions, which significantly contribute to climate change (Atedhor, 2023; Kondratyev & Varotsos, 1995). Improving energy efficiency lessens reliance on fossil fuels (reducing air pollution and conserving natural resources) and results in cost savings for consumers and businesses (Herring, 2006; Omer, 2008). Global energy demand has been rising steadily because of population growth, urbanisation, and industrialisation, especially electricity demand due to the increased use of electronic devices and electric vehicles (Adams & Shachmurove, 2008; Mesarić & Krajcar, 2015; Zhao & Zhang, 2018). Regarding energy consumption, there has been a noticeable shift from fossil fuels like coal, oil, and natural gas towards sustainable energy sources like solar, wind, and hydropower. Figure 5.1 shows types of renewable energies, which include solar energy, wind energy, geothermal energy, hydropower, and biomass energy. These types of energies do not have adverse effects on the environment. The pace of this shift is not fast enough to meet international climate goals, so there's a pressing need for enhanced energy efficiency measures (Azarpour et al., 2013; Tripathi et al., 2016).

5.1.1 Environmental and Economic Impacts of Energy Use

The environment experiences significant environmental effects due to energy usage, which causes both global warming and climate change. The environmental release of carbon dioxide (CO_2) along with multiple GHGs, as defined by Yoro and Daramola (2020), results in extreme weather events and rising sea levels while disrupting ecosystems (Fagbemi et al., 2023). The emission processes from energy production facilities endanger human health through the development of respiratory and cardiovascular diseases (Al-Kindi et al., 2020; Kim et al., 2011). Energy consumption creates unfavourable financial outcomes and impacts systems. Energy development's catalyst effect drives industrial manufacturing, transportation systems, and domestic needs. The associated consumption expenses generate operational costs for businesses while placing a financial burden on both households and businesses. Managing

DOI: 10.1201/9781003633853-5

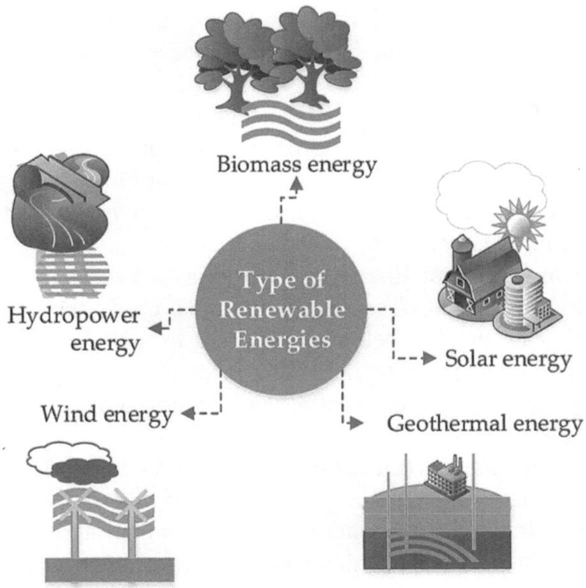

FIGURE 5.1 Renewable Sources of Energy (Avtar et al., 2019).

environmental pollution alongside health dangers stemming from energy production leads to significant economic expenses. To combat the unfortunate impacts, energy efficiency improvements serve as a fundamental solution because they minimise requirements for energy while simultaneously reducing pollutants and energy costs and raising security standards for energy supplies. (Dincer & Rosen, 1999; Hassan et al., 2024; Javaid, Sajid, et al., 2024; Ogbolumani & Nwulu, 2020).

5.1.2 Energy Efficiency

To achieve the equivalent results, people and systems need reduced amounts of energy, which defines energy efficiency. Implementing technological processes and systems helps decrease the amount of energy needed for applications in heating and cooling along with lighting and transportation (Chai & Yeo, 2012; Chel & Kaushik, 2018; Uniyal et al., 2024). Reduced energy consumption and total costs happen when energy efficiency minimises unnecessary energy waste. Each residential, commercial, and industrial sector, as well as transportation, has distinctive energy efficiency needs, so proper solutions must be developed to achieve maximum results. (Geller et al., 2006; Sorrell, 2015; Tanaka, 2008; Trianni et al., 2021; Zhou et al., 2020)

5.1.3 Energy Efficiency Concepts

Several key concepts bring about energy efficiency, each contributing to the overall goal of reducing energy use. These include energy intensity, conservation, performance indicators, life-cycle cost analysis (LCCA), and net-zero energy buildings.

1. **Energy Intensity**: Energy intensity measures a nation's economic energy efficiency, calculated by the energy consumed per unit of gross domestic product (Baloch et al., 2020; Liu et al., 2024; Lu et al., 2024; Ramanathan, 2006). Lower energy intensity indicates higher energy efficiency, meaning that although you deliver similar production output, you do so with less energy. Likewise, higher energy intensity indicates lower energy efficiency, meaning you're delivering the same production output with increased energy (Aydin & Esen, 2018).
2. **Energy Conservation**: Energy conservation is a different concept from energy efficiency, even when people mix them up. People practice energy conservation by making deliberate changes to their behaviour alongside specific operational practices, and these include turning off lights when not needed as well as lowering heating and cooling usage, and more (Tyagi et al., 2020)
3. **Energy Performance Indicators (EPIs)**: EPIs represent measurement instruments that help organisations monitor and assess their energy performance over time. All these entities including organisations, governments, and household users, apply these indicators to assess energy efficiency application success and discover areas that require improvement (Entrop et al., 2010; Geraldi & Ghisi, 2020). EPIs generate useful patterns of energy consumption to guide decisions about improving comprehensive energy efficiency as described in Belpoliti et al. (2024) and Lourenço et al. (2014).
4. **LCCA**: The method of LCCA involves determining the full budget required to own an energy-efficient investment. The examination of both upfront expenses and future savings under this approach enables complete comprehension of the financial aspects connected to energy efficiency measures. The evaluation of complete financial effects on the economy through LCCA allows businesses to decide which energy-efficient methods should be implemented. (Hajare & Elwakil, 2020; Motuziene et al., 2016).
5. **Net Zero Energy Buildings**: Net zero energy buildings are buildings that produce an amount of energy equal to their annual consumption. These structures produce equal energy by merging sustainability-powered methods with energy conservation practices. Net zero energy buildings showcase the highest achievement in energy-efficient construction by proving how buildings can reduce their environmental impact without sacrificing operational efficiency (Ahmed et al., 2022).

5.1.4 Benefits of Energy Efficiency

The practice of energy efficiency produces advantages through economic and environmental elements with social benefits. All parties are, including individuals as well as businesses and public entities, stand to benefit through significant improvements across multiple domains when they implement energy-efficient technologies and practices. Figure 5.2 shows the benefits of energy efficiency. The following benefits are present due to increased efficiency: energy savings, reduced GHG emissions, help in energy delivery, asset values, resource management and industry productivity.

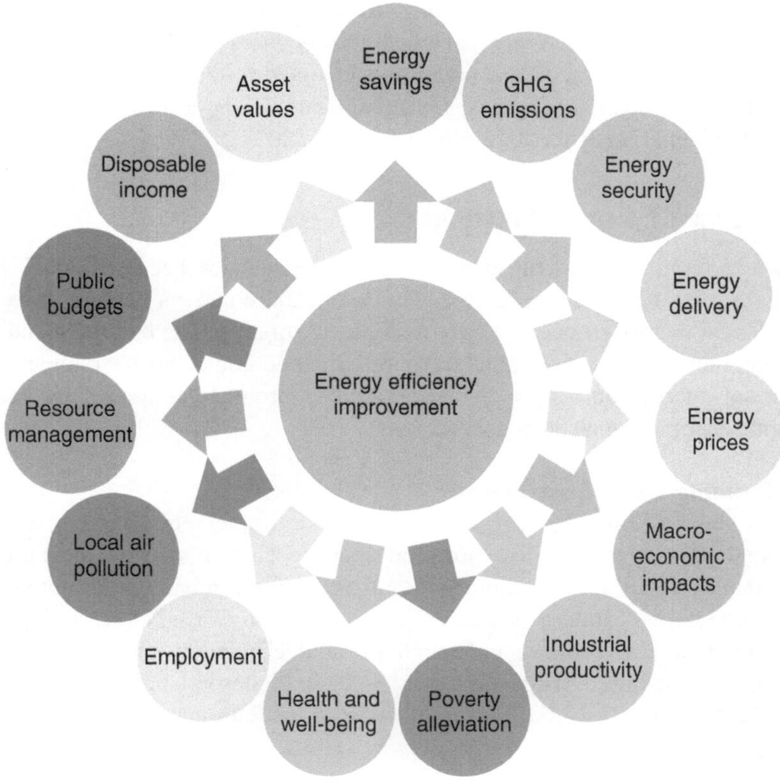

FIGURE 5.2 Advantages of Energy Efficiency (Avram et al., 2018).

Applying energy-efficient measures enhances resource values and leads to successful industrial production and proper resource management. (Avram et al., 2018).

1. **Cost savings**: Operational expenses, together with energy bills, decrease when energy consumption is reduced (Abdelaziz et al., 2011; Darshan et al., 2022; Greening et al., 2000). Installing advanced energy-efficient systems enables improved resource management to decrease operational expenses (Paramati et al., 2022). Energy-efficient appliances and lighting systems allow organisations to decrease their energy bills because they maintain lower energy costs over multiple years.
2. **Environmental benefits**: The fight against climate change depends heavily on energy efficiency because it delivers significant environmental advantages (Paramati et al., 2022). Reducing GHG emissions alongside decreased carbon footprints during all energy production and consumption stages enables energy-efficient methods to protect the environment, according to (An et al., 2023; Sarfraz, 2021).
3. **Enhanced energy security**: Energy efficiency protects the electricity supply from interruptions and market price changes (Fleck & Anatolitis, 2023)

while encouraging nations to move toward renewable power sources such as solar power, wind energy, and hydropower energy (Amir & Khan, 2022). Energy efficiency methods both protect limited resources such as coal and oil, together with natural gas, and strengthen the long-term sustainability of energy (Jhariya et al., 2022; Wang & Azam, 2024).

5.1.5 Energy Efficiency Strategies

The world requires powerful energy efficiency strategies because global energy needs are growing simultaneously as environmental concerns become more critical. Efficient energy strategies target multiple sectors, such as buildings and infrastructure, industrial processes, and transportation operations. Such strategies deliver substantial energy reductions with decreased environmental impact by adopting full-scale energy management approaches.

1. **Building and Infrastructure**
   The building and infrastructure sector implements passive design methods for proper direction and shading elements to enable natural air movement, eliminating the requirement for mechanical climate control systems. Modern insulation items applied to walls, along with roofs and windows, produce enhanced thermal efficiency and help prevent energy wastage (Arumugam et al., 2022; Figueroa-Lopez et al., 2021; Mushtaha et al., 2021). The application of insulation stands as an energy

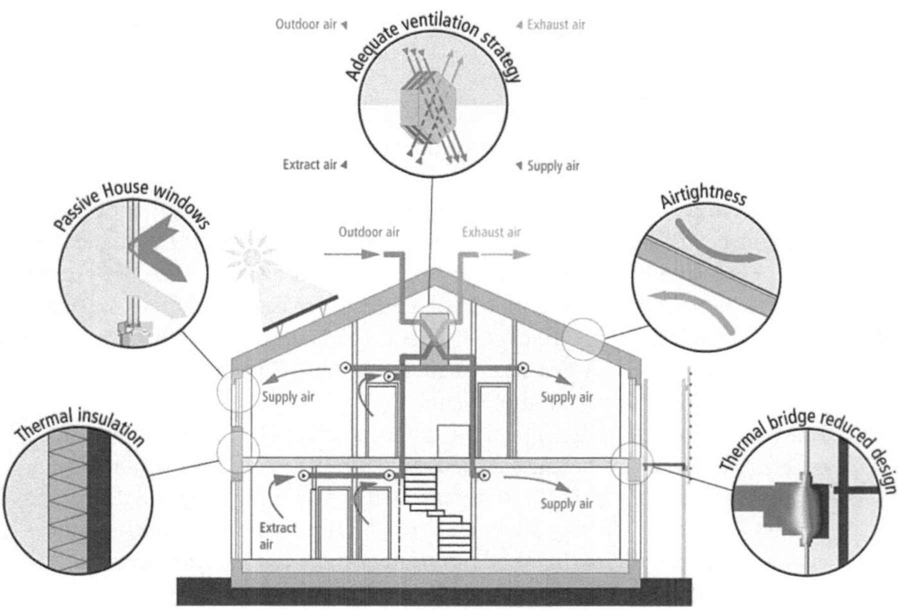

FIGURE 5.3 Reduction of Energy Use in a House through Insulation (Popescu et al., 2021).

efficiency strategy, as illustrated in Figure 5.3 (Popescu et al., 2021). The combination of optimal HVAC system operations from advanced equipment choices, automatic controls, and scheduled maintenance schedules achieves superior efficiency (Bae et al., 2021). Advanced lighting systems incorporating LEDs and smart technologies that monitor occupancy and preferences enable building staff to reduce their power bills significantly through enhanced energy savings. (Chew et al., 2017; Pandharipande & Newsham, 2018). Figure 5.4 shows the working principle and structure of LEDs. All of these ensure energy efficiency in buildings and infrastructures.

2. **Industrial Processes**

Energy Efficiency in Industries requires the automation of manufacturing processes like production, equipment operation, tracking, and evaluation of process data to minimise energy waste, improve productivity, and enhance energy performance (Menghi et al., 2019). Electric motors are standard in industrial applications and make use of a huge amount of energy, which is why, to cut energy usage and enhance the overall system efficiency, technologies like high-efficiency motors, variable frequency drives, and motor management systems are used (Menghi et al., 2019; Van Rhyn & Pretorius, 2015). Figure 5.5 shows an exploded view drawing of an electric motor that shows its components (Boopathi et al., 2022). Many industrial processes generate excess heat. Rather than letting it go to waste, it can be recovered and utilised to offset energy demand elsewhere in the facility. Waste heat recovery systems gather and reuse thermal energy from exhaust gases, process streams, and equipment cooling systems for heating, preheating, or power generation purposes. (Abdelkareem et al., 2022; Loni et al., 2021). This helps industries save energy and cost simultaneously and reduce GHG emissions. Figure 5.6 shows the principle of operation for a waste heat recovery system.

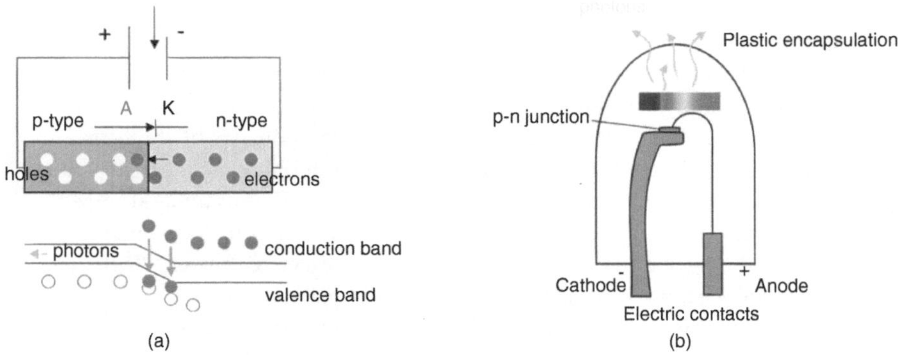

FIGURE 5.4 (a) Working principle and (b) Structure of a Light-Emitting Diode (LED) (Loi et al., 2020).

FIGURE 5.5 Electric Motor and Its Components (Boopathi et al., 2022).

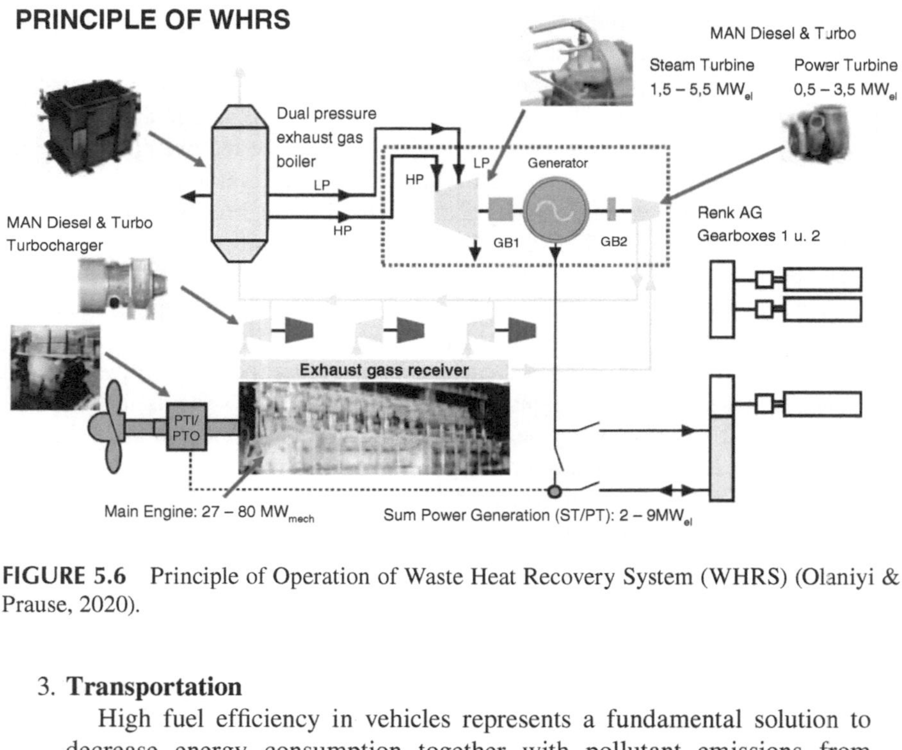

FIGURE 5.6 Principle of Operation of Waste Heat Recovery System (WHRS) (Olaniyi & Prause, 2020).

3. **Transportation**

High fuel efficiency in vehicles represents a fundamental solution to decrease energy consumption together with pollutant emissions from vehicles. The evolution of automobile technology consists of three facets: hybrid and electric powertrain systems, along with lightweight construction materials and aerodynamic engineering principles. Environmental impacts can be minimised by adopting renewable fuels, which include biofuels as well as hydrogen and compressed natural gas (Bicer & Dincer, 2018). The investment in public transit services, including buses, trains, and others, allows individuals to use alternatives to private transportation, thus lowering both fuel usage and traffic delays (Potter, 2003). Smart logistics technologies, including route planning software and tracking systems as well as vehicle telematics, ensure that delivery scheduling, combined with

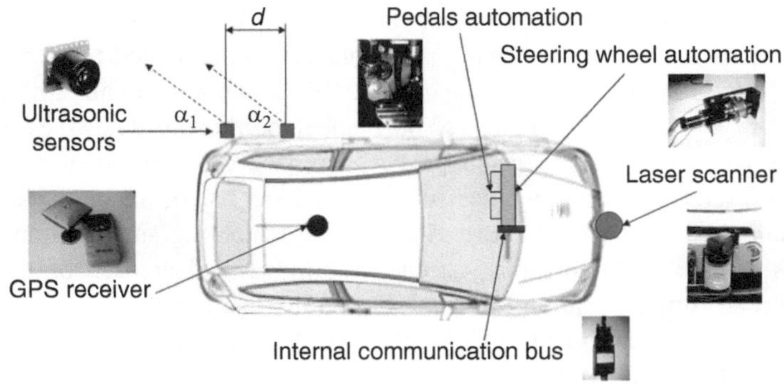

FIGURE 5.7 Live Vehicle Tracking System (Jiménez et al., 2014).

load consolidation and vehicle routing, operates at higher efficiency levels (Bechtsis et al., 2017; Comi & Russo, 2022). By adopting these logistic systems, companies decrease fuel usage and transportation expenses and enhance supply chain reliability and customer contentment.

Building implementations of efficiency tactics within industrial systems and transportation frameworks serve as essential keys to obtaining sustainable energy security and productive economic growth while reducing the environmental risks of a low-carbon transition. These methods work to enhance energy efficiency throughout different industries through resource optimisation, which reduces losses and increases operational yield. The food, energy, and water (FEW) Nexus 4.0 framework depends on energy efficiency to create equilibrium between resource management and environmental sustainability (Jiménez, et al., 2014).

5.2 ENERGY OPTIMISATION TECHNIQUES

The rising need for continuous energy delivery makes advanced optimisation techniques necessary for sustainable performance improvements. The main methods for energy optimisation consist of Energy Management Systems (EMS), Smart Grids, and Energy Storage and Demand Response Programs (Ogbolumani & Nwulu, 2020). The strategies deploy complex technological systems that track energy consumption and implement renewable power integration and dynamic usage management. Organisations that apply these optimisation strategies experience diminished energy expenses while boosting operational performance and achieving better environmental sustainability in line with the FEW Nexus 4.0 principles.

A. Energy Management Systems

Software-based tools operate through tracking and implementation of building energy usage improvements together with facilities and industrial process management. Home and facility infrastructure connect with these systems, enabling them to retrieve real-time data from embedded sensors and metres (Nematchoua et al., 2019; Oikonomou et al., 2009). Such systems interpret energy metre and sensor data in real-time to detect operational wastage patterns. Energy-saving strategies are implemented through these tools, according to (Nematchoua et al., 2019; Oikonomou et al., 2009). The five core concepts of EMS focus on permanent monitoring, data assessments, performance evaluation, and support system decisions. (Bumblauskas et al., 2017). Established tracking, decision support, and performance tracking, as fundamental elements of these systems. Implementing such systems provides benefits through improved energy efficiency, decreased operational costs, and superior equipment reliability while meeting energy regulations. (Trianni et al., 2019) In the FEW Nexus 4.0 context, the energy management system is a vital requirement for maximising interlocked systems' energy utilisation. Figure 5.8 shows the components of energy policy and strategy for an energy management system. The energy management system includes leadership activities, planning and evaluation, and operational activities within an organisational context and perfection (Machnik-Slomka et al., 2024).

B. **Smart Grids and Energy Storage**

They are advanced electricity distribution systems utilising digital communication, monitoring, and control technologies to enhance electricity generation, transmission, and consumption (Samad & Annaswamy, 2017). These systems enable bidirectional interaction between utilities and consumers, facilitating real-time energy management, responsive demand

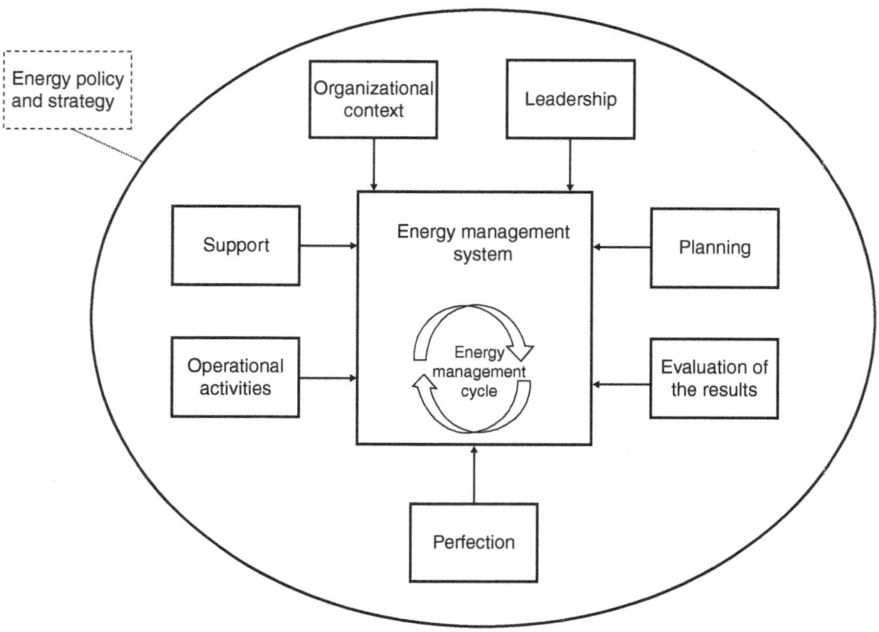

FIGURE 5.8 Energy Management System (Machnik-Slomka et al., 2024).

strategies, and the integration of renewable energy sources (Dileep, 2020; Estebsari et al., 2022). Smart grid technologies, including advanced metering infrastructure and analytics, improve grid reliability, efficiency, and robustness (Dileep, 2020; Panda & Das, 2021). Figure 5.9 shows the supply and demand sides of the smart grid system. The supply side comprises PV farms, wind farms, and thermal power plants, while the demand side comprises offices, homes, and industrial companies. Nuclear power plants and electric vehicles supply and demand smart grid energy (Çelik et al., 2024).

Energy storage systems complement smart grids by enhancing grid flexibility and stability, which is crucial for adopting sustainable energy sources (Alawad et al., 2023; Newbery, 2018). Batteries like lithium-ion, lead-acid, and flow batteries store electricity for later use, cushioning the effect of fluctuations in supply and demand and offering backup power during disruptions (Ajibola & Ogbolumani, 2024; Alawad et al., 2023). Thermal energy storage systems employ phase-change materials or sensible heat storage to absorb and discharge thermal energy, enabling efficient heating, cooling, and power generation (Parameshwaran et al., 2012). Integrating energy storage into smart grid infrastructure enhances grid reliability, optimises renewable energy utilisation, and lowers peak demand, leading to cost savings and environmental advantages. Figure 5.10 shows the various technologies and applications of energy storage systems.

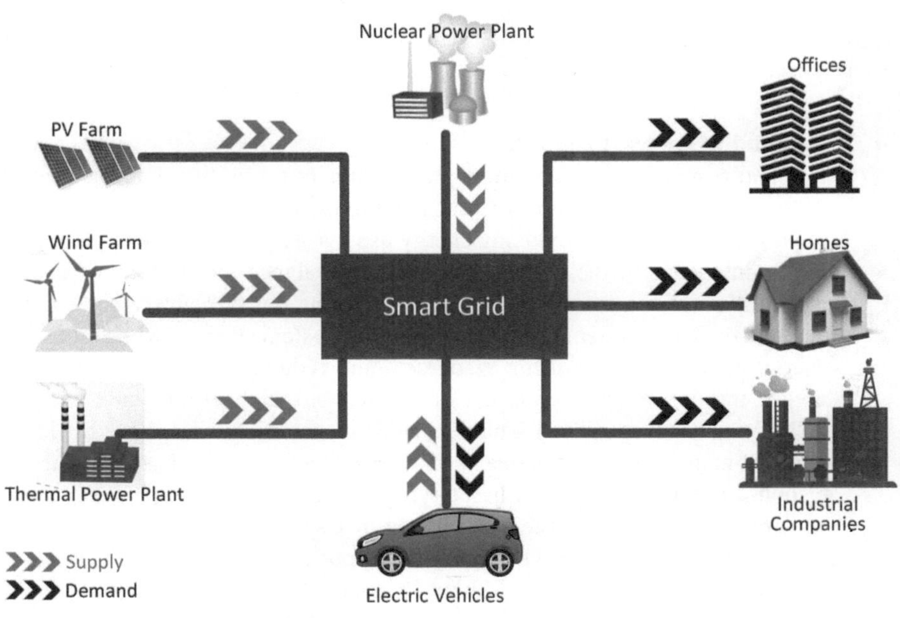

FIGURE 5.9 Supply and Demand Sides of a Smart Grid (Çelik et al., 2024).

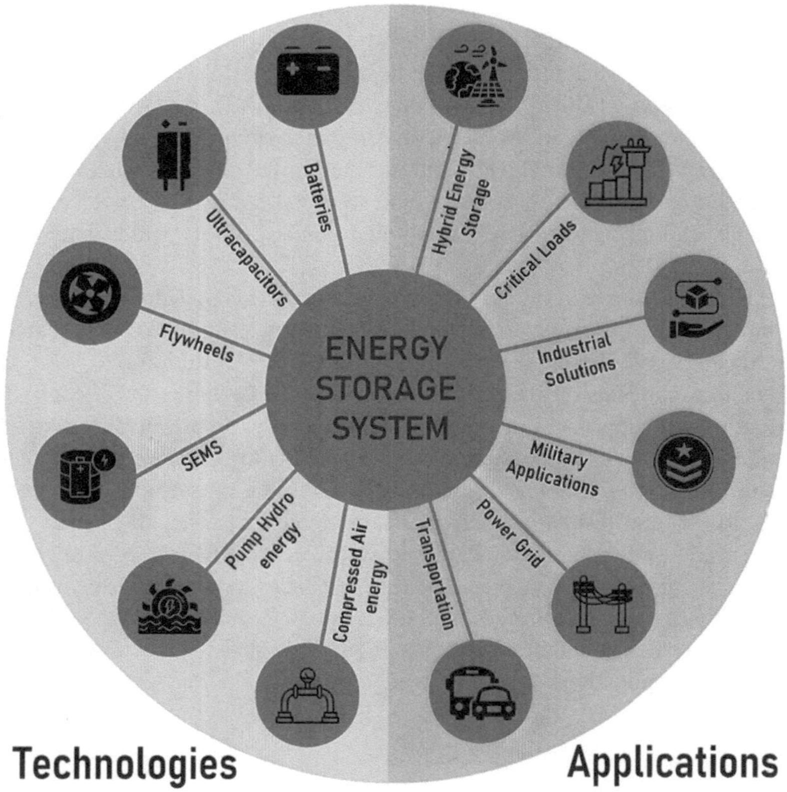

FIGURE 5.10 Technologies and Applications of Energy Storage System (Aghmadi & Mohammed, 2024).

C. **Demand Response Programs**

Demand response programs motivate consumers to modify their electricity consumption in reaction to grid conditions, pricing signals, or utility requests. By shifting or reducing energy use during peak demand or supply constraints periods, the participants can help balance the grid, avoid grid instability, and reduce the need for costly peaking power plants. Accessible demand response systems enable better power system stability while cutting energy expenses, maximising resources, and reducing the environmental impact through reduced emission of pollutants. (Michael Olatunde et al., 2024; Vasques et al., 2018). The demand response program can be divided into two major categories, namely, time-based and incentive-based demand response programs. The time-based type of demand response program can be further divided into three types: Time of Use Rate, Real-Time Pricing, and Critical Peak Pricing. There are three types of incentive-based demand response programs: Capacity Market Program, Ancillary Service markets, and Direct Load Control (Deltetto et al., 2021). Figure 5.11 shows these categories of demand response programs.

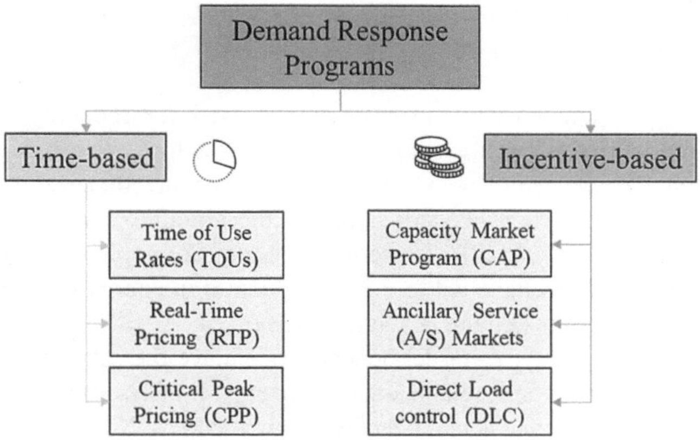

FIGURE 5.11 Categories of Demand Response Programs (Deltetto et al., 2021).

5.3 APPLICATION OF FEW NEXUS 4.0

The application of FEW Nexus 4.0 plays a crucial role in enhancing efficient energy use by recognising the interconnectedness of food, energy, and water resources and leveraging advanced technologies and integrated management practices to optimise energy use across these sectors.

A. **Internet of Things and Smart Grids**

Smart grids, enhanced by Internet of Things (IoT) technologies, are essential for achieving energy efficiency within the FEW Nexus framework (Michael Olatunde et al., 2024). In energy management, IoT optimises energy usage, improves efficiency, and reduces costs through IoT-enabled sensors, metres, and actuators (Michael Olatunde et al., 2024). These sensors and smart metres deliver live data on energy usage, enabling more accurate demand forecasting and load balancing, which reduces energy waste and ensures that energy supply precisely matches demand (Ahmad & Zhang, 2021). The use of smart grids in Denmark has optimised the allocation of electricity, minimised energy losses and wastage, and integrated renewable energy sources more effectively (Javaid et al., 2018).

B. **Artificial Intelligence and Machine Learning**

AI and ML algorithms help identify equipment faults and performance degradation before they lead to costly failures. Additionally, AI models can predict equipment failures. As a result, organisations can plan maintenance, thereby lessening costs and maximising energy use (Ahmad et al., 2021); (Donald Jambol et al., 2024). AI can optimise pump operations, aeration processes, and chemical savings in water treatment facilities, saving significant energy (Gu et al., 2023). Veolia Water Technologies & Solutions implemented an AI-driven system called "Hubgrade" to optimise the management of water, energy, and waste. It also reduces energy costs, minimises water leakages, and increases energy efficiency by capitalising on real-time data and predictive analytics (*Veolia Deploys a Digital Solution Based on Artificial Intelligence*, n.d.).

C. **Efficient Irrigation Systems**

Energy-efficient irrigation systems are vital for reducing the energy required to pump and distribute water in agriculture. Precision irrigation technologies, powered by AI and IoT power ensure that water is applied only where and when needed, reducing the amount of water and energy used (Abioye et al., 2020). These technologies are applied in Israel's agricultural sector and have led to a 20% reduction in energy consumption for irrigation purposes (Zhang & Lazin, 2023).

Energy efficiency: using less energy to achieve the same service or outcome is critical for sustainable energy systems. It offers significant environmental, economic, and social benefits within the FEW Nexus 4.0 framework. However, despite these advantages, energy efficiency initiatives still face multiple challenges. The successful advancement of food, water, and energy systems requires continuous innovation and improvement in energy efficiency technologies and processes to overcome existing obstacles. The advancement of energy efficiency efforts requires active collaboration between public authorities, commercial enterprises, and civil organisations. Partnerships between entities create the ability to spread best practices and establish successful strategies that promote cleaner and healthier sustainable energy operations. Through this collective effort, more work positions become available as financial gains increase and power bills decline. Establishing collaboration between governments, industries, and civil society produces benefits such as reduced GHG emissions alongside environmental protection, coupled with increased resilience of energy systems. The outcomes of this collaboration correspond to the fundamental targets of the FEW Nexus 4.0, which solves interconnected issues across all three sectors at once.

5.4 SUMMARY

This chapter presented the implementation of the FEW Nexus 4.0 for Energy Efficiency approaches. The text emphasises the importance of lowering energy-related environmental effects, including air pollution and climate change, together with extreme weather events, because energy consumption continues to increase. An analysis of the basics of energy efficiency and their dual impact on the environment and economics is provided in this chapter. The strategies, along with optimisation techniques for energy efficiency, used the FEW Nexus 4.0 as part of the approach. The FEW Nexus 4.0 was thoroughly described, and its application methods for maintaining energy efficiency while minimising environmental and economic impacts were discussed.

REFERENCES

Abdelaziz, E. A., Saidur, R., & Mekhilef, S. (2011). A review on energy saving strategies in industrial sector. *Renewable and Sustainable Energy Reviews*, *15*(1), 150–168. https://doi.org/10.1016/J.RSER.2010.09.003

Abdelkareem, M. A., Maghrabie, H. M., Sayed, E. T., Kais, E. C. A., Abo-Khalil, A. G., Radi, M. Al, Baroutaji, A., & Olabi, A. G. (2022). Heat pipe-based waste heat recovery systems: Background and applications. *Thermal Science and Engineering Progress*, *29*, 101221. https://doi.org/10.1016/J.TSEP.2022.101221

Abioye, E. A., Abidin, M. S. Z., Mahmud, M. S. A., Buyamin, S., Ishak, M. H. I., Rahman, M. K. I. A., Otuoze, A. O., Onotu, P., & Ramli, M. S. A. (2020). A review on monitoring and advanced control strategies for precision irrigation. *Computers and Electronics in Agriculture*, *173*, 105441. https://doi.org/10.1016/J.COMPAG.2020.105441

Adams, F. G., & Shachmurove, Y. (2008). Modeling and forecasting energy consumption in China: Implications for Chinese energy demand and imports in 2020. *Energy Economics*, *30*(3), 1263–1278. https://doi.org/10.1016/J.ENECO.2007.02.010

Aghmadi, A., & Mohammed, O. A. (2024). Energy storage systems: Technologies and high-power applications. *Batteries*, *10*(4), 141. https://doi.org/10.3390/BATTERIES10040141

Ahmad, T., & Zhang, D. (2021). Using the internet of things in smart energy systems and networks. *Sustainable Cities and Society*, *68*, 102783. https://doi.org/10.1016/J.SCS.2021.102783

Ahmad, T., Zhang, D., Huang, C., Zhang, H., Dai, N., Song, Y., & Chen, H. (2021). Artificial intelligence in sustainable energy industry: Status Quo, challenges and opportunities. *Journal of Cleaner Production*, *289*, 125834. https://doi.org/10.1016/J.JCLEPRO.2021.125834

Ahmed, A., Ge, T., Peng, J., Yan, W. C., Tee, B. T., & You, S. (2022). Assessment of the renewable energy generation towards net-zero energy buildings: A review. *Energy and Buildings*, *256*, 111755. https://doi.org/10.1016/J.ENBUILD.2021.111755

Ajibola, O. A., & Ogbolumani, O. A. (2024). Smart battery storage integration in an IoT-based solar-powered waste management system. *Journal of Digital Food, Energy & Water Systems*, *5*(2), 17–42. https://doi.org/10.36615/HREWQ861

Alawad, S. M., Mansour, R. Ben, Al-Sulaiman, F. A., & Rehman, S. (2023). Renewable energy systems for water desalination applications: A comprehensive review. *Energy Conversion and Management*, *286*. https://doi.org/10.1016/j.enconman.2023.117035

Al-Kindi, S. G., Brook, R. D., Biswal, S., & Rajagopalan, S. (2020). Environmental determinants of cardiovascular disease: Lessons learned from air pollution. *Nature Reviews Cardiology*, *17*(10), 656–672. https://doi.org/10.1038/s41569-020-0371-2

Amir, M., & Khan, S. Z. (2022). Assessment of renewable energy: Status, challenges, COVID-19 impacts, opportunities, and sustainable energy solutions in Africa. *Energy and Built Environment*, *3*(3), 348–362. https://doi.org/10.1016/J.ENBENV.2021.03.002

An, B. W., Liu, W., Basang, T. X., Li, C. Y., & Xiao, Y. (2023). Energy and air? The impact of energy efficiency improvement on air quality in China. *Environmental Science and Pollution Research*, *30*(38), 89661–89675. https://doi.org/10.21203/rs.3.rs-1830685/v1

Arumugam, P., Ramalingam, V., & Vellaichamy, P. (2022). Effective PCM, insulation, natural and/or night ventilation techniques to enhance the thermal performance of buildings located in various climates – A review. *Energy and Buildings*, *258*, 111840. https://doi.org/10.1016/J.ENBUILD.2022.111840

Atedhor, G. O. (2023). Greenhouse gases emissions and their reduction strategies: Perspectives of Africa's largest economy. *Scientific African*, *20*, e01705. https://doi.org/10.1016/J.SCIAF.2023.E01705

Avram, V., Calu, D. A., Dumitru, V. F., Dumitru, M., Glăvan, M. E., & Jinga, G. (2018). The institutionalization of the consistency and comparability principle in the European companies. *Energies*, *11*(12), 3456. https://doi.org/10.3390/EN11123456

Avtar, R., Sahu, N., Aggarwal, A. K., Chakraborty, S., Kharrazi, A., Yunus, A. P., Dou, J., & Kurniawan, T. A. (2019). Exploring renewable energy resources using remote sensing and GIS—A review. *Resources*, *8*(3), 149. https://doi.org/10.3390/RESOURCES8030149

Aydin, C., & Esen, Ö. (2018). Does the level of energy intensity matter in the effect of energy consumption on the growth of transition economies? Evidence from dynamic panel threshold analysis. *Energy Economics*, *69*, 185–195. https://doi.org/10.1016/J.ENECO.2017.11.010

Azarpour, A., Suhaimi, S., Zahedi, G., & Bahadori, A. (2013). A review on the drawbacks of renewable energy as a promising energy source of the future. *Arabian Journal for Science and Engineering*, *38*(2), 317–328. https://doi.org/10.1007/S13369-012-0436-6

Bae, Y., Bhattacharya, S., Cui, B., Lee, S., Li, Y., Zhang, L., Im, P., Adetola, V., Vrabie, D., Leach, M., & Kuruganti, T. (2021). Sensor impacts on building and HVAC controls: A critical review for building energy performance. *Advances in Applied Energy*, *4*, 100068. https://doi.org/10.1016/J.ADAPEN.2021.100068

Baloch, Z. A., Tan, Q., Iqbal, N., Mohsin, M., Abbas, Q., Iqbal, W., & Chaudhry, I. S. (2020). Trilemma assessment of energy intensity, efficiency, and environmental index: Evidence from BRICS countries. *Environmental Science and Pollution Research*, *27*(27), 34337–34347. https://doi.org/10.1007/S11356-020-09578-3

Bechtsis, D., Tsolakis, N., Vlachos, D., & Iakovou, E. (2017). Sustainable supply chain management in the digitalisation era: The impact of automated guided vehicles. *Journal of Cleaner Production*, *142*, 3970–3984. https://doi.org/10.1016/J.JCLEPRO.2016.10.057

Belpoliti, V., Yahia, M. W., Saleem, A. A., & Nassif, R. (2024). Energy consumption of UAE public schools. Mapping of a diversified sector assessing typology, conditions, and educational systems. *Energy and Buildings*, *320*. https://doi.org/10.1016/j.enbuild.2024.114599

Bicer, Y., & Dincer, I. (2018). Life cycle environmental impact assessments and comparisons of alternative fuels for clean vehicles. *Resources, Conservation and Recycling*, *132*, 141–157. https://doi.org/10.1016/J.RESCONREC.2018.01.036

Boopathi, N. G., Muthuraman, M. S., Palka, R., Wardach, M., Prajzendanc, P., Gundabattini, E., Rassiah, R. S., & Solomon, D. G. (2022). Modeling and simulation of electric motors using lightweight materials. *Energies*, *15*(14), 5183. https://doi.org/10.3390/EN15145183

Bumblauskas, D., Gemmill, D., Igou, A., & Anzengruber, J. (2017). Smart Maintenance Decision Support Systems (SMDSS) based on corporate big data analytics. *Expert Systems with Applications*, *90*, 303–317. https://doi.org/10.1016/J.ESWA.2017.08.025

Çelik, M., Lashab, Ö., Bayındır, A., Vasquez, K. Ç., Inci, M., Çelik, Ö., Lashab, A., Ça, K., Gatay Bayındır, ˇ, Vasquez, J. C., & Guerrero, J. M. (2024). Power system integration of electric vehicles: A review on impacts and contributions to the smart grid. *Applied Sciences*, *14*(6), 2246. https://doi.org/10.3390/APP14062246

Chai, K. H., & Yeo, C. (2012). Overcoming energy efficiency barriers through systems approach—A conceptual framework. *Energy Policy*, *46*, 460–472. https://doi.org/10.1016/J.ENPOL.2012.04.012

Chel, A., & Kaushik, G. (2018). Renewable energy technologies for sustainable development of energy efficient building. *Alexandria Engineering Journal*, *57*(2), 655–669. https://doi.org/10.1016/J.AEJ.2017.02.027

Chew, I., Karunatilaka, D., Tan, C. P., & Kalavally, V. (2017). Smart lighting: The way forward? Reviewing the past to shape the future. *Energy and Buildings*, *149*, 180–191. https://doi.org/10.1016/J.ENBUILD.2017.04.083

Comi, A., & Russo, F. (2022). Emerging information and communication technologies: The challenges for the dynamic freight management in city logistics. *Frontiers in Future Transportation*, *3*, 887307. https://doi.org/10.3389/FFUTR.2022.887307

Darshan, A., Girdhar, N., Bhojwani, R., Rastogi, K., Angalaeswari, S., Natrayan, L., & Paramasivam, P. (2022). Energy audit of a residential building to reduce energy cost and carbon footprint for sustainable development with renewable energy sources. *Advances in Civil Engineering*, *2022*(1), 4400874. https://doi.org/10.1155/2022/4400874

Deltetto, D., Coraci, D., Pinto, G., Piscitelli, M. S., & Capozzoli, A. (2021). Exploring the potentialities of deep reinforcement learning for incentive-based demand response in a cluster of small commercial buildings. *Energies*, *14*(10), 2933. https://doi.org/10.3390/EN14102933

Dileep, G. (2020). A survey on smart grid technologies and applications. *Renewable Energy*, *146*, 2589–2625. https://doi.org/10.1016/J.RENENE.2019.08.092

Dincer, I., & Rosen, M. A. (1999). Energy, environment and sustainable development. *Applied Energy*, *64*(1–4), 427–440. https://doi.org/10.1016/S0306-2619(99)00111-7

Donald Jambol, D., Olatoye Sofoluwe, O., Ukato, A., Joshua Ochulor, O., & Author, C. (2024). Transforming equipment management in oil and gas with AI-Driven predictive maintenance. *Computer Science & IT Research Journal*, *5*(5), 1090–1112. https://doi.org/10.51594/CSITRJ.V5I5.1117

Entrop, A. G., Brouwers, H. J. H., & Reinders, A. H. M. E. (2010). Evaluation of energy performance indicators and financial aspects of energy saving techniques in residential real estate. *Energy and Buildings*, *42*(5), 618–629. https://doi.org/10.1016/J.ENBUILD.2009.10.032

Estebsari, A., Mazzarino, P. R., Bottaccioli, L., & Patti, E. (2022). IoT-enabled real-time management of smart grids with demand response aggregators. *IEEE Transactions on Industry Applications*, *58*(1), 102–112. https://doi.org/10.1109/TIA.2021.3121651

Fagbemi, F., Oke, D. F., & Fajingbesi, A. (2023). Climate-resilient development: An approach to sustainable food production in sub-Saharan Africa. *Future Foods*, *7*, 100216. https://doi.org/10.1016/J.FUFO.2023.100216

Figueroa-Lopez, A., Arias, A., Oregi, X., & Rodríguez, I. (2021). Evaluation of passive strategies, natural ventilation and shading systems, to reduce overheating risk in a passive house tower in the north of Spain during the warm season. *Journal of Building Engineering*, *43*, 102607. https://doi.org/10.1016/J.JOBE.2021.102607

Fleck, A. K., & Anatolitis, V. (2023). Achieving the objectives of renewable energy policy – Insights from renewable energy auction design in Europe. *Energy Policy*, *173*, 113357. https://doi.org/10.1016/J.ENPOL.2022.113357

Geller, H., Harrington, P., Rosenfeld, A. H., Tanishima, S., & Unander, F. (2006). Polices for increasing energy efficiency: Thirty years of experience in OECD countries. *Energy Policy*, *34*(5), 556–573. https://doi.org/10.1016/J.ENPOL.2005.11.010

Geraldi, M. S., & Ghisi, E. (2020). Building-level and stock-level in contrast: A literature review of the energy performance of buildings during the operational stage. *Energy and Buildings*, *211*. https://doi.org/10.1016/j.enbuild.2020.109810

Greening, L. A., Greene, D. L., & Difiglio, C. (2000). Energy efficiency and consumption—The rebound effect—A survey. *Energy Policy*, *28*(6–7), 389–401. https://doi.org/10.1016/S0301-4215(00)00021-5

Gu, Y., Li, Y., Yuan, F., & Yang, Q. (2023). Optimization and control strategies of aeration in WWTPs: A review. *Journal of Cleaner Production*, *418*, 138008. https://doi.org/10.1016/J.JCLEPRO.2023.138008

Hajare, A., & Elwakil, E. (2020). Integration of life cycle cost analysis and energy simulation for building energy-efficient strategies assessment. *Sustainable Cities and Society*, *61*, 102293. https://doi.org/10.1016/J.SCS.2020.102293

Hassan, Q., Khadom, A. A., Algburi, S., Al-Jiboory, A. K., Sameen, A. Z., Alkhafaji, M. A., Mahmoud, H. A., Awwad, E. M., Mahood, H. B., Kazem, H. A., Salman, H. M., & Jaszczur, M. (2024). Implications of a smart grid-integrated renewable distributed generation capacity expansion strategy: The case of Iraq. *Renewable Energy*, *221*. https://doi.org/10.1016/j.renene.2023.119753

Herring, H. (2006). Energy efficiency—A critical view. *Energy*, *31*(1), 10–20. https://doi.org/10.1016/J.ENERGY.2004.04.055

Jiménez, F., Naranjo, J. E., Gómez, O., & Anaya, J. J. (2014). Vehicle tracking for an evasive manoeuvres assistant using low-cost ultrasonic sensors. *Sensors*, *14*(12), 22689–22705. https://doi.org/10.3390/s141222689

Javaid, A., Sajid, M., Uddin, E., Waqas, A., & Ayaz, Y. (2024). Sustainable urban energy solutions: Forecasting energy production for hybrid solar-wind systems. *Energy Conversion and Management*, *302*. https://doi.org/10.1016/j.enconman.2024.118120

Javaid, N., Hafeez, G., Iqbal, S., Alrajeh, N., Alabed, M. S., & Guizani, M. (2018). Energy efficient integration of renewable energy sources in the smart grid for demand side management. *IEEE Access*, *6*, 77077–77096. https://doi.org/10.1109/ACCESS.2018.2866461

Jhariya, M. K., Banerjee, A., & Meena, R. S. (2022). Importance of natural resources conservation: Moving toward the sustainable world. *Natural Resources Conservation and Advances for Sustainability*, 3–27. https://doi.org/10.1016/B978-0-12-822976-7.00027-2

Jiménez, F., Naranjo, J. E., Gómez, O., & Anaya, J. J. (2014). Vehicle tracking for an evasive manoeuvres assistant using low-cost ultrasonic sensors. *Sensor*, *14*(12), 22689–22705. https://doi.org/10.3390/S141222689

Kim, K. H., Jahan, S. A., & Kabir, E. (2011). A review of diseases associated with household air pollution due to the use of biomass fuels. *Journal of Hazardous Materials*, *192*(2), 425–431. https://doi.org/10.1016/J.JHAZMAT.2011.05.087

Kondratyev, K. Y., & Varotsos, C. (1995). Atmospheric greenhouse effect in the context of global climate change. *Il Nuovo Cimento C*, *18*(2), 123–151. https://doi.org/10.1007/BF02512015

Liu, M., Xia, C., Lan, H., Gao, Z., Yu, X., Wang, L., Liang, X., & Wu, Y. (2024). Driving toward sustainable cities: The interplay between Chinese emerging corporate ESG performance and climate finance in achieving low-carbon development. *Urban Climate*, *55*. https://doi.org/10.1016/j.uclim.2024.101918

Loi, M., Villani, A., Paciolla, F., Mulè, G., & Paciolla, C. (2020). Challenges and opportunities of Light-Emitting Diode (LED) as key to modulate antioxidant compounds in plants. *A Review. Antioxidants*, *10*(1), 42. https://doi.org/10.3390/ANTIOX10010042

Loni, R., Najafi, G., Bellos, E., Rajaee, F., Said, Z., & Mazlan, M. (2021). A review of industrial waste heat recovery system for power generation with organic rankine cycle: Recent challenges and future outlook. *Journal of Cleaner Production*, *287*, 125070. https://doi.org/10.1016/J.JCLEPRO.2020.125070

Lourenço, P., Pinheiro, M. D., & Heitor, T. (2014). From indicators to strategies: Key Performance Strategies for sustainable energy use in Portuguese school buildings. *Energy and Buildings*, *85*, 212–224. https://doi.org/10.1016/J.ENBUILD.2014.09.025

Lu, F., Ma, F., & Hu, S. (2024). Does energy consumption play a key role? Re-evaluating the energy consumption-economic growth nexus from GDP growth rates forecasting. *Energy Economics*, *129*. https://doi.org/10.1016/j.eneco.2023.107268

Machnik-Slomka, J., Pawlowska, E., Klosok-Bazan, I., & Goňo, M. (2024). Evaluation of the energy management system in water and wastewater utilities in the context of sustainable development—A case study. *Energies*, *17*(19), 5014. https://doi.org/10.3390/EN17195014

Menghi, R., Papetti, A., Germani, M., & Marconi, M. (2019). Energy efficiency of manufacturing systems: A review of energy assessment methods and tools. *Journal of Cleaner Production*, *240*, 118276. https://doi.org/10.1016/J.JCLEPRO.2019.118276

Mesarić, P., & Krajcar, S. (2015). Home demand side management integrated with electric vehicles and renewable energy sources. *Energy and Buildings*, *108*, 1–9. https://doi.org/10.1016/J.ENBUILD.2015.09.001

Michael Olatunde, T., Chukwudi Okwandu, A., Oluwajuwonlo Akande, D., & Queen Sikhakhane, Z. (2024). The impact of smart grids on energy efficiency: A Comprehensive review. *Engineering Science & Technology Journal*, *5*(4), 1257–1269. https://doi.org/10.51594/ESTJ.V5I4.1016

Motuziene, V., Rogoža, A., Lapinskiene, V., & Vilutiene, T. (2016). Construction solutions for energy efficient single-family house based on its life cycle multi-criteria analysis: A case study. *Journal of Cleaner Production*, *112*, 532–541. https://doi.org/10.1016/J.JCLEPRO.2015.08.103

Mushtaha, E., Salameh, T., Kharrufa, S., Mori, T., Aldawoud, A., Hamad, R., & Nemer, T. (2021). The impact of passive design strategies on cooling loads of buildings in temperate climate. *Case Studies in Thermal Engineering*, *28*, 101588. https://doi.org/10.1016/J.CSITE.2021.101588

Nematchoua, M. K., Yvon, A., Roy, S. E. J., Ralijaona, C. G., Mamiharijaona, R., Razafinjaka, J. N., & Tefy, R. (2019). A review on energy consumption in the residential and commercial buildings located in tropical regions of Indian Ocean: A case of Madagascar Island. *Journal of Energy Storage*, *24*. https://doi.org/10.1016/j.est.2019.04.022

Newbery, D. (2018). Shifting demand and supply over time and space to manage intermittent generation: The economics of electrical storage. *Energy Policy*, *113*, 711–720. https://doi.org/10.1016/j.enpol.2017.11.044

Ogbolumani, O. A., & Nwulu, N. (2020). Integrated appliance scheduling and optimal sizing of an autonomous hybrid renewable energy system for agricultural food production. *Lecture Notes in Mechanical Engineering*, 651–660. https://doi.org/10.1007/978-981-15-5753-8_60

Oikonomou, V., Becchis, F., Steg, L., & Russolillo, D. (2009). Energy saving and energy efficiency concepts for policy making. *Energy Policy*, *37*(11), 4787–4796. https://doi.org/10.1016/J.ENPOL.2009.06.035

Olaniyi, E. O., & Prause, G. (2020). Investment analysis of waste heat recovery system installations on ships' engines. *Journal of Marine Science and Engineering*, *8*(10), 811. https://doi.org/10.3390/JMSE8100811

Omer, A. M. (2008). Energy, environment and sustainable development. *Renewable and Sustainable Energy Reviews*, *12*(9), 2265–2300. https://doi.org/10.1016/J.RSER.2007.05.001

Panda, D. K., & Das, S. (2021). Smart grid architecture model for control, optimization and data analytics of future power networks with more renewable energy. *Journal of Cleaner Production*, *301*, 126877. https://doi.org/10.1016/J.JCLEPRO.2021.126877

Pandharipande, A., & Newsham, G. R. (2018). Lighting controls: Evolution and revolution. *Lighting Research & Technology*, *50*(1), 115–128. https://doi.org/10.1177/1477153517731909

Paramati, S. R., Shahzad, U., & Doğan, B. (2022). The role of environmental technology for energy demand and energy efficiency: Evidence from OECD countries. *Renewable and Sustainable Energy Reviews*, *153*, 111735. https://doi.org/10.1016/J.RSER.2021.111735

Parameshwaran, R., Kalaiselvam, S., Harikrishnan, S., & Elayaperumal, A. (2012). Sustainable thermal energy storage technologies for buildings: A review. *Renewable and Sustainable Energy Reviews*, *16*(5), 2394–2433. https://doi.org/10.1016/J.RSER.2012.01.058

Popescu, L. L., Popescu, R., & Catalina, T. (2021). Improving the energy efficiency of an existing building by dynamic numerical simulation. *Applied Sciences*, *11*(24), 12150. https://doi.org/10.3390/APP112412150

Potter, S. (2003). *Transport Energy and Emissions: Urban Public Transport*. *4*, 247–262. https://doi.org/10.1108/9781786359513-013

Ramanathan, R. (2006). A multi-factor efficiency perspective to the relationships among world GDP, energy consumption and carbon dioxide emissions. *Technological Forecasting and Social Change*, *73*(5), 483–494. https://doi.org/10.1016/J.TECHFORE.2005.06.012

Samad, T., & Annaswamy, A. M. (2017). Controls for smart grids: Architectures and applications. *Proceedings of the IEEE*, *105*(11), 2244–2261. https://doi.org/10.1109/JPROC.2017.2707326

Sarfraz, M. (2021). *Green Technologies to Combat Air Pollution*. 143–161. https://doi.org/10.1007/978-3-030-70509-1_9

Sorrell, S. (2015). Reducing energy demand: A review of issues, challenges and approaches. *Renewable and Sustainable Energy Reviews*, *47*, 74–82. https://doi.org/10.1016/J.RSER.2015.03.002

Tanaka, K. (2008). Assessment of energy efficiency performance measures in industry and their application for policy. *Energy Policy*, *36*(8), 2887–2902. https://doi.org/10.1016/J.ENPOL.2008.03.032

Trianni, A., Cagno, E., & Accordini, D. (2019). Energy efficiency measures in electric motors systems: A novel classification highlighting specific implications in their adoption. *Applied Energy*, *252*, 113481. https://doi.org/10.1016/J.APENERGY.2019.113481

Trianni, A., Cagno, E., Dolšak, J., & Hrovatin, N. (2021). Implementing energy efficiency measures: Do other production resources matter? A broad study in Slovenian manufacturing small and medium-sized enterprises. *Journal of Cleaner Production*, *287*. https://doi.org/10.1016/j.jclepro.2020.125044

Tripathi, L., Mishra, A. K., Dubey, A. K., Tripathi, C. B., & Baredar, P. (2016). Renewable energy: An overview on its contribution in current energy scenario of India. *Renewable and Sustainable Energy Reviews*, *60*, 226–233. https://doi.org/10.1016/J.RSER.2016.01.047

Tyagi, R., Vishwakarma, S., Singh, K. K., & Syan, C. (2020). *Low-Cost Energy Conservation Measures and Behavioral Change for Sustainable Energy Goal*. 1–13. https://doi.org/10.1007/978-3-319-71057-0_155-1

Uniyal, S., Lodhi, M. S., Pawar, Y., Thakral, S., Garg, P. K., Mukherjee, S., & Nautiyal, S. (2024). Passive solar heated buildings for enhancing sustainability in the Indian Himalayas. *Renewable and Sustainable Energy Reviews*, *200*. https://doi.org/10.1016/j.rser.2024.114586

Van Rhyn, P., & Pretorius, J. H. C. (2015). Utilising high and premium efficiency three phase motors with VFDs in a public water supply system. *International Conference on Power Engineering, Energy and Electrical Drives, 2015-September*, 497–502. https://doi.org/10.1109/POWERENG.2015.7266367

Vasques, T. L., Moura, P., & de Almeida, A. (2018). A review on energy efficiency and demand response with focus on small and medium data centers. *Energy Efficiency*, *12*(5), 1399–1428. https://doi.org/10.1007/S12053-018-9753-2

Veolia Deploys a Digital Solution Based on Artificial Intelligence. (n.d.). Retrieved August 15, 2024, from https://www.veolia.com/en/our-media/press-releases/veolia-worldwide-deployment-digital-solution-based-artificial-water-energy-waste

Wang, J., & Azam, W. (2024). Natural resource scarcity, fossil fuel energy consumption, and total greenhouse gas emissions in top emitting countries. *Geoscience Frontiers*, *15*(2), 101757. https://doi.org/10.1016/J.GSF.2023.101757

Yoro, K. O., & Daramola, M. O. (2020). CO2 emission sources, greenhouse gases, and the global warming effect. *Advances in Carbon Capture: Methods, Technologies and Applications*, 3–28. https://doi.org/10.1016/B978-0-12-819657-1.00001-3

Zhang, X., & Lazin, F. A. (2023). The agricultural cooperation between China and Israel – Case study of projects in Shandong province. *Asian Journal of Middle Eastern and Islamic Studies*, *17*(1), 106–126. https://doi.org/10.1080/25765949.2023.2190615

Zhao, P., & Zhang, M. (2018). The impact of urbanisation on energy consumption: A 30-year review in China. *Urban Climate*, *24*, 940–953. https://doi.org/10.1016/J.UCLIM.2017.11.005

Zhou, X., Xu, Z., & Xi, Y. (2020). Energy conservation and emission reduction (ECER): System construction and policy combination simulation. *Journal of Cleaner Production*, *267*. https://doi.org/10.1016/j.jclepro.2020.121969

6 Application of FEW Nexus 4.0 in Water

6.1 WATER MANAGEMENT AND TREATMENT STRATEGIES

Water scarcity is a worldwide concern that has been, and is still, affecting billions of people worldwide. Many have acknowledged that many nations worldwide are stepping into a seasonal period of extreme water scarcity due to various reasons, including the consistent rise of the human population, climate change, wastage of water, and the increase in water consumption per capita around the world (Belhassan, 2021; Ogbolumani & Nwulu, 2021, 2022). According to the United Nations, over 2 billion persons occupy nations that are dealing with severe water scarcity. By 2025, half of the global population might be dwelling in areas experiencing water scarcity (du Plessis, 2017). Water scarcity is a problem with many sides that is not only influenced by water availability but also its quality and access to clean drinking water, and can cause numerous economic consequences (Guarino, 2017; Gude, 2017), such as hindering industrial operations, reducing agricultural yields, and negatively affecting the production of energy, among other things (Belhassan, 2021; Zarei et al., 2020). This is why sustainable water management is crucial to tackle these problems and guarantee there's a stable supply of water all over the world; this can be done through the application of the Reduce, Reuse, Recycle, Restore, and Recover system (Yadav et al., 2022). Figure 6.1 shows the use of the 5R principle in water resource management. These are as follows: Reduce losses by boosting water efficiency, reuse water that needs minimal treatment, recycle resources and wastewater, restore water to its source at the same or better quality

6.1.1 Objectives of Water Management and Treatment

Water management and treatment are of utmost importance if we want to resolve the challenges related to global water scarcity, quality, supply, and distribution, among various others. Its primary aim is to ensure that everyone can access safe and clean water, to ensure the environment is well-protected, and to support social and economic development (Adeel, 2017).

A. Clean and safe drinking water

Worldwide, numerous persons lack access to pure and safe consumable water, with some not having access to any water (Sultana, 2018). Due to this, they have no choice but to survive on contaminated water, which could

DOI: 10.1201/9781003633853-6

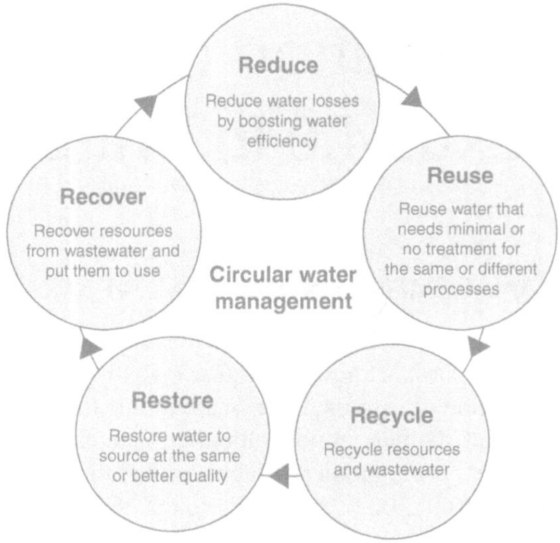

FIGURE 6.1 Circular Water Management—5R Principles (Konkol et al., 2024).

cause numerous life-threatening illnesses, including cholera, dysentery, typhoid, and hepatitis A (Qadri & Faiq, 2020; Rath, 2021). The inability to access clean and safe water can cause severe dehydration and malnutrition, and when it isn't adequately treated, it can also lead to death (Ngina, 2017). Based on the report by the Third World Academy of Sciences, polluted/dirty water is killing more people than cancer, AIDS, wars, or accidents (Ngina, 2017). The human population globally is rising, and so is the number of people who do not have access to uncontaminated water. Hence, supplying enough uncontaminated water by removing contaminants and pathogens through effective water management and treatment processes is necessary.

B. Support economic growth

This is a key objective of water management and treatment. Water is essential for economic activities such as agricultural, industrial, and energy production, which is why it aims to support economic growth by making available a reliable water supply for these activities while ensuring equity in water access (Bayu et al., 2020). Effective water management systems increase agricultural productivity through efficient irrigation practices, support industrial processes by providing the needed resources for manufacturing and cooling, and facilitate energy production by allowing for hydroelectric power generation (Majeed et al., 2023; Moreno-Pizani, 2021). Water infrastructures can attract investment, create job opportunities, and improve community resilience against economic fluctuations (Pokhrel et al., 2021). Water management and treatment help create an environment supporting long-term economic development.

C. Economic and social implications

When natural water sources aren't protected, and the production of water is destabilised, it could lead to dire consequences such as the deterioration of water quality due to pollution from industrial, agricultural, and domestic activities, and this can cause the water to become contaminated (Li et al., 2021). It can also hinder agricultural production, energy production, and even the collapse of aquatic ecosystems, and many more (Bashir et al., 2020). Water management and treatment aim to safeguard these raw water sources, like groundwater, rivers, lakes, oceans, and many more, to reduce pollution, preserve water bodies and ecosystems, and maintain balance within the water cycle and biodiversity.

Water Management and Treatment aims to protect water bodies, reduce waste, prevent water pollution, and reduce waste for use in energy production. It also seeks to ensure that the current generation has access to contaminant-free and safe drinking water, and that, without losing quality, water is equally available to future generations (Issaoui et al., 2022; Mujtaba et al., 2024; Nguyen et al., 2019; Nti et al., 2023)

6.1.2 Water Demand and Usage

Water demand and usage globally are constantly on the rise due to several conditions, including, but not restricted to, population growth, climate change, lifestyle changes, and hygiene (Borowski & Borowski, 2020; Hanjra & Qureshi, 2010). From Figure 6.2, the global water consumption shows a pattern similar to population levels. The uncertainly worsens because of climate changes impacts. The development of sufficient water

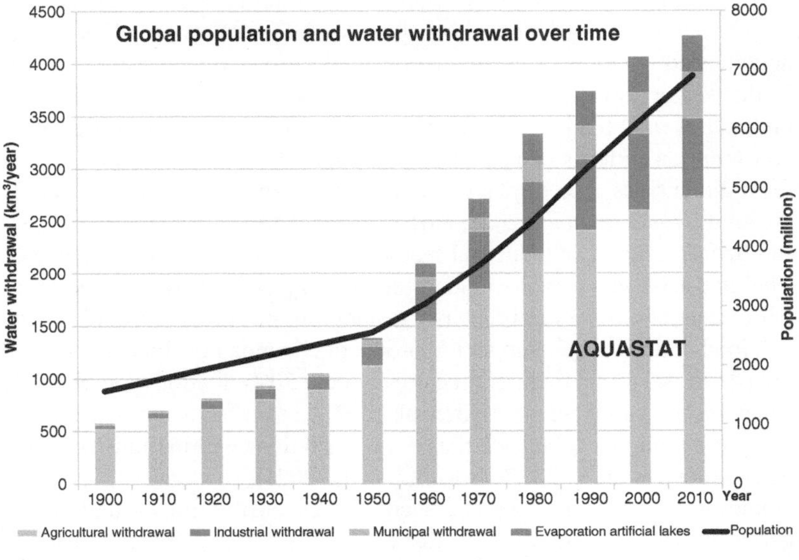

FIGURE 6.2 Worldwide Water Consumption and Population Growth Percentage since 1990 (Zisopoulou & Panagoulia, 2021).

treatment plans requires complete knowledge of water consumption patterns together with the identification of water-consuming sectors (Anderson et al., 2019; Badham et al., 2019; Hoffmann et al., 2020; Kundzewicz et al., 2018; Leal Filho et al., 2022)

The following section investigates water consumer behaviour across the agricultural sector, industrial sector, and residential sector. Water demand keeps growing worldwide because the human population increases, climate change, increasing urbanisation, and societal behaviour patterns combine with improved hygiene needs. Water use management needs effective implementation through the FEW sectors, which include energy, food production, and water services.

The total water consumption worldwide is dominated by agricultural needs, with a share of 70% that includes irrigation and livestock farming, aquaculture, and every aspect of agriculture. Manufacturing processes alongside cooling and cleaning require 20% of global water resources which the industrial sector consumes. Water resource sustainability needs proper water management to address mounting demands in these three sectors. Water consumption by the domestic sector remains lower than that in both the agriculture sector and the industrial sector. Presently, domestic use accounts for 10% of global water consumption while people utilise it for basic needs, including drinking water, cooking, household cleaning, and garden maintenance. Water management strategies need to prioritise the effective distribution of water while meeting the needs of the domestic and non-domestic sectors through responsible consumption practices.

A. Agricultural Sector: Water usage in agriculture constitutes the main water consumer since it absorbs 70% of surface water and groundwater resources. Surface water, along with groundwater sources, provide all of the total water used. The agricultural sector consumes water through various operations, starting with land treatments and continuing with irrigation water requirements. The agricultural industry uses irrigation as one of its primary activities to enhance crop production while supporting livestock growth. In this sector, animals drink water, the cleansing of farm tools, harvesting procedures, and temperature maintenance operations require water. Water consumption patterns in agriculture are affected by some factors, including farm dimensions type of plant, site placement, and geological attributes of the land (Monteiro et al., 2021; Parihar et al., 2019).
B. Industrial Sector: The industrial sector is also among the major water users. The industrial sector ranks among the primary users of water. The principal consumers of water are energy generation industries alongside chemical, food and beverage, and textile production companies. Industries. These industries as a group maintain the use of about 20% of worldwide accessible water resources (Weerasooriya et al., 2021). Manufacturing and cleaning using water happen alongside the cooling functions within industrial facilities (Bailone et al., 2022; Ravi Kumar et al., 2021).
C. Domestic Sector: The domestic sector is responsible for around 10% of global water consumption (Weerasooriya et al., 2021). Water is used for household activities such as drinking, cooking, washing clothes and dishes, flushing toilets, and irrigating lawns and gardens.

6.1.3 Water Management Strategies

Global water demand keeps rising, one of the main reasons being the ever-increasing population growth, which is why addressing water's demand and supply aspect is important. Water management strategies involve controlling and monitoring the movement of water resources to enable efficient use of water, reduce water scarcity, and protect the natural ecosystem (Siddique, 2022; Srivastav et al., 2021). The following sections talk about the key water management strategies:

A. Integrated Water Resources Management

Set up in the 1930s, Integrated water resource management tackles effective water management from a technological point of view while also factoring in social goals (Meran et al., 2021). Integrated water resources management maximises economic and social welfare by creating and managing water, land, and all associated materials together to improve economic and social well-being equitably without jeopardising the sustainable development of crucial habitats (Meran et al., 2021; Merrey et al., 2005). Implementation of integrated water management systems demands connections between local communities, national governments, and sectorial groups, according to (Galvez & Rojas, 2019). The Water Framework Directive from the European Union stands as one of the integrated water resource management successes, having established water management standards for member states (Blomquist et al., 2007; Giakoumis & Voulvoulis, 2018).

B. Water Demand Management

The Water Demand Management scheme is a national strategy to control effective water resource management by altering customer usage patterns. Strategies and systems designed to enhance water efficiency while maximising the availability of water resources operate globally. The purpose is to realise both equitable water usage and maximum utilisation. The scheme implements water conservation practices and raises public education and awareness levels while implementing economic measures based on water pricing (Koop et al., 2019; Tortajada et al., 2019).

1. Water conservation techniques: These are methods used to reduce the amount of water consumed in agriculture, industries, and households (Elahi et al., 2022; Geetha Varma, 2022; Tortajada et al., 2019). These techniques include using low-flow fixtures, efficient irrigation systems, and implementing leak detection and repair programmes (Z. Wu et al., 2020). Another way to conserve water is through water recycling (Radcliffe & Page, 2020). For example:
   a. Using unclean water to water plants
   b. Filtering, disinfecting, and using settling tanks on unclean water so they can be used for flushing toilets and even laundry.
   c. Rainwater can be collected and stored for irrigation, and after treatment, it can even be used for drinking and household use.
2. Public awareness and education: A large percentage of citizens are unaware of the threats to water quality, how their water usage

negatively and positively affects the world at large, and how much water they use, especially indirectly, while producing goods and services. Teaching the community about the importance of water conservation— through campaigns, school programmes, community workshops, and more, can lead to a reduction in their daily personal water use (Seelen et al., 2019).

C. Water Supply Management

Water supply entails providing water to the public, communities, organisations, industries, and various others for different purposes, such as drinking, sanitation, irrigation, and more (Bazaanah & Mothapo, 2024). Due to the increasing water demand, it has become harder to address the people's or industry's needs, making Water Supply Management (WSM) a necessity. WSM is the management of water supply through strategy, development, distribution, and regulation of water resources to guarantee an equitable and long-lasting water supply for the people, and this is done through infrastructure development, efficient distribution systems, leak detection, and repair, among many others (Bazaanah & Mothapo, 2024; Guta et al., 2022; Loucks & van Beek, 2017; Tadesse et al., 2013)

1. Infrastructure development: The development of sustainable water infrastructure requires planning and management activities to promote accessibility, combined with provisions for affordability and maintenance of high-quality standards (Aleixo et al., 2019; Thacker et al., 2019). The infrastructure allows the storage and transport of water to satisfy needs at different locations. Maintaining infrastructure such as dams, reservoirs, and pipelines forms part of infrastructure development efforts for steady water supply (Aleixo et al., 2019).
2. Efficient distribution systems: The continuous updates and care of water distribution systems constitute efficient distribution systems. Businesses benefit from efficient distribution systems because these systems cut down losses while also delivering better results (Razavi et al., 2019). The water sector uses modernised pipelines with smart metering technology and advanced monitoring solutions (Priyanka et al., 2021).
3. Leak detection and repair: Leak detection and quick repair represent a vital procedure. Upgraded monitoring systems allow swift detection of water leaks in distribution systems. The prompt upkeep of the damaged system can prevent water loss and avoid further situation deterioration. Rapid response procedures lower expenses while decreasing the amount of wasted resources.

The increase in population and various external influences lead to an ongoing expansion of water demand, making effective water resource management vital. The Integrated Water Resources Management process optimises water resources alongside related resources by upholding social, economic, and environmental perspectives. A successful water management system needs both coordination between different sectors and multi-level governmental participation to maintain sustainability. The use of techniques for water-saving installations along with water recycling enables water demand management to achieve maximum water usage while minimising consumption. At the same time, public awareness is essential because it teaches people how to conserve water effectively while spreading understanding of its significance. The focus of WSM is to deliver reliable water from proper water infrastructure systems. Such infrastructure development along with reservoir maintenance and pipeline management forms the core elements of this system. A leak detection programme should be implemented as part of the distribution system for preventing water loss as it identifies and repairs leaks effectively. These water use methods combined offer solutions for increasing water usage and maintaining sustainable operations.

6.1.4 Water Treatment Process

Water contamination affects millions worldwide and generates serious health threats that lead to mortality, making proper water treatment is essential. Water treatment creates safe drinking water that meets quality requirements to decrease waterborne disease-related illness and mortality among people. Three basic stages constitute the water treatment process beginning with primary, then moving to secondary, and finishing with tertiary treatment. Each of these stages is essential in ensuring clean water.

A. Primary Treatment:

The first stage in the water treatment process is primary treatment, which concentrates on getting rid of large particles and matter such as sand, grit, suspended organic matter, and many others. This is where the physical method of water treatment, which involves removing water impurities without changing or modifying its chemical structure, comes into play. Some of the standard physical methods include screening, sedimentation, and filtering.

I. Screening:

Screening is the process of removing wastewater of large debris and objects. It is an essential stage of water treatment. Depending on the source of the water being screened, it removes objects such as wood, paper, plastic, metals, rubbish, and many others. This makes it the first step in the water treatment process because it protects the further treatment stages from large objects that could cause clogging and improves their efficiency (Ahmed et al., 2022; Enfrin et al., 2019). Screening is divided into fine, medium, and coarse screening, with under 10mm, 10–40mm, and 40mm+ spacing between the bars, respectively. The size of debris that can be screened out of the water is dependent on the size of the screening bar. The smaller the bar, the smaller the debris that can be screened out. There are also scenarios where multiple screening processes can be used together to remove solids of various sizes. After the water is screened, the removed debris can be disposed of in landfill sites. Figure 6.3 shows screening at a water outlet dam.

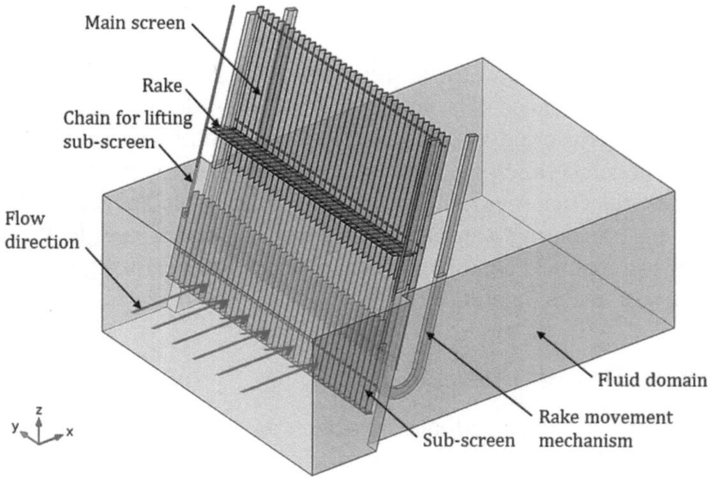

FIGURE 6.3 A 3D Representation of the Wastewater Screening Mechanism (Akhtar et al., 2023).

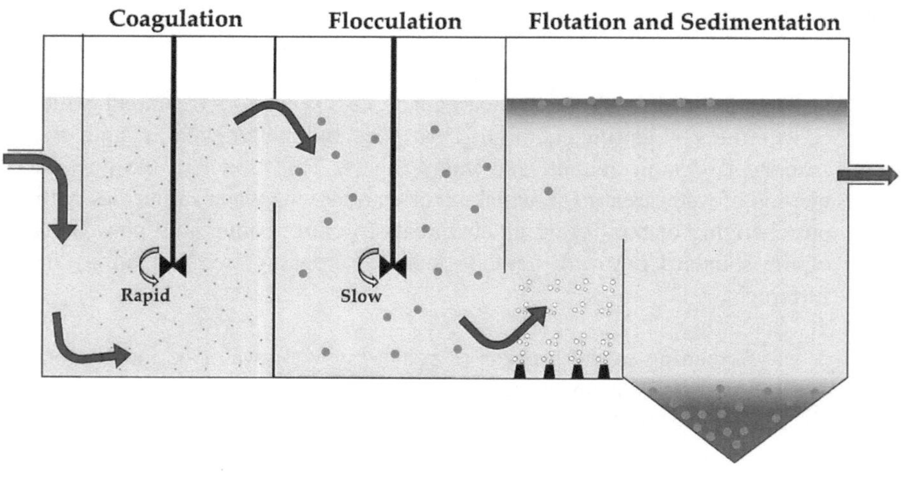

FIGURE 6.4 Sedimentation Method for Water Filtration (Fernandes et al., 2024).

1. Primary Sedimentation:

   Sedimentation involves removing suspended solids by allowing the water to sit in tanks or basins and causing the particles to settle at the bottom under the effect of gravity. It is one of the common methods used before filtration (Hurst, 2019). There are several ways through which sedimentation can be applied, and this includes horizontal and radial flow, inclined plate, ballasted floc, and floc blanket (*Quebec Septic Tank Regulations – Septic Guide*, n.d.). Figure 6.4 shows the primary processes involved in sedimentation, including: coagulation, flocculation, flotation, and sedimentation.

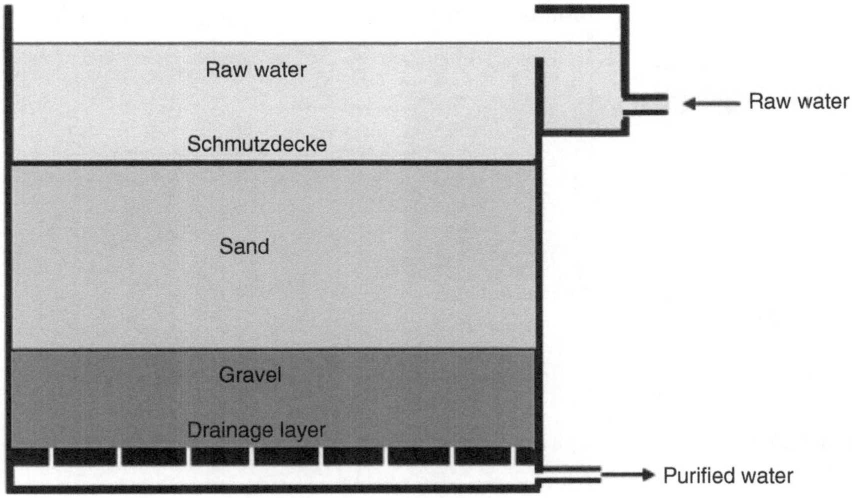

FIGURE 6.5 Schematic Diagram of a Sand Filter (Abdiyev et al., 2023).

2. Filtration:

   Filtration is the process of treating water by passing it through filters with varying pore sizes that are made from various materials, such as sand, charcoal, and gravel, to remove fragments and impurities, some of which include dust, chemicals, and bacteria. Figure 6.5 shows a schematic diagram of a sand filter.

   i. Secondary Treatment:

      Secondary treatment occurs after the physical treatment and involves removing the soluble organic matter that escaped the primary treatment and removing more of the suspended solids to facilitate the further treatment stages (Azanaw et al., 2022; Tang & Hadibarata, 2021). In other words, it purifies water through biological processes, removing the residual organics and suspended solids (Machineni, 2019). It aims to further clean the water by removing dissolved and suspended biological matter. This process typically involves the activated sludge process and a trickling filter.

      1. Activated Sludge Process:

         This widely used biological wastewater treatment method thoroughly mixes microorganisms like bacteria, fungi, and protozoa with the organic matter under conditions that stimulate their growth and remove waste materials. When the mixture settles, it forms a mass of microbes called activated sludge, which is either recycled or undergoes further treatment (Singh et al., 2021). Figure 6.6 shows a schematic diagram of an activated sludge process. It consists of two sections: a biological tank and a sedimentation tank.

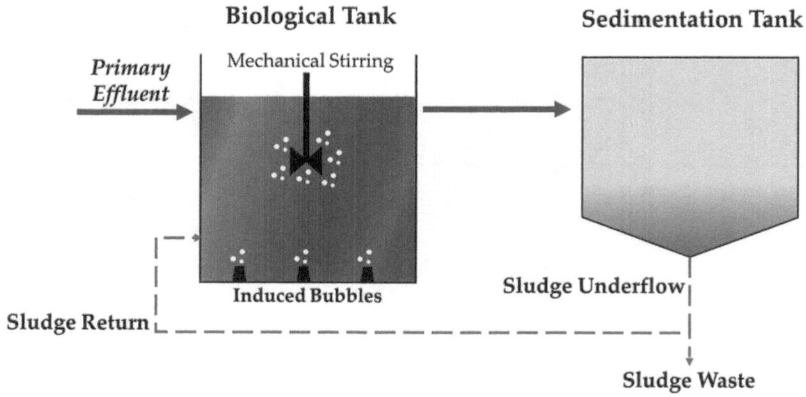

FIGURE 6.6 Activated Sludge Process (Vatten – Wikiskola, n.d.).

2. Trickling Filters:

   Trickling filter (also known as biofilter) is a type of biological treatment system and is one of the most commonly used (Loh et al., 2022). It consists of a fixed porous bed made up of different porous materials (like rocks, lava, ceramic, plastic media, gravel, and many more) and wastewater is distributed over it. As the wastewater trickles down through the bed, a microbial biofilm grown on the top layer of the bed helps to degrade waste, and then treated water is collected at the lower part of the filter. Unlike the activated sludge process, trickling filters are simple to operate, resistant to toxic and shock loads, and require low energy to function (Ali et al., 2017)

   i. Tertiary Treatment

      This is the last stage of the water treatment process, which focuses on polishing the water to remove any remaining contaminants and improve its quality to meet specific usage standards. Some of the processes used in tertiary treatment include advanced filtration and disinfection.

1. Advanced Filtration:

   Advanced filtration is a process that removes contaminants and pollutants that weren't effectively removed during the earlier treatment stages in order to improve water quality (Ahmed et al., 2021; Saravanan et al., 2021).

2. Disinfection:

   This is the last step in the water treatment process and makes use of methods such as chlorination, ozonation, or ultraviolet radiation to kill any remaining pathogens and make water safe for drinking (Srivastav et al., 2020; Tripathi & Hussain, 2022).

Water treatment ensures water purity, reduces water scarcity and assists in conservation efforts. It consists of the primary, secondary and tertiary treatment stages, each of which plays a crucial role in meeting safe and quality water standards and reducing waterborne diseases. Primary treatment, the first stage, aims to remove large particles from water through physical processes that don't change the water's chemical properties. Screening, a key part of the primary stage, rids the water of debris like wood, plastic, and metal to prevent damage and blockage in the following treatment stages. After screening comes sedimentation. In sedimentation, water is held in tanks or basins so solids can settle out, with methods like horizontal flow and inclined plate sedimentation used based on the specific needs for water treatment. The last part of the primary treatment is filtration. This process includes passing water through materials like sand, charcoal and gravel to remove smaller particles, dust and bacteria.

After the primary treatment comes the secondary treatment, which further purifies the water by removing soluble organic matter and additional suspended solids. It does this through biological processes. One of these is the activated sludge process, a standard secondary treatment method, where microorganisms like bacteria, fungi and protozoa mix with organic matter to stimulate their growth. This mixture forms a mass called activated sludge, which is either recycled or further treated. A trickling filter is another technique used in the secondary treatment stage. This technique uses a fixed porous bed of rocks or plastic, through which wastewater trickles through allowing the microbial biofilms to degrade waste, after which treated water is collected at the bottom of the filter. Tertiary treatment, the final stage, focuses on removing remaining contaminants and improving water quality to meet specific usage standards. This stage often employs advanced filtration techniques to eliminate pollutants not effectively removed in earlier stages.

6.2 APPLICATION OF FEW NEXUS 4.0 IN WATER MANAGEMENT AND TREATMENT

Applying food, energy, and water (FEW) Nexus 4.0 in water management and treatment is very important because water is crucial to the food and energy sectors and because of the interconnectedness of these three sectors. Furthermore, the use of innovative technologies like the Internet of Things (IoT), Artificial Intelligence (AI), and big data makes the management and treatment of water more efficient, curbing the issues of water scarcity and unsafe or contaminated water, among many others (Maroju et al., 2023)

1. **Internet of Things and Real-Time Monitoring**

   IoT sensors are essential in managing water today by monitoring water quality and volume in real time. These IoT sensors are used to detect contaminants, measure water levels, and continuously monitor the performance of water treatment facilities (Salam, 2020). Real-time data monitoring enables IoT devices to detect issues in managing and treating water so they can be addressed immediately, ensuring that water remains safe for consumption and reducing operational costs (Gohil et al., 2021). This technology has been implemented in various cities and countries, and one such example is the city of Singapore. By applying the IoT-based water management systems, it has reduced the occurrence of waterborne diseases (Matheri et al., 2023; Salam, 2020).

2. **Artificial intelligence and Machine Learning**

   AI and Machine Learning frameworks are used in the predictive maintenance of water treatment infrastructures by evaluating data from several sensors and previous maintenance records (Adeleke et al., 2023; Cheng et al., 2020; Hallaji et al., 2022) This enables them to project when the equipment is likely to fail or needs servicing, and to schedule time for maintenance, thereby reducing downtime, expanding the lifespan of facilities used in water treatment, and ensuring a continuous supply of clean water (Gunasekaran et al., Lee et al., 2020; Roosefert Mohan et al., 2021). The utility company Thames Water in the United Kingdom successfully used AI to foretell equipment failures in its wastewater treatment, which led to a reduction in unplanned outages and the lower maintenance costs (Gray, 2022; McMillan, 2024).

3. **Big Data Analytics**

   Big data analytics is used in the comprehensive analysis of data collected from water management systems to identify inefficiencies in water usage and treatment processes by analysing patterns and trends in the data collected (Barja-Martinez et al., 2021) The information gathered can be used to optimise operations like adjusting treatment processes to different water quality conditions or maximising energy use in water distribution networks (Barja-Martinez et al., 2021), (Wan et al., 2022) In California, data analytics has been used to effectively manage the state's extensive water resources, especially during periods of drought (Balti et al., 2020).

4. **Integrated Resource Management**

   Food, energy, and water are critical resources that are interconnected, and the FEW Nexus 4.0 emphasises the connection between water management and food and energy systems in achieving resource efficiency and sustainability (Verma et al., 2024). For example, processed wastewater can be utilised in agricultural irrigation, thereby reducing the demand for freshwater resources (Singh, 2021). Energy can also be recovered from wastewater treatment processes through technologies like anaerobic digestion, which produces biogas that can be used to generate heat or electricity (Campello et al., 2021). In the Netherlands, the Waternet utility implements a system that not only treats water but also produces biogas from sewage sludge, which is then used to generate electricity (Nkuna et al., 2024).

6.3 SUMMARY

This chapter focused on applying the FEW Nexus 4.0 strategies in the water sector to ensure it is managed and treated effectively. One of the strategies is the Reduce, Reuse, Recycle, Restore, and Recover system; a system that ensures water sustainability and prevents wastage. The chapter explained the objectives of water management and treatment, with one of the most crucial being to ensure clean and safe drinking water. It also explained how the FEW Nexus 4.0 aids the water sector's objectives of incorporating Industry 4.0s technologies in water management and highlighted the different stages of water treatment.

REFERENCES

Abdiyev, K., Azat, S., Kuldeyev, E., Ybyraiymkul, D., Kabdrakhmanova, S., Berndtsson, R., Khalkhabai, B., Kabdrakhmanova, A., & Sultakhan, S. (2023). Review of slow sand filtration for raw water treatment with potential application in less-developed countries. *Water*, *15*(11), 2007. https://doi.org/10.3390/W15112007

Adeel, Z. (2017). *Water Security as the Centerpiece of the Sustainable Development Agenda* (pp. 25–43). https://doi.org/10.1007/978-3-319-50161-1_2

Adeleke, I. A., Nwulu, N. I., & Ogbolumani, O. A. (2023). A hybrid machine learning and embedded IoT-based water quality monitoring system. *Internet of Things*, *22*, 100774. https://doi.org/10.1016/J.IOT.2023.100774

Ahmed, S. F., Mehejabin, F., Momtahin, A., Tasannum, N., Faria, N. T., Mofijur, M., Hoang, A. T., Vo, D. V. N., & Mahlia, T. M. I. (2022). Strategies to improve membrane performance in wastewater treatment. *Chemosphere*, *306*, 135527. https://doi.org/10.1016/J.CHEMOSPHERE.2022.135527

Ahmed, S. F., Mofijur, M., Nuzhat, S., Chowdhury, A. T., Rafa, N., Uddin, M. A., Inayat, A., Mahlia, T. M. I., Ong, H. C., Chia, W. Y., & Show, P. L. (2021). Recent developments in physical, biological, chemical, and hybrid treatment techniques for removing emerging contaminants from wastewater. *Journal of Hazardous Materials*, *416*, 125912. https://doi.org/10.1016/J.JHAZMAT.2021.125912

Akhtar, S., Memon, S. A., Chae, H. B., Choi, D. W., & Park, C. W. (2023). Hydrostructural phenomena in a wastewater screening channel with an ascendable sub-screen using the arbitrary lagrangian–eulerian approach. *Applied Sciences*, *14*(1), 76. https://doi.org/10.3390/APP14010076

Aleixo, B., Pena, J. L., Heller, L., & Rezende, S. (2019). Infrastructure is a necessary but insufficient condition to eliminate inequalities in access to water: Research of a rural community intervention in Northeast Brazil. *Science of The Total Environment*, *652*, 1445–1455. https://doi.org/10.1016/J.SCITOTENV.2018.10.202

Ali, I., Khan, Z. M., Peng, C., Naz, I., Sultan, M., Ali, M., Mahmood, M. H., & Niaz, Y. (2017). Identification and elucidation of the designing and operational issues of trickling filter systems for wastewater treatment. *Polish Journal of Environmental Studies*, *26*(6), 2431–2444. https://doi.org/10.15244/PJOES/70627

Anderson, E. P., Jackson, S., Tharme, R. E., Douglas, M., Flotemersch, J. E., Zwarteveen, M., Lokgariwar, C., Montoya, M., Wali, A., Tipa, G. T., Jardine, T. D., Olden, J. D., Cheng, L., Conallin, J., Cosens, B., Dickens, C., Garrick, D., Groenfeldt, D., Kabogo, J., … Arthington, A. H. (2019). Understanding rivers and their social relations: A critical step to advance environmental water management. *Wiley Interdisciplinary Reviews: Water*, *6*(6), e1381. https://doi.org/10.1002/WAT2.1381

Azanaw, A., Birlie, B., Teshome, B., & Jemberie, M. (2022). Textile effluent treatment methods and eco-friendly resolution of textile wastewater. *Case Studies in Chemical and Environmental Engineering*, *6*, 100230. https://doi.org/10.1016/J.CSCEE.2022.100230

Badham, J., Elsawah, S., Guillaume, J. H. A., Hamilton, S. H., Hunt, R. J., Jakeman, A. J., Pierce, S. A., Snow, V. O., Babbar-Sebens, M., Fu, B., Gober, P., Hill, M. C., Iwanaga, T., Loucks, D. P., Merritt, W. S., Peckham, S. D., Richmond, A. K., Zare, F., Ames, D., & Bammer, G. (2019). Effective modeling for integrated water resource management: A guide to contextual practices by phases and steps and future opportunities. *Environmental Modelling & Software*, *116*, 40–56. https://doi.org/10.1016/J.ENVSOFT.2019.02.013

Bailone, R. L., Borra, R. C., Fukushima, H. C. S., & Aguiar, L. K. (2022). Water reuse in the food industry. *Discover Food*, *2*(1), 1–17. https://doi.org/10.1007/S44187-021-00002-4

Balti, H., Ben Abbes, A., Mellouli, N., Farah, I. R., Sang, Y., & Lamolle, M. (2020). A review of drought monitoring with big data: Issues, methods, challenges and research directions. *Ecological Informatics*, *60*, 101136. https://doi.org/10.1016/J.ECOINF.2020.101136

Barja-Martinez, S., Aragüés-Peñalba, M., Munné-Collado, Í., Lloret-Gallego, P., Bullich-Massagué, E., & Villafafila-Robles, R. (2021). Artificial intelligence techniques for enabling big data services in distribution networks: A review. *Renewable and Sustainable Energy Reviews*, *150*, 111459. https://doi.org/10.1016/J.RSER.2021.111459

Bashir, I., Lone, F. A., Bhat, R. A., Mir, S. A., Dar, Z. A., & Dar, S. A. (2020). Concerns and threats of contamination on aquatic ecosystems. *Bioremediation and Biotechnology: Sustainable Approaches to Pollution Degradation*, 1–26. https://doi.org/10.1007/978-3-030-35691-0_1

Bayu, T., Kim, H., & Oki, T. (2020). Water governance contribution to water and sanitation access equality in developing countries. *Water Resources Research*, *56*(4), e2019WR025330. https://doi.org/10.1029/2019WR025330

Bazaanah, P., & Mothapo, R. A. (2024). Sustainability of drinking water and sanitation delivery systems in rural communities of the Lepelle Nkumpi Local Municipality, South Africa. *Environment, Development and Sustainability*, *26*(6), 14223–14255. https://doi.org/10.1007/S10668-023-03190-4

Belhassan, K. (2021). Water scarcity management. In *Advanced Sciences and Technologies for Security Applications* (pp. 443–462). https://doi.org/10.1007/978-3-030-76008-3_19

Blomquist, W., Haisman, B., Dinar, A., & Bhat, A. (2007). Australia: Murray-Darling Basin. *Integrated River Basin Management through Decentralization*, 65–82. https://doi.org/10.1007/978-3-540-28355-3_4

Borowski, P. F., & Borowski, P. F. (2020). Nexus between water, energy, food and climate change as challenges facing the modern global, European and Polish economy. *AIMS Geosciences*, *6*(4), 397–421. https://doi.org/10.3934/GEOSCI.2020022

Campello, L. D., Barros, R. M., Tiago Filho, G. L., & dos Santos, I. F. S. (2021). Analysis of the economic viability of the use of biogas produced in wastewater treatment plants to generate electrical energy. *Environment, Development and Sustainability*, *23*(2), 2614–2629. https://doi.org/10.1007/S10668-020-00689-Y

Cheng, J. C. P., Chen, W., Chen, K., & Wang, Q. (2020). Data-driven predictive maintenance planning framework for MEP components based on BIM and IoT using machine learning algorithms. *Automation in Construction*, *112*, 103087. https://doi.org/10.1016/J.AUTCON.2020.103087

du Plessis, A. (2017). Global water scarcity and possible conflicts. *Springer Water*, 45–62. https://doi.org/10.1007/978-3-319-49502-6_3

Elahi, E., Khalid, Z., & Zhang, Z. (2022). Understanding farmers' intention and willingness to install renewable energy technology: A solution to reduce the environmental emissions of agriculture. *Applied Energy*, *309*, 118459. https://doi.org/10.1016/J.APENERGY.2021.118459

Enfrin, M., Dumée, L. F., & Lee, J. (2019). Nano/microplastics in water and wastewater treatment processes – Origin, impact and potential solutions. *Water Research*, *161*, 621–638. https://doi.org/10.1016/J.WATRES.2019.06.049

Fernandes, J., Ramísio, P. J., & Puga, H. (2024). A comprehensive review on various phases of wastewater technologies: Trends and future perspectives. *Eng*, *5*(4), 2633–2661. https://doi.org/10.3390/ENG5040138

Galvez, V., & Rojas, R. (2019). Collaboration and integrated water resources management: A literature review. *World Water Policy*, *5*(2), 179–191. https://doi.org/10.1002/WWP2.12013

Geetha Varma, V. (2022). *Water-Efficient Technologies for Sustainable Development. 6*, 101–128. https://doi.org/10.1016/B978-0-323-91838-1.00009-9

Giakoumis, T., & Voulvoulis, N. (2018). The transition of EU water policy towards the water framework directive's integrated river basin management paradigm. *Environmental Management, 62*(5), 819–831. https://doi.org/10.1007/S00267-018-1080-Z

Gohil, J., Patel, J., Chopra, J., Chhaya, K., Taravia, J., & Shah, M. (2021). Advent of big data technology in environment and water management sector. *Environmental Science and Pollution Research, 28*(45), 64084–64102. https://doi.org/10.1007/S11356-021-14017-Y

Gray, A. (2022). *Optimisation of a Water Company's Waste Pumping Asset Base with a Focus on Energy Reduction*. https://bura.brunel.ac.uk/handle/2438/26697

Guarino, A. S. (2017). The economic implications of global water scarcity. *Research in Economics and Management, 2*(1), 51. https://doi.org/10.22158/REM.V2N1P51

Gude, V. G. (2017). Desalination and water reuse to address global water scarcity. *Reviews in Environmental Science and Biotechnology, 16*(4), 591–609. https://doi.org/10.1007/S11157-017-9449-7

Gunasekaran, K., Boopathi, S., Gunasekaran, K., & Boopathi, S. (1 C.E.). *Artificial Intelligence in Water Treatments and Water Resource Assessments* (pp. 71–98). *https://Services.Igi-Global.Com/Resolvedoi/Resolve.Aspx?Doi=10.4018/978-1-6684-6791-6.Ch004*; https://doi.org/10.4018/978-1-6684-6791-6.CH004

Guta, D. D., Tola, Y. M., Damene, S., & Pai, S. (2022). Examining factors influencing household participation in rural community based potable water supply projects in Becho woreda, Central Ethiopia. *Sustainable Water Resources Management, 8*(1). https://doi.org/10.1007/S40899-021-00595-9

Hallaji, S. M., Fang, Y., & Winfrey, B. K. (2022). Predictive maintenance of pumps in civil infrastructure: State-of-the-art, challenges and future directions. *Automation in Construction, 134*, 104049. https://doi.org/10.1016/J.AUTCON.2021.104049

Hanjra, M. A., & Qureshi, M. E. (2010). Global water crisis and future food security in an era of climate change. *Food Policy, 35*(5), 365–377. https://doi.org/10.1016/J.FOODPOL.2010.05.006

Hoffmann, S., Feldmann, U., Bach, P. M., Binz, C., Farrelly, M., Frantzeskaki, N., Hiessl, H., Inauen, J., Larsen, T. A., Lienert, J., Londong, J., Lüthi, C., Maurer, M., Mitchell, C., Morgenroth, E., Nelson, K. L., Scholten, L., Truffer, B., & Udert, K. M. (2020). A Research Agenda for the future of urban water management: Exploring the potential of nongrid, small-grid, and hybrid solutions. *Environmental Science and Technology, 54*(9), 5312–5322. https://doi.org/10.1021/ACS.EST.9B05222

Hurst, C. J. (2019). *Options for Providing Microbiologically Safe Drinking Water*. 185–260. https://doi.org/10.1007/978-3-030-16775-2_8

Issaoui, M., Jellali, S., Zorpas, A. A., & Dutournie, P. (2022). Membrane technology for sustainable water resources management: Challenges and future projections. *Sustainable Chemistry and Pharmacy, 25*, 100590. https://doi.org/10.1016/J.SCP.2021.100590

Konkol, I., Kuligowski, K., Szafranowicz, P., Vorne, V., Reinikainen, A., Effelsberg, N., Christensen, M. L., Svensson, M., Zviedris, J., Dvarioniene, J., & Cenian, A. (2024). Review of the seasonal wastewater challenges in Baltic Coastal Tourist Areas: Insights from the NURSECOAST-II Project. *Sustainability (Switzerland), 16*(22), 9890. https://doi.org/10.3390/SU16229890

Koop, S. H. A., Van Dorssen, A. J., & Brouwer, S. (2019). Enhancing domestic water conservation behaviour: A review of empirical studies on influencing tactics. *Journal of Environmental Management, 247*, 867–876. https://doi.org/10.1016/J.JENVMAN.2019.06.126

Kundzewicz, Z. W., Krysanova, V., Benestad, R. E., Hov, Piniewski, M., & Otto, I. M. (2018). Uncertainty in climate change impacts on water resources. *Environmental Science & Policy, 79*, 1–8. https://doi.org/10.1016/J.ENVSCI.2017.10.008

Leal Filho, W., Totin, E., Franke, J. A., Andrew, S. M., Abubakar, I. R., Azadi, H., Nunn, P. D., Ouweneel, B., Williams, P. A., & Simpson, N. P. (2022). Understanding responses to climate-related water scarcity in Africa. *Science of The Total Environment*, *806*, 150420. https://doi.org/10.1016/J.SCITOTENV.2021.150420

Lee, J., Ni, J., Singh, J., Jiang, B., Azamfar, M., & Feng, J. (2020). Intelligent maintenance systems and predictive manufacturing. *Journal of Manufacturing Science and Engineering, Transactions of the ASME*, *142*(11). https://doi.org/10.1115/1.4047856

Li, P., Karunanidhi, D., Subramani, T., & Srinivasamoorthy, K. (2021). Sources and consequences of groundwater contamination. *Archives of Environmental Contamination and Toxicology*, *80*(1), 1. https://doi.org/10.1007/S00244-020-00805-Z

Loh, Z. Z., Zaidi, N. S., Yong, E. L., Syafiuddin, A., Boopathy, R., & Kadier, A. (2022). Current status and future research trends of biofiltration in wastewater treatment: A bibliometric review. *Current Pollution Reports*, *8*(3), 234–248. https://doi.org/10.1007/S40726-022-00224-9

Loucks, D. P., & van Beek, E. (2017). Water resources planning and management: An overview. In *Water Resource Systems Planning and Management* (pp. 1–49). https://doi.org/10.1007/978-3-319-44234-1_1

Machineni, L. (2019). Review on biological wastewater treatment and resources recovery: Attached and suspended growth systems. *Water Science and Technology*, *80*(11), 2013–2026. https://doi.org/10.2166/WST.2020.034

Majeed, Y., Khan, M. U., Waseem, M., Zahid, U., Mahmood, F., Majeed, F., Sultan, M., & Raza, A. (2023). Renewable energy as an alternative source for energy management in agriculture. *Energy Reports*, *10*, 344–359. https://doi.org/10.1016/J.EGYR.2023.06.032

Maroju, R. G., Choudhari, S. G., Shaikh, M. K., Borkar, S. K., & Mendhe, H. (2023). Application of artificial intelligence in the management of drinking water: A narrative review. *Cureus*, *15*(11). https://doi.org/10.7759/CUREUS.49344

Matheri, A. N., Belaid, M., Njenga, C. K., & Ngila, J. C. (2023). Water and wastewater digital surveillance for monitoring and early detection of the COVID-19 hotspot: Industry 4.0. *International Journal of Environmental Science and Technology*, *20*(1), 1095–1112. https://doi.org/10.1007/s13762-022-03982-7

McMillan, L. (2024). Artificial intelligence–enabled self-healing infrastructure systems. *Doctoral Thesis, UCL (University College London)*.

Meran, G., Siehlow, M., & von Hirschhausen, C. (2021). Integrated water resource management: Principles and applications. In *The Economics of Water: Rules and Institutions* (23–121). Springer. https://doi.org/10.1007/978-3-030-48485-9_3

Merrey, D. J., Drechsel, P., De Vries, F. W. T. P., & Sally, H. (2005). Integrating "livelihoods" into integrated water resources management: Taking the integration paradigm to its logical next step for developing countries. *Regional Environmental Change*, *5*(4), 197–204. https://doi.org/10.1007/S10113-004-0088-5

Monteiro, A., Santos, S., & Gonçalves, P. (2021). Precision agriculture for crop and livestock farming—brief review. *Animals*, *11*(8), 2345. https://doi.org/10.3390/ANI11082345

Moreno-Pizani, M. A. (2021). Water management in agricultural production, the economy, and venezuelan society. *Frontiers in Sustainable Food Systems*, *4*, 624066. https://doi.org/10.3389/FSUFS.2020.624066

Mujtaba, G., Shah, M. U. H., Hai, A., Daud, M., & Hayat, M. (2024). A holistic approach to embracing the United Nation's Sustainable Development Goal (SDG-6) towards water security in Pakistan. *Journal of Water Process Engineering*, *57*, 104691. https://doi.org/10.1016/J.JWPE.2023.104691

Ngina, J. (2017). *Gender Implications of Household Inaccessibility to Safe Water and Improved Sanitation in Kaptembwo, Nakuru County*. https://41.89.96.81:8080/xmlui/handle/123456789/1485

Nguyen, T. T., Ngo, H. H., Guo, W., Wang, X. C., Ren, N., Li, G., Ding, J., & Liang, H. (2019). Implementation of a specific urban water management – Sponge City. *Science of The Total Environment*, *652*, 147–162. https://doi.org/10.1016/J.SCITOTENV.2018.10.168

Nkuna, S. G., Olwal, T. O., Chowdhury, S. D., & Ndambuki, J. M. (2024). A review of wastewater sludge-to-energy generation focused on thermochemical technologies: An improved technological, economical and socio-environmental aspect. *Cleaner Waste Systems*, *7*, 100130. https://doi.org/10.1016/J.CLWAS.2024.100130

Nti, E. K., Cobbina, S. J., Attafuah, E. E., Senanu, L. D., Amenyeku, G., Gyan, M. A., Forson, D., & Safo, A. R. (2023). Water pollution control and revitalization using advanced technologies: Uncovering artificial intelligence options towards environmental health protection, sustainability and water security. *Heliyon*, *9*(7), e18170. https://doi.org/10.1016/J.HELIYON.2023.E18170

Ogbolumani, O. A., & Nwulu, N. I. (2021). Multi-objective optimisation of constrained food-energy-water-nexus systems for sustainable resource allocation. *Sustainable Energy Technologies and Assessments*, *44*, 100967. https://doi.org/10.1016/j.seta.2020.100967

Ogbolumani, O. A., & Nwulu, N. I. (2022). A food-energy-water nexus meta-model for food and energy security. *Sustainable Production and Consumption*, *30*, 438–453. https://doi.org/10.1016/J.SPC.2021.12.019

Parihar, S. S., Saini, K., Lakhani, G. P., Jain, A., Roy, B., Ghosh, S., & Aharwal, B. (2019). Livestock waste management: A review. *Journal of Entomology and Zoology Studies*, *7*(3), 384–393.

Pokhrel, S. R., Chhipi-Shrestha, G., Hewage, K., & Sadiq, R. (2021). Sustainable, resilient, and reliable urban water systems: Making the case for a "one water" approach. *30*(1), 10–29. https://doi.org/10.1139/ER-2020-0090

Priyanka, E. B., Thangavel, S., & Gao, X. Z. (2021). Review analysis on cloud computing based smart grid technology in the oil pipeline sensor network system. *Petroleum Research*, *6*(1), 77–90. https://doi.org/10.1016/J.PTLRS.2020.10.001

Qadri, R., & Faiq, M. A. (2020). Freshwater pollution: Effects on aquatic life and human health. *Fresh Water Pollution Dynamics and Remediation*, 15–26. https://doi.org/10.1007/978-981-13-8277-2_2

Quebec Septic Tank Regulations – Septic Guide. (n.d.). Retrieved January 30, 2025, from https://septiclearn.blogspot.com/2021/12/quebec-septic-tank-regulations.html

Radcliffe, J. C., & Page, D. (2020). Water reuse and recycling in Australia—History, current situation and future perspectives. *Water Cycle*, *1*, 19–40. https://doi.org/10.1016/J.WATCYC.2020.05.005

Rath, S. (2021). Microbial contamination of drinking water. *Water Pollution and Management Practices*, 1–17. https://doi.org/10.1007/978-981-15-8358-2_1

Ravi Kumar, K., Krishna Chaitanya, N. V. V., & Sendhil Kumar, N. (2021). Solar thermal energy technologies and its applications for process heating and power generation – A review. *Journal of Cleaner Production*, *282*, 125296. https://doi.org/10.1016/J.JCLEPRO.2020.125296

Razavi, S. E., Rahimi, E., Javadi, M. S., Nezhad, A. E., Lotfi, M., Shafie-khah, M., & Catalão, J. P. S. (2019). Impact of distributed generation on protection and voltage regulation of distribution systems: A review. *Renewable and Sustainable Energy Reviews*, *105*, 157–167. https://doi.org/10.1016/J.RSER.2019.01.050

Roosefert Mohan, T., Preetha Roselyn, J., Annie Uthra, R., Devaraj, D., & Umachandran, K. (2021). Intelligent machine learning based total productive maintenance approach for achieving zero downtime in industrial machinery. *Computers & Industrial Engineering*, *157*, 107267. https://doi.org/10.1016/J.CIE.2021.107267

Salam, A. (2020). Internet of Things in water management and treatment. *Internet of Things*, 273–298. https://doi.org/10.1007/978-3-030-35291-2_9

Saravanan, A., Senthil Kumar, P., Jeevanantham, S., Karishma, S., Tajsabreen, B., Yaashikaa, P. R., & Reshma, B. (2021). Effective water/wastewater treatment methodologies for toxic pollutants removal: Processes and applications towards sustainable development. *Chemosphere*, *280*, 130595. https://doi.org/10.1016/j.chemosphere.2021.130595

Seelen, L. M. S., Flaim, G., Jennings, E., & De Senerpont Domis, L. N. (2019). Saving water for the future: Public awareness of water usage and water quality. *Journal of Environmental Management*, *242*, 246–257. https://doi.org/10.1016/J.JENVMAN.2019.04.047

Siddique, I. (2022). Sustainable water management in urban environments. *SSRN Electronic Journal*. https://doi.org/10.2139/SSRN.4885908

Singh, A. (2021). A review of wastewater irrigation: Environmental implications. *Resources, Conservation and Recycling*, *168*, 105454. https://doi.org/10.1016/J.RESCONREC.2021.105454

Srivastav, A. L., Dhyani, R., Ranjan, M., Madhav, S., & Sillanpää, M. (2021). Climate-resilient strategies for sustainable management of water resources and agriculture. *Environmental Science and Pollution Research*, *28*(31), 41576–41595. https://doi.org/10.1007/S11356-021-14332-4

Srivastav, A. L., Patel, N., & Chaudhary, V. K. (2020). Disinfection by-products in drinking water: Occurrence, toxicity and abatement. *Environmental Pollution*, *267*, 115474. https://doi.org/10.1016/J.ENVPOL.2020.115474

Sultana, F. (2018). Water justice: Why it matters and how to achieve it. *Water International*, *43*(4), 483–493. https://doi.org/10.1080/02508060.2018.1458272

Tadesse, A., Bosona, T., & Gebresenbet, G. (2013). Rural water supply management and sustainability: The case of Adama Area, Ethiopia. *Journal of Water Resource and Protection*, *05*(02), 208–221. https://doi.org/10.4236/JWARP.2013.52022

Tang, K. H. D., & Hadibarata, T. (2021). Microplastics removal through water treatment plants: Its feasibility, efficiency, future prospects and enhancement by proper waste management. *Environmental Challenges*, *5*, 100264. https://doi.org/10.1016/J.ENVC.2021.100264

Thacker, S., Adshead, D., Fay, M., Hallegatte, S., Harvey, M., Meller, H., O'Regan, N., Rozenberg, J., Watkins, G., & Hall, J. W. (2019). Infrastructure for sustainable development. *Nature Sustainability*, *2*(4), 324–331. https://doi.org/10.1038/s41893-019-0256-8

Tortajada, C., González-Gómez, F., Biswas, A. K., & Buurman, J. (2019). Water demand management strategies for water-scarce cities: The case of Spain. *Sustainable Cities and Society*, *45*, 649–656. https://doi.org/10.1016/J.SCS.2018.11.044

Tripathi, S., & Hussain, T. (2022). Water and wastewater treatment through Ozone-based technologies. In *Development in Wastewater Treatment Research and Processes: Removal of Emerging Contaminants from Wastewater through Bio-Nanotechnology* (pp. 139–172). https://doi.org/10.1016/B978-0-323-85583-9.00015-6

Verma, N., Talwar, P., Upadhyay, A., Singh, R., Lindenberger, C., Pareek, N., Sarangi, P. K., Zorpas, A. A., & Vivekanand, V. (2024). Food-energy-water Nexus in compliance with sustainable development goals for integrating and managing the core environmental verticals for sustainable energy and circular economy. *Science of The Total Environment*, *930*, 172649. https://doi.org/10.1016/J.SCITOTENV.2024.172649

Wan, X., Kuhanestani, P. K., Farmani, R., & Keedwell, E. (2022). Literature review of data analytics for leak detection in water distribution networks: A focus on pressure and flow smart sensors. *Journal of Water Resources Planning and Management*, *148*(10), 03122002. https://doi.org/10.1061/(ASCE)WR.1943-5452.0001597

Weerasooriya, R. R., Liyanage, L. P. K., Rathnappriya, R. H. K., Bandara, W. B. M. A. C., Perera, T. A. N. T., Gunarathna, M. H. J. P., & Jayasinghe, G. Y. (2021). Industrial water conservation by water footprint and sustainable development goals: A review. *Environment, Development and Sustainability, 23*(9), 12661–12709. https://doi.org/10.1007/S10668-020-01184-0

Wu, Z., Wu, Z., Li, H., Zhang, X., & Jiang, M. (2020). Developing a strategic framework for adopting water-saving measures in construction projects. *Environmental Geochemistry and Health*, *42*(3), 955–968. https://doi.org/10.1007/s10653-019-00407-2

Yadav, M., Vashisht, B. B., Jalota, S. K., Kumar, A., & Kumar, D. (2022). *Sustainable Water Management Practices for Intensified Agriculture*. 131–161. https://doi.org/10.1007/978-3-031-12059-6_8

Zarei, Z., Karami, E., & Keshavarz, M. (2020). Co-production of knowledge and adaptation to water scarcity in developing countries. *Journal of Environmental Management*, *262*, 110283. https://doi.org/10.1016/J.JENVMAN.2020.110283

Zisopoulou, K., & Panagoulia, D. (2021). An in-depth analysis of physical blue and green water scarcity in agriculture in terms of causes and events and perceived amenability to economic interpretation. *Water*, *13*(12), 1693. https://doi.org/10.3390/W13121693

7 Application of FEW Nexus 4.0 in Food, Energy, and Water Systems

7.1 SYSTEMS THINKING AND INTEGRATION

The concept of systems thinking and integration plays crucial roles in the context of the FEW nexus, especially in addressing the complex interdependencies and challenges associated with these interconnected systems (food–energy–water) (Ogbolumani & Nwulu, 2021). Figure 7.1 illustrates, pictorially, systems thinking in FEW Nexus 4.0.

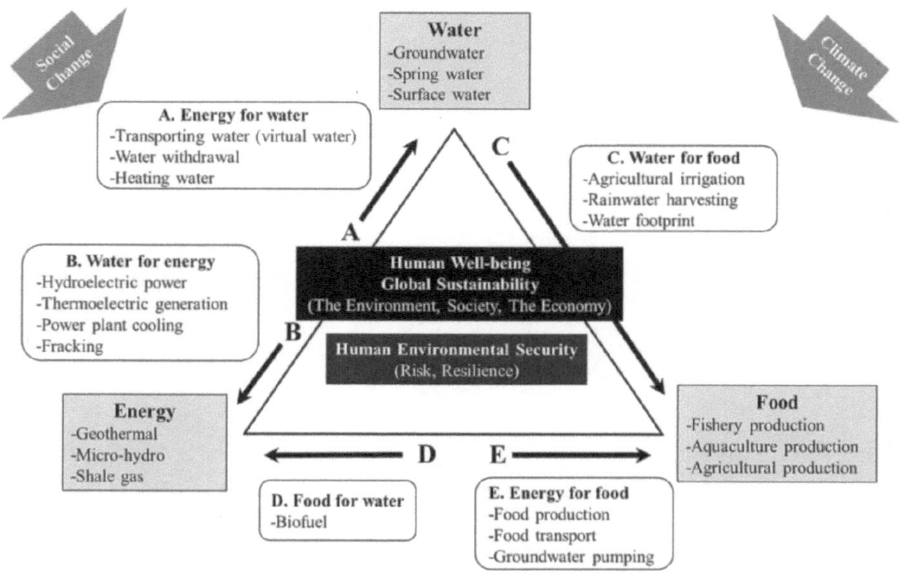

FIGURE 7.1 Systems Thinking and Integration in FEW 4.0 (Endo et al., 2015).

DOI: 10.1201/9781003633853-7

There exists quite a good number of applications of systems thinking and integration in FEW Nexus 4.0. Some of them include:

- **Understanding Interconnections**: The understanding of the intricate relationships and connection loops between food, energy and water systems is facilitated by systems thinking. Systems Thinking also provides some insights into the implications of FEW across sectors and strategies for implementation in a single sector. It allows stakeholders to identify and understand how changes in one system (e.g., water availability) can filter through and affect others (e.g., food production and energy generation) (Ogbolumani & Nwulu, 2022).
- **Integrated Resource Management**: Systems thinking and integration describe integrated procedures that facilitate the use of integrated approaches in managing resources across the FEW nexus. Instead of approaching them in isolation, it encourages strategies that address holistic capture of synergies and trade-offs between food production, energy generation, and water supply considered in optimising resource use and enhancing resiliency to shocks and stresses (Laspidou et al., 2020; Ogbolumani & Nwulu, 2021, 2024). Figure 7.2 illustrates systems thinking and an integrated approach in integrated resource management in the FEW nexus.
- **Policy and Decision-Making**: The global community, through systems thinking, can make multi-sector integrated approaches to complex trade-offs and challenges and also increase resource security through governance that incorporates the stakeholders of the FEW nexus arenas in all the decision-making processes. This will also strengthen and empower

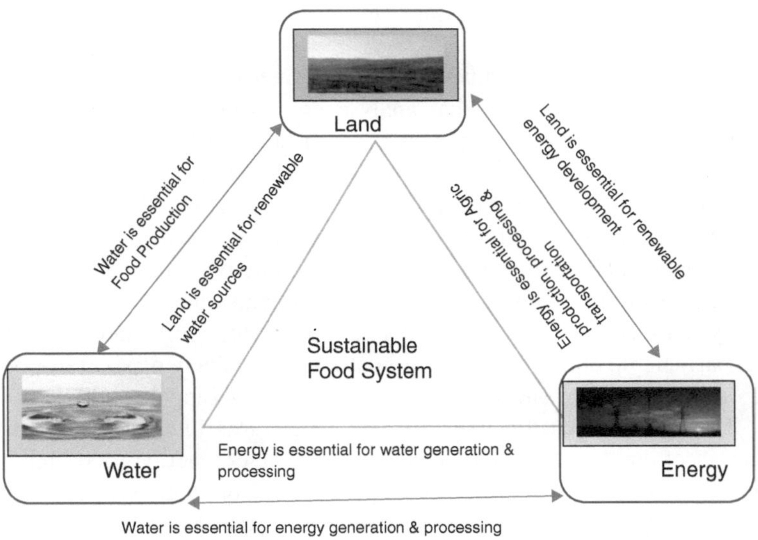

FIGURE 7.2 Systems Approach Illustration (Cho et al., 2023).

analyses and management decisions with its appropriate tools (Namany et al., 2019). The tools must address the complications of such a system by providing adequate details that could reflect the essential trends while offering data for nexus analysis in an informed and quantified manner. The perspective taken by the nexus analysis appears to lose effectiveness as a means of policy-making when the environment is one where various groups of stakeholders are involved with conflicting interests and objectives and who themselves are limited because of complex institutional dynamics and politics (Bai & Sarkis, 2022). Systems-thinking provides decision-makers in management teams with a data-driven framework to analyse the impacts of policies and interventions across the FEW nexus. It facilitates the evaluation of policy options in terms of their immediate effects and broader implications for sustainability, equity, and long-term resilience (Stringer et al., 2018).

- **Planning and Resilience of Scenario**: Scenario planning exercises are supported by systems thinking and integration. These exercises explore alternative opportunities in the future and their implications for the FEW nexus. By taking careful consideration of various scenarios, for example, climate change impacts, population growth, and technological advancements, stakeholders can come up with robust strategies that will improve resilience and adaptability across food, energy, and water systems (Laspidou et al., 2020).
- **Collaboration Across Sectors**: Since its inception in 2011, the water-food-energy-ecosystem nexus approach has emerged as a crucial framework of sustainable development and for advancing Sustainable Development Goals (SDGs) commonly referred to as the SDGs. The approach has evolved from policy and development issues to focus on the common interlinkages among water, food, energy, and ecosystems (Jain et al., 2023).

In 2013, within the framework of the Water Convention, work started to develop methodologies that analyse cross-sectoral linkages in transboundary cross-sectoral linkages in transboundary settings, managing trade-offs, harnessing synergies, and building cooperation and compromise among different uses of water across many sectors. The Task Force on the Water-Food-Energy-Ecosystems nexus was established to oversee the development and application of the methodology with an explicit mandate to facilitate understanding of interlinkages among water, food, energy, and ecosystems. So far, the methodology has been successfully undertaken with the partnership of the national administrations in reviewing six transboundary river basins and one shared aquifer (Strasser & Lipponen, 2021).

This development thus signals a move towards integrated approaches that accommodate the interdependencies of water, energy, food, and ecosystems to effect SDGs through collaborative and mutually supportive management strategies. Integrating systems thinking across various sectors is essential for addressing complex challenges in the food, energy, and water nexus. Systems thinking creates an avenue that fosters collaboration among multiple stakeholders, some of whom include policymakers, scientists, practitioners, and communities. This collaboration enables stakeholders to develop viable solutions that balance the competing demands and maximise the combined benefits of food security, energy access, and water availability (Mensah, 2019).

- **Technological and Innovation Pathways**: The idea of systems thinking and integration facilitates the exploration of technological innovations and other technology-driven pathways that promote sustainable development within the FEW nexus. The adoption of innovative solutions and technologies such as renewable energy technologies, bio-innovations, digital agro-technologies, water-efficient agricultural practices, E-commerce, and sustainable food production systems that contribute to resource efficiency and environmental sustainability are encouraged by systems thinking and integration (Ogbolumani & Nwulu, 2020). Even though these technologies are subject to policy changes and regulations across multiple regions of the world, they will still go a long way in addressing challenges found along the agricultural value chain. Biological and digital technologies afford new possibilities for complex issues emerging in the FEW nexus. This, however, underlines the necessity for an appropriate degree of preparedness to exploit these scenarios and establish a suitable environment for developing, implementing, and using the product (Benfica et al., 2023). Figure 7.3 shows a pictorial representation of the impact of technological and innovation pathways driven by systems thinking and integration in FEW 4.0.

  Another opportunity that system thinking and integration opens regarding technology is agricultural R&D investment. Agricultural research and innovation, mainly through funding by the public sector, has strategic importance in increasing innovation within food systems. Part of this funding goes to research that seeks to solve societal problems and could spur the local private sector to implement solutions (Singh et al., 2022). However, in developing countries, specifically Nigeria, Nepal, Yemen, Jordan, and Bolivia, there has been a decline in funding to the public sector in recent

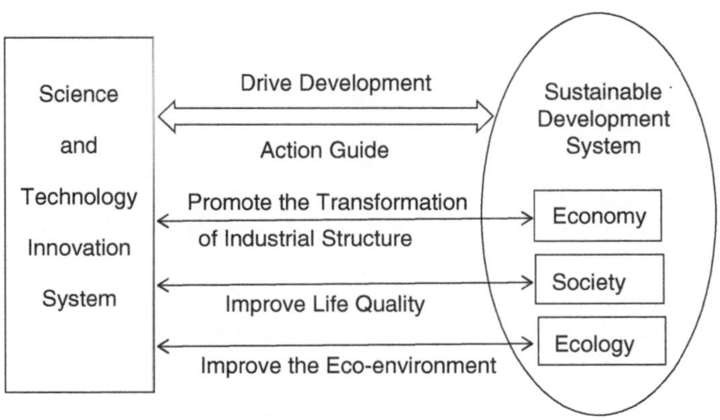

FIGURE 7.3 Technological and Innovation Pathways of System Thinking and Integration in the FEW nexus (Sun et al., 2023).

decades. In contrast, to private-sector funding for agricultural research has, on the other hand, been on the rise. This change has transformed the global agriculture research system (Malec et al., 2024). More precisely, international agricultural R&D investments have shifted from developed countries to large emerging ones, including Brazil, China, and India. These countries have transformed into dependable producers and even policymakers in the agriculture field. Most of the research institutes in developing countries, on the other hand, still encountered significant deficits in terms of capital expenditure on agricultural R&D and human capital as well as institutional endowments. Thus, these countries have not posted increased agricultural productivity. Developing countries characterise certain specific gaps in agricultural research and development systems that influence their potential to exploit the benefits of bio-innovations and digital technologies (Benfica et al., 2023). These include:

a. Investment Gaps: Countries with weak agricultural research systems account for only a meagre portion, and this is about 3% of the overall investment in R&D in developing countries. Such a low investment limits the scope for meaningful research and technology adoption.
b. Productivity Challenges: In developing countries with lagging research systems, agriculture growth is slower than in countries where research systems are at an average or advanced stage; productivity challenges are more evident in developing countries. In this sense, targeted investments in R&D in agriculture would support faster output growth and enhanced resilience.
c. Global Disparities: The poorest countries with weak research systems have a far higher incidence of extreme poverty, while countries with more developed research systems have much more significant investments in R&D and have achieved much higher agricultural productivity levels in the long term.

This would require targeted efforts and commitment to raising investment in agricultural research capacity, involving research quality and productivity, and promoting innovation adoption. Indeed, these efforts could potentially trigger momentous gains but also economic development in food security, economic development, and sustainability in developing countries (Benfica et al., 2023). A nation's ability to invest in agricultural research and development is dependent upon many factors other than just the value of that nation's agricultural GDP. There are certain critical factors that influence a nation's ability to invest in this area; some of them are its economy's total size, its income level, and its access to applicable technology spillovers from other countries, which are far more advanced. In other words, these factors collectively determine a nation's ability to absorb investment in agricultural research and innovation and agricultural productivity and growth. In this sense, knowledge and full utilisation of these broader economic and technological contexts become equally crucial for investments in agricultural R&D to be effective and sustainable in achieving agricultural development goals. Aside from the agricultural

R&D investments gap, other gaps needed to be filled include institutional capacity gap and human capital gap (Benfica et al., 2023).

- **Education and Capacity Building**: Considerable knowledge has been nurtured to provide consumers with safe food with good nutritional value. However, this knowledge often fails to reach food professionals and consumers effectively in real life. Indeed, to improve worldwide food and water safety and sustainably bring better consumer nutrition, this gained knowledge has to leave these paper trails. It has to get imbibed within each stakeholder's minds and hearts and spread throughout the world. Systems thinking would enhance learning and capacity building of stakeholders involved in FEW nexus management, enhancing their understanding of the complex interactions and dependencies between food, energy, and water systems and how this informs decision-making and adaptive management practices. Despite the increased knowledge, improved laws and policies and other tools such as hazard analysis and critical control points, barriers to food safety still exist. Situations persist worldwide with botulism episodes in various countries and salmonellosis with food products. One of the critical issues highlighted is the lack of emphasis on professionalisation and professional development throughout the food chain, especially in the culinary sector. More often than not, training efforts are considered as a matter of response rather than as a course of action. Thus, there is an urgent need for a meaningful and encompassing training program in food safety. This program should target all the food handlers and caregivers who handle and prepare food. The objective is to ensure that people know what they should or should not do to avoid an incident or mitigate risk before it occurs. Minimally informed, distorted risk perception and insufficient power of decision have been rated as the leading causes of food-borne disease outbreaks (Wegener et al., 2018)

  Enhancing appreciation and action on food safety ensures that the knowledge being passed is understood theoretically and practically by those handling foods. This change entails creating an authentic pledge among staff regarding food safety practices as their duty. Energy is also vital in guaranteeing food safety in terms of storage and energy generation. Adequate research and knowledge should be made available on managing the interdependencies between food/water and energy so as not to grow one aspect at the expense of the other. In all, enhancing conditions of food and water safety as well as nutrition requires collaborative approaches that would engage various stakeholders at different levels, all of whom should be armed with appropriate education and training for their respective tasks (Gurinovic et al., 2020)

7.2 CIRCULAR ECONOMY AND ACHIEVING ZERO WASTE

The FEW nexus highlights the interrelatedness and interdependencies between food, energy, and water systems and resources. Previously, the management process involved in these sectors was handled independently. However, the nexus approach realises that activities surrounding a given sector can dramatically affect other sectors. For example, production in agriculture calls for water, and treatment, as well as distribution of these agro-products, requires energy. This section focuses on how this nexus applies to the circular economy and achieving zero waste.

Though nexus thinking and circular economy can be favourable approaches for improving the sustainability of resources, the FEW nexus thinking focuses more on reducing inefficiencies and waste that may arise in the use of resources (Ajibola & Ogbolumani, 2024; Ogbolumani et al., 2025). It aims to do this by understanding and optimising their interdependence. On the other hand, the circular economy tries to reduce waste to its barest minimum through reusing, recycling, and recovering materials. These complementary strategies try to improve resilience, manage resources, and preserve the environment (Parsa et al., 2021). The circular economy is another progressive method of resource management, which aims at minimising use, recycling, and recovering to maintain the usage and value of products (Fernandes & Cunha Marques, 2023; Hoosain et al., 2023; Sauvé et al., 2021). The concept of circular economy changes the perspective of consumers, markets, and natural resources in a way that encourages sustainable and resource utilisation practices (Liu et al., 2021; Tzanakakis et al., 2023)

The idea behind the circular economy is to avoid the depletion of requisite resources, and with this, it strives to support continued economic development (Liu et al., 2021; Tzanakakis et al., 2023). The circular economy is also viewed as a systemic solution for reacting to challenges related to consumption and production patterns (Salminen et al., 2022). Since the regenerative concept of a circular economy is oriented to decreasing emissions, using renewable energy sources, and minimising waste by the closed-loop system and continual reutilisation of resources and materials, it is considered an integrative approach to reacting to challenges relating to consumption and production patterns (Nika et al., 2021).

Recent times have witnessed a shift towards a circular economy, which is likely to improve efficient water utilisation, given the current scarcity of water supply plaguing some parts of the world (Tzanakakis et al., 2023). They indicated that of the circular economy initiatives surveyed, 66% selected water and sanitation as an important sector for circular economy practices, followed by waste management at 78%. The primary strategy remains waste valorisation. Waste valorisation aims to convert waste into essential resources to advance the sustainability of the environment and establish the circular economy model, as identified by (Tanveer et al., 2022). All these efforts are important in advancing efficient and sustainable use of resources, attenuating possible adverse effects from more conventional linear economy approaches.

The option of closing the loops found in water utilisation and management can be associated with considerable economic, environmental, and social benefits. From the ecological point of view, water reclamation can take part in ameliorating environmental conditions by substituting abstraction, reducing anthropogenic pressures, and alleviating the sensitive ecosystems from wastewater treatment plant (WWTP) discharges (Tzanakakis et al., 2023). Economically, this prolongs the lifetime of valued water resources and improves resource efficiency, thus reducing the heavy reliance on basic resources such as energy, water, phosphorus, and nitrogen compounds for secondary activities (Tzanakakis et al., 2023). This can be very important in growing sustainable water management, but in a more traditional setting, it fosters broader resource sustainability in our contemporary world.

Water sector-based circular economy strategies for agriculture have the advantage of reusing water and treating wastewater (Al Hamedi et al., 2023; Fernandes & Cunha Marques, 2023). This will help solve water shortage issues while raising the local food output in agriculture. This makes the method crucial for conserving groundwater supplies because when treated wastewater is used for irrigation, it reduces dependency on freshwater sources. Moreover, since the water is cleaned, the chances of contamination are close to zero. Technologies for wastewater reclamation reduce the demand for freshwater resources by 70% because they work with the existing effluent for farming and solve the global issues of groundwater supplies (Al Hamedi et al., 2023).

Also, a study sought to model specific frameworks for introducing new technologies for interconnected FEW systems in some newly built cities situated in Wadi-Dara, Egypt. This approach entailed using a concealed dam to attain a closed-loop system, which, in this context, regarded the interdependency of energy and water. The study connected water and energy in a closed loop (Abdelzaher et al., 2023). Current energy systems in the world today are still heavily dependent on sources that emit carbon. By extension, this implies that water treatment, which uses energy, is also a carbon-intensive process. Aside from being important potential producers of energy, these carbon- and energy-intensive wastewater treatment processes can also become major recyclers of metals and organic materials, thereby contributing to the attainment of the circular economy and other broader SDGs related to the nexus of FEW, sanitation, and carbon reduction. Rani and his team examined energy-extracting technologies from wastewater alongside their techno-economic viability, environmental benefits, and engineering performance. Their study brought forth four relevant strategies to achieve energy self-sufficiency and near net-zero carbon emissions in the water sector; they include: (1) process energy efficiency improvement; (2) recovery of energy from sewage which has been treated; (3) maximisation of on-site renewable energy sources; (4) improvement of biogas and innovative, decentralised energy-water supply solutions. These strategies would improve the energy self-sufficiency and sustainability of the water sector and its environmental performance (Rani et al., 2022).

According to a study (Samberger, 2022), for SDGs to be realised, there has to be decarbonisation of energy through scaling up renewable energy and efficiency in the use of energy. Other important measures in meeting this goal include establishing a working circular economy, conserving ecosystems and biological diversity, shifting to Green and healthy agricultural practices, promoting electric vehicle use, and supporting green energy development. All these steps will collectively help achieve the SDGs and the UN 2050 target for zero-net carbon emission objectives because they bring sustainability to several sectors of the economy.

Based on studies from Rezvani Ghomi et al. (2021), adopting a circular economy system enables a rational raw materials life cycle. It promotes sustainable economic growth for the environment and businesses. Circular economy differs from linear economy in ownership as it seeks to conserve resources with items and materials being used optimally through repair, recycling, reuse, and renewable energy. Using wastewater in irrigation is useful in cutting costs related to fertilisers and managing the steadily rising water demands within water-scarce countries. Due to the nutritional value of fertilisers, which are usually expensive, and which present in wastewater, there will be a reduction in the usage of commercial fertilisers, which are usually expensive and require energy to produce. The main idea behind shifting to a circular economy is that it focuses on recovering and recycling materials (Lahlou et al., 2021).

The circular economy has given rise to new technologies used to manage water, recycle water, and search for new sources to recover water, especially treated wastewater in urban areas, to rebalance demand and supply (Valenti & Toscano, 2021). Certain organic materials, cellulose, nutrients, and biopolymers coming from wastewater treatment procedures may, in turn, be used to develop high-strength bio-composites and energy forms (Valenti & Toscano, 2021). On a global scale, demands for organic matter and renewable energy are constantly rising. Consequently, waste like sewage sludge are increasingly seen as valuable products. Anaerobic digestion plays a significant role in waste reduction in disposal and the generation of bio-methane, a raw material used in electricity production. The second important part of WWTPs is sewage sludge, which also becomes the substrate for conventional energy production techniques and soil fertilisation or improvement (Valenti & Toscano, 2021).

7.3 WASTE MANAGEMENT AND FEW 4.0 NEXUS

The issue of food wastage worldwide is enormous, with enormous amounts of edible food being thrown away yearly. The need to help reduce global warming has increased the interest in renewable energy sources. Converting food waste into energy taps into the potential within this food-water-energy nexus (Goodwin, 2023). Figure 7.4 shows a pictorial relationship between food wastage and its resultant impact on the globe.

For biological beings, survival and sustainability are directly connected to the intrinsic relationship between food, water, and energy. Food production requires about 30% of our energy resources and accounts for up to 70% of freshwater usage (Food and Agriculture Organization of the United Nations, 2014). Enhancing efficiency in food production and, no less importantly, being prudent in our relationship with food consumption are tasks of paramount necessity and urgency.

The FEW nexus is an important concept globally because these three vital resources are intricately interlinked; therefore, the effect in one area spreads onto others. With an ever-increasing population, supply pressure is mounted on food, energy, and water, which increases simultaneously. This demands new ways of ensuring that humanity has serviceable and balanced management and sharing of these resources, both now and in the future. Food waste is commonly defined as food that has been contaminated or spoilt, and it is usually rejected after it becomes spoiled or stale, most commonly in the consumer stage that exists in the food chain (UNEP, n.d.). Annually, it is estimated that between 1.3 and 2 million tonnes, or roughly a third of the entire food cultivated, or approximately a third of the whole food cultivated, is wasted, according to FAO of the United Nations. However, there is a need for better ways of managing resources with the increase in population all over the world (Norouzi & Kalantari, 2020).

It is essential to understand food wastage regarding the efficiency and sustainability of the FEW nexus by being a vital resource. Part of the solution to the problems

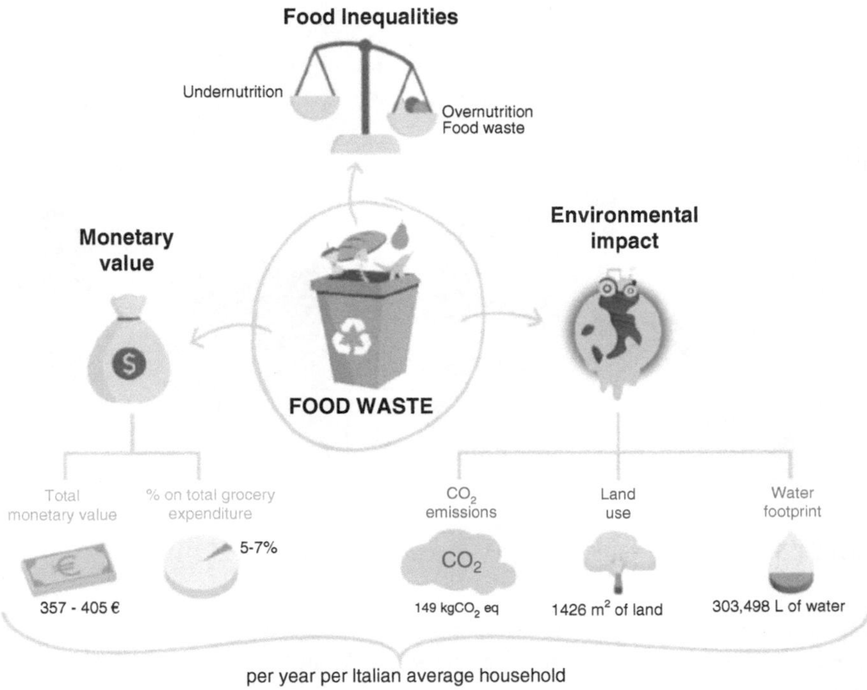

FIGURE 7.4 Food Wastage and Global Impact (Antonelli et al., 2024).

associated with the current constraint of resources can be achieved through food waste recycling for energy and water. This sustainable approach takes the pressure off conventional resource production systems because waste from food can be harnessed and used in energy generation processes and water. Some of the technologies involved with these systems include reverse osmosis, UV sterilisation, filtration, and distillation, which are aimed at purifying water both in homes and businesses (Sridhar et al., 2021). Other energy-producing methods, like hydropower generating stations, nuclear power stations, and solar power farms, influence the general architecture of resource management and sustainability. Hence, strategies like these enhance waste reduction and resource efficiency, supporting environmental and economic dimensions of sustainability under the SDGs agenda (Siaw et al., 2022). Figure 7.5 shows a pictorial relationship existing between food waste and sustainability.

Current waste conversion systems utilise anaerobic digesters that take organic waste foods and turn them into energy. These systems use what is termed a co-digestion process. The principle of co-digestion refers to a process in which various types of organic wastes fed into the digester are degraded. The presented approach by Siaw et al. (2022) improves the parameters of the anaerobic digestion process, which can be used to transform different kinds and specifications of naturally occurring material, such as food waste into biogas, the renewable energy source based on methane, which is effectively used for heating and electricity production and/or as a fuel for the car. Co-digestion also becomes optimal in terms of

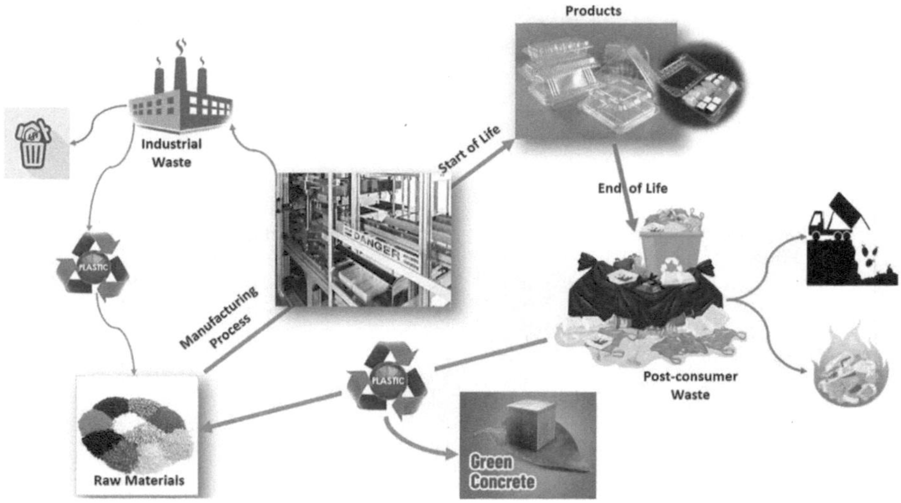

FIGURE 7.5 Waste and Sustainability (Mohammadhosseini et al., 2021).

energy recovery and beneficial environmental effects resulting from reduced landfill disposal and associated greenhouse gas emissions (Siaw et al., 2022).

The food waste recycling process proposed by Siaw et al. (2022) begins with sorting through wastes and setting aside non-biodegradable components to isolate the targeted food waste. Non-biodegradable wastes cannot break down naturally through oxygen and microbial actions (Kumari & Bauddh, 2023). These include glass and plastic waste. Other food wastes considered non-ingestible include items such as eggshells and parts of plants torn out while harvesting; these are also separated using this process. Together with other biodegradable waste, these non-ingestible food items can undergo co-digestion with green waste and animal manure. Upon completing this process, the waste undergoes mechanical compression to recover moisture that should be used for irrigation. Figure 7.6 shows a picture of the mechanical form of an expeller press, with labels indicating its various components. The mechanical device utilised was derived from the structure employed for oil extraction, specifically the expeller press. The press can extract as much as 80% of the oil, and before pressing, the seed is subjected to heat (Bogaert et al., 2018). However, this will not occur with the food waste, as it will not be heated before pressing. Some nutrients would be destroyed by heating food, undermining the objective of collecting the nutrients for irrigation (Siaw et al., 2022).

The feeder integrated with the system feeds food wastes into a mechanical press. These wastes are ground by the screw shaft inside the pressing cage due to friction. The liquids coming out of the grinded waste are tapped as it can be extracted out of the pressing cage due to its tiny pores that allow the liquid to escape into a collection chamber and prevent the solid material to pass through. Residual waste materials refer to the dried portion of food waste. The wastewater contains some hazardous microorganisms, so it is not fit for direct irrigation. Therefore, this wastewater must be treated in order to obtain some form of nutrient-rich solution fit for irrigation.

FIGURE 7.6 Mechanical Expeller Press (Screw Oil Press, n.d.).

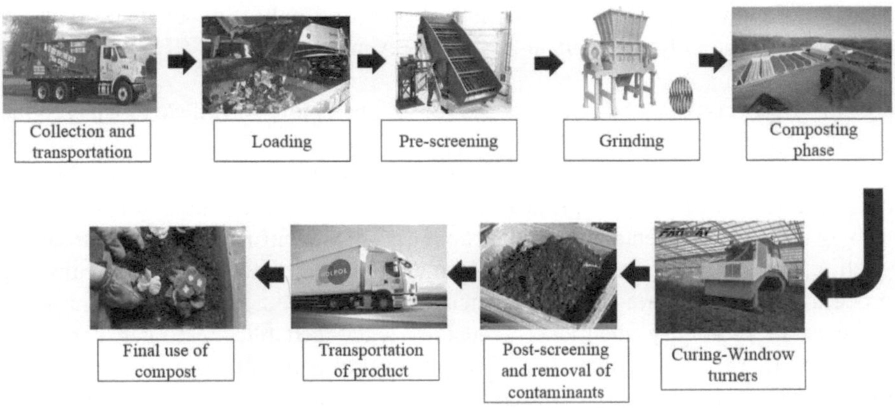

FIGURE 7.7 Flowchart of Food Waste Recycling Process (Al-Rumaihi et al., 2020).

Subsequently, the dehydrated food waste is put into an anaerobic digester system for it to decompose and for methane production. The anaerobic digester is an oxygen-free tank containing microorganisms that assist decomposition. Before being fed into the digester, the dried material is uniformly crushed. The substrate is kept at of about 37°C or 38°C (Farm Energy, n.d.) and continually agitated.

There are three phases involved in the anaerobic digestion process that occurs within the digester: hydrolysis of enzyme, acetogenesis process, and gas generation. In the enzymatic process of hydrolysis, the microorganisms break down any dried material and turn them into soluble compounds. In acetogenesis, the bacteria break down the above derivative into acids, carbon dioxide, hydrogen, and ammonia. Lastly, the methanogens act upon the solution and produce carbon dioxide and methane gas. Methanogens are catalysts in the creation of methane (Enzmann et al., 2018). A flowchart explaining this system is shown in Figure 7.7.

The system can achieve wastewater extraction in the following ways:

The proposed expeller press is fabricated using oil seed extraction technologies, working with mechanical compression and friction to extract between 68% and 80% liquid content from food wastes. Heat generation is intended during compression in the pressing cage; it typically reaches temperatures between 60°C and 100°C. Accordingly, running the expeller press in atmospheric conditions is recommended to avoid probable nutritional breakdowns in the extracted liquid (Goodnature, n.d.). The expeller press is powered by a single unit of 20 horsepower three-phase motor and can make approximately 5–6 tonnes of liquid daily at optimum operation (Siaw et al., 2022)

There are two ways of extracting liquids: one, suggested here, is mechanical pressing, and another is extraction with the help of solvents, such as chemical extraction (Keneni et al., 2021). Some examples of what can be used to extract liquids include the chemical solvents ethanol, n-heptane, n-hexane, and diethyl ether. It has been known that, among the other solvents compared, n-hexane can extract oil better. There are, however, serious problems with the process involving chemical solvents, like air pollution or possibly dangerous contents of the extracted liquid. These troubles are circumvented by the possibility of resorting only to mechanical pressing. An option to use only mechanical pressing prevents these issues (Khan, Mondal, & Hoque, 2016).

This system is designed in the process described below to achieve biogas production in the following way:

Various elements must be carefully examined in the layout of an anaerobic process digester because they significantly affect the efficient production and quantity of biogas. These elements broadly fall under process control systems and feedstock quality. The key considerations include maintaining certain pH levels within the process digester, certain nutrient particulates of the feedstock, particle mass, the presence of inhibitory chemicals, feedstock load, feed rate, hydrogen concentration, and others (Zhang et al., 2021). The pH level directly affects methane yield; for instance, excessive formation of acetic acid during acetogenesis decreases pH levels, resulting in increased acidity and reduced methane generation. Contrary to anticipation, increasing feed loads do not generally correspond to increased methane yield. Studies have found that proteins generate the maximum methane production, sometimes reaching up to 69.5% (Talan et al., 2021). Conducting biochemical methane potential analysis assists in identifying suitable substrates and determining necessary process parameters. Once relevant substrates and parameters are discovered, experimental layout methods such as the Taguchi process, the use of neural networks, and response surface technology may also be applied. Approaches such as these successfully optimise system efficiency, decrease costs, and enhance overall performance (Sridhar et al., 2021).

The food waste recycling system proposed by Siaw et al. (2022) should be utilised in the most efficient manner for agriculture and energy production-friendly ways. Assisted by a mechanical presser, it extracts minerals like calcium compounds, potassium-based minerals, zinc compounds, and iron minerals, along with vitamins B and C, into a concentrated solution that can be applied in watering farmlands to provide good health and growth to the plants, hence reducing water demand

(Cooper et al., 2018). Anaerobic digestion reworks food waste and dehydrates into biogas. Biogas is a renewable energy source that could be used to power farm works and generate electricity or as fuel for vehicles. The resulting digestate from the process may also be further used as organic fertilisers that increase the fertility of soils. In other words, using this new way to harness energy from food wastes would maximise the recycling of nutrients to foster sustainable agriculture and minimise impacts on the environment.

Briefly, FEW Nexus 4.0 is a conceptual framework for the attainment of sustainability goals through the integration of attaining sustainability goals through integrating interrelated systems in food, energy, and water with an emphasis on the context of circular economy and zero-waste programs. As regards the circular economy, it allows economical and efficient use and recycling of resources between these interlinked systems. It promotes strategies such as:

- Resource Efficiency: Making sure that food, energy, and water are used optimally so that wastes are minimal.
- Waste Valorisation: The utilisation of waste production and coming up with ways and means of reclaiming these wastes for valuable use through processes like biogas production from anaerobic digestion or compost production from waste materials for nutrient recycling.
- Closed-Loop Systems: Such systems as production systems that could integrate wastes from one sector (for example, in food processing) to be used as inputs to another industry, such as energy or as fertilisers.
- Technological Integration: Implementing smart technologies for resources, such as smart sensors, the Internet of things, and data analytics, to monitor and coordinate actions and performances.
- Policy and Governance: Application of circular economy policies and governing such as encouraging long-life large-scale products and waste-to-energy programs.

Systems thinking and integration make provision for the emergence of the FEW Nexus 4.0 challenges to be met by gaining an appreciation of interconnections, coordination, integrated resource management, evidence-based decision-making, adapting to shocks and stressors, enhancing coordination, embracing new technologies, and creating stakeholder capacity. These approaches are required to assist in achieving both the universal SDGs and improved health of global communities. FEW Nexus 4.0 aims to reduce food, energy, and water waste by enhancing resource efficiency and considering environmental concerns. It deals with the application of holistic and cross-sector approaches consisting of partnerships in designing and developing programmes that provide answers to the challenges of the current century, including climatic changes and limited resources.

7.4 SUMMARY

While the previous chapters talked about the individual application of FEW Nexus 4.0 in each sector, this chapter discusses the application of FEW Nexus 4.0

in the food, energy, and water systems as an interconnected system. It explains the importance of systems thinking and the circular economy in the sustainability of these resources. It also explains how nexus thinking is different from the circular economy and how they complement each other. This chapter talks about how FEW Nexus 4.0 uses its technologies and systems to manage waste, with one of the systems being to convert food waste into energy. It also talks about how this is carried out.

REFERENCES

Abdelzaher, M. A., Farahat, E. M., Abdel-Ghafar, H. M., Balboul, B. A. A., & Awad, M. M. (2023). Environmental policy to develop a conceptual design for the water–energy–food nexus: A case study in Wadi-Dara on the Red Sea Coast, Egypt. *Water (Switzerland), 15*(4). https://doi.org/10.3390/W15040780

Ajibola, O. A., & Ogbolumani, O. A. (2024). Smart battery storage integration in an IoT-based solar-powered waste management system. *Journal of Digital Food, Energy & Water Systems, 5*(2), 17–42. https://doi.org/10.36615/HREWQ861

Al Hamedi, F. H., Kandhan, K., Liu, Y., Ren, M., Jaleel, A., & Alyafei, M. A. M. (2023). Wastewater irrigation: A promising way for future sustainable agriculture and food security in the United Arab Emirates. *Water (Switzerland), 15*(12). https://doi.org/10.3390/W15122284

Al-Rumaihi, A., McKay, G., Mackey, H. R., & Al-Ansari, T. (2020). Environmental impact assessment of food waste management using two composting techniques. *Sustainability, 12*(4), 1595. https://doi.org/10.3390/su12041595

Antonelli, M., Giordano, C., Chiriacò, M. V., Casari, S., Cadel, E., Chen, P.-J., Magnani, A., Pizzileo, G., Falasconi, L., Alboni, F., & Cicatiello, C. (2024). Assessing the monetary value and environmental impact of household food waste in Italy. *Sustainability, 16*(23), 10614. https://doi.org/10.3390/su162310614

Bai, C., & Sarkis, J. (2022). The water, energy, food, and sustainability nexus decision environment: A multistakeholder transdisciplinary approach. *IEEE Transactions on Engineering Management, 69*(3), 656–670. https://doi.org/10.1109/TEM.2019.2946756

Benfica, R., Chambers, J., Koo, J., Nin-Pratt, A., Falck-Zepeda, J., Stads, G. J., & Arndt, C. (2023). Food system innovations and digital technologies to foster productivity growth and rural transformation. *Science and Innovations for Food Systems Transformation*, 421–437. https://doi.org/10.1007/978-3-031-15703-5_22

Bogaert, L., Mathieu, H., Mhemdi, H., & Vorobiev, E. (2018). Characterization of oilseeds mechanical expression in an instrumented pilot screw press. *Industrial Crops and Products, 121*, 106–113. https://doi.org/10.1016/J.INDCROP.2018.04.039

Cho, M. A., Ramoelo, A., & Sibanda, L. M. (2023). Exploring the integration of the land, water, and energy nexus in sustainable food systems research through a socio-economic lens: A systematic literature review. *Sustainability, 15*(23), 16528. https://doi.org/10.3390/su152316528

Cooper, K. A., Quested, T. E., Lanctuit, H., Zimmermann, D., Espinoza-Orias, N., & Roulin, A. (2018). Nutrition in the bin: A nutritional and environmental assessment of food wasted in the UK. *Frontiers in Nutrition, 5*, 343653. https://doi.org/10.3389/FNUT.2018.00019

Endo, A., Burnett, K., Orencio, P., Kumazawa, T., Wada, C., Ishii, A., Tsurita, I., & Taniguchi, M. (2015). Methods of the water-energy-food nexus. *Water, 7*(10), 5806–5830. https://doi.org/10.3390/w7105806

Enzmann, F., Mayer, F., Rother, M., & Holtmann, D. (2018). Methanogens: Biochemical background and biotechnological applications. *AMB Express, 8*(1), 1–22. https://doi.org/10.1186/S13568-017-0531-X

Fernandes, E., & Cunha Marques, R. (2023). Review of water reuse from a circular economy perspective. *Water (Switzerland), 15*(5). https://doi.org/10.3390/W15050848

Food and Agriculture Organization of the United Nations. (2014). *The Water-Energy-Food Nexus A New Approach in Support of Food Security and Sustainable Agriculture*. https://openknowledge.fao.org/items/c171bc10-9788-4f3d-878a-a137be150c71

Goodwin, L. (2023). *The Global Benefits of Reducing Food Loss and Waste, and How to Do It*. https://www.wri.org/insights/reducing-food-loss-and-food-waste

Gurinovic, M., Milesevic, J., Zekovic, M., Kadvan, A., Ranic, M., & Glibetic, M. (2020). Capacity development in food and nutrition in central and Eastern Europe: A decade of achievements. *Food Policy*, *96*, 101850. https://doi.org/10.1016/J.FOODPOL.2020.101850

Hoosain, M. S., Paul, B. S., Doorsamy, W., & Ramakrishna, S. (2023). The influence of circular economy and 4IR technologies on the climate–water–energy–food nexus and the SDGs. *Water (Switzerland), 15*(4). https://doi.org/10.3390/W15040787

Jain, S. K., Sikka, A. K., & Alam, M. F. (2023). Water-energy-food-ecosystem nexus in India—A review of relevant studies, policies, and programmes. *Frontiers in Water*, *5*, 1128198. https://doi.org/10.3389/FRWA.2023.1128198

Keneni, Y. G., Bahiru, L. A., & Marchetti, J. M. (2021). Effects of different extraction solvents on oil extracted from Jatropha seeds and the potential of seed residues as a heat provider. *Bioenergy Research*, *14*(4), 1207–1222. https://doi.org/10.1007/S12155-020-10217-5

Khan, M. H. U., Mondal, D., & Hoque, S. (2016). Design and construction of oil expeller press with structural analysis of screw with ansys. *PDm*, *50*(4). https://www.researchgate.net/publication/315111267_Design_and_Construction_of_Oil_Expeller_Press_with_Structural_Analysis_of_Screw_with_Ansys

Kumari, K., & Bauddh, K. (2023). Phytoremediation of inorganic contaminants from the aquatic ecosystem using Eichhornia crassipes. *Metals in Water*, 353–368. https://doi.org/10.1016/B978-0-323-95919-3.00001-X

Lahlou, F. Z., Mackey, H. R., & Al-Ansari, T. (2021). Wastewater reuse for livestock feed irrigation as a sustainable practice: A socio-environmental-economic review. *Journal of Cleaner Production*, *294*, 126331. https://doi.org/10.1016/J.JCLEPRO.2021.126331

Laspidou, C. S., Mellios, N. K., Spyropoulou, A. E., Kofinas, D. T., & Papadopoulou, M. P. (2020). Systems thinking on the resource nexus: Modeling and visualisation tools to identify critical interlinkages for resilient and sustainable societies and institutions. *Science of The Total Environment*, *717*, 137264. https://doi.org/10.1016/J.SCITOTENV.2020.137264

Liu, Q., Yang, L., & Yang, M. (2021). Digitalisation for water sustainability: Barriers to implementing circular economy in smart water management. *Sustainability (Switzerland), 13*(21). https://doi.org/10.3390/SU132111868

Malec, K., Rojík, S., Maitah, M., Abdu, M., & Abdullahi, K. T. (2024). Impact of investments in agricultural innovation on food security in sub-Saharan Africa. *Heliyon*, *10*(16), e35913. https://doi.org/10.1016/J.HELIYON.2024.E35913

Mechanical Extraction Processing Technology for Biodiesel – Farm Energy. (n.d.). Retrieved August 24, 2024, from https://farm-energy.extension.org/mechanical-extraction-processing-technology-for-biodiesel/

Mensah, J. (2019). Sustainable development: Meaning, history, principles, pillars, and implications for human action: Literature review. *Cogent Social Sciences*, *5*(1). https://doi.org/10.1080/23311886.2019.1653531

Mohammadhosseini, H., Alyousef, R., & Md. Tahir, M. (2021). Towards sustainable concrete composites through waste valorisation of plastic food trays as low-cost fibrous materials. *Sustainability*, *13*(4), 2073. https://doi.org/10.3390/su13042073

Namany, S., Al-Ansari, T., & Govindan, R. (2019). *Sustainable Energy, Water and Food Nexus Systems: A Focused Review of Decision-Making Tools for Efficient Resource Management and Governance*. https://doi.org/10.1016/j.jclepro.2019.03.304

Nika, C. E., Expósito, A., Kisser, J., Bertino, G., Oral, H. V., Dehghanian, K., Vasilaki, V., Iacovidou, E., Fatone, F., Atanasova, N., & Katsou, E. (2021). Validating circular performance indicators: The interface between circular economy and stakeholders. *Water (Switzerland)*, *13*(16), 2198. https://doi.org/10.3390/W13162198

Norouzi, N., & Kalantari, G. (2020). The food-water-energy nexus governance model: A case study for Iran. *Water-Energy Nexus*, *3*, 72–80. https://doi.org/10.1016/J.WEN.2020.05.005

Ogbolumani, O. A., & Nwulu, N. (2020). Integrated appliance scheduling and optimal sizing of an autonomous hybrid renewable energy system for agricultural food production. In *Lecture Notes in Mechanical Engineering* (pp. 651–660). https://doi.org/10.1007/978-981-15-5753-8_60

Ogbolumani, O. A., & Nwulu, N. I. (2021). Multi-objective optimisation of constrained food-energy-water-nexus systems for sustainable resource allocation. *Sustainable Energy Technologies and Assessments*, *44*, 100967. https://doi.org/10.1016/j.seta.2020.100967

Ogbolumani, O. A., & Nwulu, N. I. (2022). A food-energy-water nexus meta-model for food and energy security. *Sustainable Production and Consumption*, *30*, 438–453. https://doi.org/10.1016/J.SPC.2021.12.019

Ogbolumani, O. A., & Nwulu, N. I. (2024). Environmental impact assessment for a meta-model-based food-energy-water-nexus system. *Energy Reports*, *11*, 218–232. https://doi.org/10.1016/j.egyr.2023.11.033

Ogbolumani, O. A., & Waste, Y. W. O. R. D. S. (2025). Intelligent waste management optimization through machine learning analytics. *Journal of Science Research and Reviews*, *2*(1), 7–26. https://doi.org/10.70882/JOSRAR.2025.V2I1.25

Parsa, A., Van De Wiel, M. J., & Schmutz, U. (2021). Intersection, interrelation or interdependence? The relationship between circular economy and nexus approach. *Journal of Cleaner Production*, *313*, 127794. https://doi.org/10.1016/J.JCLEPRO.2021.127794

Rani, A., Snyder, S. W., Kim, H., Lei, Z., & Pan, S. Y. (2022). Pathways to a net-zero-carbon water sector through energy-extracting wastewater technologies. *Npj Clean Water*, *5*(1). https://doi.org/10.1038/S41545-022-00197-8

Rezvani Ghomi, E., Khosravi, F., Tahavori, M. A., & Ramakrishna, S. (2021). Circular economy: A comparison between the case of Singapore and France. *Materials Circular Economy*, *3*(1), 1–12. https://doi.org/10.1007/S42824-020-00016-W

Salminen, J., Määttä, K., Haimi, H., Maidell, M., Karjalainen, A., Noro, K., Koskiaho, J., Tikkanen, S., & Pohjola, J. (2022). Water-smart circular economy – Conceptualisation, transitional policy instruments and stakeholder perception. *Journal of Cleaner Production*, *334*. https://doi.org/10.1016/J.JCLEPRO.2021.130065

Samberger, C. (2022). The role of water circularity in the food-water-energy nexus and climate change mitigation. *Energy Nexus*, *6*, 100061. https://doi.org/10.1016/J.NEXUS.2022.100061

Sauvé, S., Lamontagne, S., Dupras, J., & Stahel, W. (2021). Circular economy of water: Tackling quantity, quality and footprint of water. *Environmental Development*, *39*, 100651. https://doi.org/10.1016/J.ENVDEV.2021.100651

Siaw, M. N. K., Oduro-Koranteng, E. A., & Dartey, Y. O. O. (2022). Food-energy-water nexus: Food waste recycling system for energy. *Energy Nexus*, *5*, 100053. https://doi.org/10.1016/J.NEXUS.2022.100053

Singh, R. B., Paroda, R. S., & Dadlani, M. (2022). *Science, Technology and Innovation*. 213–250. https://doi.org/10.1007/978-981-19-0763-0_8

Sridhar, A., Kapoor, A., Senthil Kumar, P., Ponnuchamy, M., Balasubramanian, S., & Prabhakar, S. (2021). Conversion of food waste to energy: A focus on sustainability and life cycle assessment. *Fuel*, *302*. https://doi.org/10.1016/J.FUEL.2021.121069

Strasser, L. de., & Lipponen, Annukka. (2021). *Solutions and Investments in the Water-Food-Energy-Ecosystems Nexus : A Synthesis of Experiences in Transboundary Basins*. United Nations.

Stringer, L. C., Quinn, C. H., Le, H. T. V., Msuya, F., Pezzuti, J., Dallimer, M., Afionis, S., Berman, R., Orchard, S. E., & Rijal, M. L. (2018). A new framework to enable equitable outcomes: Resilience and nexus approaches combined. *Earth's Future*, *6*(6), 902–918. https://doi.org/10.1029/2017EF000694

Sun, L., Wang, Z., & Yang, L. (2023). Research on the dynamic coupling and coordination of science and technology innovation and sustainable development in Anhui Province. *Sustainability*, *15*(4), 2874. https://doi.org/10.3390/su15042874

Talan, A., Tiwari, B., Yadav, B., Tyagi, R. D., Wong, J. W. C., & Drogui, P. (2021). Food waste valorization: Energy production using novel integrated systems. *Bioresource Technology*, *322*, 124538. https://doi.org/10.1016/J.BIORTECH.2020.124538

Tanveer, M., Khan, S. A. R., Umar, M., Yu, Z., Sajid, M. J., & Haq, I. U. (2022). Waste management and green technology: Future trends in circular economy leading towards environmental sustainability. *Environmental Science and Pollution Research*, *29*(53), 80161–80178. https://doi.org/10.1007/S11356-022-23238-8

Tzanakakis, V. A., Capodaglio, A. G., & Angelakis, A. N. (2023). Insights into global water reuse opportunities. *Sustainability*, *15*(17), 13007. https://doi.org/10.3390/SU151713007

Valenti, F., & Toscano, A. (2021). A gis-based model to assess the potential of wastewater treatment plants for enhancing bioenergy production within the context of the water–energy nexus. *Energies*, *14*(10). https://doi.org/10.3390/EN14102838

Wegener, J., Fong, D., & Rocha, C. (2018). Education, practical training and professional development for public health practitioners: A scoping review of the literature and insights for sustainable food system capacity-building. *Public Health Nutrition*, *21*(9), 1771–1780. https://doi.org/10.1017/S1368980017004207

Zhang, J., Gu, D., Chen, J., He, Y., Dai, Y., Loh, K. C., & Tong, Y. W. (2021). Assessment and optimization of a decentralized food-waste-to-energy system with anaerobic digestion and CHP for energy utilization. *Energy Conversion and Management*, *228*, 113654. https://doi.org/10.1016/J.ENCONMAN.2020.113654

8 Sustainability Metrics and Assessments For FEW Nexus 4.0

8.1 SUSTAINABILITY IN FOOD, ENERGY AND WATER SYSTEMS

Urban centres are the hub of consumption of food, water, and energy; most of the sustainability challenges are created in the cities. Much of the existing literature underlines the need for a more integrated approach to Food, Energy and Water (FEW) nexus within policy and decision-making frameworks, but there is still a lack of reliable assessment systems that, at least in principle, could identify how governance strategies might improve the provision of such basic services. Looking forward, various studies have proposed an indicator system to better capture the diversity of sustainability issues associated with the food, energy, and water nexus at the urban level. Results from these studies will eventually help to facilitate implementation strategies for optimised urban sustainability efforts (Yuan et al., 2021). Figure 8.1 shows a framework for the relationship between a crop production system and the FEW nexus system.

The United Nations described sustainable development as "development that meets the needs of the present without compromising the ability of future generations to meet their own needs" (Basiago, 1995). This principle drew the attention of the whole world and spurred many theories and policies to guide society toward a sustainable future (Al-Saidi & Elagib, 2017; Parnell, 2016). The centrality of cities in this debate is underlined by the fact that nearly 50% of the world's population is already residing in cities, with the continuance of the urbanisation trend driven by aspirations for improved living standards and economic opportunities (Hall & Pfeiffer, 2013; Henderson, 2010). This means we must recognise that cities are not just passive spaces for these processes but can themselves, through their sheer scale, help attain sustainability goals. This argument has been advanced recently by the authors cited in (Saladini et al., 2018) and more recently by the authors cited in (Momblanch et al., 2019). There is nothing wrong with urban living, provided it is done right. Urban living tackles problems coming from within the cities and also those that are associated with city expansion. Most cities today are built for commercial purposes; however, people often find it difficult to use cities naturally because they often get used to using them excessively(Yuan et al., 2017).

In cities, basic needs of life like FEW are very important. Food is vital to stop hunger; with it, everyone is sure of being satisfied and, by extension, can maintain

DOI: 10.1201/9781003633853-8

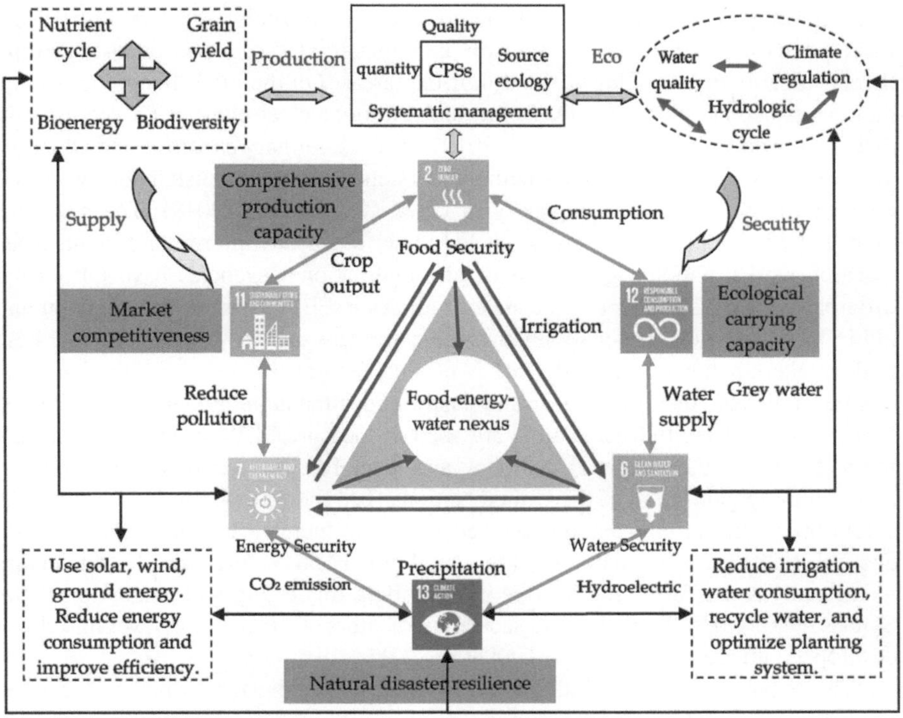

FIGURE 8.1 Framework for the Relationship between Crop Production System and Food, Energy and Water Nexus (Guo et al., 2024).

good health and vitality. The availability of food ensures that crops are grown the right way (Crush & Frayne, 2011). Energy helps us get cheap, steady, and green power (Kaygusuz, 2012). Water is an essential resource that we need to keep safe (Biswas & Tortajada, 2010). The need to focus on how cities can sustain food, energy, and water is growing because these areas are very important for healthy city life (Leck et al., 2015). This focus is also important for planning future cities (Verma & Raghubanshi, 2018). However, there's actually no single way to look at how cities grow in a green way. This is because every city faces its own set of challenges and situations (Artmann et al., 2019). Without a common way to compare, it's hard to find and share the best ways to make cities more sustainable.

While researchers generally agree that the FEW nexus concept ought to be integrated into governance, discussions regarding its present status and general trends are ongoing. In the FEW nexus systems, some authors have carried out research that addresses issues in the areas of optimisation, challenges and opportunities, synergies and trade-offs, integrative approaches, and model simulations in studies. The gaps in these studies indicate a pressing need to engage in a FEW nexus approach towards attaining sustainability (Dargin et al., 2019; Mercure et al., 2019; Ogbolumani & Nwulu, 2020, 2021, 2022, 2024; Özcan et al., 2025; Sivakumar, 2021; Wu et al., 2021; Yuan et al., 2019; Zhang et al., 2019).

According to the authors cited in reference (Covarrubias & Boas, 2020), the FEW nexus marks a milestone towards greening food systems in cities regarding the sustainable provisioning of food in urban areas. On the other hand, knowledge on governance in cities is relatively low and without a clear niche in the mainstream discussions on the nexus (Feng et al., 2019). Interdisciplinary expertise and collaborative efforts are important in enhancing understanding of FEW sustainability knowledge due to the intrinsic complexity involved (Covarrubias, 2019). The quest for more knowledge on sustainability has led to serious academic research. Numerous research works, national governments, and international agencies have introduced different yardsticks to identify the state of affairs in FEW systems development and aid in FEW sustainability assessments. Indicators are the key instruments that help policymakers better understand what is needed for governance and how decisions can be improved; they offer clear and easily understandable guidance. It is important to note that multiple indicators are used to measure FEW sustainability at different scales, be they regional, national, and/or global. For example, the Food and Agriculture Organisation of the United Nations (FAO) developed food security indicators based on the information received from 113 member countries (FAO, IFAD and WFP, 2013). Also, the Committee on World Food Security drafted indicators to measure/assess food security that provide data for testing and evaluation (ESA SOFA-Team, 2015). Specific food security indicators at the national level have been developed and are in place (Cleveland et al., 2015).

With regard to energy sustainability, the IEA uses indicators which are oriented towards satisfying total energy demand as well as towards managing the price risk of supply (Tsai, 2010). Studies have reviewed the indicators and the trends in energy policies that are being implemented for future sustainable energy development at the national level in countries like Taiwan (Chuang & Ma, 2013; Martchamadol & Kumar, 2012), Thailand (Serwaa Mensah et al., 2014), Ghana (Machado et al., 2015), Brazil (Hannan et al., 2018), Malaysia (Gnansounou, 2008), etc. The same also applies at regional and global levels (Mendes et al., 2011). Authors like Badea et al. (2011) and Cleveland et al. (2015) addressed challenges related to uncertainty in the determination of energy-related indicators. Mendes et al. (2011) discussed how indicators developed for addressing energy uncertainties can be defined.

Authors in cited the reference (Sullivan et al., 2003) developed a Water Poverty Index, which uses associated performance and management issues of household water consumption to try and measure the water pressure at home and in the community. The significance of energy efficiency indicators and hydraulics for improved water supply service management is highlighted by international case studies (Vilanova et al., 2015).

Household, economy, city, river, and community are the five dimensions for securing water, as described in the Asian Water Development Outlook (Asian Water Development Outlook, 2020). Even with all of these advancements, seamless indicators are still required to direct the long-term growth of the food, energy, and water nexus, particularly in the global space (Yuan & Lo, 2020).

The emergence of the term "FEW sustainability" in the policy of various international bodies has brought about various definitions. Reviewing a specific definition that should be built on a sound literature review becomes crucial (Yuan & Lo, 2020). In their research, they considered the significance of FEW sustainability as the provision of affordable, reliable, and sufficient quantity of Food, Energy, and Water compared to spatial demands. For global applicability, the study conducted by Yuan and Lo (2020) proposes an approach for the development of Sustainable Development Goals (SDG)-based indicators; this approach is briefly explained below:

- Firstly, there is a need to identify sustainability indicators, generally confined to Food-Energy-Water domains. The approach developed here selects all indicators of the SDG related to FEW: Thirteen for food sustainability, six for energy sustainability, and eleven for water sustainability.
- Their studies attempted to explore the feasibility of the Universal FEW Sustainability (LIFEWAY) and frame its relevance across countries.
- These data were scrutinised to ascertain if they qualify for inclusion into the established database. The fourteen indicators identified for use in the study by (Yuan & Lo, 2020) will contribute to the reduction of the number of indicators, ensuring that the data are available. In addition, a proper time horizon was set within which the monitoring processes can be followed. If possible, indicators whose information or functions are redundant, e.g., prevalence of undernourishment, stunting, and malnutrition, are condensed into one.

Table 8.1 shows the basic indicators along with their corresponding measuring units.

There are 12 indicators in LIFEWAY, each of them is briefly described in the following section.

TABLE 8.1
Twelve Indicators for LIFEWAY based on the Sustainable Development Goals (SDGs), (Yuan & Lo, 2020)

Part of Nexus Under Consideration	Indicator	Unit
Food Sustainability	Incidence of undernourishment	% of population
	Cereal yield	kg per hectare
	Arable land	hectares per person
	Proportion of local breeds being at risk	% of total local breeds
Energy Sustainability	Access to electricity	% of population
	Access to clen fuels and technologies for cooking	% of population
	Energy use	Constant 2011 PPP $ per kg of oil equivalent
	Renewable energy share	% of total final energy consumption
Water Sustainability	People using at least basic drinking water services	% of population
	People using at least basic sanitation services	% of population
	Renewable internal freshwater resources	Cubic metres per capita
	Water body extent	% of total land area

8.2 LIFEWAY INDICATORS

8.2.1 Cases of Undernourishment

This is one of the major indicators describing the food insecurity problem within a population. Good nutrition is relevant for work productivity, health, and learning ability (Özilgen, 2018). The malnutrition case rate trends are falling in sub-Saharan Africa and South Asia (Rosegrant & Cline, 2003). This indicator reports the nutritional status of a country's members.

8.2.2 Cereal Production

This indicator, which includes wheat, rice, maize, barley, oats, rye, millet, sorghum, buckwheat, and mixed grains, calculates production yield in kilograms per hectare of cultivated land area hectare. The dry grain yields are used in the production data. In 2018, 320 billion kilograms of these crops were harvested globally. Some projections indicate that this crop production has to expand to 440 billion kg by 2050 for reasons related to food security. An important and crucial agricultural indicator of improved food sustainability is crop production efficiency (Alston et al., 2009; Rosegrant & Cline, 2003).

8.2.3 Arable Land

This indicator measures the most productive type of land for raising permanent crops and pastures. Research carried out by policymakers in the early 2000s suggests that there is an increase in the amount of land used for agriculture (Foley et al., 2005). However, when land remains fallow owing to changes in human behaviour toward farming, it affects the ecosystem, the productivity of the land, and our food security. As a result, it's critical to preserve productive farms. This indicator helps estimate the amount of food that humanity is losing out on, together with the previous two.

8.2.4 Share of Local Breeds at Risk

The Food and Agriculture Organization defines this indicator as the proportion of threatened livestock breeds. The metric becomes relevant to nutrition and biodiversity because the genetic material from food-producing animals forms an integral part of the naturally based platform underpinning long-term food systems, biodiversity, and farming. Resilience in the face of new social, economic, and environmental conditions requires this measure to be carefully considered.

8.2.5 Access to Reliable Electricity

This indicator is important as it shows what fraction of people who live in cities have access to clean and affordable electricity. According to data published by IEA, 1.3 billion people don't have reliable power, and 85% of them live in rural areas (World Energy Outlook 2011 – Analysis – IEA, n.d.). It is essential to ensure everyone gets cheap, eco-friendly electricity. This will help businesses greatly in the long run.

8.2.6 Access to Eco-Friendly Fuels and Technologies for Cooking and Other Domestic Heating Purposes

This indicator expresses the proportion of a population having access to modern cooking solutions. Kerosene continues to be a very important energy source in developing regions, including Africa and Central Asia (World Energy Outlook 2011 – Analysis – IEA, n.d.). Still, concerning WHO guidelines, kerosene cannot be rated or ranked as a clean cooking fuel. Approximately 2.7 billion people still lack access to clean fuels and technologies for cooking. Monitoring energy sustainability at the national level is essential in fostering modern solutions for cooking (Farabi-Asl et al., 2019).

8.2.7 Utilisation of Energy

This indicator is very important in terms of the creative use of energy in an eco-friendly way. It refers to the PPP GDP per kilogram of oil equivalent for energy use. Efficient energy use is an essential part of a sustainable energy supply. It creates a tight link between the production of energy and economic development by establishing how efficiently people can access products and services and also guiding them on how to use energy effectively and efficiently (Yuan & Lo, 2020).

8.2.8 Share of Energy Derived from Renewables

To fight climate change, the European Union set a binding goal of using 20% green energy by 2020 (Liu, 2014). Renewable energy comes from the sun, the wind, water, tides, geothermal heat, biofuels, and the part of waste that can be used again. These sources are very important in many places because they cut down on greenhouse gas emissions, make energy sources more diverse, and make countries less reliant on fossil fuel markets (Lund, 2007; Turner, 1999). When you put together Indicators 5, 6, and 7, energy use shows how important the energy industry is.

8.2.9 People Using at Least Basic Drinking Water Services

This indicator expresses the proportion of the population using at least essential services for drinking water, thereby covering both basic and safely managed drinking water services. Essential drinking water sources include delivery points such as piped water, public taps, boreholes or tube wells, protected dug wells and springs, rainwater, and bottled water. Access to water positively affects global health and prosperity. This is in consideration of the fact that, according to WHO, communities that lack essential water services face high risks of various diseases, which accounts for thousands of deaths each year (Gleick, 2003; WHO & Unisef, 2000). This scenario greatly affects the world's health.

8.2.10 People Using at Least Basic Sanitation Services

This indicator refers directly to basic sanitation, ensuring safe and effective disposal of human, animal, and insect waste, either in situ or through collection and conveyance, treatment, and disposal off-site. Basic sanitation facilities include flush toilets connected to piped sewer systems, septic tanks, or pit latrines. This indicator is designed to measure the proportion of people using basic sanitation services exclusively. Inadequate separation of excreta from human contact, compounded by inaccessibility to basic water services, can result in huge health risks and a number of deaths (Yuan & Lo, 2020).

8.2.11 Renewable Internal Freshwater Resources

Renewable internal freshwater resources include renewable sources, such as flows within a country and groundwater recharge by rainfall. This variable is therefore highly linked with variables on land use and land cover, particularly agricultural lands, forests, and wetlands, which are key contributors to creating such resources (Yuan & Lo, 2020).

8.2.12 Water Body Size

Over 70% of the Earth is covered with water, and most of this water is found in oceans, seas, and notable rivers; only a minute portion of the Earth's water is used for terrestrial purposes. With less than 1% of Earth's water, terrestrial waterways contribute significantly to the global water cycle and climate. They have an incredibly high influence on both life and the environment. A subset of the global hydrological cycle originates in reservoirs, surface water bodies, and intermittently flooded areas that majorly affect water sustainability and climate change. Other indicators supporting the freshwater usage indicator are Indicators 9 and 11, associated with the management of ongoing water resources, assisting in scrutinising efficiency performance in a water system. Studies by Yuan and Lo (2020) considered 12 SDG metrics as showing FEW sustainability in forty-two countries, representing both the developed and developing world. This study assigned equal weights to all indicators. Data sources used in this study were obtained from the World Bank's World Development Indicators database and the United Nations Global SDG Indicators Database. Their studies assessed data from 2014 since it had maximum coverage of FEW sustainability worldwide. These normalisation formulae guarantee that all indices are scored from a base of 0 to 100. This then allows for direct comparison between indicators.

In the context of food sustainability, both food-energy malnutrition and food inadequacy indicators are important as they give a holistic view of the food and nutrition status across different countries. Considering the sustainability of energy, it is noted that issues such as access to electricity and clean energy within the sector cannot be ignored. As for the evaluation framework, water sustainability involves indicators concerning access to drinking water and sanitation services.

The assessments and metrics relevant for sustainability in the context of understanding, managing, and governing a transforming nexus of FEW Nexus 4.0 enable an understanding of new systems complexities and interrelationships between the food, energy, and water systems in relation to sustainable development.

8.3 SUMMARY

This chapter discussed governance and its role in efficiently managing resources across food, water, and energy. Cities are the focus of the FEW Nexus 4.0s systems and strategies due to the percentage of people residing there, and this chapter explained how cities could improve sustainability across the food, energy, and water nexus systems. It explained how crucial the FEW nexus resources are for current and future cities. This chapter briefly describes the 12 indicators in LIFEWAY based on the SDGs.

REFERENCES

Al-Saidi, M., & Elagib, N. A. (2017). Towards understanding the integrative approach of the water, energy and food nexus. *Science of The Total Environment*, *574*, 1131–1139. https://doi.org/10.1016/J.SCITOTENV.2016.09.046

Alston, J. M., Beddow, J. M., & Pardey, P. G. (2009). Agricultural research, productivity, and food prices in the long run. *Science*, *325*(5945), 1209–1210. https://doi.org/10.1126/SCIENCE.1170451

Artmann, M., Kohler, M., Meinel, G., Gan, J., & Ioja, I. C. (2019). How smart growth and green infrastructure can mutually support each other—A conceptual framework for compact and green cities. *Ecological Indicators*, *96*, 10–22. https://doi.org/10.1016/J.ECOLIND.2017.07.001

Asian Water Development Outlook 2020. (2020). https://doi.org/10.22617/SGP200412-2

Badea, A. C., Rocco S., C. M., Tarantola, S., & Bolado, R. (2011). Composite indicators for security of energy supply using ordered weighted averaging. *Reliability Engineering and System Safety*, *96*(6), 651–662. https://doi.org/10.1016/J.RESS.2010.12.025

Basiago, A. D. (1995). Methods of defining 'sustainability.' *Sustainable Development*, *3*(3), 109–119. https://doi.org/10.1002/SD.3460030302

Biswas, A. K., & Tortajada, C. (2010). Future water governance: Problems and perspectives. *International Journal of Water Resources Development*, *26*(2), 129–139. https://doi.org/10.1080/07900627.2010.488853

Chuang, M. C., & Ma, H. W. (2013). An assessment of Taiwan's energy policy using multi-dimensional energy security indicators. *Renewable and Sustainable Energy Reviews*, *17*, 301–311. https://doi.org/10.1016/j.rser.2012.09.034

Cleveland, D. A., Carruth, A., & Mazaroli, D. N. (2015). Operationalizing local food: Goals, actions, and indicators for alternative food systems. *Agriculture and Human Values*, *32*(2), 281–297. https://doi.org/10.1007/S10460-014-9556-9

Covarrubias, M. (2019). The nexus between water, energy and food in cities: Towards conceptualizing socio-material interconnections. *Sustainability Science*, *14*(2), 277–287. https://doi.org/10.1007/S11625-018-0591-0

Covarrubias, M., & Boas, I. (2020). The making of a sustainable food city in Barcelona: Insights from the water, energy, and food urban nexus. *Journal of Integrative Environmental Sciences*, *17*(2), 1–19. https://doi.org/10.1080/1943815X.2019.1675715

Crush, J. S., & Frayne, G. B. (2011). Urban food insecurity and the new international food security agenda. *Development Southern Africa*, *28*(4), 527–544. https://doi.org/10.1080/0376835X.2011.605571

Dargin, J., Daher, B., & Mohtar, R. H. (2019). Complexity versus simplicity in water energy food nexus (WEF) assessment tools. *Science of The Total Environment*, *650*, 1566–1575. https://doi.org/10.1016/J.SCITOTENV.2018.09.080

ESA SOFA-Team. (2015). The State of Food and Agriculture 2015 in brief. In *Crítica Literaria* (Vol. 9). FAO . https://openknowledge.fao.org/handle/20.500.14283/i4953e

FAO, IFAD and WFP. (2013). *The State of Food Insecurity in the World 2013. The Multiple Dimensions of Food Security*. Rome, FAO.

Farabi-Asl, H., Taghizadeh-Hesary, F., Chapman, A., Bina, S. M., & Itaoka, K. (2019). *Energy Challenges for Clean Cooking in Asia, the Background, and Possible Policy Solutions*. https://www.econstor.eu/handle/10419/222774

Feng, C., Qu, S., Jin, Y., Tang, X., Liang, S., Chiu, A. C. F., & Xu, M. (2019). Uncovering urban food-energy-water nexus based on physical input-output analysis: The case of the Detroit Metropolitan Area. *Applied Energy*, *252*, 113422. https://doi.org/10.1016/J.APENERGY.2019.113422

Foley, J. A., DeFries, R., Asner, G. P., Barford, C., Bonan, G., Carpenter, S. R., Chapin, F. S., Coe, M. T., Daily, G. C., Gibbs, H. K., Helkowski, J. H., Holloway, T., Howard, E. A., Kucharik, C. J., Monfreda, C., Patz, J. A., Prentice, I. C., Ramankutty, N., & Snyder, P. K. (2005). Global consequences of land use. *Science*, *309*(5734), 570–574. https://doi.org/10.1126/SCIENCE.1111772

Gleick, P. H. (2003). Global freshwater resources: Soft-path solutions for the 21st century. *Science*, *302*(5650), 1524–1528. https://doi.org/10.1126/SCIENCE.1089967

Gnansounou, E. (2008). Assessing the energy vulnerability: Case of industrialised countries. *Energy Policy*, *36*(10), 3734–3744. https://doi.org/10.1016/J.ENPOL.2008.07.004

Guo, B., Cheng, T., Sun, T., Tong, Y., & Nian, S. (2024). Identifying the complexity of the food-energy-water nexus based on emergy analysis in crop production systems of China. *Land*, *13*(11), 1756. https://doi.org/10.3390/land13111756

Hall, P., & Pfeiffer, U. (2013). *Urban Future 21 : A Global Agenda for Twenty-First Century Cities*. https://doi.org/10.4324/9781315011523

Hannan, M. A., Begum, R. A., Abdolrasol, M. G., Hossain Lipu, M. S., Mohamed, A., & Rashid, M. M. (2018). Review of baseline studies on energy policies and indicators in Malaysia for future sustainable energy development. *Renewable and Sustainable Energy Reviews*, *94*, 551–564. https://doi.org/10.1016/J.RSER.2018.06.041

Henderson, J. V. (2010). Cities and development. *Journal of Regional Science*, *50*(1), 515–540. https://doi.org/10.1111/J.1467-9787.2009.00636.X

Kaygusuz, K. (2012). Energy for sustainable development: A case of developing countries. *Renewable and Sustainable Energy Reviews*, *16*(2), 1116–1126. https://doi.org/10.1016/J.RSER.2011.11.013

Leck, H., Conway, D., Bradshaw, M., & Rees, J. (2015). Tracing the water–energy–food nexus: Description, theory and practice. *Geography Compass*, *9*(8), 445–460. https://doi.org/10.1111/GEC3.12222

Liu, G. (2014). Development of a general sustainability indicator for renewable energy systems: A review. *Renewable and Sustainable Energy Reviews*, *31*, 611–621. https://doi.org/10.1016/J.RSER.2013.12.038

Lund, H. (2007). Renewable energy strategies for sustainable development. *Energy*, *32*(6), 912–919. https://doi.org/10.1016/J.ENERGY.2006.10.017

Machado, P. G., Picoli, M. C. A., Torres, L. J., Oliveira, J. G., & Walter, A. (2015). The use of socioeconomic indicators to assess the impacts of sugarcane production in Brazil. *Renewable and Sustainable Energy Reviews*, *52*, 1519–1526. https://doi.org/10.1016/J.RSER.2015.07.127

Martchamadol, J., & Kumar, S. (2012). Thailand's energy security indicators. *Renewable and Sustainable Energy Reviews*, *16*(8), 6103–6122. https://doi.org/10.1016/j.rser.2012.06.021

Mendes, C., De Souza, L. S., Kalid, R., Esquerre, K., & Kiperstok, A. (2011). Assessment of the uncertainty associated with the energy indicator. *Renewable and Sustainable Energy Reviews*, *15*(6), 3156–3164. https://doi.org/10.1016/J.RSER.2011.04.031

Mercure, J. F., Paim, M. A., Bocquillon, P., Lindner, S., Salas, P., Martinelli, P., Berchin, I. I., de Andrade Guerra, J. B. S. O., Derani, C., de Albuquerque Junior, C. L., Ribeiro, J. M. P.,

Knobloch, F., Pollitt, H., Edwards, N. R., Holden, P. B., Foley, A., Schaphoff, S., Faraco, R. A., & Vinuales, J. E. (2019). System complexity and policy integration challenges: The Brazilian energy-water-food nexus. *Renewable and Sustainable Energy Reviews*, *105*, 230–243. https://doi.org/10.1016/J.RSER.2019.01.045

Momblanch, A., Papadimitriou, L., Jain, S. K., Kulkarni, A., Ojha, C. S. P., Adeloye, A. J., & Holman, I. P. (2019). Untangling the water-food-energy-environment nexus for global change adaptation in a complex Himalayan water resource system. *Science of The Total Environment*, *655*, 35–47. https://doi.org/10.1016/J.SCITOTENV.2018.11.045

Ogbolumani, O. A., & Nwulu, N. (2020). Integrated appliance scheduling and optimal sizing of an autonomous hybrid renewable energy system for agricultural food production. In *Lecture Notes in Mechanical Engineering* (pp. 651–660). https://doi.org/10.1007/978-981-15-5753-8_60

Ogbolumani, O. A., & Nwulu, N. I. (2021). Multi-objective optimisation of constrained food-energy-water-nexus systems for sustainable resource allocation. *Sustainable Energy Technologies and Assessments*, *44*, 100967. https://doi.org/10.1016/j.seta.2020.100967

Ogbolumani, O. A., & Nwulu, N. I. (2022). A food-energy-water nexus meta-model for food and energy security. *Sustainable Production and Consumption*, *30*, 438–453. https://doi.org/10.1016/J.SPC.2021.12.019

Ogbolumani, O. A., & Nwulu, N. I. (2024). Environmental impact assessment for a meta-model-based food-energy-water-nexus system. *Energy Reports*, *11*, 218–232. https://doi.org/10.1016/j.egyr.2023.11.033

Özcan, Z., Köksal, M. A., & Alp, E. (2025). An integrated modeling approach to assess water-energy nexus in a semi-arid watershed. *Environmental Modelling & Software*, *185*, 106326. https://doi.org/10.1016/J.ENVSOFT.2025.106326

Özilgen, M. (2018). Nutrition and production related energies and exergies of foods. *Renewable and Sustainable Energy Reviews*, *96*, 275–295. https://doi.org/10.1016/J.RSER.2018.07.055

Parnell, S. (2016). Defining a global urban development agenda. *World Development*, *78*, 529–540. https://doi.org/10.1016/J.WORLDDEV.2015.10.028

Rosegrant, M. W., & Cline, S. A. (2003). Global food security: Challenges and policies. *Science*, *302*(5652), 1917–1919. https://doi.org/10.1126/SCIENCE.1092958

Saladini, F., Betti, G., Ferragina, E., Bouraoui, F., Cupertino, S., Canitano, G., Gigliotti, M., Autino, A., Pulselli, F. M., Riccaboni, A., Bidoglio, G., & Bastianoni, S. (2018). Linking the water-energy-food nexus and sustainable development indicators for the Mediterranean region. *Ecological Indicators*, *91*, 689–697. https://doi.org/10.1016/J.ECOLIND.2018.04.035

Serwaa Mensah, G., Kemausuor, F., & Brew-Hammond, A. (2014). Energy access indicators and trends in Ghana. *Renewable and Sustainable Energy Reviews*, *30*, 317–323. https://doi.org/10.1016/J.RSER.2013.10.032

Sivakumar, B. (2021). Water-energy-food nexus: Challenges and opportunities. *Stochastic Environmental Research and Risk Assessment*, *35*(1), 1–2. https://doi.org/10.1007/S00477-020-01927-5

Sullivan, C. A., Meigh, J. R., Giacomello, A. M., Fediw, T., Lawrence, P., Samad, M., Mlote, S., Hutton, C., Allan, J. A., Schulze, R. E., Dlamini, D. J. M., Cosgrove, W., Delli Priscoli, J., Gleick, P., Smout, I., Cobbing, J., Calow, R., Hunt, C., Hussain, A., … Steyl, I. (2003). The water poverty index: Development and application at the community scale. *Natural Resources Forum*, *27*(3), 189–199. https://doi.org/10.1111/1477-8947.00054

Tsai, W. T. (2010). Energy sustainability from analysis of sustainable development indicators: A case study in Taiwan. *Renewable and Sustainable Energy Reviews*, *14*(7), 2131–2138. https://doi.org/10.1016/J.RSER.2010.03.027

Turner, J. A. (1999). A realizable renewable energy future. *Science*, *285*(5428), 687–689. https://doi.org/10.1126/SCIENCE.285.5428.687

Verma, P., & Raghubanshi, A. S. (2018). Urban sustainability indicators: Challenges and opportunities. *Ecological Indicators*, *93*, 282–291. https://doi.org/10.1016/J.ECOLIND.2018.05.007

Vilanova, M. R. N., Magalhães Filho, P., & Balestieri, J. A. P. (2015). Performance measurement and indicators for water supply management: Review and international cases. *Renewable and Sustainable Energy Reviews*, *43*, 1–12. https://doi.org/10.1016/J.RSER.2014.11.043

WHO, & Unicef. (2000). Global water supply and sanitation assessment 2000 report. *Water Supply*, *87*. https://scholar.google.com/scholar?hl=en&btnG=Search&q=intitle:Global+Water+Supply+and+Sanitation+Assessment+2000+Report0

World Energy Outlook 2011 – Analysis – IEA. (n.d.). Retrieved August 25, 2024, from https://www.iea.org/reports/world-energy-outlook-2011

Wu, L., Elshorbagy, A., Pande, S., & Zhuo, L. (2021). Trade-offs and synergies in the water-energy-food nexus: The case of Saskatchewan, Canada. *Resources, Conservation and Recycling*, *164*, 105192. https://doi.org/10.1016/J.RESCONREC.2020.105192

Yuan, D., Yau, Y., & li, R. (2017). *Urban Village Renewal in China: From State-Led to Self-Organized Land Readjustment* (pp. 209–231). https://www.researchgate.net/profile/Dinghuan-Yuan/publication/324976634_Urban_Village_Renewal_in_China_From_State-Led_to_Self-Organized_Land_Readjustment/links/5aeec9250f7e9b01d3e1c897/Urban-Village-Renewal-in-China-From-State-Led-to-Self-Organized-Land-Readjustment.pdf

Yuan, M. H., Chiueh, P. Te, & Lo, S. L. (2019). Understanding synergies and trade-offs between water and energy production at landfill sites. *Science of The Total Environment*, *687*, 152–160. https://doi.org/10.1016/J.SCITOTENV.2019.06.075

Yuan, M. H., Chiueh, P. Te, & Lo, S. L. (2021). Measuring urban food-energy-water nexus sustainability: Finding solutions for cities. *Science of The Total Environment*, *752*, 141954. https://doi.org/10.1016/J.SCITOTENV.2020.141954

Yuan, M. H., & Lo, S. L. (2020). Developing indicators for the monitoring of the sustainability of food, energy, and water. *Renewable and Sustainable Energy Reviews*, *119*, 109565. https://doi.org/10.1016/J.RSER.2019.109565

Zhang, P., Zhang, L., Chang, Y., Xu, M., Hao, Y., Liang, S., Liu, G., Yang, Z., & Wang, C. (2019). Food-energy-water (FEW) nexus for urban sustainability: A comprehensive review. *Resources, Conservation and Recycling*, *142*, 215–224. https://doi.org/10.1016/J.RESCONREC.2018.11.018

9 Policy and Governance Frameworks For FEW Nexus 4.0

9.1 FOOD, ENERGY, AND WATER NEXUS 4.0 GOVERNANCE FRAMEWORKS

The food, energy, and water (FEW) nexus concept illustrates reconnecting the interrelated essential components of life, such as food, energy, and water, with some dependencies among these components for advanced survivors. Figure 9.1 illustrates the FEW nexus system that shows the interdependencies of the elements, showing the inflow within and between them. However, it should be emphasised that for our survival in this world and for anything constructive to be achieved, we should find innovations to establish appropriate guidelines that contribute to the synergies in extracting, distributing, marketing, and applying these components (Ogbolumani & Nwulu, 2021). The nexus approach does not perceive these relations as separate entities. Still, as components of the whole, recent years have seen the use of this technique to investigate the relationship between food, energy, and water (Hülsmann et al., 2019; Purwanto et al., 2021a). It is also important to point out at this position that the nexus approach aims at resolving the issues of synergy between food,

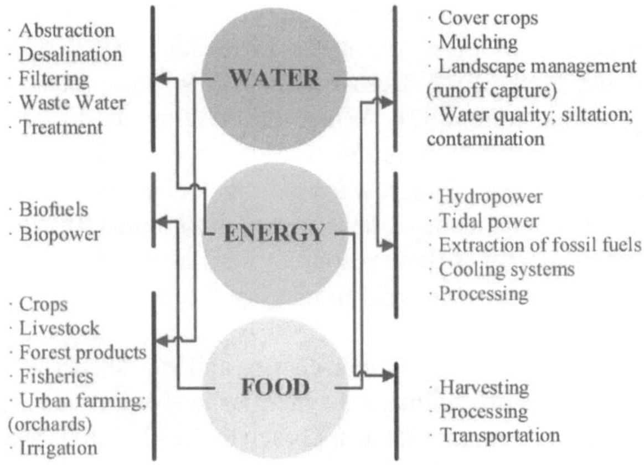

FIGURE 9.1 The Food, Energy, and Water Nexus System with Its Interdependencies (Bian & Liu, 2021).

DOI: 10.1201/9781003633853-9

energy, and water resources development so that problems are non-existent and that the expansion or sustainability of specific resources accumulated from such sectors cross-feed into other sectors (Bian & Liu, 2021). Policy analysts, especially those who will be dealing with natural resources, must be able to come up with a comprehensive understanding of the economic, environmental, political, and other related factors (Nhamo et al., 2020; Ogbolumani & Nwulu, 2022; Rosales-Asensio et al., 2020).

The 2011 Bonn Conference concerning the FEW nexus discussed the security of these resources. Discussions at this conference revolved around integrated management and the principle of governance of these vital resources and analysing the complex policy changes that may arise in managing the synergy that exists between them, which may also include the challenges that may affect them (Orimoloye et al., 2021; Saidmamatov et al., 2020).

The FEW nexus approach, in simple terms, stresses that the food, energy, and water systems must be jointly managed in such a way as to take into consideration their interlinkages and impacts on society at large if the attainment of sustainable development goals is to be achieved amidst the complexity of the management of resources in the global context (Albrecht et al., 2018). The FEW nexus approach enhances the security of these vital resources by better integrating their management across sectors in terms of greater effectiveness, reduced risk of trade-offs, and more synergy through better governance. Guidance was given at the 2011 World Economic Forum meeting on adopting this approach, emphasising the interrelation of food, energy, and water systems and some policy strategies to handle them more effectively (Holger Hoff, 2011). This concept gained prominence after the 2011 World Economic Forum meeting.

There are significant gaps in the existing FEW nexus frameworks that govern these resources, and subsequent studies have supported this observation. Closing these gaps will go a long way in managing these resources. It is also worth noting that the FEW nexus significantly contributes to formulating policies for managing natural resources and climate change (Albrecht et al., 2018; Endo et al., 2017). Governance of these resources has a way of placing attention on increasing demands that arise from changes occurring in society over time and how these changes impact the availability of these resources in the environment, especially water (Bakhshianlamouki et al., 2020; Busayo et al., 2019; Dyosi et al., 2021; Mpandeli et al., 2018; Orimoloye et al., 2021). When the FEW nexus is effectively managed, the effective use of water, which is the core of these resources, is well integrated into various sectors, thus building resilience in water basins. Bodies like integrated water resources management use a linear approach for managing these resources. In contrast, FEW nexus manages these resources nonlinearly, thus providing feedback to various sectors actively using and managing these resources, especially water (Al-Saidi, 2017; Hack, 2015).

The nexus approach can improve water, energy, and food security through increased efficiency, reduced trade-offs, enhanced synergies, and improved sector governance (Holger Hoff, 2011). Such an integrated approach to the FEW nexus is often perceived to achieve green economic growth by maximising the effectiveness of resource use (Tantoh et al., 2020). Notably, gaps still exist on how the FEW nexus will be implemented and its policy relevance, especially within decision-making processes. Seeds of the FEW nexus must be germinated into policy and decision-making at all levels by integrating outcomes from relevant studies if this framework is to be implemented effectively.

Research becomes core in quantifying critical interlinkages and interactions of the FEW nexus, thereby bridging knowledge gaps, identifying trade-offs, and examining potential synergies after scientific research (Tantoh et al., 2020; Wichelns, 2017). This necessitates an evidence-based approach, which is very important in developing countries, where sustainable management of these resources is a requirement apart from being a means of damage minimisation at the sectoral level to achieve broader developmental goals.

Consequently, the FEW nexus is a promising way to control and implement water, energy, and food resources policy. Given the difficulties inherent in realising sustainable development, scientific research must be employed in decision-making support. The FEW nexus governance model brings together extreme weather conditions, ill-planned urban expansion, large-scale terrorist attacks, geopolitical instability, implications of climate change, and infrastructural problems. These risks directly and seriously affect the process of water, energy, and food management by placing threats at local and global levels, which further identifies interferences among hazards, resource management, and global uncertainties. With efficient policies established around the FEW nexus, threats like this can be adequately handled with minimal shocks to the afflicted clime.

9.1.1 Governance Structure of FEW Nexus 4.0

FEW nexus 4.0 governance embraces the frameworks, policies, and institutions that oversee and regulate the interaction within and between the food, energy, and water systems. The principal components of this, according to (Srigiri & Dombrowsky, 2021), include:

Integrated Approaches: A good number of governments and organisations are adopting integrated approaches within their governance structures, considering how interlinked FEW systems are. This involves the development of policies and strategies that take into notice trade-offs or synergies among the food, energy, and water sectors.

Policy Coherence: Any path to sustainable FEW nexus 4.0 requires that policies be coherent across different sectors and scales. Several studies investigate the cases of food security, access to energy, or availability of water as goals that must be aligned with larger sustainability goals based on their contribution to overall well-being.

Multilevel Governance: When it comes to the FEW nexus, just a few of the challenges associated with the nexus can be effectively tackled without good governance at all levels, ranging from the local level to the global level. Yet, some sustainable practices are most suitably implemented at local government levels. At the same time, to effectively handle trans-regional issues related to shared water, food, and/or energy trade, international cooperation is greatly needed.

9.1.2 Economic and Environmental Impacts of FEW Nexus 4.0

Understanding FEW nexus 4.0 means understanding the impacts and changes in all three sectors: food, energy, or water, together with broader links to the environment and society that drive economic outcomes. Some of the key impacts, according to Zarei et al. (2021), include:

Environmental Impacts: Through the nexus approach, one gets an insight into environmental impacts such as water use efficiency in agriculture, greenhouse gas emissions from energy production, and the ecological effects of land-use changes for food production.

Social Impacts: The socioeconomic dimensions include impacts on livelihoods, food security, and access to energy and water services. In many cases, the poor bear the worst brunt of resource scarcity and environmental degradation.

Economic Impacts: Economic assessments include estimating the costs and benefits that accrue because of the interventions of the nexus; for example, renewable energy investments, gains in water-use efficiency, and sustainable agriculture practices.

9.1.3 Future Consideration of FEW Nexus 4.0

Looking ahead, authors like Purwanto et al. (2021b) and Kalvani and Celico (2023) have discussed some directions that could be followed to improve understanding and management of FEW Nexus 4.0. Their points are organised as follows:

Improving Metrics and Indicators: One needs to build resilient measures for metrics and indicators that would accommodate the complexity of interactions within this nexus. This involves indicators such as resource-use efficiency, resilience to climate change, and social equity about access to FEW services.

Data and Modelling Enhancement: There is an ultimate requirement to enhance modelling capabilities that can effectively support data collection and project future trends and impacts within the nexus. Integrated assessment models, which are applied in simulating scenarios to inform policy, are increasingly being used.

Innovation and Technology: Improving sustainability in the FEW nexus can, therefore, be driven by technological innovations in smart agriculture, renewable energy technologies, and water-efficient practices.

Capacity Building and Knowledge Sharing: Capacity-building among stakeholders at the policy, research, and community levels is essential. This also includes sharing best practices, lessons learned, and collaboration across sectors and regions.

The arrangement of the food–energy–water security governance model is illustrated in Figure 9.2, which describes the interactions between these essential resources

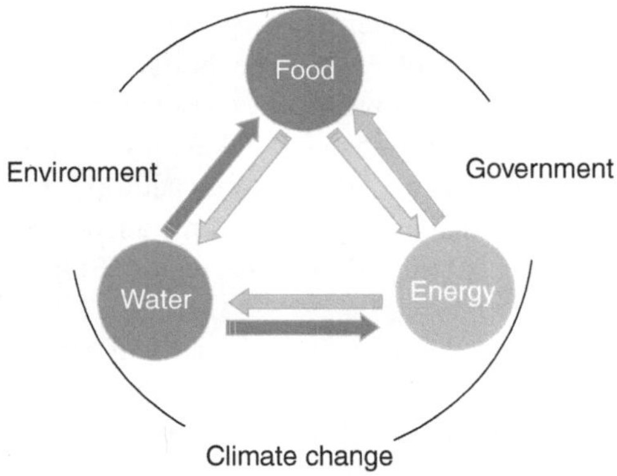

FIGURE 9.2 Governance Structure of Food, Energy, and Water Nexus (Bian & Liu, 2021).

and the externality affecting them. These challenges are well captured within the framework of the governance model in this context, as it seeks to achieve more integrated solutions that can be imaginative to manage water, energy, and food systems simultaneously. FEW nexus framework aims to improve the outcome of resource management processes by acknowledging and solving such relations (de Amorim et al., 2018).

The challenges facing the nexus must be identified and addressed to achieve sustainable development goals via the FEW nexus. These pressing challenges include managing regulatory frameworks addressing specific gaps in the nexus's governance structure and creating an enabling environment that fosters collaboration and cooperation amongst various stakeholders at different levels. It is also worth noting that moving from traditional methods of managing these resources to more innovative approaches will undoubtedly add complexity to managing these essential resources (Urbinatti et al., 2020). Implementing these new and creative approaches will open up more growth and economic opportunities within the FEW nexus. Some of these opportunities, which will be opened through innovative approaches to managing the FEW nexus, include new job opportunities, especially in areas where water is lacking or in excess. However, these opportunities are limited and may only benefit a few workers with certain skill sets in the long run, while unemployment will persist in other sectors where resources in the FEW nexus are not in maximum use (Markantonis et al., 2019; Nhamo et al., 2020).

The unavailability of global data threatens the implementation of an effective governance structure for the FEW nexus. The European Economic Area has tried to improve this by initiating the Shared Environmental Information System, which attempts to build a harmonised database of water-related activities, publications, events, and the like involved in the process. In most instances, however, the developing nations contribute little data, making the studies on the FEW nexus very challenging at both the sectoral and intersectoral levels (Boluwade, 2021). There are also cases where countries would not like to disclose some critical data considered strategically sensitive.

The scarcity of arable land and water resources also contributes to the challenges that restrict the implementation of FEW. The building of greenhouses and the provision of water tankers for the movement of water for greenhouse farming are two techniques that have been proposed to solve these problems (Adedeji et al., 2020; Adetoro et al., 2020; Orimoloye et al., 2021). Strong natural forces, such as cyclones' constant rainfall and other meteorological circumstances, could, however, seriously hinder the long-term feasibility of these strategies (Boluwade, 2021; Martinez et al., 2018).

The lack of holistic databases capturing the impacts of extreme climate events as experienced on the FEW nexus is a significant challenge. In this line of argumentation, it can be pointed out that very minimal quantitative data or metrics are available to assess the effect of this nature in a systemic way. The available research is scant on the specific impacts of climate change on the FEW, coupled with very few attempts to give quantitative indicators from this perspective (Boluwade, 2021; Martinez et al., 2018). While setting up greenhouses and transporting water are done, how far these alternates could be resilient to the impacts of extreme climate-induced events remains unknown, with very little data or research conducted on this aspect. These knowledge gaps must be properly addressed to have strategies that would sustainably manage the resources of water, energy, and food under the impacts of climate change.

Scholarly works on FEW nexus research reveal governance gaps that may disconnect the nexus concept from the decision-making and policy processes it intends to influence (Weitz et al., 2017). One visible leaning so far across the literature assessed is the fragmentation of policies between the water, food, and energy sectors. This is problematic for several intertwined reasons represented by the different perspectives. One of the first views from these four perspectives is that such failure will further aggravate resource scarcity and foster more conflict. Policies operating in isolation may eventually compete for the same limited resources or fail to identify synergies to improve overall efficiency (Jha, 2021).

9.1.4 Governance Challenges within the Food-Energy-Water Nexus

Governance at the nexus means managing plurality and cross-sectoral interactions of policies. While there is recognition of the strong interlinkages between the water, energy, and food sectors, and undoubtedly much attention has been paid recently to the nexus by development agencies, policy analysts, and stakeholders, a silo mentality persists (Bhaduri et al., 2015; Terrapon-Pfaff et al., 2018). FEW challenges remain compartmentalised within sectoral boundaries, often by isolated projects and policy decisions. This kind of policymaking in watertight compartments cannot ensure water, food, and energy security and sustainability (Jha, 2021). For instance, most agricultural policies and fertiliser subsidy schemes are tailored independently of their interlinkages with the energy and water sectors. Similarly, the energy subsidy offered to farmers might not consider the repercussions on groundwater depletion.

However, effective governance over the nexus means that the values of the different stakeholder groups and interest groups must be credibly reflected openly and transparently. Integration with the private sector, bureaucratic structures, and informal networks will also be required. Bhattacharyya et al. (2015) discussed the results of the water-energy-food nexus in all 15 nations of the Economic Community of West African States (ECOWAS) (Bhattacharyya et al., 2015). These countries represent a population of over 300 million, which could double by 2030. Out of these populations, about 58% are rural, while 42% are below the poverty line. While raising an interest in the FEW nexus concept among the ECOWAS member countries, success in implementation varies greatly. Most countries are still stuck on the sectoral approach to policymaking, and integrated approaches are rare. This governance continues to be centralised and top-down, with little or no decentralised, bottom-up mobilisation of initiatives. Even in cases where decentralisation has been encouraged, corresponding authorities are not fully delegated within ECOWAS (Jha, 2021).

The governance of the nexus requires solid political will to enable its effective execution on the ground. India is the world's highest extractor and user of groundwater, based on 19.8 million agricultural wells and tube wells (Addressing Groundwater Depletion: Lessons from India, the World's Largest User of Groundwater | Independent Evaluation Group, n.d.; India Seeks to Arrest Its Alarming Decline in Groundwater, n.d.; Bhaduri et al., 2015; Shah, 2009). While groundwater has always contributed much towards irrigation in India, over-exploitation commenced through intensive use during the Green Revolution, which began in the 1970s. Through this over-exploitation, many states in northern and western parts of India are experiencing acute depletion (Mukherji & Das, 2014).

Policies were brought into force to marshal these challenges and limit the over-extraction of groundwater. For instance, West Bengal adopted a modality of policy whereby "charging for groundwater was based on metered consumption rather than a flat rate" (Shah, 2009). This has been effective in checking the falling groundwater levels. Besides, in the northern and western states of the country, different electricity feeders for agriculture are installed, which are different from domestic feeders and those for industrial purposes. This separation helps regulate and control the over-extraction of groundwater (Shah, 2009).

National-level policies can very strongly affect the local communities. For instance, dam construction may drastically alter local stakeholders' access to water. Hydropower development in the transboundary river basins generates connected cross-sectional challenges and policy issues across different levels (Bhaduri et al., 2015). The limited influence of a water-focused organisation on hydropower development and debated issues related to authority in international development and hydropower policies were considered in the case of the Xayaburi Dam in the Lower Mekong Basin. The Xayaburi Dam project illustrates FEW nexus issues at the transboundary level, with seasonal variations of water release and conflicting demands for irrigation and energy production causing huge conflicts that negatively impacted local communities (Hensengerth, 2015). It becomes very important to guard the rights of local stakeholders concerning water use (Bhaduri et al., 2015).

An assessment methodology of implementing the WEF nexus at the level of a river basin was developed, and the tool was applied to a case study related to the Duero River basin in Spain (Mayor et al., 2015). This methodology helped reveal trade-offs and synergies, evaluate the political coordination needed, and thus provided options for solutions or ways to improve the management of water, energy, and food resources in the region. Some of the critical issues brought out through the case study included the rise in energy costs to irrigated agriculture by farming becoming more mechanised, water treatment capacity remaining limited, and new demands for water coming up for hydraulic fracturing of oil and gas. Also, it was highlighted that the use of fresh water for oil and gas exploration is growing equally, specifically in areas already under some water stress. Resources must be co-optimised so that, with reduced freshwater withdrawals, sufficient energy demand can be met, and exploitation and depletion of the remainder of the freshwater resources are avoided (Mayor et al., 2015).

According to Mayor et al. (2015),

> the WEF nexus can be considered as a framework that helps connect, coordinate and reinforce individual water, energy and food management policies by paying special attention to the identification, understanding, and characterisation of interconnections and tradeoffs, to build aligned and coherent strategies/actions that permit synergies. This should be applicable both within the political sphere, at either national, regional or local scales, and at a business or even household scale, where resource management and optimisation is also required.
>
> *(Mayor et al., 2015)*

Significant implementation gaps and challenges exist for multi-sectoral WEF nexus approaches that are not easy to implement and then translate into policy. There is a need for considerable efforts to integrate WEF nexus perspectives into policy frameworks so that its application can be scaled up across different communities, regions, and environments. Moreover, the added complexities of climate change are aggravating the nexus-related governance problems. Hence, linkages between, and vulnerabilities of, water, energy, and food resources are very important in gaining a better understanding. It entails interdisciplinary research since assessing climate impacts on these resources and their linked ecological processes is challenging (Liu et al., 2021).

In the absence of sufficient studies, it is still somewhat of a challenge to fully review WEF nexus interplays and quantify the benefits and costs associated with strategies that enhance synergies among sectors (Liu et al., 2021).

9.1.5 Tools for FEW Implementations

Several dedicated assessment tools have been developed to guide the implementation of the WEF nexus at different scales.

For example, multi-scale integrated analysis of societal and ecosystem metabolism (MuSIASEM framework) offers a comprehensive description of the dynamics of water, food, and energy systems, integrating factors like labour and hours of human activity into its assessments. Figure 9.3 shows a practical implementation of MuSIASEM Grammar.

Figure 9.4 shows the water-food nexus assessment tools in agriculture. The models analysed here are: Life Cycle Assessment, Common Agricultural Policy Regional Impact, Global Food and Water System, Soil and Water Assessment Tool, Water Evaluation and Planning system, and Soil Water Atmosphere Plant (Corona-López et al., 2021)

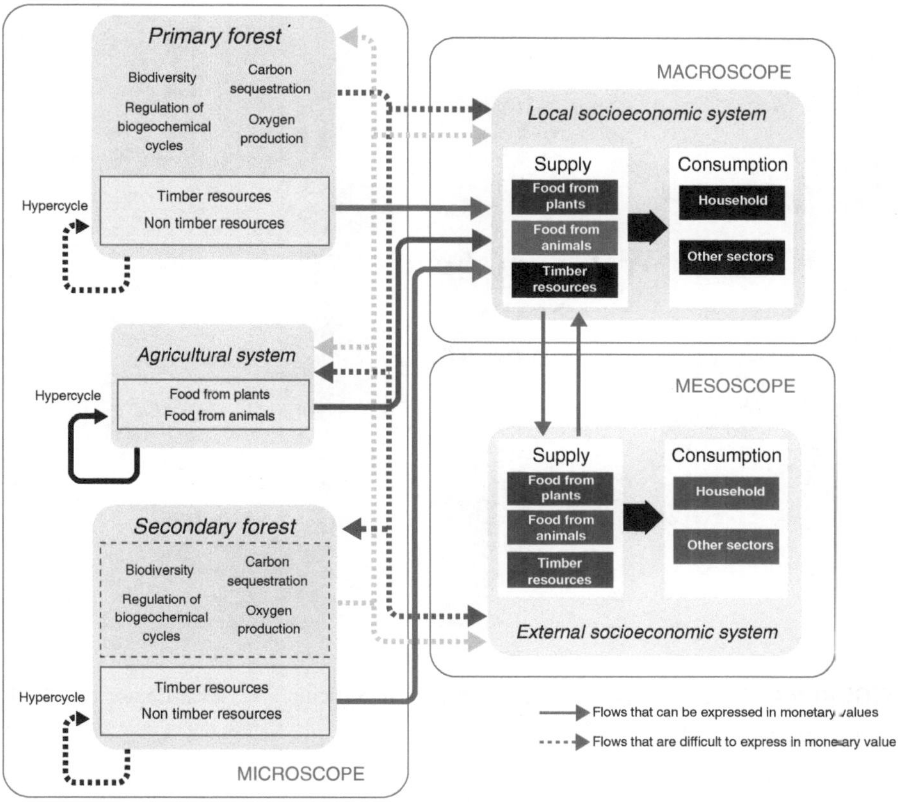

FIGURE 9.3 MuSIASEM Grammar in Agriculture (Cadillo-Benalcazar et al., 2022).

The Water, Land, Energy and Food (WLEF) nexus provides a network approach to studying the interlinkages between water, land, energy, and food. These resources are closely related to reacting to external social, economic, and environmental systems. The WLEF is characterised by a temporal-spatial distribution, which is a result of the different natural, economic, and social conditions that occur in different regions and at different times, such as climate and hydrology; a pictorial view of the WLEF model is shown in Figure 9.5.

Digital models also serve to help frame the WEF nexus. One model is the national outlook model from Australia's Commonwealth Scientific and Industrial Research Organization. This model is very data-intensive and thus only applicable in Australia due to the high costs involved. The WEF nexus Tool 2.0 is another tool; however, it requires manual input in the sense that user input is provided to visualise nexus goals somewhat more interactively; it was applied in Qatar for agricultural crop production, though with limited scope regarding livestock and processed food concerns (Daher & Mohtar, 2015).

Similarly, Foreseer is a tool applied in the United Kingdom to create scenarios according to the supply of natural resources and integrate different models to build localised services. Still, challenges abound in scaling these kinds of

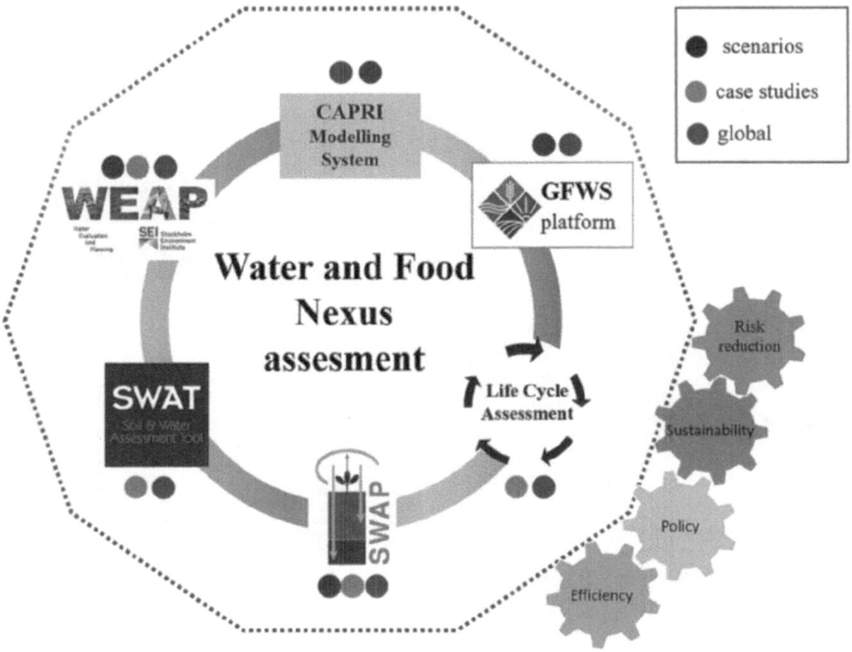

FIGURE 9.4 Water-Food Nexus Assessment Tools in Agriculture (Corona-López et al., 2021).

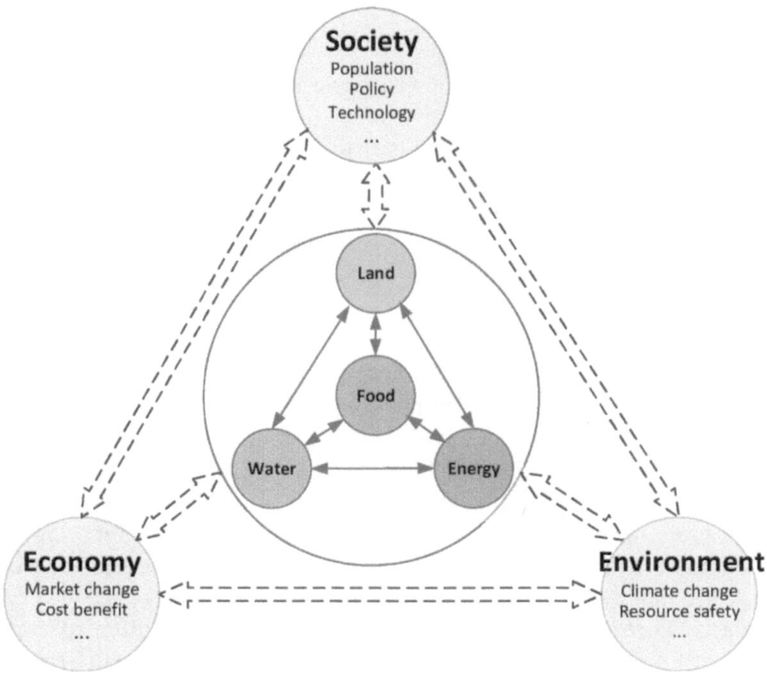

FIGURE 9.5 Water, Land, Energy and Food (WLEF) Nexus Framework (Chen et al., 2023).

tools down to more localised levels due to data limitations (Foreseer of Future Resources | University of Cambridge, n.d.).

Solutions to scale the WEF nexus emphasise the need to engage political systems and stakeholders in conjunction with technical variables. It first needs two main ingredients: Setting up inclusive institutions empowered by horizontal and vertical coordination mechanisms that allow them to generate nexus-driven policy. Finally, an appeal is made for more transparency and free data provision to enable better decisions based on evidence in a nexus framework (Jha, 2021).

The policy and governance frameworks of the WEF nexus 4.0 reflect an evolution in how societies decide to manage water, energy, and food resources in a much more connected way. This one is oriented to more integration, innovation, and adaptability concerning contemporary challenges such as climate change, increasing population, and urbanisation. WEF nexus 4.0 centres on the idea that conventional approaches focused on sectors cannot achieve sustainability and resilience in resource management, but rather that it calls for comprehensive policies that "consider the complex interplays and trade-offs among the water, energy, and food systems." In this respect, governance frameworks are specially developed to enhance coordination among sectors and levels of governance, from local to global, to promote coherence in decision-making and resource allocation.

Notable aspects of the governance and policy frameworks of the Water-Energy-Food nexus 4.0 are as follows:

Joint Planning and Decision-Making: The policies in FEW 4.0 look out for integrated planning. Food, energy, and water are very close sectors with clear synergies to maximise resource use, minimise conflict, and build greater cooperation.

Innovation and Technology: FEW nexus 4.0 cannot be without embracing technological innovation and creative ideas. Technologies applicable here are digital models, data analytics, and smart monitoring systems to increase resilience and efficiency.

Adaptive Governance: They can adapt to dynamic changes and uncertainties, notably climate change, by implementing adaptive governance frameworks. They call for adaptability and learning, which enable changes in policies, given fresh data and suggestions emanating from various stakeholders.

Inclusive Decision-Making: It has strongly emphasised the need for inclusivity in governance processes by giving a voice to concerns from many local stakeholders at the local level, such as local communities, businesses, and civil society organisations. They become more effective in practice due to the feeling of ownership toward the policy.

Resilience and Sustainability: FEW nexus 4.0 is dedicated to developing resilience and sustainability in managing natural resources. This includes the augmentation of the long-term resiliency of FEW systems, among other measures aimed at reducing the risk related to resource shortage, climatic variability, and environmental deterioration.

These vital resources will now be managed through integrated, adaptable, and inclusive methods thanks to the overall shift to FEW nexus 4.0. If societies solve the governance gaps and adopt technology breakthroughs, they can be better positioned to handle the complexity of interrelated resource systems and accomplish sustainable development goals.

9.2 SUMMARY

This chapter explains the governance structure that oversees food, energy, and water systems, and the principal frameworks of this structure are integrated approaches, policy coherence, and multilevel governance. It explains why understanding the environmental, social, and economic impacts and changes in the food, energy, and water sectors is crucial to understanding the FEW nexus 4.0. Considering that the governance structure for FEW nexus 4.0 is under development and constantly being refined, it comes with its challenges, and this is further explained in this chapter. Future directions that'd aid the understanding and management of FEW nexus 4.0 are also discussed in this chapter.

REFERENCES

Addressing Groundwater Depletion: Lessons from India, the World's Largest User of Groundwater | Independent Evaluation Group. (n.d.). Retrieved September 2, 2024, from https://ieg.worldbankgroup.org/blog/addressing-groundwater-depletion-lessons-india-worlds-largest-user-groundwater

Adedeji, O., Olusola, A., James, G., Shaba, H. A., Orimoloye, I. R., Singh, S. K., & Adelabu, S. (2020). Early warning systems development for agricultural drought assessment in Nigeria. *Environmental Monitoring and Assessment*, *192*(12). https://doi.org/10.1007/S10661-020-08730-3

Adetoro, A. A., Abraham, S., Paraskevopoulos, A. L., Owusu-Sekyere, E., Jordaan, H., & Orimoloye, I. R. (2020). Alleviating water shortages by decreasing water footprint in sugarcane production: The impacts of different soil mulching and irrigation systems in South Africa. *Groundwater for Sustainable Development*, *11*, 100464. https://doi.org/10.1016/J.GSD.2020.100464

Albrecht, T. R., Crootof, A., & Scott, C. A. (2018). The water-energy-food nexus: A systematic review of methods for nexus assessment. *Environmental Research Letters*, *13*(4), 043002. https://doi.org/10.1088/1748-9326/AAA9C6

Al-Saidi, M. (2017). Conflicts and security in integrated water resources management. *Environmental Science & Policy*, *73*, 38–44. https://doi.org/10.1016/J.ENVSCI.2017.03.015

Bakhshianlamouki, E., Masia, S., Karimi, P., van der Zaag, P., & Sušnik, J. (2020). A system dynamics model to quantify the impacts of restoration measures on the water-energy-food nexus in the Urmia lake Basin, Iran. *Science of the Total Environment*, *708*, 134874. https://doi.org/10.1016/J.SCITOTENV.2019.134874

Bhaduri, A., Ringler, C., Dombrowski, I., Mohtar, R., & Scheumann, W. (2015). Sustainability in the water–energy–food nexus. *Water International*, *40*(5–6), 723–732. https://doi.org/10.1080/02508060.2015.1096110

Bhattacharyya, S., Bugatti, M. N., & Bauer, M. H. (2015). A bottom-up approach to the nexus of energy, food and water security in the Economic Community of West African States (ECOWAS) region. *The Nexus Network*. Retrieved May 15, 2025, from https://thenexusnetwork.org/wp-content/uploads/2014/08/Bhattacharyya-thinkpiece_2015.pdf

Bian, Z., & Liu, D. (2021). A comprehensive review on types, methods and different regions related to water–energy–food nexus. *International Journal of Environmental Research and Public Health*, *18*(16), 8276. https://doi.org/10.3390/IJERPH18168276

Boluwade, A. (2021). Impacts of climatic change and database information design on the water-energy-food nexus in water-scarce regions. *Water-Energy Nexus*, *4*, 54–68. https://doi.org/10.1016/J.WEN.2021.03.002

Busayo, E. T., Kalumba, A. M., & Orimoloye, I. R. (2019). Spatial planning and climate change adaptation assessment: Perspectives from Mdantsane Township dwellers in South Africa. *Habitat International*, *90*, 101978. https://doi.org/10.1016/J.HABITATINT.2019.04.005

Cadillo-Benalcazar, J. J., Silva-Macher, J. C., & Salinas, N. (2022). Applying the multi-scale integrated analysis of societal and ecosystem metabolism (MuSIASEM) to characterize the society–agriculture–forest system: The case of Huayopata, Cuzco (Peru). *Environment, Development and Sustainability*, 1–24. https://doi.org/10.1007/S10668-022-02457-6

Chen, A., Hao, Z., Wang, R., Zhao, H., Hao, J., Xu, R., & Duan, H. (2023). Cultivated land sustainable use evaluation from the perspective of the water–land–energy–food nexus: A case study of the major grain-producing regions in Quzhou, China. *Agronomy*, *13*(9), 2362. https://doi.org/10.3390/AGRONOMY13092362

Corona-López, E., Román-Gutiérrez, A. D., Otazo-Sánchez, E. M., Guzmán-Ortiz, F. A., & Acevedo-Sandoval, O. A. (2021). Water–food nexus assessment in agriculture: A systematic review. *International Journal of Environmental Research and Public Health*, *18*(9), 4983. https://doi.org/10.3390/IJERPH18094983

Daher, B. T., & Mohtar, R. H. (2015). Water–energy–food (WEF) nexus Tool 2.0: Guiding integrative resource planning and decision-making. *Water International*, *40*(5–6), 748–771. https://doi.org/10.1080/02508060.2015.1074148

de Amorim, W. S., Valduga, I. B., Ribeiro, J. M. P., Guazzelli Williamson, V., Krauser, G. E., Magtoto, M. K., & de Andrade Guerra, J. B. S. O. (2018). The nexus between water, energy, and food in the context of the global risks: An analysis of the interactions between food, water, and energy security. *Environmental Impact Assessment Review*, *72*, 1–11. https://doi.org/10.1016/J.EIAR.2018.05.002

Dyosi, M., Kalumba, A. M., Magagula, H., Zhou, L., & Orimoloye, I. R. (2021). Drought conditions appraisal using geoinformatics and multi-influencing factors. *Environmental Monitoring and Assessment*, *193*(6). https://doi.org/10.1007/S10661-021-09126-7

Endo, A., Tsurita, I., Burnett, K., & Orencio, P. M. (2017). A review of the current state of research on the water, energy, and food nexus. *Journal of Hydrology: Regional Studies*, *11*, 20–30. https://doi.org/10.1016/J.EJRH.2015.11.010

Foreseer of Future Resources | University of Cambridge. (n.d.). Retrieved August 25, 2024, from https://www.cam.ac.uk/research/news/foreseer-of-future-resources

Hack, J. (2015). Application of payments for hydrological ecosystem services to solve problems of fit and interplay in integrated water resources management. *Water International*, *40*(5–6), 929–948. https://doi.org/10.1080/02508060.2015.1096122

Hensengerth, O. (2015). Where is the power? Transnational networks, authority and the dispute over the Xayaburi Dam on the Lower Mekong Mainstream. *Water International*, *40*(5–6), 911–928. https://doi.org/10.1080/02508060.2015.1088334

Hoff, H. (2011). *Understanding the Nexus*. Background Paper for the Bonn2011 Conference: The Water, Energy and Food Security Nexus. Stockholm Environment Institute, Stockholm. https://www.sei.org/publications/understanding-the-nexus/

Hülsmann, S., Sušnik, J., Rinke, K., Langan, S., van Wijk, D., Janssen, A. B., & Mooij, W. M. (2019). Integrated modelling and management of water resources: The ecosystem perspective on the nexus approach. *Current Opinion in Environmental Sustainability*, *40*, 14–20. https://doi.org/10.1016/J.COSUST.2019.07.003

India Seeks to Arrest Its Alarming Decline in Groundwater. (n.d.). Retrieved September 2, 2024, from https://blogs.worldbank.org/en/endpovertyinsouthasia/india-seeks-arrest-its-alarming-decline-groundwater

Jha, D. (2021). Implementation and governance issues in the water-energy-food nexus. *International Journal of High School Research*, *3*(3), 21–27. https://doi.org/10.36838/V3I3.21

Kalvani, S. R., & Celico, F. (2023). The water–energy–food nexus in European countries: A review and future perspectives. *Sustainability (Switzerland)*, *15*(6). https://doi.org/10.3390/SU15064960

Liu, Q., Yang, L., & Yang, M. (2021). Digitalisation for water sustainability: Barriers to implementing circular economy in smart water management. *Sustainability (Switzerland)*, *13*(21). https://doi.org/10.3390/SU132111868

Markantonis, V., Reynaud, A., Karabulut, A., El Hajj, R., Altinbilek, D., Awad, I. M., Bruggeman, A., Constantianos, V., Mysiak, J., Lamaddalena, N., Matoussi, M. S., Monteiro, H., Pistocchi, A., Pretato, U., Tahboub, N., Tunçok, I. K., Ünver, O., Van Ek, R., Willaarts, B., … Bidoglio, G. (2019). Can the implementation of the water-energy-food nexus support economic growth in the Mediterranean region? The current status and the way forward. *Frontiers in Environmental Science*, *7*(JUL), 409800. https://doi.org/10.3389/FENVS.2019.00084

Martinez, P., Blanco, M., & Castro-Campos, B. (2018). The water–energy–food nexus: A fuzzy-cognitive mapping approach to support nexus-compliant policies in Andalusia (Spain). *Water*, *10*(5), 664. https://doi.org/10.3390/W10050664

Mayor, B., López-Gunn, E., Villarroya, F. I., & Montero, E. (2015). Application of a water–energy–food nexus framework for the Duero river basin in Spain. *Water International*, *40*(5–6), 791–808. https://doi.org/10.1080/02508060.2015.1071512

Mpandeli, S., Naidoo, D., Mabhaudhi, T., Nhemachena, C., Nhamo, L., Liphadzi, S., Hlahla, S., & Modi, A. T. (2018). Climate change adaptation through the water-energy-food nexus in Southern Africa. *International Journal of Environmental Research and Public Health*, *15*(10), 2306. https://doi.org/10.3390/IJERPH15102306

Mukherji, A., & Das, A. (2014). The political economy of metering agricultural tube wells in West Bengal, India. *Water International*, *39*(5), 671–685. https://doi.org/10.1080/02508060.2014.955408

Nhamo, L., Mabhaudhi, T., Mpandeli, S., Dickens, C., Nhemachena, C., Senzanje, A., Naidoo, D., Liphadzi, S., & Modi, A. T. (2020). An integrative analytical model for the water-energy-food nexus: South Africa case study. *Environmental Science & Policy*, *109*, 15–24. https://doi.org/10.1016/J.ENVSCI.2020.04.010

Ogbolumani, O. A., & Nwulu, N. I. (2021). Multi-objective optimisation of constrained food-energy-water-nexus systems for sustainable resource allocation. *Sustainable Energy Technologies and Assessments*, *44*, 100967. https://doi.org/10.1016/j.seta.2020.100967

Ogbolumani, O. A., & Nwulu, N. I. (2022). A food-energy-water nexus meta-model for food and energy security. *Sustainable Production and Consumption*, *30*, 438–453. https://doi.org/10.1016/J.SPC.2021.12.019

Orimoloye, I. R., Belle, J. A., & Ololade, O. O. (2021). Exploring the emerging evolution trends of disaster risk reduction research: A global scenario. *International Journal of Environmental Science and Technology*, *18*(3), 673–690. https://doi.org/10.1007/S13762-020-02847-1

Purwanto, A., Sušnik, J., Suryadi, F. X., & de Fraiture, C. (2021a). Quantitative simulation of the water-energy-food (WEF) security nexus in a local planning context in indonesia. *Sustainable Production and Consumption*, *25*, 198–216. https://doi.org/10.1016/J.SPC.2020.08.009

Purwanto, A., Sušnik, J., Suryadi, F. X., & de Fraiture, C. (2021b). Water-energy-food nexus: Critical review, practical applications, and prospects for future research. *Sustainability (Switzerland)*, *13*(4), 1–18. https://doi.org/10.3390/SU13041919

Rosales-Asensio, E., de la Puente-Gil, Á., García-Moya, F. J., Blanes-Peiró, J., & de Simón-Martín, M. (2020). Decision-making tools for sustainable planning and conceptual framework for the energy–water–food nexus. *Energy Reports*, *6*, 4–15. https://doi.org/10.1016/J.EGYR.2020.08.020

Saidmamatov, O., Rudenko, I., Pfister, S., & Koziel, J. (2020). Water-Energy-Food nexus framework for promoting regional integration in Central Asia. *Water (Switzerland)*, *12*(7). https://doi.org/10.3390/W12071896

Shah, T. (2009). Climate change and groundwater: India's opportunities for mitigation and adaptation. *Environmental Research Letters*, *4*(3). https://doi.org/10.1088/1748-9326/4/3/035005

Srigiri, S. R., & Dombrowsky, I. (2021). *Governance of the water-energy-food nexus for an integrated implementation of the 2030 Agenda: Conceptual and methodological framework for analysis* (No. 2/2021). Discussion Paper. https://www.econstor.eu/handle/10419/228835

Tantoh, H. B., Leonard, L., & Simatele, M. D. (2020). Strengthening the scaffolds of community flexibility: Policy and institutional response to the rural water supply and sustainability challenge. *African Geographical Review*, *39*(3), 208–223. https://doi.org/10.1080/19376812.2019.1694045

Terrapon-Pfaff, J., Ortiz, W., Dienst, C., & Gröne, M. C. (2018). Energising the WEF nexus to enhance sustainable development at local level. *Journal of Environmental Management*, *223*, 409–416. https://doi.org/10.1016/J.JENVMAN.2018.06.037

Urbinatti, A. M., Dalla Fontana, M., Stirling, A., & Giatti, L. L. (2020). 'Opening up' the governance of water-energy-food nexus: Towards a science-policy-society interface based on hybridity and humility. *Science of The Total Environment*, *744*, 140945. https://doi.org/10.1016/J.SCITOTENV.2020.140945

Weitz, N., Strambo, C., Kemp-Benedict, E., & Nilsson, M. (2017). Closing the governance gaps in the water-energy-food nexus: Insights from integrative governance. *Global Environmental Change*, *45*, 165–173. https://doi.org/10.1016/J.GLOENVCHA.2017.06.006

Wichelns, D. (2017). The water-energy-food nexus: Is the increasing attention warranted, from either a research or policy perspective? *Environmental Science & Policy*, *69*, 113–123. https://doi.org/10.1016/J.ENVSCI.2016.12.018

Zarei, S., Bozorg-Haddad, O., Kheirinejad, S., & Loáiciga, H. A. (2021). Environmental sustainability: A review of the water-energy-food nexus. *Aqua Water Infrastructure, Ecosystems and Society*, *70*(2), 138–154. https://doi.org/10.2166/AQUA.2020.058

10 Cultural and Societal Factors in Technology Adoption in FEW Nexus 4.0

10.1 CULTURAL AND SOCIETAL INFLUENCES ON TECHNOLOGY ADOPTION

The adoption of FEW 4.0 technologies cannot be regarded simply as a matter of technical feasibility but is deeply connected to the cultural and societal settings in which such technologies will be deployed. Traditional practices and belief systems in food production, energy use, and water management can significantly influence the adoption of new technologies to a large and considerable extent. For example, communities committed to traditional farming methods may resist implementing precision agriculture practices, even if those practices are far more efficient. Similarly, cultural practices that are deeply ingrained regarding water sources and their importance may hinder the acceptance of advanced water purification or conservation technologies. Effective technology implementation must be understood in light of these pre-existing practices and beliefs (Raj et al., 2020).

Social norms and values also play a critical role. When technologies relate to being an individualistic society or a collectivist one, the latter could influence the priorities given to technologies. While individualistic societies may readily embrace technologies that benefit the individual, community needs and solutions may be favoured in the case of collectivist societies. A further determinant is the risk tolerance within the culture. Relatively risk-averse cultures may not adopt new technologies as quickly as others, and this is further complicated if risks to health, the environment, or social relations are perceived. Additionally, society's notion of gender and the dynamics of gender play a role in technology adoption. This is based on the belief that some technologies are traditionally defined as exclusively masculine or feminine, hence regulating access and determining the beneficiaries (Theis et al., 2018).

Most importantly, technology adoption is linked to livelihoods and economic structures. Inequalities may exist in the access to FEW 4.0 technologies, further exacerbating social divides. In such contexts, affordability and digital literacy are significant concerns. New technologies might disrupt the ways people lead their lives. Technology will continue to pose opportunities and challenges; for example,

DOI: 10.1201/9781003633853-10

the automation of agricultural activities can increase the volume of food production, but this might result in a reduction of agricultural jobs (Bag et al., 2021). Success will also emanate from market dynamics, i.e. consumer demand, investment, and regulatory frameworks guiding the implementation of FEW 4.0 technologies.

Lastly, governance and policy act as drivers for technology adoption. Public trust in government and other institutions creates conditions that favour easy acceptance of new technologies. Transparency and community participation in decision-making processes help develop a feeling of confidence, and responsibility regarding the implementing new technologies. Critically, community engagement in the design and implementation of FEW 4.0 solutions is paramount. This concerns the appropriateness, functionality, and benefits of technologies within the needs and aspirations of their respective communities so that their introduction may be achieved with success and sustainability (Lee et al., 2013; Ogbolumani & Nwulu, 2021).

10.1.1 Cultural Beliefs and Energy 4.0 Adoption

The energy sector is transforming and shifting into the era of Energy 4.0. In this most advanced phase comes a suite of leading-edge technologies: the Internet of Things (IoT), Artificial Intelligence (AI), big data analytics, and blockchain. All these converge to revolutionise how we generate, distribute, and consume energy. These converging technologies drive the building of more decentralised and sustainable power systems that have improved operational efficiency and are more resilient (Ogbolumani & Nwulu, 2024). Energy 4.0 involves smart grids and renewable energy resources, including solar and wind power, along with advanced digital energy management systems (Nwulu & Damisa, 2023). All this aims to make the energy landscape more dynamic and responsive. However, whether these promising technological solutions for solving global energy challenges are effectively adopted is not only a matter of economic rationales or technical feasibility. Great opportunities like what has been referred to as Energy 4.0 are closely linked with cultural beliefs, social norms, and the value system in society. What is most technically appropriate may find resistance when it challenges certain areas of cultural practice or societal expectations (Marcon et al., 2022). Understanding the cultural context is thus quite crucial to achieving successful technology integration.

For example, a community heavily attached to traditional energy sources might refuse to move away from these energies toward renewable technologies, even in cases where this technology promises some long-term economic and ecological gain. Furthermore, social norms relating to energy use, such as attitudes about the waste of energy or the desirability of energy independence, strongly influence the degree to which smart grid technologies and energy efficiency measures are adopted. The values that a community embraces, for instance, economic growth, environmental protection, or social equity, go a long way in determining what kind of energy technologies will be acceptable and desirable. Besides pointing out such cultural factors, it is important to know how cultural values could be overcome toward wider acceptance of Energy 4.0. This could include targeted communication strategies addressing particular cultural issues, community outreach that builds trust and promote understanding, and educational programmes that outline the advantages of new technologies in ways that are understandable in their cultural context (Clean Energy Technology Adoption, Cultural Differences and Basic Human Dispositions, 2021).

By recognising and working within the cultural dimensions of energy technology uptake, we clear the path toward a more sustainable and equitable energy future wherein benefits from Energy 4.0 are accrued by communities worldwide. That calls for moving beyond a pure emphasis on technological matters to a more holistic one that recognises the complex interplay between technology, culture, and society (Govindarajan & Ganesh, 2022).

10.1.1.1 Overview of Energy 4.0

Energy 4.0 is a sea change that has hit the energy sector. It is a reflection of broader changes wrought by the Fourth Industrial Revolution. It's all about integrating more digital technologies like IoT, AI, big data analytics, and blockchain into every part of the energy value chain (Ogbolumani et al., 2025). That translates into smart grids-upgraded electrical grids that can be monitored and controlled in real time, hence being more efficient, dependable, and resilient (Ajibola & Ogbolumani, 2024). Energy 4.0 also favours decentralisation by enabling distributed generation sources in the form of renewable energy. The concept of Energy 4.0 gives more power to consumers, thus enabling them to have better control over their energy use. Energy 4.0 also aims to offer a more sustainable energy future by increasing renewable energy integration, optimising energy consumption, and reducing environmental impacts. It should foster innovation and new business opportunities in the sector (Ahmad et al., 2022; Ajibola & Ogbolumani, 2024).

10.1.1.2 The Role of Cultural Beliefs in Energy 4.0 Adoption

The acceptance of Energy 4.0 technologies depends greatly on cultural beliefs, since they impact people's opinions about new technology. These ideas affect people's attitudes toward energy use, environmental behaviours, and their readiness to accept contemporary technical systems (D'Cruz et al., 2022). For example, some societies might have a strong inclination toward conventional energy sources and a certain opposition to more modern, sophisticated systems, including smart grids or renewable energy technology (Ogbolumani & Nwulu, 2020). In other civilisations, the focus on environmental sustainability could make them more receptive to inventions suggesting better, more efficient means of generating energy.

People's daily interactions with technology also reflect their cultural values. In societies where people strongly believe in creativity and technical development, there will be a high rate of acceptance, and the application of Energy 4.0 technologies could be faster. Societies with more conservative or traditional values may require more time to be persuaded to accept technology (Rasa & Laherto, 2022).

Ultimately, though, the diffusion of Energy 4.0 technologies relates not only to new tools but also to knowledge and respect for the cultural settings in which every person on Earth forms their ideas and interactions with energy. It is these characteristics and cultural features that energy businesses and policymakers who want integration must be in contact with so as to ensure that new technologies connect to the beliefs, behaviours, and needs of the communities they wish to serve (Goggins et al., 2022).

10.1.1.3 Environmental Ethics and Cultural Perception of Sustainability

There are circumstances where environmental ethics are closely related to a profound cultural bond with nature, and sustainability has indeed constituted the bedrock of diverse civilisations' behaviours, notably with regard to renewable energy. Where one passionately believes in living in harmony with the environment, people are likely to look upon the natural world not only as a resource to be exploited but also as something to be valued and preserved. It calls for responsibility to the environment by safeguarding its security for the next generation. Thus, it pays attention to energy methods that support such care. In communities of this character, renewable energy systems are more than merely technological answers; solar, wind, and other green technologies form part of the cultural ideals of ecological balance and long-term environmental health. The concept of the coexistence of society with nature then becomes a strong driving force toward adopting these sources of clean energy, reflecting a commitment toward less harm to the environment and building a better future. For certain cultures, however, sustainability is not about technology but a way of life deeply rooted in ethical principles. This connectedness with nature, in turn, makes more and more people accept renewable energy and practices of sustainability as extensions of their ethical will and cultural identity. These cultures can thus be forerunners in the transition toward cleaner and more sustainable forms of energy, boosting solutions favourable for the planet and future generations (Allum et al., 2008; Ogbolumani & Nwulu, 2024; Olorunfemi et al., 2022, 2023; Stephenson, 2018).

10.1.1.4 Indigenous Knowledge and Renewable Energy

The philosophical framework that Indigenous people maintain regarding environmental respect aligns directly with renewable power principles. Solar and wind energy remain natural, harmonious elements according to many Indigenous cultures, which believe they should be used respectfully. The cultural focus on natural resource preservation by indigenous communities in Latin American regions has encouraged them to become early adopters of solar power projects (Edenhofer et al., 2011).

10.1.1.5 Growth-Oriented Cultures

Culture-based priorities for industrial expansion alongside economic development prevent sustainability from becoming a critical factor. Societies exclude renewable energy adoption unless policy requirements create substantial financial reasons for change. Traditional fossil fuel preferences are preferred in developing nations because they represent industrial development and national economic expansion (Carrillo, 2015).

10.1.1.6 Religious Beliefs and Symbolism in Energy Perception

Therefore, religious beliefs can be a factor that influences people's perception and acceptance of energy technologies. Most religious tenets have been influential in determining the values people hold in relation to the environment and in utilising natural resources. For example, some religious traditions emphasise stewardship of the earth, thus advocating for responsible and sustainable use of resources. These can allow the fostering or use of energy technologies that consider the importance of environmental care, such as renewable energy, which is conceptualised as being in harmony with nature (Ogbolumani & Nwulu, 2020).

Certain energy religious symbolism within various cultural settings also carries added meaning in the way energy technologies have been conceptualised, such as cases of light linked to enlightenment or spiritual clarity; thus, the harnessing of the sun for energy may come to the fore as particularly compelling. Also, religious traditions prioritising overall sustainability and just sharing will especially applaud those energy technologies contributing to the advancement of justice and care on a larger scale.

On the other hand, religious views might also conflict with some technologies in the energy field and act as barriers to the use of particular technologies, especially those perceived as running against religious doctrine or cultural practice. For example, some communities might have resistance when they are presented with new technologies. Certain people believe that new technology will disrupt their conventional ways of life, which some may perceive as unnatural.

Finally, religious beliefs and symbolism link to people's perception and acceptance of energy technologies. These should be understood from a cultural and spiritual viewpoint, enabling policymakers, businesses, and innovators to come up with effective energy solutions that are culturally and ethically acceptable to various communities.

10.1.1.7 Solar Energy as a Divine Gift

Some cultural traditions connect solar power with divine elements that bring forth life energy because of its sacred status. The symbolic relationship between solar energy and natural power helps cultures accept solar energy as an auspicious natural source. Parts of India regard solar energy as a sustainable and sacred source after linking it with their sun deity, Surya (Clean Energy Technology Adoption, Cultural Differences and Basic Human Dispositions, 2021).

In other cultures, the sun is regarded as a manifestation of heavenly forces that bring warmth, light, and food to all living beings; consequently, the energy emanating from it becomes something sacred. Given this, the exploitation of solar energy aligns better with a fundamental spiritual or ethical commitment to using natural resources by humanity without harming the environment. It's not only about technology, but it's also tapping into that cosmic resource that has always been there and has formed part of the divine order of the planet. In this way, it may also create feelings of reverence and duty for the adoption of solar energy, seeing it as a practical tool but also linked morally with spiritual principles of sustainability, stewardship, and care for the land.

Solar energy, in this context, is a source of energy that is more than a renewable resource; it is a means to maintain balance with nature, emphasising the harmonious interaction between human growth and the natural environment. As such, many communities may embrace solar technology for its environmental and economic benefits and because it connects with their religious or spiritual ideals.

10.1.1.8 Traditional Fuels and Religious Rituals

specific populations connect wood and coal as traditional energy sources to religious ceremonies and rituals cultural traditional practices. The associations function as a barrier that opposes modern energy solutions. The people living in rural Africa and Asia use firewood and biomass not just for energy generation but also to perform essential cultural activities, including preparing communal meals and conducting religious ceremonies with them. Communities might view contemporary energy alternatives with doubt when they need to replace their traditional sources (Goggins et al., 2022). In this context, energy is discussed under the following subtopics:

10.1.1.9 Cultural Norms and Resistance to Modernisation

The acceptance of new energy technologies by communities depends heavily upon their existing cultural beliefs about modernisation. While those who have had contact with the Western world, where emerging technologies are not just innovations but part of their culture, tend to accept new technologies, core naturalists and believers in ancient tradition are likely to resist any appearance of new technologies (Evans & Britt, 2023; Gui & MacGill, 2018).

10.1.1.10 Preservation of Traditional Lifestyles

Rural communities worldwide preserve their traditional ways of life with outstanding commitment. Technologies that disturb traditional life patterns encounter notable opposition from the affected communities. The combination of loss of control for energy usage and external monitoring concerns prompts remote small communities to reject smart metres and automated energy management systems.

10.1.1.11 Trust in Technology and Institutions

Trust maintains its essential position in the process of adopting new technologies. People living in nations with minimal trust in public institutions, and international technology suppliers tend to resist these systems. Acceptance of government-backed energy projects faces resistance within specific regions of Eastern Europe and parts of Africa, which impedes smart grid deployments and other Energy 4.0 technology implementations.

10.1.1.12 Societal Influence on Energy 4.0 Adoption

Societal influence is highly relevant to implementing Energy 4.0 technologies. For example, some people shape their attitudes and values regarding behaving properly toward energy transition. In addition, religious and cultural beliefs can create a favourable setting for accepting innovative energy transition options that will be highly affected by societal values, such as social norms, economic context, and civic commitment. In societies that strongly support innovation, sustainability, and environmental responsibility, the implementation of Energy 4.0 systems will be fast. On the other hand, in communities that are dominated by traditional energy practices or where people are resistant to change, advanced energy solutions might face challenges. These factors are necessary to understand how the society around us influences the promotion of Energy 4.0 in ways that fit the values and needs of diverse societies. In this context, the following points are worth examining:

10.1.1.13 Influence of Community Leaders

Community leaders maintain responsibility for protecting cultural standards through their influence on public sentiments. The fair and regular participation of leaders will help them support Energy 4.0 technological initiatives, which will promote better community reception. Rural Japanese communities adopted renewable energy solutions like solar farms and wind projects after their leaders promoted them (Clean Energy Technology Adoption, Cultural Differences and Basic Human Dispositions, 2021).

10.1.1.14 Media and Public Perception

European countries demonstrate slow smart grid acceptance because media reports create public concerns about expensive technologies and technical hazards (Jørgensen et al., 2025). Organisations need to eliminate cultural obstacles that prevent Energy 4.0 implementation. A comprehensive solution to cultural barriers includes education programmes, community outreach initiatives, legislative changes, and communications that respect cultural elements.

10.1.1.15 Education and Awareness Campaigns

Through educational instruction, operators can eliminate false beliefs about Energy 4.0 technology products while demonstrating their advantageous aspects within local cultural contexts. The acceptance rate of solar energy systems increases when their connection to environmental ethics and cultural value systems is properly communicated to the population.

10.1.1.16 Community Engagement and Participatory Approaches

Central participation in energy project planning and implementation enables respect for regional cultural values. Smart energy solutions gain better community acceptance when participants actively co-design those solutions, as the process instils ownership and minimises resistance (Adamos et al., 2024).

10.1.1.17 Policy Support and Incentives

The implementation of Energy 4.0 technologies is based on cultural belief systems within the community. Identifying energy-related cultural beliefs represents a mandatory step for creating successful adoption strategies regarding renewable energy together with smart grids and digital energy management systems. Government programmes that prioritise cultural factors in policy development, along with educational outreach plans and community programmes, help stakeholders establish acceptance of modern technologies. We need to handle these cultural obstacles to unlock the full potential of Energy 4.0 for sustainable and equitable energy prospects (Goggins et al., 2022).

10.1.1.18 Sociocultural Barriers in Agri-Food Industry 4.0

The Agri-Food Industry 4.0 constitutes a technological advancement for agriculture and food production through precision farming, IoT systems, AI systems, blockchain traceability, and automation technologies. The technologies introduced present several advantages that boost production levels while protecting both food security and environmental sustainability. While the advantages are obvious, there are multiple factors that make the technology adoption process complex (Kumar et al., 2022).

Agri-Food Industry 4.0 technologies experience limited adoption because of sociocultural barriers, which include conventional farming methods and people's unwillingness to adapt to existing social arrangements and specific cultural beliefs. The exploration in this article details sociocultural barriers to technology adoption in the agri-food sector while presenting potential solutions to these adoption challenges.

10.1.1.19 Understanding Agri-Food Industry 4.0

The application of sophisticated technologies enables the digital evolution of agricultural and food manufacturing techniques known as the Agri-Food Industry 4.0. Some key technologies in this space include:

- The technology enables optimal crop production alongside resource optimisation through data acquired from sensors, drones, and GPS systems.
- The implementation of automated systems efficiently manages planting and harvesting tasks, along with packaging functions to increase operational efficiency as well as minimise expenses through reduced workforce requirements (Fuentes-Peñailillo et al., 2024; Javaid et al., 2022).
- Blockchain technology establishes transparent and secure food supply chain tracking, which builds confidence between farmers and their end-market consumers (Apeh & Nwulu, 2025).
- The utilisation of AI enables users to analyse extensive datasets to generate yield predictions, identify disease, and provide strategic recommendations (Fuentes-Peñailillo et al., 2024; Javaid et al., 2022).

The implementation of these modern technologies depends on diverse social customs that vary between different geographical areas and communities (Senturk et al., 2023).

10.1.1.20 Key Sociocultural Barriers

Some social and cultural barriers influence the spread of the Agri-Food Industry 4.0, especially in the rural parts of the world. Some of them are briefly explained:

Resistance to Change and Technological Fear: A substantial sociocultural obstacle exists in the acceptance of Agri-Food Industry 4.0 technologies because older farmers and rural residents strongly resist the implementation of new technology. People resist technology adoption primarily because they fear the loss of traditional farm management systems and practices, and they question the authenticity of new technological solutions.

Traditional Farming Practices: Different farming practices have been transmitted within families for many generations. People living in various communities maintain these traditional practices because they uphold their cultural heritage and reflect their cultural identity. Automation and data analytics systems cause apprehension among traditional practitioners because they fear the loss of established customs (Kumar et al., 2022).

Subsistence farmers throughout the African and Asian regions prefer to use established farming methods rather than adopt new technological solutions that might increase yield and efficiency levels.

Lack of Technological Literacy: The lack of technological aptitude for operating sophisticated equipment exists among a considerable numbers of farmers. A lack of technological capabilities makes farmers increasingly anxious about new technology systems because they perceive the devices as inaccessible and unstable.

Social Structures and Influence: Rural community dwellers base their responses to new technologies primarily on their social and community interactions. This, in turn, affects their rate of acceptance of these new technologies. The opposite applies to urban community dwellers.

Influence of Social Networks: The agricultural community depends on both peer advice and community leader guidance when deciding to adopt new practices. Members of the community who reject Agri-Food 4.0 technologies may influence other residents to avoid these innovations (Senturk et al., 2023).

The Chinese social network system known as guanxi is essential for acquiring both resources as well as new technologies in China. People lacking strong social connections often face exclusion from the benefits of innovation.

Socio-Economic Divides: The acceptance of new practices among farmers is impacted by the wealth disparities that exist between them. Commercial farming groups have better access to affordable technology investments than those operating smallholder farms. The agricultural industry functions as two separate systems, which increases the growing distance between modern and traditional farming operations.

Cultural Perceptions of Food and Technology: The relationship between cultural perceptions of food and technology is complex and multifaceted. Food carries deep cultural, social, and emotional meanings that go far beyond basic sustenance; indeed, traditions, rituals, and identities have long been linked with food. Therefore, this naturally sets up varied reactions toward technology in the food system (Park & Widyanta, 2022). Whereas some cultures might embrace the technology to increase the production or preservation of food, others might view such innovations with scepticism and opt to retain more traditional ways, perhaps viewing technological interventions with suspicion. Besides issues related to food safety, transparency, the "naturalness" of food, ethical concerns regarding animal welfare, and environmental impact are part of the shaping cultural perceptions (Henchion et al., 2022). Examples of such varied attitudes towards organic food, genetically modified organisms (GMOs), and food taboos are manifold and indicate the many ways in which cultural beliefs influence food choices and acceptance of new technologies. Such a complex landscape must be negotiated by considering the particular cultural context in which the food technology is being introduced, by being open in one's communication and by showing respect for traditional food habits.

Genetically Modified Organisms: The general public shows wariness toward genetic crop technology due to beliefs that these developments are both artificial and dangerous for human health. The same resistance to GMOs exists for all biotechnology applications that Agri-Food 4.0 employs, including AI-guided plant breeding and cultured meat production (Lucht, 2015). The people of Europe display intense opposition to GMOs as they worry about their effects on both health and the environment (Stapleton (2016). The same mindset prevents people from adopting new technologies in food-related applications.

Perceptions of Natural vs. Artificial: Societies worldwide show a tendency to prefer natural food production systems rather than man-made strategies that appear artificial or industrialised. People tend to develop doubts about precision agriculture and automated farming systems based on their perceptions (Senturk et al., 2023).

Policy and Regulatory Barriers: The adoption of sociocultural factors into the system is directly impacted by the existing policy framework together with regulatory conditions. A policy that favours the integration of technology into agri-systems will lead to a massive expansion of the Agri-Food Industry 4.0 across the board. The reverse is the case for communities that implement policies that don't favour technology integration in the agricultural space.

Lack of Supportive Policies: Several countries today do not support policies that promote Agri-Food 4.0 technology adoption. The support system directed toward traditional farming practices receives more funding than does investments in innovative solutions. Developing country farmers encounter bureaucracy and corruption, which prevent them from obtaining smart farming equipment subsidies.

Inconsistent Standards and Regulations: The implementation of technology adoption faces challenges because inconsistent regulations make data ownership, traceability, and AI ethical standards unclear, which causes fear of adoption among users (Kumar et al., 2022).

10.1.1.21 Strategies to Overcome Sociocultural Barriers

There are some strategies that, when employed, can reduce these sociocultural barriers discussed above. Some of them include:

Education and Capacity Building: Trained farmers with access to resources become more capable of adopting Agri-Food 4.0 solutions by overcoming their technological illiteracy and gaining confidence.

Non-profit organisations, along with government entities, should provide farmers with demonstrations and workshops to show them the advantages of adopting technology in agriculture.

Academic institutions, working in partnership, enable studies and the development of specific solutions for agricultural communities in particular regions.

Participatory Approach and Community Engagement: When farmers actively contribute to the Agri-Food Industry 4.0 technology planning process, it leads to solutions that match regional requirements and local farming culture. Predevelopment involvement of farmers through co-design enables them to build a sense of ownership while decreasing their resistance to new technologies. Local community champions should be identified and provided with specialised training to serve as advocates who will promote new technologies while generating trust and adoption.

Government Support and Policy Reforms: Government support helps overcome sociocultural barriers by designing favourable policies that motivate people to adopt technology. The provision of financial support to smallholder farmers through subsidies helps lower their expenses related to adopting innovative technologies.

Acceptance of Agri-Food Industry 4.0 technology goes beyond technical matters because it relates fundamentally to social and cultural aspects. These technologies receive varied receptions based on traditional practice adoption, social structures, cultural beliefs, and existing regulatory standards. Implementation needs a complete strategy that unites educational sessions, policy backing, and culturally mindful communication to overcome these barriers. Stakeholders who establish trust through local engagement will ensure Agri-Food Industry 4.0 benefits reach every farmer to develop a sustainable agri-food sector alongside resilience and equity (Kumar et al., 2022).

10.1.1.22 Social Acceptance of Smart Water 4.0 Management Technologies

Water management sectors in society benefit from revolutionary solutions that emerged during the Fourth Industrial Revolution. Smart Water 4.0 technologies combine the IoT together with AI alongside data analytics tools to boost water management through real-time monitoring systems and predictive maintenance capabilities and automated system control. High-tech solutions seek to improve resource management by limiting water loss while delivering better resource planning and combating worldwide water shortage (Glória et al., 2020). Public acceptance alongside societal engagement determines how much Smart Water 4.0 will succeed as a technological innovation despite its potential. The reception of these innovative technologies depends on how well communities trust them and their beliefs, economic diversity, and regulatory frameworks. This paper analyses what affects social acceptance rates for Smart Water 4.0 technology and proposes methods to boost acceptance levels.

10.1.1.23 What Are Smart Water 4.0 Technologies?

Current water management efforts implement digital technologies as the backbone of Smart Water 4.0 to achieve more efficient and sustainable water operations. The most common innovations include:

- Smart Water Metres operate in real-time to monitor and report water usage, which enables users to understand their consumption behaviour.
- AI-driven water quality monitoring systems have demonstrated their ability to identify and anticipate water quality problems, which helps public health personnel tackle risks early on.

- IoT-based leak detection Systems function as advanced monitoring systems that detect drainage network leaks to decrease water loss in ageing infrastructure.
- Automated wastewater treatment systems utilise real-time operating adjustments to maximise efficiency and lower environmental impact.

The implementation of these modern technologies offers substantial advantages but society remains unsure about their acceptance. Multiple cultural, along with economic and social elements, demand attention to guarantee successful technology adoption.

10.1.1.24 Factors Influencing Social Acceptance

1. Cultural and Historical Perspectives on Water:

   Many nations attribute cultural value, along with symbolic meaning, to water that affects their response to new water management systems.
2. Water as a Communal Resource:

   The local people in developing nations manage water as a communal resource under traditional social practices. Various communities may oppose new smart water management systems because these technologies seem to encourage privatisation and external control.

   Traditional Indian rural areas practice water management through customs and religious rituals, which form the bedrock of their water operations. The rejection of smart water metres is likely in these communities because they represent opposition to conventional resource management methods. The people of Africa share a common heritage in water resources, while they see water service digitisation as a foreign intervention (Liu et al., 2021).
3. Religious and Symbolic Significance of Water:

   Failure in technology adoption arises because religious and cultural beliefs shape particular perspectives about water resources. Islamic communities emphasise moral duties of water conservation that can support sustainable technologies when cultural sensitivity is maintained in their presentation.
4. Trust and Transparency:

   Society needs people to trust both technological systems and governmental institutions in order to accept them, and research has found that trust depends on how transparent information appears, while fairness is a factor, as well as the amount of public involvement in the decision-making process.
5. Data Privacy and Security Concerns:

   Smart water technologies gather broad water utilisation data, while corresponding institutions face challenges regarding personal water usage data management. Water usage data availability causes concern due to possible profit exploitation or unfair pricing methods. People are less likely to adopt new systems when there is insufficient disclosure about data management practices, because such secrecy diminishes public confidence.
6. Public Engagement and Awareness:

   The implementation of Smart Water 4.0 technologies becomes more socially acceptable through community programmes that spread knowledge about these systems. The integration of public education efforts with straightforward water conservation information, presented by PUB Singapore, has enabled successful smart water solution implementation in the country (Ramos et al., 2019).

7. Socio-Economic Barriers:

   The cost of participating in and embracing Smart Water 4.0 applications varies directly based on current economic inequalities between different groups.

8. Cost and Affordability:

   The high initial expenditure needed to implement Smart Water 4.0 products presents a barrier to their adoption in lower-income population areas. Transformer operations require expensive updates for existing infrastructure, which makes adoption difficult for selected municipalities. Low-income families face increased financial hardships from paying additional rates for smart water metres, together with maintenance services, which creates resistance.

9. Digital Divide and Inequity:

   The unequal distribution of technology access exacerbates disparities in water management. Urban wealthier areas gain more advantages from technological applications because less fortunate rural settlements, along with marginalised regions, face the risk of lagging behind in adoption.

10.1.1.25 Political and Regulatory Challenges

1. Policy and Governance Issues:

   Smart water technologies require the development of an appropriate political framework to maintain transparency measures and protect data integrity while ensuring affordable pricing. The absence of clear regulations for data management involving smart water systems across various nations creates uncertainty among providers and consumers (Abiri et al., 2023). Real-time data-driven water pricing through dynamic models raises public controversy when implemented without proper regulatory frameworks and effective public communication (Zarreh et al., 2024).

2. Political Influence and Corruption:

   The presence of high corruption rates together with political manipulation causes people in these areas to have minimal trust in technological advancements. People frequently doubt that smart water solutions will function as tools for political manipulation instead of serving the public welfare (Adam & Fazekas, 2021)

10.1.1.26 Strategies for Enhancing Social Acceptance

1. Community Engagement and Participatory Design:

   The design process, along with implementation of smart water solutions, becomes more successful when communities actively participate because their needs and cultural preferences are integrated into the initiative. Communities that engage in co-design activities boost their acceptance and ownership of modern technologies through participatory design practices. Facilitating learning sessions that allow people to directly experience smart water devices removes uncertainty while creating confidence in the technology.

2. Transparent Communication and Education Campaigns:

   Massive adoption rates for Smart Water 4.0 will result from properly targeted educational campaigns that spread awareness about its benefits throughout the public. The marketing initiatives must display substantial benefits like water conservation and better water quality, yet they need to reassure people about privacy protection and cost-effectiveness (Gupta et al., 2020).

3. Ensuring Equitable Access:

   Public access to smart water technologies requires policymakers to establish supportive financial programmes, which should be combined with government-private partnership initiatives. Government subsidies directed toward low-income families create a way for them to afford smart water technology. Cooperation between public entities and private businesses reduces cost and improves accessibility of services.

   Social acceptance of Smart Water 4.0 technologies serves as a key requirement for establishing sustainable water management programmes. Public understanding of Smart Water 4.0 technologies is shaped by cultural beliefs in conjunction with trust elements, economic disparities, and regulatory systems. The implementation of Smart Water 4.0 requires policymakers and technology developers to use a human-centred framework, which honours cultural sensitivities alongside transparent operations and equal opportunities for access.

   Organisations dedicated to public education, together with supporting inclusive policies, will lead to extensive adoption. The resolution of current obstacles will enable Smart Water 4.0 achieve sustainable water resource protection for upcoming generations alongside equitable distribution.

10.1.1.27 Socio-Economic Factors and Agrivoltaic Technology Adoption

The joining of photovoltaic solar power with agricultural systems, known as agrivoltaics, has emerged as an effective, sustainable approach to combining energy production with food production. The factors determining agrivoltaic adoption appear differently across geographic areas because of local socio-economic and cultural conditions. According to a report in the Renewable and Sustainable Energy Reviews journal, report agrivoltaic initiatives deal with issues stemming from land conflicts between different users and economic dissimilarities between groups (Giri & Mohanty, 2022).

The concern about traditional farming space shrinking causes farmers from certain rural areas to reject agrivoltaic systems. The lack of equal financial resources prevents small-scale farmers from purchasing these advanced technologies. The acceptance of agrivoltaics faces obstacles because of how societies handle proprietorship alongside agriculture. The deep historical connection between culture and land ownership leads people to resist changes in existing land management systems through technological interventions.

The solution to sociocultural obstacles requires policymakers to develop inclusive financial systems through subsidies, cooperative business models, and public-private partnerships. The implementation process of agrivoltaic projects gains acceptance and equitable benefits through active community participation from the start of the project design (Giri & Mohanty, 2022).

10.1.1.28 Cultural Factors in Consumption Patterns and Sustainable Practices

Variant consumer actions play a primary role in determining how people adopt Food 4.0, Energy 4.0, and Water 4.0 technologies. People's approaches to new technologies depend on their culture-based norms together with their dietary practices and life patterns. Food technology is popular for plant-based and lab-grown meat products because Western society shows increased interest in environmental and health concerns. Cultures in Argentina and Brazil strongly maintain meat consumption during their traditions and social ceremonies, so alternative protein products encounter strong opposition (Banbury et al., 2012).

People from Scandinavia who strongly value the environment have introduced smart grids while adopting energy-efficient home appliances. The areas heavily reliant on fossil fuel industries tend to resist the implementation of quick transitions toward renewable power systems.

Global consumption patterns demonstrate that cultural backgrounds create obstacles to behaviour change, thus impacting how well new technological solutions contribute to achieving sustainability. Programmes combining education with culture-specific awareness messages and incentives will accelerate sustainable consumption habits among different cultural groups (Banbury et al., 2012; Goggins et al., 2022).

The implementation of Food 4.0, Energy 4.0, and Water 4.0 technologies faces social-cultural barriers rather than purely technical ones. Food 4.0, together with Energy 4.0 and Water 4.0, receive diverse responses because cultural systems and economic differences and government structures, and social traditions influence them. These innovations will achieve their full success potential through policy-making leadership, which incorporates sensitive cultural considerations together with business and research initiatives.

This also helps to engage local communities in decision-making. These technologies should properly align with prevailing cultural standards and social structures. This addresses economic barriers through inclusive policies. Public trust will increase as organisations practice transparent, ethical governance. Through improved cultural inclusivity, these sustainability measures based on technology will achieve significant success in food, energy, and water system management.

10.2 IMPACT OF CLIMATE CHANGE ON FOOD, ENERGY, AND WATER SYSTEM 4.0 GLOBALLY

Several economic sectors, along with societal environments, face profound consequences from the global climate change challenge, particularly affecting the food, water, and energy systems. The multiple sectors continue to merge in ways that make them important for human development as well as environmentally sustainable practices (Field et al., 2014). These sectors are set to obtain new opportunities for solving global issues thanks to "System 4.0" which brings together AI and IoT technologies and big data analytics. Understanding this interconnected network of systems requires complete comprehension of their risks alongside opportunities, together with necessary adaptive strategies in order to achieve sustainable global development as climate change continues to impact these networks.

This text delves into the convergence of climate change with evolving trends in food, energy, and water systems by investigating climate change effects on how these systems function, as well as their productivity and resilience. The study reviews contemporary System 4.0 technological research along with its protective measures against climate change impacts and the difficulties they pose during this period of climate change (Pardoe et al., 2018).

10.2.1 Climate Change and Its Impacts on Global Food Systems

Food systems remain highly susceptible to climate change because it worsens existing threats faced by food security, alongside risks in agriculture and nutrition systems. The growing temperatures, together with inconsistent rainfall, extreme weather phenomena, and changing ecosystems, degrade food production, processing, and distribution systems (Bocchiola et al., 2019).

Multiple scientific studies have established how climate change reduces agricultural output and agricultural production. Staple crops within food-deprived regions will experience decreased production because of rising temperatures, modified rainfall, and intensified droughts. New pests and diseases that result from changing climate conditions endanger both crops and livestock production. The regions of sub-Saharan Africa, Southeast Asia and South America will face substantial yield decreases, which will create obstacles to satisfying international food requirements.

Climate-based changes negatively impact all aspects of food logistics between production and consumption. Temperature increases, while extreme weather disrupts supply chains and leads to spoilage throughout transportation. The scarcity of water, which arises from climate change consequences disrupts irrigation practices that subsequently affect both agricultural production and available food. Published research confirms that climate change causes food systems to suffer price hikes and supplies to decline, which reduces accessibility for food security and equitable access (Bocchiola et al., 2019).

System 4.0 technologies applied in agricultural operations through precision agriculture, combined with smart irrigation, and AI-managed supply chain optimisation, provide potential remedies to defend food systems against climate change effects. The IoT, through its use of sensors, tracks soil conditions alongside crop health and meteorological data, which allows farmers to enhance their resource management capabilities. The implementation of this strategy leads to stronger crop resistance against climate-based stress factors while simultaneously reducing negative environmental effects. AI strengthens predictive systems that estimate crop yields, so governments alongside organisations may plan for shortages resulting from climate change (Esham et al., 2018).

10.2.2 Climate Change and Its Impact on Global Energy Systems

Climate change produces dual effects on energy systems by influencing both how much energy is used and how it is provided to consumers. Energy production, together with distribution and consumption processes experience major impacts because of increasing temperatures, alongside varying weather patterns, and more frequent severe weather occurrences (Cronin et al., 2018).

Renewable energy resources, including hydropower and wind-based facilities, experience increased challenges from climate change effects in their supply operations. The transformation of precipitation patterns, together with modifications in wind direction, result in reduced water levels needed for hydropower and impaired wind farm operations (Kara & Şahin, 2023; Singh et al., 2022). Fossil fuel electricity production faces increased difficulties because of heatwave conditions that decrease the operational efficiency of power generation facilities (Understanding the Climate-Water-Energy-Food Nexus and Streamlining Water-Related Policies – European Commission, n.d.).

Both the quantity of energy needed and the way energy is distributed come under the influence of climate change. The global rise in temperature boosts building cooling needs, particularly in hot areas, thus boosting energy usage during heatwaves. During intense periods of cold weather, the need for heating energy production typically rises. The frequent shifts in energy demand place excessive stress on present-day power networks and cause power outages alongside rising energy expenses (Taseska et al., 2012).

When integrating System 4.0 technologies such as smart grids, and blockchain-based trading platforms, and energy storage capabilities, into the energy sector, these tools improve the efficiency and durability of power systems. The monitoring capability enabled by smart grids, empowered by AI and big data analytics, permits the optimisation of energy distribution in real time, reducing energy disruptions during climate-induced weather events. Batteries and hydrogen storage systems, together with other energy storage technologies enable better management of renewable power when intermittent sources fail, thus maintaining a steady energy supply during climate-related interruptions (Edenhofer et al., 2011).

10.2.3 Climate Change and Its Impact on Global Water Systems

The pressure on water systems from climate change produces adverse effects that spread across human health, agriculture, and natural ecosystems. Temperature rises, together with pattern modifications in rainfall systems, cause major changes in water resource availability and quality distribution patterns (Glória et al., 2020).

The occurrence of more severe and common droughts as a result of climate change creates widespread water shortages throughout various regions. The combination of high rainfall events brings about flooding that results in water contamination, thus negatively affecting water quality. Water supply for drinking and irrigation, together with hydropower production, undergoes changes because of these environmental modifications, thus affecting food and energy systems (Gopikrishnaa, 2016).

The combined effect of elevated temperatures and droughts leads to worse water body pollution which produces toxic algal blooms that damage water quality.

Waterborne disease occurrences become more frequent because of precipitation pattern changes, while water treatment facilities, together with sanitation systems, experience increased strain. Changing water availability and quality create severe problems with access to clean water and sanitation facilities in developing nations.

Water management professionals are finding useful applications for System 4.0 technologies to tackle climate change problems. The integration of IoT with irrigation systems delivers enhanced water optimisation through continuous monitoring of soil wetness, temperature data, and meteorological predictions. AI-driven algorithms enable forecasting of water needs for better dam operation, which leads to enhanced distribution systems that reduce water loss. Machine learning provides two important capabilities, as it enables identification of pollutants and ensures effective water treatment through monitoring mechanisms (*How AI Can Optimize Water Usage in Agriculture, n.d.*).

The three interconnected systems which make up food, energy, and water show progressive impacts between them through modifications to any single system element. In addition to agricultural effects, water scarcity creates food shortages, and energy shortages break down water-pumping operations and irrigation systems. The improper management of agricultural operations increases energy consumption rates and reduces water efficiency levels. The challenges brought by climate change need solutions which analyse their complex interdependent nature Ogbolumani & Nwulu, 2022; *Understanding the Climate-Water-Energy-Food Nexus and Streamlining Water-Related Policies – European Commission, n.d.*).

Data System 4.0 represents a promising solution for food-energy-water system connections by uniting IoT systems with big data through AI technologies and additional system 4.0 capabilities. Real-time demand monitoring through integrated systems enables better resource management, which decreases total waste. Machine learning models with AI capabilities predict how climate factors will influence the interactive food-energy-water systems, thus allowing for proactive strategic planning (Ogbolumani & Adekoya, 2025). Through blockchain technologies, operators can monitor resource use across various sectors to obtain transparent and efficient management (Katel et al., 2022).

Climate change affects global food, energy, and water systems through major processes that lead to technological advancements for future assessment. The next generation of system technology (System 4.0) demonstrates capabilities to improve industrial sectors through the development of sustainable solutions that adapt to climate-related disturbances. System 4.0 technologies need a strategic evaluation of food-energy-water system cross-relationships, while following implementation practices that build fairness and universal access, in addition to stability and robustness. Sustainable development in the face of climate change impacts depends heavily on ongoing research efforts dedicated to technological advancement.

10.3 SOCIAL, ECONOMIC, AND ENVIRONMENTAL BENEFITS OF FOOD, ENERGY, AND WATER NEXUS 4.0

The FEW Nexus 4.0 refers to the interlinkages between these three critical resources and the need for integrated systems that leverage advanced technologies to optimise their use and sustainability. The FEW Nexus 4.0 incorporates emerging technologies such as AI-powered, IoT, big data and blockchain. This creates a digital framework that enables food, energy, and water systems to be run more efficiently and sustainably (Cho et al., 2023).

This section examines features of the FEW Nexus 4.0 and its resultant social, economic, and environmental advantages. It evaluates how technology affects the sustainability aspects across interconnected systems.

10.3.1 Social Benefits of the FEW Nexus 4.0

The social benefits of FEW Nexus 4.0 aim to enhance community welfare by providing equal food, energy and water resources to all people. The technological innovation within these sectors demonstrates potential toward three outcomes: enhancing social equity, decreasing poverty levels, and stimulating inclusive growth patterns. Advanced technologies included in the FEW Nexus 4.0 provide increased resource availability for food and water and energy, particularly for distant or underdeveloped locations (Cho et al., 2023).

The adoption of IoT technology through irrigation control systems enables remote farmers to enhance resource efficiency and boost crop output, which protects their food supply. The implementation of solar mini-grids, coupled with blockchain energy trading platforms, allows rural areas to acquire affordable renewable electricity. The infrastructure development barriers that exist can be overcome by these systems leading to social inclusion and diminishing inequality (Adeleke et al., 2023; Ogbolumani & Mabaso, 2023).

The implementation of AI along with IoT technologies in disaster mitigation and climate adaptation programmes builds community resilience against climate-triggered emergencies to reach improved community strength. The community obtains efficient disaster response through proper preparedness against droughts, floods, and heatwaves that reduce interruptions in food, water, and energy supplies. The FEW Nexus 4.0 provides capabilities that help communities strengthen their capacity to adapt to future disruptions, thus securing their social stability and wellness over long periods (Cho et al., 2023). Community members receive training about advanced technology utilisation within FEW Nexus 4.0 (Capacity Building and Empowerment) programmes, which allows them to utilise their resources effectively. Farmers who use precision agriculture tools will learn to operate these systems to maximise their crop yields through reduced water usage and energy consumption. Natural resource ownership is enhanced when people gain better livelihoods through this technology. Community involvement in managing local resources becomes more possible through efforts that implement data-driven platforms.

10.3.2 Economic Benefits of the FEW Nexus 4.0

The FEW Nexus version 4.0 generates favourable economic results through technological implementations in food, energy, and water resource management, enabling better efficiency and enhanced productivity, as well as lower costs. This new technological progress creates opportunities for business development, creating employment, and generating economic diversity (Hoosain et al., 2023). The FEW Nexus 4.0 enables improved efficiency across all three sectors of food, energy, and water during its operation. Precise farming methods enabled by IoT tools, together with AI algorithms, help users manage water resources, and energy supply, and fertilisers effectively, thus lowering operating expenses while producing greater output. The same energy optimisation benefits of smart grids and energy storage occur in electricity distribution networks. These technological systems enable their application, resulting in extended affordability benefits for both individuals, businesses, and governments. The FEW Nexus 4.0 solution improves agricultural output by helping farmers monitor their water systems and pests and observe their crops for enhanced economic growth. AI predictive analytics helps farmers predict crop yields, which allows them to forecast market demand to increase their farming profitability (Hoosain et al., 2023).

Moreover, organic farming practices and aquaculture are implemented while observing less input of water and pesticides leading to more durable and sustainable agricultural practices, thereby facilitating an increase in the rural economy. Utilisation of the technology of System 4.0 across the FEW Nexus can be a breakthrough into a new era of occupations and interconnections amongst different spheres of work. The areas on the topic that I introduced consist of AI development, renewable energy, smart agriculture, and water management. That is to say, apart from the environmental benefits, the greening of the economies of such countries will also be a result of the synergy among green technologies and green industries. This approach will give rise to a fairly intricate web of job possibilities that include green entrepreneurship, and government and private sector support for local cases of sustainable development.

10.3.3 Environmental Advantages of the FEW Nexus 4.0

One of the main roles of the FEW Nexus 4.0 is to minimise the side effects created when dealing with these global issues such as resource constraints, ecosystem destruction, and climate change. Through technological innovation, we can improve and optimise the long-term sustenance of nature, thus providing a hopeful answer to the climate change issue (Zarei et al., 2021).

The FEW Nexus 4.0 has been developed to employ more sustainable technologies that will allow the use of resources to be minimised but will ensure safe and healthy food growth and adequate water and energy supply. Thus, an example of this might be a high-tech precision irrigation system using IoT sensors, which saves water waste and preserves local water resources. The implementation of AI-based energy optimisation in buildings and industries results in a decrease of the energy consumption of the energy system thereby largely reducing the carbon emissions of the energy system (Zarei et al., 2021).

The FEW Nexus 4.0 can be seen as a solution that conserves resources and does more to ensure that climate change remains minimal if not reversed. Using renewable energy sources such as wind and solar power helps to solve the problem of reliance on fossil fuels and as a result, reduces greenhouse gas emissions in environmental settings. Sustainable agricultural practices have been grounded in data-driven platforms, thus not only reducing soil degradation but also lowering greenhouse gases from farming activities.

The FEW Nexus 4.0 might be a tool for modifying and preserving ecosystems. Nowadays, employing modern agricultural practices needs less land for agriculture and thus prevents deforestation and habitat destruction. Meanwhile, the prudent use of energy and water is a must for the environment to be clean and for the population to grow in number, so biodiversity may remain healthy (Zarei et al., 2021).

10.4 FOOD-ENERGY-WATER NEXUS SYNERGETIC INTERACTION

Taking advantage of the synergies in the FEW Nexus would greatly help the environment. For example, the replacement of water desalination or irrigation systems with solar or wind power can greatly reduce the carbon footprints when they are being used to drive these processes. Just like sustainable food production practices like agroforestry or precision agriculture which can also promote both water conservation and carbon sequestration (Cho et al., 2023).

The social, economic, and environmental benefits of integrating technologies that improve the management of food, energy, and water resources are achieved in the FEW Nexus 4.0. This interconnectedness can help redistribute natural resources and minimise wastage. It also yields also efficiency and diversification, and subsidies become a source of employment and, therefore, sustainable industries are fostered. This nexus is environmentally friendly, promotes the efficient use of resources, reduces climate change, and protects ecosystems. In a world where climate change is present, and resources are getting scarcer, the FEW Nexus 4.0 is probably the best way forward to a better and sustainable future (David, Aigbavboa, et al., 2024a, 2024b; David et al., 2022; Liu et al., 2018).

10.5 SUMMARY

Significant changes have occurred in the food energy and water sectors because of the industrial revolution 4.0 through the implementation of digital innovations, including AI, the IoT, blockchain and smart solutions. The alliance of modern innovations constitutes the elements that form Food 4.0 along with its counterparts Energy 4.0 and Water 4.0. Technology implementation needs factors that surpass basic economic viability and technical possibilities before making any decisions. The seamless integration of digital technologies within the food energy and water nexus needs some degree of expertise in technology and also some level of acceptance regarding the culture of the people who the nexus is going to benefit. The culture of the people plays a vital role in determining if a specific type of technology will be accepted and utilised to drive development. A good understanding of how culture influences the FEW Nexus 4.0, especially in harnessing it for positive development, is necessary to exploit the potential of the FEW Nexus 4.0. Community-wide cultural identity and societal beliefs drive how industrial sectors embrace new technological solutions throughout their operational areas.

REFERENCES

Abiri, R., Rizan, N., Balasundram, S. K., Shahbazi, A. B., & Abdul-Hamid, H. (2023). Application of digital technologies for ensuring agricultural productivity. *Heliyon*, *9*(12), e22601. https://doi.org/10.1016/J.HELIYON.2023.E22601

Adam, I., & Fazekas, M. (2021). Are emerging technologies helping win the fight against corruption? A review of the state of evidence. *Information Economics and Policy*, *57*, 100950.

Adamos, G., Caucci, S., Charpentier, L., Chediek, J., Krol, D., & Laspidou, C. (2024). Exploring nexus policy insights for water-energy-food resilient communities. *Sustainability Nexus Forum*, *31*(1), 69–82. https://doi.org/10.1007/S00550-024-00534-0

Adeleke, I. A., Nwulu, N. I., & Ogbolumani, O. A. (2023). A hybrid machine learning and embedded IoT-based water quality monitoring system. *Internet of Things*, *22*, 100774. https://doi.org/10.1016/J.IOT.2023.100774

Ahmad, T., Zhu, H., Zhang, D., Tariq, R., Bassam, A., Ullah, F., AlGhamdi, A. S., & Alshamrani, S. S. (2022). Energetics systems and artificial intelligence: Applications of industry 4.0. *Energy Reports*, *8*, 334–361. https://doi.org/10.1016/J.EGYR.2021.11.256

Ajibola, O. A., & Ogbolumani, O. A. (2024). Smart battery storage integration in an IoT-based solar-powered waste management system. *Journal of Digital Food, Energy & Water Systems*, *5*(2), 17–42. https://doi.org/10.36615/HREWQ861

Allum, N., Sturgis, P., Tabourazi, D., & Brunton-Smith, I. (2008). Science knowledge and attitudes across cultures: A meta-analysis. *Public Understanding of Science*, *17*(1), 35–54. https://doi.org/10.1177/0963662506070159

Apeh, O. O., & Nwulu, N. I. (2025). Improving traceability and sustainability in the agri-food industry through blockchain technology: A bibliometric approach, benefits and challenges. *Energy Nexus*, *17*, 100388. https://doi.org/10.1016/J.NEXUS.2025.100388

Bag, S., Gupta, S., & Kumar, S. (2021). Industry 4.0 adoption and 10R advance manufacturing capabilities for sustainable development. *International Journal of Production Economics*, *231*, 107844. https://doi.org/10.1016/J.IJPE.2020.107844

Banbury, C., Stinerock, R., & Subrahmanyan, S. (2012). Sustainable consumption: Introspecting across multiple lived cultures. *Journal of Business Research*, *65*(4), 497–503. https://doi.org/10.1016/J.JBUSRES.2011.02.028

Bocchiola, D., Brunetti, L., Soncini, A., Polinelli, F., & Gianinetto, M. (2019). Impact of climate change on agricultural productivity and food security in the Himalayas: A case study in Nepal. *Agricultural Systems*, *171*, 113–125. https://doi.org/10.1016/J.AGSY.2019.01.008

Carrillo, F. J. (2015). Knowledge-based development as a new economic culture. *Journal of Open Innovation: Technology, Market, and Complexity*, *1*(2), 1–17. https://doi.org/10.1186/S40852-015-0017-5

Cho, M. A., Ramoelo, A., & Sibanda, L. M. (2023). Exploring the integration of the land, water, and energy nexus in sustainable food systems research through a socio-economic lens: A systematic literature review. *Sustainability*, *15*(23), 16528. https://doi.org/10.3390/SU152316528

Clean Energy Technology Adoption, Cultural Differences and Basic Human Dispositions. (2021). https://doi.org/10.32657/10356/151797

Cronin, J., Anandarajah, G., & Dessens, O. (2018). Climate change impacts on the energy system: A review of trends and gaps. *Climatic Change*, *151*(2), 79–93. https://doi.org/10.1007/S10584-018-2265-4

David, L. O., Aigbavboa, C., Nwulu, N., & Adepoju, O. O. (2024a). Application of digital technologies of the fourth industrial revolution (4IR) in water, energy, and food (WEF) nexus. *Green Energy and Technology*, Part F3772, 287–317. https://doi.org/10.1007/978-3-031-72833-4_9

David, L. O., Aigbavboa, C., Nwulu, N., & Adepoju, O. O. (2024b). General description of water, energy, and food resources. *Green Energy and Technology*, Part F3772, 21–61. https://doi.org/10.1007/978-3-031-72833-4_2

David, L. O., Nwulu, N. I., Aigbavboa, C. O., & Adepoju, O. O. (2022). Integrating fourth industrial revolution (4IR) technologies into the water, energy & food nexus for sustainable security: A bibliometric analysis. *Journal of Cleaner Production*, *363*, 132522. https://doi.org/10.1016/J.JCLEPRO.2022.132522

D'Cruz, P., Du, S., Noronha, E., Parboteeah, K. P., Trittin-Ulbrich, H., & Whelan, G. (2022). Technology, megatrends and work: Thoughts on the future of business ethics. *Journal of Business Ethics*, *180*(3), 879–902. https://doi.org/10.1007/S10551-022-05240-9

Edenhofer, O., Madruga, R. P., Sokona, Y., Seyboth, K., Matschoss, P., Kadner, S., Zwickel, T., Eickemeier, P., Hansen, G., Schlömer, S., & von Stechow, C. (2011). Renewable energy sources and climate change mitigation: Special report of the intergovernmental panel on climate change. *Renewable Energy Sources and Climate Change Mitigation: Special Report of the Intergovernmental Panel on Climate Change*, 1–1075. https://doi.org/10.1017/CBO9781139151153

Esham, M., Jacobs, B., Rosairo, H. S. R., & Siddighi, B. B. (2018). Climate change and food security: A Sri Lankan perspective. *Environment, Development and Sustainability*, *20*(3), 1017–1036. https://doi.org/10.1007/S10668-017-9945-5

Evans, M. I., & Britt, D. W. (2023). Resistance to change. *Reproductive Sciences*, *30*(3), 835–853. https://doi.org/10.1007/S43032-022-01015-9

Field, C. B., Barros, V. R., Dokken, D. J., Mach, K. J., Mastrandrea, M. D., Bilir, T. E., Chatterjee, M., Ebi, K. L., Estrada, Y. O., Genova, R. C., Girma, B., Kissel, E. S., Levy, A. N., MacCracken, S., Mastrandrea, P. R., & White, L. L. (2014). Climate change 2014 – Impacts, adaptation and vulnerability: Part A: Global and sectoral aspects: Working group II contribution to the IPCC fifth assessment report. In *Climate Change 2014 Impacts, Adaptation and Vulnerability: Part A: Global and Sectoral Aspects* (pp. 1–1131). https://doi.org/10.1017/CBO9781107415379

Fuentes-Peñailillo, F., Gutter, K., Vega, R., & Silva, G. C. (2024). Transformative technologies in digital agriculture: Leveraging internet of things, remote sensing, and artificial intelligence for smart crop management. *Journal of Sensor and Actuator Networks*, *13*(4), 39. https://doi.org/10.3390/JSAN13040039

Giri, N. C., & Mohanty, R. C. (2022). Design of agrivoltaic system to optimize land use for clean energy-food production: A socio-economic and environmental assessment. *Clean Technologies and Environmental Policy*, *24*(8), 2595–2606. https://doi.org/10.1007/S10098-022-02337-7

Glória, A., Sebastião, P., Dionísio, C., Simões, G., & Cardoso, J. (2020). Water management for sustainable irrigation systems using internet-of-things. *Sensors*, *20*(5), 1402. https://doi.org/10.3390/S20051402

Goggins, G., Rau, H., Moran, P., Fahy, F., & Goggins, J. (2022). The role of culture in advancing sustainable energy policy and practice. *Energy Policy*, *167*, 113055. https://doi.org/10.1016/J.ENPOL.2022.113055

Gopikrishnaa, S. M. (2016). Advanced irrigation systems impacting sustainability and reducing water needs—role of ICT in irrigation. *Advances in Intelligent Systems and Computing*, *409*, 415–426. https://doi.org/10.1007/978-981-10-0135-2_41

Govindarajan, H. K., & Ganesh, L. S. (2022). Integrating energy governance and environmental justice: Role of renewable energy. *Renewable Energy Focus*, *43*, 24–36. https://doi.org/10.1016/J.REF.2022.08.006

Gui, E. M., & MacGill, I. (2018). Typology of future clean energy communities: An exploratory structure, opportunities, and challenges. *Energy Research & Social Science*, *35*, 94–107. https://doi.org/10.1016/J.ERSS.2017.10.019

Gupta, A. D., Pandey, P., Feijóo, A., Yaseen, Z. M., & Bokde, N. D. (2020). Smart water technology for efficient water resource management: A review. *Energies*, *13*(23), 6268. https://doi.org/10.3390/EN13236268

Henchion, M. M., De Backer, C. J. S., Hudders, L., & O'Reilly, S. (2022). Ethical and sustainable aspects of meat production; consumer perceptions and system credibility. In *New Aspects of Meat Quality: From Genes to Ethics*, Second Edition, (pp. 829–851). https://doi.org/10.1016/B978-0-323-85879-3.00001-5

Hoosain, M. S., Paul, B. S., Doorsamy, W., & Ramakrishna, S. (2023). The influence of circular economy and 4IR technologies on the climate–water–energy–food nexus and the SDGs. *Water*, 15(4), 787. https://doi.org/10.3390/W15040787

How AI Can Optimize Water Usage in Agriculture. (n.d.). Retrieved February 8, 2025, from https://keymakr.com/blog/precision-irrigation-how-ai-can-optimize-water-usage-in-agriculture/

Javaid, M., Haleem, A., Singh, R. P., & Suman, R. (2022). Enabling flexible manufacturing system (FMS) through the applications of industry 4.0 technologies. *Internet of Things and Cyber-Physical Systems*, *2*, 49–62. https://doi.org/10.1016/J.IOTCPS.2022.05.005

Jørgensen, B. N., Gunasekaran, S. S., & Ma, Z. G. (2025). Impact of EU Laws on AI Adoption in Smart Grids: A Review of Regulatory Barriers, Technological Challenges, and Stakeholder Benefits. *Energies*, 18(12), 3002. https://doi.org/10.3390/en18123002

Kara, T., & Şahin, A. D. (2023). Implications of climate change on wind energy potential. *Sustainability*, *15*(20), 14822. https://www.mdpi.com/2071-1050/15/20/14822

Katel, S., Mandal, H. R., Subedi, D., Koirala, S., Timsina, S., Poudel, A., & Koirala, G. P. (2022). Climate smart agriculture for food security, adaptation, and migration: A review. *Turkish Journal of Agriculture – Food Science and Technology*, *10*(8), 1558–1564. https://doi.org/10.24925/TURJAF.V10I8.1558-1564.5162

Kumar, A., Mangla, S. K., & Kumar, P. (2022). Barriers for adoption of Industry 4.0 in sustainable food supply chain: A circular economy perspective. *International Journal of Productivity and Performance Management*, *73*(2), 385–411. https://doi.org/10.1108/IJPPM-12-2020-0695

Lee, S. G., Trimi, S., & Kim, C. (2013). The impact of cultural differences on technology adoption. *Journal of World Business*, *48*(1), 20–29. https://doi.org/10.1016/J.JWB.2012.06.003

Liu, J., Hull, V., Godfray, H. C. J., Tilman, D., Gleick, P., Hoff, H., Pahl-Wostl, C., Xu, Z., Chung, M. G., Sun, J., & Li, S. (2018). Nexus approaches to global sustainable development. *Nature Sustainability*, *1*(9), 466–476. https://doi.org/10.1038/s41893-018-0135-8

Liu, Q., Yang, L., & Yang, M. (2021). Digitalisation for water sustainability: Barriers to implementing circular economy in smart water management. *Sustainability*, *13*(21), 11868. https://doi.org/10.3390/SU132111868

Marcon, É., Soliman, M., Gerstlberger, W., & Frank, A. G. (2022). Sociotechnical factors and industry 4.0: An integrative perspective for the adoption of smart manufacturing technologies. *Journal of Manufacturing Technology Management*, *33*(2), 259–286. https://doi.org/10.1108/JMTM-01-2021-0017

Lucht, J. M. (2015). Public acceptance of plant biotechnology and GM crops. *Viruses*, *7*(8), 4254–4281. https://doi.org/10.3390/v7082819.

Nwulu, N., & Damisa, U. (2023). Energy 4.0: Concepts and applications. *Energy* 4.0. https://doi.org/10.1063/9780735425163

Ogbolumani, O. A., & Adekoya M. (2025). Intelligent waste management optimization through machine learning analytics. *Journal of Science Research and Reviews*, *2*(1), 7–26. https://doi.org/10.70882/JOSRAR.2025.V2I1.25

Ogbolumani, O. A., & Mabaso, B. (2023). An IoT-based hydroponic monitoring and control system for sustainable food production. *Journal of Digital Food, Energy & Water Systems*, *4*(2), 106–140.

Ogbolumani, O. A., & Nwulu, N. (2020). Integrated appliance scheduling and optimal sizing of an autonomous hybrid renewable energy system for agricultural food production. In *Lecture Notes in Mechanical Engineering* (pp. 651–660). https://doi.org/10.1007/978-981-15-5753-8_60

Ogbolumani, O. A., & Nwulu, N. I. (2021). Multi-objective optimisation of constrained food-energy-water-nexus systems for sustainable resource allocation. *Sustainable Energy Technologies and Assessments*, *44*, 100967. https://doi.org/10.1016/j.seta.2020.100967

Ogbolumani, O. A., & Nwulu, N. I. (2022). A food-energy-water nexus meta-model for food and energy security. *Sustainable Production and Consumption*, *30*, 438–453. https://doi.org/10.1016/J.SPC.2021.12.019

Ogbolumani, O. A., & Nwulu, N. I. (2024). Environmental impact assessment for a meta-model-based food-energy-water-nexus system. *Energy Reports*, *11*, 218–232. https://doi.org/10.1016/j.egyr.2023.11.033

Olorunfemi, B. O., Nwulu, N. I., & Ogbolumani, O. A. (2023). Solar panel surface dirt detection and removal based on arduino color recognition. *MethodsX*, *10*, 101967. https://doi.org/10.1016/J.MEX.2022.101967

Olorunfemi, B. O., Ogbolumani, O. A., & Nwulu, N. (2022). Solar panels dirt monitoring and cleaning for performance improvement: A systematic review on smart systems. *Sustainability*, *14*(17), 10920. https://doi.org/10.3390/SU141710920

Pardoe, J., Conway, D., Namaganda, E., Vincent, K., Dougill, A. J., & Kashaigili, J. J. (2018). Climate change and the water–energy–food nexus: Insights from policy and practice in Tanzania. *Climate Policy*, *18*(7), 863–877. https://doi.org/10.1080/14693062.2017.1386082

Park, E., & Widyanta, A. (2022). Food tourism experience and changing destination foodscape: An exploratory study of an emerging food destination. *Tourism Management Perspectives*, *42*, 100964. https://doi.org/10.1016/J.TMP.2022.100964

Raj, A., Dwivedi, G., Sharma, A., Lopes de Sousa Jabbour, A. B., & Rajak, S. (2020). Barriers to the adoption of industry 4.0 technologies in the manufacturing sector: An inter-country comparative perspective. *International Journal of Production Economics*, *224*, 107546. https://doi.org/10.1016/J.IJPE.2019.107546

Ramos, H. M., McNabola, A., López-Jiménez, P. A., & Pérez-Sánchez, M. (2019). Smart water management towards future water sustainable networks. *Water*, *12*(1), 58. https://doi.org/10.3390/W12010058

Rasa, T., & Laherto, A. (2022). Young people's technological images of the future: Implications for science and technology education. *European Journal of Futures Research*, *10*(1), 1–15. https://doi.org/10.1186/S40309-022-00190-X

Senturk, S., Senturk, F., & Karaca, H. (2023). Industry 4.0 technologies in agri-food sector and their integration in the global value chain: A review. *Journal of Cleaner Production*, *408*, 137096. https://doi.org/10.1016/J.JCLEPRO.2023.137096

Singh, R., Bhattarai, N., Prajapati, A., & Shakya, S. R. (2022). Impact of variation in climatic parameters on hydropower generation: A case of hydropower project in Nepal. *Heliyon*, *8*(12). https://www.sciencedirect.com/science/article/pii/S2405844022035289

Stapleton, P. A. (2016). From Mad Cows to GMOs: The side effects of modernization. *European Journal of Risk Regulation*, 7(3), 517–531. https://doi.org/10.1017/S1867299X0000605X

Stephenson, J. (2018). Sustainability cultures and energy research: An actor-centred interpretation of cultural theory. *Energy Research and Social Science*, *44*, 242–249. https://doi.org/10.1016/j.erss.2018.05.034

Taseska, V., Markovska, N., & Callaway, J. M. (2012). Evaluation of climate change impacts on energy demand. *Energy*, *48*(1), 88–95. https://doi.org/10.1016/J.ENERGY.2012.06.053

Theis, S., Lefore, N., Meinzen-Dick, R., & Bryan, E. (2018). What happens after technology adoption? Gendered aspects of small-scale irrigation technologies in Ethiopia, Ghana, and Tanzania. *Agriculture and Human Values*, *35*(3), 671–684. https://doi.org/10.1007/S10460-018-9862-8

Understanding the Climate-Water-Energy-Food Nexus and Streamlining Water-Related Policies – European Commission. (n.d.). Retrieved February 8, 2025, from https://research-and-innovation.ec.europa.eu/news/all-research-and-innovation-news/understanding-climate-water-energy-food-nexus-and-streamlining-water-related-policies-2021-03-19_en

Zarei, S., Bozorg-Haddad, O., Kheirinejad, S., & Loáiciga, H. A. (2021). Environmental sustainability: A review of the water–energy–food nexus. *Journal of Water Supply: Research and Technology-Aqua*, *70*(2), 138–154. https://doi.org/10.2166/AQUA.2020.058

Zarreh, M., Yaghoubi, S., & Bahrami, H. (2024). Pricing of drinking water under dynamic supply and demand based on government role: A game-theoretic approach. *Water Resources Management*, *38*(6), 2101–2133. https://doi.org/10.1007/S11269-024-03745-8

11 Case Studies and Best Practices in FEW Nexus 4.0

11.1 CASE STUDIES AND BEST PRACTICES FROM AROUND THE WORLD

The Food Energy Water (FEW) nexus concept focuses on the interlinkage and interdependence of the three vital resources of life and urges integrated approaches towards their management. Case studies and good practice examples in FEW nexus implementation can show strategies, policies, and innovations that optimise resource use, enhance sustainability, and reduce conflicts. These studies often examine holistic and collaborative frameworks for how different regions and sectors face complex challenges related to resource scarcity, the impacts of climate change, and socio-economic dynamics. Stakeholders can learn relevant lessons from examples of success regarding effective governance structures, strategies for technological innovation, and community engagement that have complemented the fostering of resilience and sustainable development in the FEW nexus (Apeh & Nwulu, 2024; David et al., 2024; Ogbolumani & Nwulu, 2020, 2021, 2022, 2024).

As the understanding of the nexus improves, it provides opportunities for societies to move away from the constraints of sectoral silos and better appreciate the connections between natural resource uses, This, in turn, enables the negotiations of compromises and potential synergies regarding competing concerns in the FEW nexus.

11.1.1 Best Practice Nexus Cases

The International Union for Conservation of Nature (2020) listed several examples of best practices case studies in the FEW nexus (*The Power of Best Practice Nexus Cases | IUCN*, n.d.). These examples are:

1. Cross-Sectoral Investment Analysis in the Zambezi Basin
2. Infrastructure Investment Trade-offs in the Volta River Basin
3. Irrigation Infrastructure Use across Sectors in Asia
4. Applying a Nexus Approach in the Senegal River Basin
5. Agribusiness as Natural Infrastructure in Southern Africa
6. Multi-stakeholder collaboration in the Rhine River Basin
7. Multi-benefit analysis of water hyacinth control in Indonesia
8. Improving inefficient use of irrigation in Indonesia

DOI: 10.1201/9781003633853-11

The following section discusses case studies of FEW Nexus 4.0 from different world regions, including Africa, Asia, Europe, North America, South America, Australia, and the Middle East.

11.2 CASE STUDIES FROM AFRICA

Case Study 1: Zambezi River Basin

Description: The Zambezi River Basin spans several countries in Southern Africa, namely Zambia, Zimbabwe, Mozambique, and Angola. It is a vital water resource supporting agriculture, hydropower generation, and ecosystem-related services.

Multi-sector integration, as in the case of the Zambezi River Basin, applies to infrastructure developments relating to hydropower, such as the Kariba Dam, that can serve energy generation and irrigation purposes simultaneously. A key precept of this approach is optimising water use while matching energy demand with agricultural needs.

Transboundary Cooperation: For the proper management of the water resources in the basin, the riparian states must consider transboundary cooperation. Countries within the basin have sought to cooperate through organisations such as the Zambezi River Authority in negotiating water allocations across borders and sharing benefits accruing from the resource.

Stakeholder Participation: Different decision-making levels involve governments at various levels, local communities, and NGOs to come up with a resource management framework that assures inclusiveness and equity. Open forums for public consultations and stakeholder forums are utilised to facilitate dialogue and reach consensus on decisions related to the nexus (*The Power of Best Practice Nexus Cases | IUCN*, n.d.).

Case Study 2: The Mekong River Basin

The Mekong River Basin covers six known countries in the Southeast Asia region, generally referred to as the Greater Mekong Sub-region: China, Myanmar, Thailand, Laos, Cambodia, and Vietnam. It is a critical resource for agriculture, fisheries, hydropower, and navigation. The construction of hydropower dams like the Nam Theun 2 and Xayaburi Dams in the Mekong Basin serves the multi-variant purposes of electricity generation, flood regulation, and irrigation. This is a balancing act between energy and the environment, energy needs, environment, and social concerns. The Mekong River Commission allows its member countries to integrate their policies on sustainable development and resource management. The strategic framework of the MRC is harnessed to strike a balance between economic growth and environmental protection, coupled with social equity. The scenario planning in the Mekong Basin considers estimations of future scenarios on dam construction, impacts of climate change, and population growth. This thus sets out the risks and opportunities that come with them to guide the adaptive management strategies aimed at building resilience through structured handling of uncertainty. The two case studies above have shown good practice in managing the FEW nexus. From successful implementation, it can be observed that sizeable transboundary river basins require integration, cooperation, stakeholder participation, policy alignment, and scenario planning to achieve sustainable development (*The Power of Best Practice Nexus Cases | IUCN*, n.d.).

The best practice cases of the nexus (Zambezi River Basin and Mekong River Basin) demonstrate how multi-purpose infrastructure can combine social and environmental benefits across sectors with one investment. That financial burdens on single ministries are reduced by sharing costs and risks, while enhanced economic returns result from multiple benefits and better management. These cases offer valuable learning to global stakeholders and spur actions and solutions toward complex issues in the governance of transboundary resources. They pinpoint a way through which the nexus can reconcile interests across different water, land, energy, food, and environmental sectors, hence underlining inter-sectoral competition, particularly within transboundary contexts. As workable building blocks globally, every nexus intervention must be context-sensitive and underpinned by dedicated capacity-building and awareness efforts. In the case of Central Asia, these cases are developed using a Stakeholder-Problem typology, which addresses region-specific challenges and invites knowledge adaptation for integrated investments and sectoral synergies to be planned. Moreover, future scenarios and institutional analysis also feed into proposals for possible nexus interventions at national and regional levels to develop tailored strategies for sustainable development (*The Power of Best Practice Nexus Cases | IUCN*, n.d.).

Case Study 3: Morocco and South Africa

The study described the Water-Energy-Food (WEF) Nexus concept in Africa, using Morocco and South Africa as the main examples. The population boom experienced throughout Africa is the primary cause of higher food, water, and energy demand, hence a greater need to conserve natural resources. Climate change and population explosion influence both reasons for this surge in demand. The WEF Nexus model is set to play the role of a significant tool for the sustainability of resources and policy coordination in the sectors, thus transforming the other interventions into success. The two African nations, Morocco and South Africa, are both grappling with the issues of water, energy, and food sustainability. As the main component, Moroccan agriculture is dominant in water use, accounting for 85% of the total, irrigating its crop; therefore, the government has an edge on food security; however, the rest is exported to very few importers, and the energy dependency weighs on the country's price. In South Africa, water scarcity, the ageing energy infrastructure, and the ageing energy infrastructure are limiting factors that have seen agricultural output in the area being restricted. In the face of these obstacles, the countries in question have been found to have no unified policies to deal with issues like the WEF Nexus. Thus, the situation appears chaotic in the FEW Nexus. Hence, the situation appears chaotic, and the resources become overexploited in the process. Climate change undoubtedly presents a big problem to the FEW nexus in the two nations through water availability, crop yield, and power production. In Morocco, periods of increasing drought and unpredictable rainfall affect rain-fed agriculture, which is the main region's farming system. South Africa is in the same boat, where severe weather conditions aggravate water scarcity and worsen energy reliability. The study indicates the need to develop new approaches that are aimed at climate-proofing the key resources and lessening the impacts of climate change on the environment. The work highlighted the need to deploy new technology in tackling WEF Nexus challenges. Solutions such as precision agriculture, the integration of renewable energy, and the application of water-saving technology are fundamental to sustain resource use. The use of remote sensing and Internet of Things (IoT) is another promising method in managing and monitoring the dynamics of nexus resources. Morocco's initiative in introducing renewable sources of energy, mainly solar and wind power, not to mention plans for building desalination plants, are some of the ways in which technology may successfully counter resource shortages. The analysis brings up solutions with the aim of designing integrated WEF Nexus frameworks that allow policy coupling and sector coordination. It is also about raising greater funds towards renewable energy, sustainable agriculture, and water infrastructure development. The idea of promoting mutual cooperation locally after bringing in technology-assisted business models cannot be overemphasised; in this regard, resources will be optimally allocated and the health of businesses will also be taken care of in the long run. For a start, both Morocco and South Africa should strive to allocate finance into these areas to meet the demand they will face while at the same time ensuring that they are sustainable and lead to economic development (El Youssfi et al., 2020).

Case Study 4: Infrastructure Investment Trade-offs in the Volta River Basin of Ghana

Making water infrastructure investment decisions in a changing climate. The research investigates political and economic aspects of river basin development throughout Ghana but focuses on a single analysis of the Kpong Multi-Purpose Storage project.

The government of Ghana has opted for constructed water infrastructure like dams and levees to lead their investment in hydropower generation and irrigation projects. Although climate change raise many concerns about environmental risks, natural infrastructure methods such as floodplains and wetlands see limited use. The government demonstrates leadership toward climate adaptation primarily through development partner influence, though its decentralised and fragmented implementation structure persists.

The WISE-UP Project represents a programme that uses natural infrastructure as a sustainable method to adapt to climate change. Through an analysis of combined natural and man-made water infrastructure methods, the project explores how different benefits like food security, biodiversity, conservation, and climate resilience can be optimised. This study inspects Ghanaian water administration dynamics while examining political engagement and economic aspects that guide decision-making through stakeholder engagement analysis.

The report includes a significant case study of the Pwalugu Multipurpose Dam project, which aims to enhance economic development throughout Ghana's northern region. Since the project requires irrigation support alongside hydropower generation and flood protection management, land-use decisions, community resettlement plans, and limited local participatory decision-making processes remain critical concerns. The examination reveals competing demands between nation-minded objectives, business needs, and community requirements.

To Ghana's water sector decision-making process experiences disparities in authority between national and local agencies, with national agencies maintaining greater authority due to political and economic interest influence. Weak coordination between policymakers, insufficient funding, and political intervention hinder the successful implementation of existing climate change and environmental sustainability policies. Public institutions prefer energy development and irrigation projects to environmental protection measures, although donor-funded climate policies guide their decisions.

Three key guidelines emerged from the report for improving sustainable water management through (1) developing extended development plans combining natural and built systems, (2) developing stronger international collaborations for water management across borders, and (3) enhancing water management knowledge among the public and institutional capacities. A strong partnership between the government and civil society organisations, together with international partners, must be created to guarantee climate-resilient and sustainable development within Ghanas water management sector (Mosello et al., 2017).

Case Study 5: Wadi-Dara on the Red Sea Coast, Egypt

A conceptual design for the WEF Nexus in a selected location on the Red Sea coast, Egypt, would be a suitable venture for Wadi-Dara. The issue of water and food scarcity is becoming more of a problem not only because of the climatic challenges in Africa. Egypt needs innovative solutions to address its increasing water and food shortages caused by its growing population and industrial capacity. The investigation has shown one of the models for sustainable development of cities through the application of the newest and most environmentally friendly technologies, such as solar energy generation, automation in agriculture, and desalination as a means of wastewater treatment that could make the community self-reliant (Abdelzaher et al., 2023).

A primary part of the proposed system is the desalination of seawater, which combines the use of beach sand filtration wells as the pre-treatment step and the utilisation of renewable energy sources such as wind and photovoltaic solar power.

This technique saves energy and protects the environment during freshwater supply. Among other confounding issues, the underground water reservoir in Wadi-Dara is the main water source, and it has the potential to support 1,000 acres of smart farming, with IoT and artificial intelligence (AI) being the technologies used for efficient resource management.

In this study, algae biomass cultivation has been discussed as a sustainable way to produce food and energy. Open pond systems and photobioreactors are microalgaes resources that will be cultivated for animal feed, biofuel, and wastewater treatment. Furthermore, apart from municipal wastewater treatment, which leads to the production of irrigation water, public lands and forests are integrated into the design. The system promotes the culture of mixotrophy, which improves ecological balance.

The project's environmental impact assessment describes how sustainable green energy, recycling of waste, and zero carbon emissions can be relied on; thus, the risk level is low. The study concluded that the improvement of the WEF Nexus in the Al-Darb Valley could become a blueprint for sustainable urban expansion in arid areas. It suggests more research and support policy for the IRM, and its long-term viability be secured and closely monitored (Abdelzaher et al., 2023).

Case Study 6: Agrivoltaic Systems Across the Water-Energy-Food Nexus in West Africa

In this work, the possible development of agrivoltaic systems (AVS) in West Africa, which can solve problems of the WEF-Environment Nexus, is examined. Today's water, energy, and food demands often compete with one another, but AVS may be a rescue technology that envelops solar energy generation with agriculture, which is the greenest way to produce food, increasing land use and making sustainability stronger. Nevertheless, research on AVS in the region remains limited.

The study states that many West African countries, including Benin, Guinea-Bissau, Sierra Leone, The Gambia, and Togo, are facing land constraints, yet they aim to use ambitious solar energy targets. To that end, the AVS technique produces the twin benefits of generating energy and providing the necessary shade for crops, improving water retention and reducing irrigation. Furthermore, it promotes solar pumps and small solar power stations for the sake of water supplies while also benefiting the agrarian sector and the economy.

However, despite all of these positives, AVS adoption is hindered by challenges like sky-high initial investment costs, technical gaps, and the need to understand relevant policies.

The survey's reviewers outline that AVS is a viable solution for energy production in the years to come, tightening the energy-food security nexus without entering land use. It encourages research, pilot projects, and public policy formulation that could effectively integrate AVS into regional development strategies. The work contributes to the discourse on sustainable development, as one of the challenges it sets forth is the management of resources in West Africa (Favi et al., 2024).

Case Study 7: The Agribusiness Ecosystem in East and Southern Africa

The contents of updated materials about this ecosystem have been examined in change distributions over the whole region of East and South Africa to see how it can become a key exponent of the transformation of the agricultural sector with existing stakeholders. Meanwhile, the agriculture sector's share of GDP and the job market is way too big to be ignored, while the existing weaknesses are due to the frequent natural disasters, disjointed market fragmentation, and the unavailability/underperformance of the main infrastructure. Continually, the study looks at agribusinesses, investors, agriculture corporates, and entrepreneur support organisations that are key drivers for sustainable food systems.

One of the key points addressed is climate-saving agriculture (CSA), which, in digital ways, develops processes, raises farmers' incomes, and makes food, water, and land systems stronger. In agritechs, the digital platforms, precision farming, and the protein production alternatives account for how rethinking agriculture value chains can be made efficient. Nonetheless, one of the difficulties these innovations face is their delivery to the end users due to high cost, with some infrastructure gaps, and the rigid or forbidding regulations. Unavailable still to holders of noncommercial lending are profitable divisions such as agribusinesses exchanging less than 3% of commercial loans.

The key message of the study is that the only way to overcome the $106 billion financial gap between SMEs in Africa and the rest of the world is through alternative financing models like borrowing from venture capital, funding impact investments, and combining these with other inclusive financial methods. Agripreneurs are often unable to start the investment process efficiently, thus requiring help in areas such as technical assistance and marketing strategies to grow their businesses.

Public officials are in the driver's seat to steer growth in the sector through better regulations, more cross-border trading, and fostering an environment that is more inclusive of both young and older women. Making supply chains robust, allowing other women to access to CSA technologies and having trade regulations across the region harmonised are the steps the report points out for long-term sustainability in agriculture.

The study clarifies that agribusinesses in East and Southern African countries offer a real opportunity to bring about economic growth and food security. Still, this potential is not likely to be realised unless it is accompanied by a set of actions on the part of all the stakeholders who align policy reforms, provision of financial support, adoption of technology, and capacity-building endeavours to establish a sector that is resilient and competitive (*THE AGRIBUSINESS ECOSYSTEM IN EAST AND SOUTHERN AFRICA Exploring the Role and Synergies of Key Stakeholders in the Space*, 2022).

Case Study 8: Simultaneous Intercropping in Sri Lanka and Malawi

Integrated food–energy systems enable bioenergy production that fulfils dual requirements for food and energy needs. The technique incorporates environmentally friendly methods to simultaneously produce food along with other types of biomass across different ecological, spatial and temporal scales through multiple-cropping systems or intermittent planting of annual crops with perennial plants, which falls under the classification of agro-forestry systems.

Agroforestry delivers farmers a set of tools that boost both production levels and soil quality and earnings while strengthening food security. Modern intercropping combines nitrogen-fixing woody trees and annual crop production on a single plot of land. *Gliricidia sepium* represents a quick-growing nitrogen-fixing leguminous tree that farmers use for both nitrogen supplementation and organic matter supply to their soil.

The addition of organic matter through this method brings significant crop yield increases to the land. In Malawi, intercropping maize, which involves the combination of maize and *Gliricidia*, resulted in greater than 5 tonnes per hectare harvest in favourable seasons yet typically delivered 3.7 tonnes per hectare. *Gliricidia sepium* trees reveal an overall yield capacity of 3.7 tonnes per hectare for annual cultivation areas without mineral fertiliser input when compared to average yields of 0.5–1.0 tonnes per hectare. 0.5–1.0 tonnes per hectare without *Gliricidia* or mineral fertiliser. Similarly, in Sri Lanka, *Gliricidia* is intercropped with coconut.

A food–energy system enabling rural electrification provides commercial service using *Gliricidia* plants as base material. The feedstock is used for electricity production. The installation site of this 1–10 MW power generation plant operates in a designated rural zone, which comprises *Gliricidia* trees and provides a sustainable agricultural solution for the smallholder maize farmers who maintain predominantly food crop cultivation.

A regular pattern of *Gliricidia* trees runs across maize fields during cropping. The *Gliricidia* trees grow wood material alongside plant foliage. A yearly crop harvest system is practised for this periodic gathering operation. The foliage is stripped from the branches for high-protein livestock fodder and/or fertiliser for the crops. The company moved cut branches from the site to serve as fuel at the power plant for electrical power generation. The power plant provides electricity to nearby rural consumers with additional production directed to the national power grid.

The *Gliricidia* systems help African farmers increase their domestic firewood supply, which shows diminishing availability throughout Africa's small agricultural regions. The sustainable system relieves women and children from the exhausting task of collecting firewood, which creates time for other income-producing activities.

This practice protects the state of natural resources from destruction. The need for firewood from public parks decreases because farmers produce their wood on their properties. The envisioned integrated food–energy sector would supply electrical power through its operation. The model combines sustainable food production with increased nutritional availability while increasing living standards and forest preservation for economic advancement (*Renewable Energy in the Water Energy: Food Nexus*, n.d.).

Case Study 9: Water-Energy-Food Nexus for Sustainable Agriculture and Food Security in Burundi

Challenges to Agriculture and Food Security in Burundi. Burundi's economic structure depends upon agriculture because it provides jobs to 84% of people and 40% of the GDP. Various obstacles like soil deterioration and climate-related effects, coupled with insufficient infrastructure and improper water management, reduce the availability of food in the region. The country holds one of the worst positions regarding global food security because more than 40% of Burundian people suffer persistent food shortages. National and regional projects for agricultural productivity enhancement face slow advancement since they must overcome resource shortages and administrative difficulties in their implementation.

The WEF Nexus Approach. The research emphasises the necessity of adopting a connected WEF Nexus strategy to enhance agricultural sustainability in Burundi. This approach emphasises sector dependencies and develops methods to boost resource utilisation. The research team used econometric measures, particularly the Autoregressive Distributed Lag and the ARIMA models, to assess agricultural yield relationships with water resources, energy allocation, land utilisation, and to construct future yield predictions.

Key Findings and Future Projections. The studies demonstrate that agricultural yield strengthens when the sectors of water, energy, and land are used together positively. The combined resources show that a 1% enhancement leads to a 3.74% increase in agricultural output. Recent forecasts predict a future decline of agricultural productivity in 2030 that will result in a decreased yield of 74.9 kg per hectare alongside a reduction of 11.9×10^4 hectares of agricultural land. According to the study, resources need better management because they reveal negative patterns.

Policy Recommendations for Sustainable Agriculture. This research proposes three methods for sustainability improvement in agriculture: organisations should enhance irrigation functionality, while using renewable energy in farming operations, and embrace modern farming technology. Policy implementations must include methods to manage land resources since these play a crucial role in handling the agricultural soil effects of urbanisation and rising population numbers. Besides improving the management of regional resources in the Nile Basin and East African Community member states, policymakers must work toward a sustained food security framework.

Conclusion and Implications for Future Research. This research highlights why applying WEF Nexus principles remains essential for Burundi in dealing with its food security problems. Distinct evidence-based policy development, sustainability-focused agricultural investments, and better technological integration processes must be established. Research moving forward should compare sustainable agriculture models across different countries to create broad knowledge about similar implementation efforts across contexts (Mperejekumana et al., 2023).

11.3 CASE STUDIES FROM EUROPE

Case Study 1: The German Rhine Basin

Floods pose significant risks for the German Rhine Basin region and present various management complications in this area. The German territory has experienced major flood occurrences even after implementing numerous flood prevention measures.

Floods accompanied by economic losses and fatalities will probably increase because of climate change, economic development, and human settlements in at-risk areas.

Technical knowledge already exists, yet ineffective flood risk management persists because institutions lack strength and their responsibilities are scattered while public awareness remains minimal. During the last four decades, the dissertation has assessed structural changes in German Rhine Basin flood management and evaluated institutional developments for enhanced resilience.

Shifts in Flood Management Approaches. German flood management practices in the past exclusively relied on the implementation of dikes and levees as their primary defence mechanisms. Public authorities have evolved their flood management system toward a risk-based strategy, including ecological floodplain restoration and water retention methods. The Flood Control Act of 2005 and Federal Water Act of 2010, implemented through legislative changes, enhanced flood prevention by defining flood-prone zones. Federal states within Germany maintain separate authorities that prevent unified flood risk management across jurisdictions.

Transboundary Cooperation and Institutional Barriers. Through an analytical study, this research examines different aspects of flood management collaboration between Germany and the Netherlands. Under Dutch management, the central government oversees safety standards, whereas German states operate independently; therefore, safety standards and funding differ between states. The International Commission for the Protection of the Rhine represents international cooperation, but bilateral collaboration faces difficulties because of institutional frameworks and policy divergences between the two countries.

Adaptive Co-Management (ACM) and Stakeholder Participation. This research examines the implementation of ACM in Germany to study its decentralised governance system, along with public participation and adaptable flood risk management. Germany has progressed in using ACM principles but continues to experience problems with government-level coordination and insufficient stakeholder participation. Communities that demonstrate more acute risk awareness tend to establish stronger preparations against future events.

Policy Recommendations for Improved Flood Resilience. Improved flood resilience requires three main elements: better connections between federal institutions and local administrations, standardised flood safety regulations between states, and public education campaigns. An extended "Rhine 2100" strategic plan connects climate adaptation with both sustainable flood management and a financial planning framework. The Rhine Basin needs better land-use control, improved risk messaging, and stronger community engagement practices to build effective flood defences.

Case Study 2: Virtual Water and Embodied Energy in Food Consumption in the Tamar Catchment, UK

Research evaluates the relationships between FEW systems within the Tamar catchment region of Southwest England. The research analyses virtual water and embodied energy through a life cycle analysis approach in this region. The research demonstrates that food imports, together with local production, create resource use, which reveals previously unknown environmental implications of food supply chains.

Resource Consumption in Local and Imported Food. Research shows that the Tamar catchment heavily depends on imported foods, which make up 51% of its total embodied energy usage and consume 88% of blue virtual water resources and 45% of all green virtual water. Animal meat products and milk, along with chicken, consume the greatest amount of resources because animal feed production alone requires 58% of all energy used and water consumed. The use of virtual blue water from rice imports becomes substantial because these crops require irrigation.

The Environmental Impact of Food Imports. Nearly all imported food originates from the European Union, where Ireland, together with the Netherlands and Spain, holds primary supplier positions. The virtual supply of blue water primarily originates in water-stressed areas, which include Pakistan and India. The Tamar catchment becomes indirectly susceptible to water shortages and trade disturbances because it depends on foreign food imports. Food production primarily utilises irrigation to the extent of 92% for blue water distribution, as water-intensive crops require substantial amounts of energy.

Policy Implications for Sustainable Food Systems. Better food, water, and energy policies must be implemented to reduce imports and enhance sustainability, according to this research. The strategies for sustainable development consist of boosting local food systems while applying water-saving farming practices and promoting diet modifications towards sustainable meal choices. The study demonstrates that government entities should carefully expand food import relationships while taking into account the environmental conditions of exporting nations to reduce potential resource hazards.

Life Cycle Assessment methods have great potential to advance FEW Nexus research. Life Cycle Assessment proves itself useful for assessing FEW connections as well as improving policy decision-making processes. Although this analysis recognises its boundaries due to missing data about UK domestic commerce and varying effects from transporting food products. Future research must combine LCA models with spatial specificity to enhance resource tracking capabilities and better decision-making processes. The research demonstrates that food production needs full FEW nexus knowledge to achieve enduring sustainability in food systems (Salmoral & Yan, 2018).

Case Study 3: The Water-Food Nexus in Italy: A Virtual Water Perspective

The Water-Food Nexus and Virtual Water in Italy. The agricultural sector of Italy depends to a great extent on both the water availability from its territory and imported water supplies. Researchers have investigated Italian farming water utilisation by using the water footprint and virtual water trade theory. Virtual water stands for all waters that enter the production of food and agricultural commodities, starting from water usage in imported goods. Italy has transformed into a virtually water-dependent food security nation by adopting increased foreign water consumption while reducing its local water usage over an extended period (Laio et al., 2020).

Italy's Virtual Water Balance and Consumption Trends. Research data from 1961 to 2013 show that Italy experienced substantial increases in virtual water imports, which outpaced its domestic water consumption for food production. The yearly virtual water consumption by Italy reached 71 km³ in 2013, while its domestic agricultural water needs stood at 66 km³. Global trade now plays a larger role in food supply because Italian agriculture continues to decline. The work demonstrates that food supply and water security in Italy remain vulnerable because of their significant dependence on global trade.

Italy's Role in the Global Virtual Water Trade. Italy imports virtual water as well as exports it, but operates as a net water importer because its dependence on foreign water supply is extensive. Italy obtains essential agricultural products such as pork and rice, together with coffee and palm oil, from four major import partners, namely France, Germany, Brazil, and Indonesia. The Italian virtual water trade functions by exporting processed food together with pasta and cheese to consumers across the United States and the United Kingdom and Mediterranean region countries.

Spatial Variability in Italy's Water Footprint. The variability of agricultural water use throughout Italy stems from different environmental conditions found among regions together with local variations in soil properties and farming crops. The regions of Piemonte and Puglia in Northern Italy extensively conduct irrigation activities, but central and southern Italy display a combination of both rain-fed cultivation and irrigation practices. Water usage within Tuscany and Lazio exhibits robust spatial variations because these regions have different farming practices. Both rice and maize cultivation remove substantial quantities of blue water through irrigation, which results in high blue water consumption within specific geographical areas.

Policy Implications and Future Challenges. The research concludes that sustainable water resource management should become an essential policy to cut foreign water dependencies and secure future food availability. A holistic approach to water management should focus on both enhancing irrigation systems and adopting water conservation techniques, and implementing virtual water assessments into trade strategies. Beyond climate and trade uncertainties, Italy should pursue combined approaches to protect agricultural sustainability while managing its foreign water supply utilisation (Laio et al., 2020).

11.4 CASE STUDIES FROM NORTH AMERICA

Case Study 1: The Water-Food-Energy Nexus Optimisation Approach to Combat Agricultural Drought: A Case Study in the United States

Climate change and population growth, together with rising economic requirements, place escalated stress on the U.S. food supply, water sources, and energy systems. Drought creates a lasting and extreme situation that impacts three vital sectors, namely agriculture, water resources, and infrastructure. Several types of drought affect the U.S. nation, including meteorological due to the absence of rain, agricultural from impacts on crops and livestock, and hydrological from reduced water levels in reservoirs and rivers, which generate widespread problems (Zhang et al., 2018).

Drought in the United States creates substantial economic destruction of billions of dollars through annual agricultural losses. Drought poses exceptional risk to three important agricultural states, including California, Texas, and the entire Midwest region. The 2012 drought, which stood as one of the worst droughts in modern times, forced the American economy to pay $30 billion in damages while driving up how much people had to pay for their food. Drought creates multiple economic struggles combined with social problems that force job loss and water inadequacy in rural areas.

The American agriculture sector suffers extensive damage when droughts reduce production numbers of basic crops such as wheat and soybeans, and corn. Food prices increase on multiple fronts when the United States experiences droughts since the nation exports substantial amounts of food products. The western United States faces acute water concerns from depleted aquifer reserves combined with diminished Rocky Mountain snowfalls that endanger extended supplies to agricultural and urban areas.

The U.S. government, alongside researchers, invests funds to develop three key elements for drought mitigation: drought-tolerant crops, together with precise farming practices and advanced irrigation equipment. The state of California leads by example through its enactment of rigorous water conservation measures that utilise water consumption limits along with ecological farming rewards. Farming communities, together with policymakers, can better prepare for drought through advanced weather prediction technologies and early warning systems.

The solution to U.S. drought challenges requires joint efforts between national departments at all levels of government together with active participation from the private sector. Policymakers from NOAA, the USDA, and state water boards collaborate to enhance drought resistance through infrastructure funding, awareness programmes, and policy reform. The rise in extreme weather occurrences demands that the United States establish sustained multi-agency coordination to defend its food security, water supplies, and economic structure (Zhang et al., 2018).

Case Study 2: A Machine Learning Approach for the Water-Energy-Food-Ecosystem Nexus: A Resilience Perspective for Sustainability in Mexico

Machine learning should be integrated as part of a systemic analysis within the Water-Energy-Food-Ecosystem (WEFE) Nexus. Machine learning technology serves as the basis for developing an optimisation model that enables studies of the WEFE Nexus while strengthening both environmental sustainability and resilience.

The method utilises regular technologies along with innovative methods to handle resources, starting from creation until distribution ends at utilisation. The model employs artificial neural networks from machine learning to forecast complex mathematical model outputs, thereby lowering computations while enhancing decision-making speed.

Case Study in Mexico and Model Evaluation. A case study in Mexico utilises the proposed model to check the status of the WEFE nexus for five specific years, beginning with 2020 and continuing through 2040. The predicted data demonstrates superior accuracy through a 0.99 coefficient of determination, along with an estimated mean square error value of 0.0007. Economic profit results from the 2020 optimisation delivery demonstrate superior resource security while striking an environmental impact equilibrium that surpasses previous models in sustainability trade-offs management.

Projected Future Trends and Challenges. Analysis indicates that environmental impact will expand substantially to 50% higher than in 2022 as resource needs escalate in the upcoming years until 2040. Strategic planning must be established to meet self-sufficiency needs because food imports will increase according to projections. The current energy production stability requires additional development of renewable energy technologies for sustainable longevity.

Policy and Planning Recommendations. The study shows how sustainable policies need integration between WEFE nexus management and machine learning insights. The findings show that enhancing financial backing for renewable power systems, together with water optimisation activities and better farming methodologies would minimise environmental harm.

Developments and Upcoming Studies. The study demonstrates how machine learning may be used to tackle difficult resource and environmental management problems. The model should be improved to take climate uncertainties into consideration, its use should be extended to other geographical areas, and more data sources should be included. Decision-making for global sustainable resource management will be further improved by enhancing predictive capacities (López-Flores et al., 2023).

Case Study 3: Why are Californian farmers adopting more (and larger) renewable energy operations?

Adoption of Renewable Energy in California's Agriculture. This research examines the factors that drive Californian farmers to implement renewable energy systems, especially solar and wind power technologies. The combination of increasing energy costs, together with government support and community interest in environmental protection functions as the main reason behind California's shift toward renewable energy systems. A double-hurdle model serves the research purpose to evaluate farmers' renewable energy adoption choices and their specific decisions regarding installation sizes.

Factors Influencing Renewable Energy Adoption. A farmer makes and installs renewable energy system decisions based on multiple variables such as electricity costs, alongside farming practices, and property dimensions, and internet connectivity status. People who practice conservation methods and own most of their land tend to choose investments in renewable energy systems. The adoption of these technologies shows a lesser likelihood among farmers who are older and those dependent primarily on farm income.

Determinants of System Size. The choice to adopt renewable energy systems depends on certain factors, but the actual system size develops from alternative criteria. Farms with expensive land value and hired managers, together with government support, tend to install larger systems. The dimensions of renewable energy systems installed do not change based on electrical cost factors.

Variation in Renewable Energy Adoption Across Farm Types. Installation rates of renewable energy technologies are more common in specific types of farms rather than general agricultural operations. The implementation of renewable energy systems in commercial applications grows larger in fruit farms, yet these systems normally remain independent and smaller for cattle farms. The adoption of renewable energy by organic farms reaches higher numbers, but these properties deploy energy systems with lower capacity compared to conventional operations.

Policy Implications and Future Considerations. Research on renewable energy elements enables policy builders to establish enhanced support programmes for agricultural renewable energy deployments. Technical incentives encouraging larger renewable energy installations will produce long-term sustainability in California's agriculture sector. Additional investigation should focus on evaluating how renewable power technologies perform economically and what impact they have on continued agricultural profitability (Beckman & Xiarchos, 2013).

11.5 CASE STUDIES FROM SOUTH AMERICA

Case Study 1: Application of a Food-Energy-Water Nexus Approach to Water Resources Management in the Colombian Andean Region

Application of the FEW Nexus in the Colombian Andean Region. Researchers study how to implement the FEW nexus strategy in water management for the Otun River Watershed (ORW), which encompasses different land uses in Colombia's Andean territory. The FEW nexus concept represents the unified nature between the three fundamental resources so they can achieve sustainable use at their optimal levels. The research shows that water availability and quality within the region are affected by three factors: climate change, alongside population growth and urban growth, thus requiring a comprehensive management strategy.

Challenges in Water Resources Management in Colombia. Multiple obstacles exist within the water management sector, according to the study, which combines water contamination problems with resource waste issues and insufficient wastewater treatment capabilities. The Pereira/Dosquebradas urban space causes significant water resource damage due to its high amounts of energy usage, along with water requirements and poor wastewater treatment programmes. The growing urban nature of the area has caused agricultural operations inside the watershed to decrease while creating a dependence on external food suppliers.

The research predicts how changing climate will affect resources while exploring future resource requirements. Water availability models created through hydrological modelling show substantial modifications for 2030–2039 because of climate change effects. The forthcoming projections show possible runoff loss at 35% and water supply reduction at 25%, which will directly impact the operation of agriculture, industry, and household water usage. The amount of water required will expand by 16%, while energy needs will surge to 30% underestimated future demand, therefore mandating essential policy modifications together with infrastructure upgrades.

Policy Recommendations. The research predicts how changing climate will affect resources while exploring future resource requirements. Water availability models created through hydrological modelling show substantial modifications for 2030–2039 because of climate change effects. The forthcoming projections show possible runoff loss at 35% and water supply reduction at 25% which will directly impact the operation of agriculture and industry and household water usage. The amount of water required will expand by 16% while energy needs will surge to 30% under the estimated future demand therefore mandating essential policy modifications together with infrastructure upgrades.

Policy Recommendations for Sustainable Resource Management. Three types of policies, along with enhanced water distribution systems, improved conservation practices, decentralised wastewater management, and urban farming leadership are recommended to enhance water sustainability in the ORW region. The implementation of FEW nexus monitoring must become part of urban management to achieve proper resource distribution and use. Additional emphasis is placed on the need for improved monitoring protocols, since they will enhance water quality assessment accuracy.

Conclusion and Broader Implications. The revolution of water resource management in Colombia's Andean region becomes possible by implementing a FEW nexus structure. The study advocates for stakeholder collaboration improvements, alongside enhanced investments in sustainable infrastructure systems, alongside strengthened climate resilience approaches. The developed framework serves as a model for other Latin American watersheds to benefit from sustainable water, energy, and food management activities worldwide (Pardo, 2022).

Case Study 2: Water-Energy-Food Security: A Nexus Perspective of the Current Situation in Latin America and the Caribbean

The Importance of the WEF Nexus in Latin America and the Caribbean (LAC). Sustainable development throughout LAC depends fundamentally on the implementation of a WEF Nexus approach. Growing resource security challenges affect this area because it depends on commodity markets for earnings, its population keeps expanding, and climate patterns are changing. The study implements a performance index system to identify security levels in water, energy, and food sectors and investigates their functional relationships. The data demonstrates the immediate requirement for ensuring sustainable progress by improving infrastructure and executing low-carbon energy transitions, alongside improved water governance (Mahlknecht et al., 2020).

Water, Energy, and Food Security Trends in LAC. The study demonstrates that LAC functions as a food-exporting nation while experiencing water scarcity throughout regions because of uneven precipitation patterns accompanied by climate variations. Hydropower is the largest power source in energy production, even though power generation still heavily depends on fossil fuels. Energy usage has continued to rise, thus producing more greenhouse gas emissions. The research reveals why agriculture requires knowledge of both water and energy resources especially in Mexico, Peru, and Caribbean areas that face water shortages.

Challenges in Implementing the WEF Nexus Approach. The LAC region shows limited integration of the WEF Nexus approach within its policy structures. The region encounters challenges because it faces poorly coordinated institutions, alongside fragmented governance bodies and deficient multi-sector planning structures. Political instability, combined with economic challenges in certain countries, makes it difficult to invest money into sustainable resource management on a long-term basis. The research sets forth a need to move away from single-sector policies by adopting holistic approaches to discover and balance relationships between food systems and water and energy sectors.

Regional Trade-Offs and Resource Management. The way LAC conducts trade operations creates substantial consequences for how its resources are utilised. The export of soybeans and beef from Brazil and Argentina produces two main environmental effects, including forest destruction and excessive water usage. According to the study, researchers emphasise the necessity of better land-use policies and sustainable farming techniques, partnered with international collaborations, for reducing environmental effects. Renewable energy development helps decrease fossil fuel usage by securing both food and water resources for the future.

Policy Recommendations and Future Directions. The research proposes updating regulatory control systems while promoting sector unity and building a sustainable infrastructure base. To improve irrigation systems, along with non-hydropower renewable energy development and sustainable food system promotion, these represent the specific recommended measures. Decisions based on an integrated WEF Nexus strategy represent the key solution for LAC to achieve sustainable resource access and environmental resilience, combined with economic durability (Mahlknecht et al., 2020).

Case Study 3: The Water-Energy-Food Nexus: An Analysis of Food Sustainability in Ecuador

WEF Nexus in Ecuador's Food Sustainability. The research evaluates Ecuador's water-energy-food relationships while underscoring the necessity of preserving resources in a sustainable manner. Food production in Ecuador constitutes a major economic sector since agricultural activities require significant amounts of water as well as energy use. The investigated study measures virtual water consumption in exports, alongside embodied energy use during food production, to demonstrate Ecuador's natural resource depletion from international trade.

Water and Energy Consumption in Ecuador's Agriculture. Ecuador's agricultural industry uses half of its total imported energy, and shrimp and fish farming stand as its most power-consuming agricultural operations. Cocoa and bananas represent major exports in Ecuador but have especially high water consumption levels, totalling 7.6 and 3.8 km^3, respectively. Water pollution in Ecuador has become an urgent issue because agricultural authors use excessive agrochemicals which pollute water resources, in addition to the problems caused by insufficient irrigation systems.

Impact of Virtual Water and Embodied Energy Exports. The research indicates that Ecuador sends many components of its virtual water, along with embodied energy, to other countries primarily through its exports of bananas, cocoa, and shrimp. The food production sectors in Ecuador use more than 50% of its complete water and energy supply to export their products to foreign markets. The sustainable management of Ecuador's natural resources becomes doubtful because intensive agricultural and aquacultural operations continue to escalate environmental pressures.

Challenges in Policy and Resource Management. Sustainable resource management faces difficulties because of poor governance, combined with ineffective water conservation policy enforcement and insufficient cross-departmental collaboration. Despite Ecuadorian constitutional protections related to food sovereignty and environmental sustainability, the country still faces problems due to inadequate water pricing and energy subsidies, which enable overconsumption. The environmental risks increase because of inadequate wastewater management techniques and continuing deforestation activities.

Recommendations for Sustainable Resource Use. To maintain food production equilibrium with resource preservation, the study suggests implementing better irrigation methods, agricultural renewable energy adoption, and strengthening environmental statutes. Successful economic sustainability depends on the expansion of the economy away from resource-based exports. Better fossil fuel management through policy integration with water systems and food regulation, along with improved monitoring will boost Ecuador's ability to withstand climate change and market volatility (Páez & Salazar, 2022).

11.6 CASE STUDIES FROM THE MIDDLE EAST

Case Study 1: The Groundwater-Energy-Food Nexus in Iran's Agricultural Sector: Implications for Water Security

This research analysed Iran's agricultural sector regarding groundwater-energy-food linkages alongside their unsustainable water usage effects. Iran's subsidised water and energy policies have brought about intense groundwater depletion beyond renewable capacity, creating destructive water shortages in the country. With its

20% workforce employed and 90% water withdrawals, the agricultural sector makes groundwater depletion the dominant cause of declining water levels across Iran. The country's efforts to achieve food self-sufficiency, which began during the post-1979 Revolution and in the Iran-Iraq War years, have intensified the misuse of ground aquifers. Heavy water and energy subsidies created conditions for agricultural inefficiency, resulting in soil salinisation and water quality deterioration across the nation. Throughout almost all the provinces of Iran, groundwater extraction surpasses sustainable volumes, while many wells serving agricultural purposes operate without registration, resulting in limited enforcement capacity. Research on all three aspects—groundwater, energy use, and land productivity—across Iran's provinces indicates problems within agronomic production systems. Most Iranian territory demonstrates poor resource productivity because the northern and western regions are exceptions to this pattern, but cereal, bean, and rice cultivation exhibit especially low efficiency. The ongoing stress on groundwater resources has not caused any changes in the size of irrigated farming regions because water availability shows no link to cultivation expansion. Water security in Iran is in grave danger due to ongoing groundwater depletion patterns. Provinces such as Tehran, Isfahan, Fars, Kerman, and Khorasan Razavi face severe groundwater vulnerability, with average annual declines of around one metre. Because of the transition to electric-powered wells, the energy requirements have increased, thus creating additional concerns about sustainability. The study demands immediate revisions to policies that need to expand groundwater surveillance, enhance agricultural productivity, and implement financial rewards to reduce excessive water consumption. If no corrective measures are implemented, groundwater depletion will result in job losses in agriculture, environmental collapse, and deteriorating food supplies (Mirzaei et al., 2019).

Case Study 2: Smart Greenhouses as The Path Towards Precision Agriculture In the Food-Energy And Water Nexus: Case Study of Qatar

This work identified the role of smart greenhouses in the shift from the traditional farming method to precision farming within the FEW nexus in Qatar. Due to Qatar's dry climate and low natural resources, greenhouse farming is the fundamental factor in increasing the local yield and obtaining food safety. Smart greenhouses apply the latest technology, such as auto climate control, AI, and live tracking, to achieve the best yields using less energy and water. The greenhouse market in Qatar is increasing; by 2025, it is expected to grow substantially. The report emphasises that traditional farming will not be able to meet the demand for increased food production and adapt to climate change and the end of resources. Smart greenhouses, which use such systems as hydroponics and advanced lighting and automation, not only provide a better alternative by enhancing water use but also reduce environmental harm. Emerging technologies such as wireless sensor networks, the IoT, and AI enable greenhouses to operate more efficiently. In addition to that, using renewable energy sources (solar panels, wind turbines) will also enhance their sustainability. The study also refers to multiple cooling and watering systems that can be applied to Qatar's specific climate and land conditions, stressing the importance of water saving (Karanisa et al., 2022). Government policies and investments are key factors that determine the success of the greenhouse industry. For example, the Qatar National Food Security Strategy aims to embrace technology for greenhouse cultivation to boost the country's independence from vegetables. The government also supports agricultural expansion using financial incentives, infrastructure development, and research collaborations. There are still some obstacles to be overcome, including initial installation costs that are too high, water scarcity, and farmers' inability to adapt to new technologies. The study suggests further research and development in eco-friendly greenhouse designs (sustainable), better automation, and the use of energy-saving technology to reduce the agricultural sector's dependence on foreign suppliers and increase long-term food security in Qatar (Karanisa et al., 2022).

Case study 3: Multi-scale analysis of the water-energy-food nexus in the Gulf region

The study aimed to gain insight into the WEF Nexus of the Gulf region with emphasis on Kuwait, Qatar, and Saudi Arabia (Siderius et al., 2020). It considered domestic resource dependencies, international trade linkages, and global sustainability challenges. It was discovered that FEW trade-offs are minimal at the domestic level. At the same time, international dependencies on food imports exacerbate the region's vulnerabilities, such as groundwater depletion in exporting countries like India and Pakistan. The study related the Gulf countries' reliance on seawater desalination to freshwater shortages, nothing that a highly energy-intensive process is the main cause, primarily powered by fossil fuels. Water scarcity is a key limiting factor in the Gulf region, and hence, the countries are forced to import their essential fruit, vegetables, and seeds from other countries. The new direction of Saudi Arabia's policy from wheat self-sustainability in 2006 has led to a decrease in national groundwater depletion. Still, it has not led to the reduction of the external water-stressed area through imports. The study explored the effects of reducing the stress of climate change on the economies of the Persian Gulf. Applying the social cost of carbon could be the factor in the fall in oil export earnings, making buying food and desalinating water economically unfeasible. This indicates the region's necessity to diversify its economic activity and use nonconventional energy sources. The work argued that while oil income had protected the Gulf nations from the immediate FEW nexus challenges, ensuring the sustainability of the resource needs an integrated resource management system, investments in renewable energy technologies, and improved agricultural strategies that involve minimum reliance on external resources (Siderius et al., 2020)

11.7 CASE STUDIES FROM ASIA

Case Study 1: Rapid Assessment of Wetland Ecosystem Services (RAWES): An example from Colombo, Sri Lanka

Importance of Wetland Ecosystem Services. Wetlands serve as essential resources to human well-being by offering four main ecosystem services that include flood management along with water purification and climate regulation and biodiversity conservation. Government officials require quick, cost-efficient tools to properly assess and integrate the benefits of wetlands into their nature conservation programmes. The Rapid Assessment of Wetland Ecosystem Services (RAWES) introduced an easy and practical solution to measure wetland product value. This method functions for a wide spectrum of wetland habitats by requiring minimal operational resources. Services under the RAWES framework are grouped based on provisioning, such as food, along with regulating through flood control, up to cultural benefits like recreation and support types, including habitat provision. Researchers used RAWES to evaluate 62 wetland sites throughout the rapidly developing Colombo Metropolitan Region because urbanisation has severely reduced local wetland areas. The analysis confirmed that local communities within the Colombo Metropolitan Region gain essential benefits from wetlands regarding food supply alongside water management and flood prevention. The sites encountered numerous adverse effects from waste disposal together with pollution while also facing threats from urban encroachment. The research demonstrates why urban planners should include wetland conservation programmes when designing urban development plans to achieve sustainable resilience. Wetlands demonstrated notable contributions to climate regulation functions as well as pollution control operations while supporting biodiversity levels. Habitat provision together with other supporting services showed the strongest relationship to ecosystem health outcomes. The research revealed that such regulating services as flood and disease control generate immediate benefits to local population groups. Service delivery becomes more effective as wetlands receive better management from the local governing bodies. The authors promote worldwide integration of RAWES as an essential tool for wetland management. The recommendations stress two aspects: inserting wetland concepts into policy systems through better enforcement of rules. Policymakers gain better decision-making capabilities through systemic understanding of complete ecosystem services delivered by wetlands for sustainable long-term use (McInnes & Everard, 2017).

Case Study 2: Understanding the Water-Food-Energy Nexus for Supporting Sustainable Food Production and Conserving Hydropower Potential in China

The work investigated the WFE nexus in China, where the primary focus is on the balance of sustainability in crop production and the conservation of hydropower. Using the grid-based crop model (PEPIC) and the hydrological model (DBH), the research determines the relationship of irrigation and crops on yield, and thus on hydropower (hydro-) production. It is concluded that the WEF Nexus in the north of China is the most severe, in terms of trade-offs, against the background of large water demands that irrigation needs and of the littleness of water availability, notably in the Yellow, Hai, and Liao River basins. The survey results demonstrate irrigation to be an indispensable factor in crop yield enhancement, particularly in arid regions, but, at the same time, there is the risk of losing up to 50% of the hydropower potential for some areas. Irrigation not only stimulates food production but also hampers the energy generation process by diverting streamflows, the consequences of which are discussed in this report. The study proposes the notion that the modernisation of irrigation systems and a relatively low correction rate of more efficient distribution justified the overextension of currently insufficient irrigation. The research provides a timely message on the integrated water-energy system and the acute situation of water use between food and energy, especially in the case of conflicts between water use for food and energy. It also stresses the importance of further research to look at irrigation optimisation, reservoir regulation inclusion, and groundwater use issues. They conclude that the integrated approach to WFE management is critical for the long-term sustainability and resource security in China (Liu et al., 2019).

Case Study 3: Multi-Scale Analysis of the Water-Energy-Food Nexus in the Gulf Region

The study aimed to gain an insight into the WEF Nexus of the Gulf region with emphasis on Kuwait, Qatar, and Saudi Arabia. It considers domestic resource dependencies, international trade linkages, and global sustainability challenges. It finds that WEF trade-offs are minimal at the domestic level while international dependencies on food imports exacerbate the region's vulnerabilities, such as groundwater depletion in the exporting countries like India and Pakistan. The work related the Gulf countries' reliance on seawater desalination to freshwater shortages, by which a highly energy-intensive process is the main cause, and it is primarily powered by fossil fuels. Water scarcity is a key limiting factor in the Gulf region, and hence, the countries are forced to import their essential fruit, vegetables, and seeds from other countries. The new direction of Saudi Arabia's policy from wheat self-sustainability in 2006 has led to a decrease in the national groundwater depletion, but it has not led to a decrease in the external water-stressed area through imports. The study was undertaken to explore the effects of lightening the load of climate change on the economies of the Persian Gulf. The application of the social cost of carbon could be the factor in a fall in oil export earnings, which could make the buying of food and desalinating water economically unfeasible. This indeed points out the region's necessity to diversify its economic activity and to use nonconventional energy sources. The authors argued that while the oil income had protected the Gulf nations from the immediate WEF challenges, ensuring the sustainability of the resource needs an integrated resource management system, investments in renewable energy technologies, and improved agricultural strategies that involve minimal reliance on external resources (Siderius et al., 2020).

Case Study 4: Water-Energy-Food Nexus Simulation in South Korea and Indonesia

The work analysed the embedded relationship that exists between water resources and energy systems and food production in South Korea together with Indonesia. By applying the WEF Nexus approach, these resources acquire equal importance as interconnected parts, which leads to better allocation and management methods. The research developed a simulation model to assess resource distribution patterns and maximise supply and demand as well as promote sustainability within regional and national territories. The study implemented Ribasim, MODSIM, and WEAP model frameworks as part of its water balance system used for resource availability and demand prediction. The model draws from historical data spanning 2000–2012 to create future projections from 2012 to 2037 that depict population growth, GDP alterations, and climate alterations. A new reliability index enables researchers to establish resource management self-sufficiency levels. According to projections, both South Korea and Indonesia will experience mounting strain on their water supply combined with food and energy resources as their populations expand together with the effects of climate change. The Korean market expects growing demands for energy and food but rapid agricultural growth in Indonesia leads to substantial water shortages. The research indicates that better approaches for resource efficiency together with integrated management must be established. The research implemented an optimisation tool which achieves maximum resource reliability and self-sufficiency while decreasing resource deficiencies. The model includes the recommendation of a trade module, which enables better resource exchange practices between inter-basin areas. Decision-making quality will improve through river basin implementation of these modules to guarantee regional sustainability. The proposed WEF Nexus model stands as a beneficial resource-planning instrument both for states and regions. The model delivers practical information to policy developers who need to optimise resource usage while ensuring environmental sustainability and building resistance against potential resource deficits. The model's application in river basin management programmes offers advanced capabilities to boost resource allocation and policy development for South Korea and Indonesia ((PDF) Water-Energy-Food Nexus Simulation in South Korea and Indonesia, n.d.).

11.7.1 Future Scenarios Possible for Central Asia

In the context of Central Asia, mainstreaming of the WEF Nexus approach should be done at a concerted level in national and regional processes. At the national level, this would involve the integration of the nexus dimensions into policies, strategies, and plans, while at the same time enhancing capacities through bottom-up approaches for regional challenges. After all, locally driven solutions can solve the nexus issues within their contexts. At the regional level, an enabling environment for dialogue and negotiation needs to be established to set a common development agenda for the Aral Sea Basin, as well as an appropriate mechanism of benefit-sharing and strengthening regulation for collaborative decision-making (*The Power of Best Practice Nexus Cases | IUCN*, n.d.).

The IUCN analysis highlights the potential competitive advantage that pertinent institutions and sectors might attain through enhancing their comprehension and awareness of intricate, interdependent cause-and-effect interactions within the WEF Nexus. Best practice cases are useful instruments for advancing this understanding because they facilitate insightful discussions and encourage well-informed policy actions and reforms across multiple stakeholders.

To summarise, the practical solutions for FEW case studies concentrate on optimising concurrent operations to minimise the possibility of conflicts across the systems. Usually, these case studies include:

i. **Multi-sector Integration**: Coordinating planning and management with goals to optimise food, energy, and water to arrange the benefits and positives in the approved settings. The approach is useful in managing resources as well as improving the efficiency of the operations in the business (Khan et al., 2022).
ii. **Multi-purpose Infrastructure**: Relay and infrastructure projects that apply in different sectors to benefit many people. For instance, infrastructure development for irrigation aiming to cost-effective energy production or the provision of water will go a long way in reducing the rebound effect (Tian et al., 2018).
iii. **Transboundary Cooperation**: To enhance the interaction between countries or geographical areas that use the same water source. This way, there are certainties that the benefits will be fairly divided and the social resources are properly utilised (Elsayed et al., 2022).
iv. **Stakeholder Engagement**: Involving stakeholders in decisions-making at various levels and from various fields of practice. This inclusion assists in the consideration of conflicts of interest and facilitation of consensus that are characteristics of FEW nexus concerned policies and projects (Bielicki et al., 2019; Dickinson, 2022).
v. **Policy Integration**: Mainstreaming the FEW nexus into national and regional policies, strategies, and plans. This integration equally ensures that nexus principles are mainstreamed within governance systems for improved resilience and sustainability (Norouzi & Kalantari, 2020).
vi. **Capacity Building**: Developing sufficient and competent local capacities that would help in confronting nexus challenges. That is, improving specialist training, sharing information, and encouraging innovation in utilisation of resources (Papas, 2024)
vii. **Scenario Planning**: The application of scenario planning as a method to develop the future tense issues and potentials of FEW nexus. Situations enable one to determine what can go wrong, and thus assist in the development of management adjustments.

11.8 CASE STUDIES FROM AUSTRALIA

Case Study 1: A Comprehensive Analysis in the Context of New South Wales, Australia

Research on water-energy nexus operations throughout New South Wales. This research investigates the interrelated connections between water and energy resources in New South Wales, Australia. Water plays a vital role in electricity generation both for coal-fired power stations and hydropower plants due to their dependence on water while electrical power is vital for all aspects of water treatment and distribution. Traditional policy development between water and energy sectors remains separate, which produces inefficient results and performance trade-offs mainly during droughts and climate change. This research fills the knowledge gap through a complete analysis of how water depends on energy and energy needs water.

Historical Evolution and Policy Reforms. The study explores historical developments in both the water and electricity industries of New South Wales, with a focus on important reforms that happened during recent decades. The research spotlights changes made at the state level during the 1980s that preceded the formation of a nationwide electricity market in the 1990s, which affected resource utilisation. The study reveals how economic efficiency has improved yet environmental and sustainability issues have received insufficient attention, which resulted in aggravated water scarcity together with heightened energy requirements.

Empirical Analysis of Water and Energy Linkages. The research presents three analytical models, including input-output analysis, along with historical assessment and price elasticity modelling to determine water-energy connectivity. The analysis reveals that coal-fired power plants, which employ high water consumption levels control most electricity production operations. The water sector, in turn, relies on energy-intensive technologies such as desalination and wastewater treatment. Trade-offs occur as a result of these connected systems, which become more problematic under conditions of climate change and rising resource requirements.

Future Scenarios and Resource Management. The study evaluates water and energy requirements for 2031 through different policy implementations as it examines technology adoption possibilities using scenario analysis. The research reveals how water solutions that operate with high electricity requirements, such as desalination, would substantially boost electrical power consumption. The research demonstrates that models employing renewable energies together with water-efficient systems produce more lasting, sustainable outcomes. The study supports infrastructure development that combines distributed systems together with performance-based designs and climate-adapted features for better risk management.

Policy Recommendations for an integrated approach. The research stresses how integrated water and energy policies should work together because they need to solve problems created by climate change as well as resource limitations and growing populations. New South Wales should prioritise three key actions to enhance its sustainable development. The first priority is to build stronger regulatory structures. The second priority involves supporting demand management initiatives. Finally, the region must invest more heavily in sustainable innovations. Holistic water-energy nexus management will improve New South Wales resource security while strengthening its economic resilience, based on findings that could help similar regions with identical concerns (Marsh, 2008).

Overall, FEW-case studies bring out the following: better performance of systems, excellent resource conditions and well-being of people associated with the efficient use of resources, reduced vulnerability to negative environmental influences, and intersectoral and inter-disciplinary management.

These case studies highlight the potential of 4IR technologies to:

i. **Enhance Food Security**: From precision agriculture using IoT sensors and AI-powered analytics to optimise crop yields and reduce waste, to blockchain technology ensuring traceability and transparency in food supply chains, the 4IR is transforming food production and distribution.
ii. **Improve Water Management**: Smart irrigation systems driven by real-time data and AI algorithms are optimising water use in agriculture, while advanced filtration and treatment technologies are improving access to clean water (How AI Can Optimise Water Usage in Agriculture, n.d.). IoT-enabled monitoring systems are also enhancing water resource management and reducing leakage in distribution networks (*Water Quality: IoT for Water Resource Management.*, n.d.).
iii. **Optimise Energy Systems**: Smart grids integrating renewable energy sources and utilising AI for demand forecasting are enhancing energy efficiency and grid stability (*Smart Grids Unleashed: How Is AI Used to Forecast Energy Consumption?*, n.d.). Furthermore, advanced materials and nanotechnology are enabling the development of more efficient energy storage and conversion technologies (*Processes | Special Issue : Advanced Materials for Energy Conversion and Storage Processes*, n.d.).
iv. **Foster Nexus Integration**: The interconnected nature of the FEW nexus demands integrated solutions. 4IR technologies, particularly AI and big data analytics, are enabling a more holistic understanding of the complex interactions within the nexus, facilitating the development of integrated management strategies that optimise resource use across all three domains.

11.9 SUMMARY

The FEW Nexus strategy has been implemented in different regions and sectors under different strategies, and this chapter talks about two of these case studies and good practice examples. These two are the Zambezi River Basin and the Mekong River Basin. It explains the strategies, policies, and frameworks used in those two case studies and how they improved sustainability and efficiency across the food, energy, and water sectors. It presents FEW Nexus case studies from Africa, Asia, Europe, North America, South America, Australia, and the Middle East. It then centres on Central Asia and the approaches that should be taken at different levels when implementing FEW Nexus. It also discusses the benefits of FEW Nexus case studies and their role in highlighting the potential of fourth industrial revolution technologies.

REFERENCES

Abdelzaher, M. A., Farahat, E. M., Abdel-Ghafar, H. M., Balboul, B. A. A., & Awad, M. M. (2023). Environmental policy to develop a conceptual design for the water–energy–food nexus: A case study in Wadi-Dara on the Red Sea Coast, Egypt. *Water*, *15*(4), 780. https://doi.org/10.3390/W15040780

Apeh, O. O., & Nwulu, N. I. (2024). The water-energy-food-ecosystem nexus scenario in Africa: Perspective and policy implementations. *Energy Reports*, *11*, 5947–5962. https://doi.org/10.1016/J.EGYR.2024.05.060

Becker, G. (2015). *Adaptive Flood Management: Institutional Challenges and Opportunities in the German Rhine Basin*. https://research.vu.nl/en/publications/adaptive-flood-management-institutional-challenges-and-opportunit

Beckman, J., & Xiarchos, I. M. (2013). Why are Californian farmers adopting more (and larger) renewable energy operations? *Renewable Energy*, *55*, 322–330. https://doi.org/10.1016/J.RENENE.2012.10.057

Bielicki, J. M., Beetstra, M. A., Kast, J. B., Wang, Y., & Tang, S. (2019). Stakeholder perspectives on sustainability in the food-energy-water nexus. *Frontiers in Environmental Science*, *7*(FEB). https://doi.org/10.3389/fenvs.2019.00007

David, L. O., Aigbavboa, C., Nwulu, N., & Adepoju, O. O. (2024). Application of digital technologies of the fourth industrial revolution (4IR) in water, energy, and food (WEF) Nexus. *Green Energy and Technology*, *Part F3772*, 287–317. https://doi.org/10.1007/978-3-031-72833-4_9

Dickinson, K. (2022, September 26). *How to Involve Stakeholders in the Food-Water-Energy Nexus*. JPI Urban Europe. https://jpi-urbaneurope.eu/how-to-involve-stakeholders-in-the-food-water-energy-nexus/

El Youssfi, L., Doorsamy, W., Aghzar, A., Cherkaoui, S. I., Elouadi, I., Faundez, A. G., & Salazar, D. R. (2020). Review of water energy food nexus in Africa: Morocco and South Africa as case studies. *E3S Web of Conferences*, *183*, 02002. https://doi.org/10.1051/E3SCONF/202018302002

Elsayed, H., Djordjevic, S., Savic, D., Tsoukalas, I., & Makropoulos, C. (2022). Water-food-energy nexus for transboundary cooperation in Eastern Africa. *Water Supply*, *22*(4), 3567–3587. https://doi.org/10.2166/ws.2022.001

Favi, S. G., Adamou, R., Godjo, T., Giri, N. C., Kuleape, R., & Trommsdorff, M. (2024). Agrivoltaic systems offer symbiotic benefits across the water-energy-food-environment nexus in West Africa: A systematic review. *Energy Research & Social Science*, *117*, 103737. https://doi.org/10.1016/J.ERSS.2024.103737

How AI *Can Optimise Water Usage in Agriculture*. (n.d.). Retrieved February 6, 2025, from https://keymakr.com/blog/precision-irrigation-how-ai-can-optimize-water-usage-in-agriculture/

Karanisa, T., Achour, Y., Ouammi, A., & Sayadi, S. (2022). Smart greenhouses as the path towards precision agriculture in the food-energy and water nexus: Case study of Qatar. *Environment Systems and Decisions*, *42*(4), 521–546. https://doi.org/10.1007/S10669-022-09862-2

Khan, Z., Abraham, E., Aggarwal, S., Ahmad Khan, M., Arguello, R., Babbar-Sebens, M., Bereslawski, J. L., Bielicki, J. M., Campana, P. E., Silva Carrazzone, M. E., Castanier, H., Chang, F. J., Collins, P., Conchado, A., Dagani, K. R., Daher, B., Dekker, S. C., Delgado, R., Diuana, F. A., … Yue, Q. (2022). Emerging themes and future directions of multi-sector nexus research and implementation. *Frontiers in Environmental Science*, *10*. https://doi.org/10.3389/fenvs.2022.918085

Laio, F., Tamea, S., & Tuninetti, M. (2020). *The Water-Food Nexus in Italy: A Virtual Water Perspective* (pp. 319–334). https://doi.org/10.1007/978-3-030-36460-1_13

Liu, W., Yang, H., Tang, Q., & Liu, X. (2019). Understanding the water-food-energy nexus for supporting sustainable food production and conserving hydropower potential in China. *Frontiers in Environmental Science*, *7*, 419479. https://doi.org/10.3389/fenvs.2019.00050

López-Flores, F. J., Sánchez-Zarco, X. G., Rubio-Castro, E., & Ponce-Ortega, J. M. (2023). A machine learning approach for optimizing the water-energy-food-ecosystem nexus: A resilience perspective for sustainability. *Environment, Development and Sustainability*, 1–29. https://doi.org/10.1007/s10668-023-04257-y

Mahlknecht, J., González-Bravo, R., & Loge, F. J. (2020). Water-energy-food security: A nexus perspective of the current situation in Latin America and the Caribbean. *Energy*, *194*, 116824. https://doi.org/10.1016/J.ENERGY.2019.116824

Marie Marsh, D.M. (2008.). *The Water-Energy Nexus: A Comprehensive Analysis in the Context of New South Wales*. University of Technology Sydney.

McInnes, R. J., & Everard, M. (2017). Rapid assessment of Wetland ecosystem services (RAWES): An example from Colombo, Sri Lanka. *Ecosystem Services*, *25*, 89–105. https://doi.org/10.1016/J.ECOSER.2017.03.024

Mirzaei, A., Saghafian, B., Mirchi, A., & Madani, K. (2019). The groundwater–energy–food nexus in Iran's agricultural sector: Implications for water security. *Water*, *11*(9), 1835. https://doi.org/10.3390/W11091835

Mosello, B., Adamtey, R., & Obuobie, E. (2017). *Making Water Infrastructure Investment Decisions in a Changing Climate: A Political Economy Study of River Basin Development in Ghana.* London: Overseas Development Institute. Retrieved from https://www.odi.org/sites/odi.org.uk/files/resource-documents/11929.pdf

Mperejekumana, P., Shen, L., Zhong, S., Muhirwa, F., Nsabiyeze, A., Nsigayehe, J. M. V., & Nyirarwasa, A. (2023). Assessing the capacity of the water–energy–food nexus in enhancing sustainable agriculture and food security in Burundi. *Sustainability*, *15*(19), 14117. https://doi.org/10.3390/SU151914117

Norouzi, N., & Kalantari, G. (2020). The food-water-energy nexus governance model: A case study for Iran. *Water-Energy Nexus*, *3*, 72–80. https://doi.org/10.1016/J.WEN.2020.05.005

Ogbolumani, O. A., & Nwulu, N. (2020). Integrated appliance scheduling and optimal sizing of an autonomous hybrid renewable energy system for agricultural food production. *Lecture Notes in Mechanical Engineering* (pp. 651–660). https://doi.org/10.1007/978-981-15-5753-8_60

Ogbolumani, O. A., & Nwulu, N. I. (2021). Multi-objective optimisation of constrained food-energy-water-nexus systems for sustainable resource allocation. *Sustainable Energy Technologies and Assessments*, *44*, 100967. https://doi.org/10.1016/j.seta.2020.100967

Ogbolumani, O. A., & Nwulu, N. I. (2022). A food-energy-water nexus meta-model for food and energy security. *Sustainable Production and Consumption*, *30*, 438–453. https://doi.org/10.1016/J.SPC.2021.12.019

Ogbolumani, O. A., & Nwulu, N. I. (2024). Environmental impact assessment for a meta-model-based food-energy-water-nexus system. *Energy Reports*, *11*, 218–232. https://doi.org/10.1016/j.egyr.2023.11.033

Páez, C. F. T., & Salazar, O. V. (2022). The water–energy–food nexus: An analysis of food sustainability in Ecuador. *Resources*, *11*(10), 90. https://doi.org/10.3390/RESOURCES11100090

Papas, Maureen. (2024). *Capacity-Building and the Water-Energy-Food Nexus: Rethinking Integration in the Asia-Pacific*. Routledge. https://www.routledge.com/Capacity-Building-and-the-Water-Energy-Food-Nexus-Rethinking-Integration-in-the-Asia-Pacific/Papas/p/book/9780367356293

Pardo, C. T. (2022). *Application of a Food-Energy-Water (FEW) Nexus Approach to Water Resources Management in the Colombian Andean Region.* https://doi.org/10.25394/PGS.19678401.V1

(PDF) Water-Energy-Food Nexus Simulation in South Korea and Indonesia. (n.d.). Retrieved February 6, 2025, from https://www.researchgate.net/publication/322636650_WATER-ENERGY-FOOD_NEXUS_SIMULATION_IN_SOUTH_KOREA_AND_INDONESIA

Processes | Special Issue : Advanced Materials for Energy Conversion and Storage Processes. (n.d.). Retrieved February 6, 2025, from https://www.mdpi.com/journal/processes/special_issues/647GWZL7CR

Renewable Energy in the Water Energy: Food Nexus. (n.d.). Retrieved February 6, 2025, from https://www.irena.org/Publications/2015/Jan/Renewable-Energy-in-the-Water-Energy--Food-Nexus

Salmoral, G., & Yan, X. (2018). Food-energy-water nexus: A life cycle analysis on virtual water and embodied energy in food consumption in the Tamar catchment, UK. *Resources, Conservation and Recycling*, *133*, 320–330. https://doi.org/10.1016/J.RESCONREC.2018.01.018

Siderius, C., Conway, D., Yassine, M., Murken, L., Lostis, P. L., & Dalin, C. (2020). Multi-scale analysis of the water-energy-food nexus in the Gulf region. *Environmental Research Letters*, *15*(9), 094024. https://doi.org/10.1088/1748-9326/AB8A86

Smart Grids Unleashed: *How is AI Used to Forecast Energy Consumption?*. (n.d.). Retrieved February 6, 2025, from https://www.datategy.net/2023/12/12/smart-grids-unleashed-how-is-ai-used-to-forecast-energy-consumption/

The Agribusiness Ecosystem in East and Southern Africa. Exploring the Role and Synergies of Key Stakeholders in the Space. (2022). https://www.cgiar.org/research/cgiar-portfolio/

The Power of Best Practice Nexus Cases | IUCN. (n.d.). Retrieved February 6, 2025, from https://iucn.org/news/eastern-europe-and-central-asia/202004/power-best-practice-nexus-cases

Tian, H., Lu, C., Pan, S., Yang, J., Miao, R., Ren, W., Yu, Q., Fu, B., Jin, F. F., Lu, Y., Melillo, J., Ouyang, Z., Palm, C., & Reilly, J. (2018). Optimizing resource use efficiencies in the food–energy–water nexus for sustainable agriculture: From conceptual model to decision support system. *Current Opinion in Environmental Sustainability*, *33*, 104–113. https://doi.org/10.1016/J.COSUST.2018.04.003

Water *Quality: IoT for Water Resource Management.* (n.d.). Retrieved February 6, 2025, from https://www.synox.io/en/smart-watering/water-quality-how-can-iot-improve-water-resource-monitoring-and-management/

Zhang, J., Campana, P. E., Yao, T., Zhang, Y., Lundblad, A., Melton, F., & Yan, J. (2018). The water-food-energy nexus optimization approach to combat agricultural drought: A case study in the United States. *Applied Energy*, *227*, 449–464. https://doi.org/10.1016/J.APENERGY.2017.07.036

12 Challenges, Opportunities, and Future Directions of FEW Nexus 4.0

12.1 CHALLENGES IN THE FOOD, ENERGY AND WATER NEXUS 4.0

The coming years will be marked by dramatic changes in the climate, environment, and infrastructure across our world. More important than anything else in these changes will be how food, energy, and water are strongly linked. Not only are these resources essential for human survival, but they also strongly shape environmental sustainability. A critical framework through which these elements come together is provided by the Food-Energy-Water (FEW) nexus. While urbanisation accelerates, cities again become focal points to which one has to pay special attention in sustainability studies. Otherwise, a holistic nexus approach can derive comprehensive analyses only in this case, which will inform sustainable urban development strategies worldwide (Heard et al., 2017). Some notable challenges and opportunities in the FEW nexus are elaborated in the following section.

12.1.1 Current Challenges in Water Management and Treatment

Water management and treatment face numerous challenges that threaten water availability and quality for future generations. Climate change is one such challenge that affects water resources. It can lead to changes in the distribution and intensity of rainfall where some areas would face extreme dry periods due to higher temperatures causing drought and water scarcity, while others would face intense downpour leading to flooding (Miyan, 2015; Watanabe et al., 2018). This affects water quality because intensified temperatures can lead to the development of toxic organisms in water bodies, making water treatment more difficult, and extreme weather events can overwhelm water treatment infrastructures, leading to contamination (Giri, 2021; Hughes et al., 2021; Li, et al., 2023; Mishra, 2023). The prolonged drought in California, worsened by climate change, impacted water availability for agriculture, urban use, and ecosystems, and to manage it, authorities applied stringent water conservation strategies and focused on alternative water solutions such as desalination (Berbel & Esteban, 2019; Medellín-Azuara et al., 2024). Pollution is another

DOI: 10.1201/9781003633853-12

challenge threatening water quality, and it comes from industries, agriculture, and domestic use in the form of pollutants like pesticides, heavy metals, herbicides, pathogens, and many more (Cañedo-Argüelles et al., 2016; du Plessis, 2018, 2019), making water unsafe for use.

Another challenge is the lack of finances to access advanced water management and treatment systems that can effectively remove contaminants and integrate new technologies into existing ones, especially in developing regions. In numerous African nations, financial and technological barriers prevent the development of effective water infrastructure, which has brought about programmes like the African Development Bank's "Africa Water Facility" aimed at providing financial support and technical assistance to improve water and sanitation services across the continent (Manda et al., 2019; Zvobgo & Do, 2020). Overcoming these challenges involves a comprehensive approach, including investment in research and development of new technologies, international cooperation, capacity-building initiatives, and sustainable financing mechanisms (Hecht, 2004; Tambo, 2016).

12.1.2 Current Challenges in Energy Efficiency and Optimisation

Despite the advancements in energy efficiency technologies, several technical challenges remain to be surmounted. Some of these include the development of non-costly energy storage solutions, improvements in grid incorporation of sustainable energy sources, and the improvement of energy-consuming processes in industry (Aneke & Wang, 2016). The cost required to adopt some of these technologies and infrastructure upgrades has also posed a challenge. Many businesses and consumers may hesitate to invest in energy-efficient solutions and technologies due to concerns about return on investment and payback periods (Chai & Yeo, 2012; Gohari et al., 2024; Imran Khan et al., 2024). In addition, developing countries and underserved communities may find it challenging to adopt energy-efficient technologies due to limited access to financing and incentives for energy management projects (Liu et al., 2024; Santopietro et al., 2024; Sarkar & Singh, 2010). Despite awareness of the importance of energy efficiency, human behaviour often prioritises short-term convenience over long-term gains and poses a challenge to energy efficiency (Alsalemi et al., 2019).

12.2 OPPORTUNITIES IN THE FEW NEXUS 4.0

12.2.1 High Growth of Urbanisation

The United Nations estimated that in 2014, 54% of the world's population was living in urban areas, which is expected to grow to 66% by 2050 (Leeson, 2018). It is known that cities are net sinks for both energy and materials (Villarroel Walker et al., 2014). Like other ecosystems, they garner their needed ecosystem services and resources from outside their political boundaries, often beyond their immediate rural surroundings (Güneralp et al., 2017). For example, cities need an area of 27 to 621 times their physical size to sustainably supply them with water (Darrel Jenerette & Larsen, 2006).

Whereas urban areas are linked to enhanced environmental impacts such as air and water pollution and reduced biodiversity, they also promise reduced contemporaneous per capita environmental impact if planned appropriately (Bibri et al., 2020). Cities bring enormous environmental effects together in a small geographic area because of their high use of resources. On the other hand, this population density may also have overriding benefits by reducing per-person environmental footprints (Dodman, 2009; Yi et al., 2017). This duality is responsible for the challenges and opportunities to fashion solutions for the sustainability of urban life and encourages innovative, customised approaches to meet them.

12.2.2 Urban Metabolism, Input-Output Models, and the Food-Energy-Water Nexus

All components of the FEW nexus are fast-growing, with billions of people globally lacking sufficient services and adequate quantity and quality of such resources. They operate within heavily regulated markets and exhibit high regional availability, variability, and strong interdependencies with climate change and the environment, among other traits (Bazilian et al., 2011). Those who lack access to the essential components of the FEW nexus tend to explore solutions using other methods, and oftentimes this exacerbates existing challenges faced by the nexus. The engineering community will continue to be under greater pressure for rigorous analysis and development of solutions to such challenges, especially with expanding urbanisation trends. Most methods developed to combat these challenges in the urban setting are based on urban metabolism.

Methods based on urban metabolism provide a framework in which inflows and outflows can be considered together with the stock and production of socio-economic and technical processes within cities (Kennedy & Hoornweg, 2012). In this way, it could provide an overview of the opportunities to synergistically improve their sustainability. For instance, in London, a metabolism study demonstrated that separating urine in sewage could recover nitrogen from food people consume, produce fertilisers to generate revenues, obtain biofuels, and even create a source of income by blending food waste with algae in wastewater treatment plants. These kinds of projects have economies of scale advantages through the nexus relationships that are not feasible at smaller scales. Therefore, cities have unique opportunities to implement new systems for improving sustainability (Apeh & Nwulu, 2024; Ogbolumani & Nwulu, 2024; Opeyemi et al., 2024).

Ramaswami et al. (2012) brought together urban metabolism and life cycle assessment under the concept of “social-ecological-infrastructural systems.” This concept is used to research transboundary impacts from infrastructure supply chains, water use, energy consumption, and greenhouse gas emissions associated with cities. This will focus on understanding the actions of policy actors and infrastructure users and measuring the risks at multiple scales, including climate change, pollution, and the urban heat island effect. It assesses stocks and flows of materials at all stages of life cycles and at all geographic scales that constitute urban metabolism (Ramaswami et al., 2012).

Input-output (IO) models, according to Miller and Blair (2009), have been increasingly used to quantify regions' environmental footprints. The models trace global supply chains linking urban systems with areas external to these urban systems, thus enabling the tracing of the transboundary environmental impacts of urban systems. The food, energy, and water production processes embed both direct environmental impacts and indirect impacts related to the inputs of production processes. IO models for food, energy, and water supply chains can be established to trace nexus-related effects and trade impacts of urban FEW nexus applications. Construction of IO tables at an urban level has remained challenging due to issues related to data unavailability (Liang & Zhang, 2011; Ogbolumani & Nwulu, 2022). Further research in estimation methodology refinement for urban IO data is accordingly warranted.

Secondly, national and local level government-level statistical systems should be built up to collect the data necessary for compiling urban IO tables. These will facilitate comprehensive analyses of FEW systems' environmental impacts at the urban scale.

12.2.3 Food, Land Use, and Energy

Seto and Ramankutty (2016) presented a brief but comprehensive review of the far-reaching interlinkages between urbanisation and food systems, but their research remains relatively unexplored across global systems. However, their review reveals how little, to date, these relations have been covered in existing literature (Seto & Ramankutty, 2016). The authors then summarise and briefly discuss the implications of these complex interrelations for the urban situation. The FEW nexus has impressive potential for resource efficiency and sustainability in food production, land use, and in the energy sector. Specifically, the junction of such systems increases resource efficiency, eases pressures on the environment, and promotes economic growth by embracing innovations such as renewable energy sources in input and land management techniques in a sustainable way. Major cost reductions, increased resilience to climate change, and new markets for technology and associated services can be unlocked within these interdependent systems through optimisation. In addition to this, the FEW nexus also offers some considerable challenges, some of which include the management of the complicated trade-offs that exist between these resources, strategies in addressing knowledge gaps, overcoming outdated or poorly formulated policies, and addressing financial issues (Ogbolumani & Mabaso, 2023; Ogbolumani & Nwulu, 2020, 2021). Successful implementation of the FEW nexus plan requires streamlined efforts among relevant stakeholders, formulation of quality policies, and a system that promotes comprehensive data sharing. This will address these challenges and promote the wide sharing of the benefits of an integrated system.

12.2.4 Urban Food Demand and Its Impact on Agricultural Land

Spatially, the increasing food demand from urban centres and its associated deforestation for agriculture are critical factors (Angelsen, 1999). Food demand from cities can result in significant environmental impact in far-away areas (Defries et al., 2010).

Urban expansion is likely to cause global loss of productive croplands by 1.8–2.4% over the next two decades, and the lost croplands will be approximately 1.77 times more productive than the average globally (D'Amour et al., 2017). Even with all the efforts to close the yield gaps, the additional food needs are still likely to lead to further expansion of agricultural land (Bajželj et al., 2014). Much of the livelihood improvement within the urban populations in developing cities can be enabled through better distribution and supply of food. This would not require any additional interventions in existing croplands.

Urban agriculture is one of the possible solutions to growing food demand, particularly in the developing world, with positive social and economic benefits for poor communities. Existing literature, however, theorises that the contributions should be viewed sceptically because of the limiting factors, which include its small contribution (Ogbolumani & Mabaso, 2023; Ogbolumani & Nwulu, 2020, 2021) to household income and overall agricultural production, as well as the insecurity of food supply due to the limitation of land in urban area. Promoting the urban agricultural system in this respect is key to enhancing the security of urban food supplies.

In developed countries, land prices and binding policy restrictions are obstacles to the scaling of urban agriculture, although it is feasible in terms of land use (Angotti, 2015). Vertical farming, then, appears as an interesting alternative with its stacked greenhouses that reduce the need for ground-level land and probably enhance the efficiency of food production (Despommier, 2011).

12.2.5 Urbanisation, Changes in Dietary Patterns, and Energy Production

Increasing incomes and urbanisation facilitate shifts toward diets with higher intakes of refined sugars, fats, oils, and meats (Tilman & Clark, 2014). Unless curbed, such dietary shifts may more than double emissions from food manufacture and land clearance, as well as contribute to an increased burden from diet-related diseases (Tilman & Clark, 2014).

In contrast, there is the possibility of utilising food waste for energy production. Anaerobic co-digestion of thickened waste activated sludge and grease interceptor waste from food establishments can also produce methane (Wang et al., 2013).

12.2.6 Water, Food, and Nutrient Flows

Urbanisation can either improve or worsen water vulnerability. On the one hand, establishing new water infrastructure and decreasing irrigated areas will lead to a drop in vulnerability. Also, an augmented population density and increased affluence tend to raise the level of vulnerability. Future trends in urban water use will therefore depend on how population density, per capita water use, climate, and dedication of water to ecosystem services in the coming future evolve (Darrel Jenerette & Larsen, 2006). There are review works on how urban water systems intersect within the energy-water-GHG nexus, citing the availability of a framework to address the dynamics between elements of the nexus as one of the major research gaps (Nair et al., 2014).

12.2.7 Dietary Shifts and Water

Among the major opportunities and challenges put forward by FEW nexus-related issues are dietary and water management changes. Indeed, shifting to a diet rich in plant-based foods and with low water use could dramatically increase water-use efficiency and shift some burden off freshwater resources (Henderson et al., 2017). It also helps to preserve healthy ecosystems, which reduces pressure on agricultural water demand and offers a few economic and environmental benefits by making the water and the energy used for food production more affordable. This can be further enhanced by promoting resilience to climate unpredictability through sustainable food systems in areas with low endowments (Merrett et al., 2003).

However, the challenges arise in the course of transitioning to the FEW nexus space. Some of these challenges include the need to create a balance between the competing resources and to overcome any cultural and behavioural barriers to dietary shifts (Hou et al., 2013).

However, this transition presents several challenges, including balancing competing resource demands and overcoming cultural and behavioural barriers to dietary change (Forkes, 2007). It may require huge investments in newer, more efficient infrastructure and new technologies for diet shift and water-efficient agriculture techniques. Greater policy integration among the food, water, and energy sectors in bringing about these changes could also be difficult, so some creativity in governance will be required. If dietary adjustments are to realise their full potential for delivering improved resource sustainability and resilience, these barriers must be overcome (Jarvie et al., 2015).

12.2.8 Water and Wastewater Management

Wastewater management relates to significant problems but also holds possible resource recycling opportunities. For example, urine separation systems can recover nitrogen and phosphorus from urban wastewater and help make the environment more sustainable compared to conventional sewage systems (Jönsson, 2002). However, their implementation requires energy consumption and infrastructure investments to achieve overall energy savings at the system level.

Treated wastewater reclamation for non-portable urban uses is an important opportunity for sustainable urban water supply (Heard & Miller, 2016; Okun, 1996). Coupled with the increasing interest in integrated approaches to urban water management, Srinivasan et al. (2010) indicate that such paradigms are less generally applicable, especially in parts of India and other countries where decentralised methods of water acquisition, such as through private wells or even rainwater harvesting, are prevalent (Srinivasan et al., 2010).

In literature, there is a focus on increasing the sustainability of urban water systems through supporting decentralised or hybrid solutions, stormwater management, rainwater harvesting, water reclamation, and source separation for wastewater (Daigger, 2009). Such approaches not only enhance the efficiency of water management but also reduce the interdependencies between water, energy, and farming.

12.2.9 Energy Generation and Water

Large volumes of energy are used by cities, and the overall processes associated with energy production are extremely water-intensive, whether for cooling in thermoelectric generation, water extraction in fossil fuel mining, or hydroelectric generation (Torcellini et al., 2003). However, water use for these purposes can certainly be reduced, particularly in companies related to thermoelectric generation (Feeley et al., 2008). Still, there is little prospect of an undisputed decoupling of water use from energy production into the foreseeable future.

12.2.10 Desalination

The water-energy nexus could be strengthened, particularly for desalination processes (Paul et al., 2016). As stress on freshwater resources increases, there will be greater reliance on seawater desalination to meet demands. While there have been numerous technological advances in the desalination process, it still requires far more energy than conventional freshwater treatment processes (Elimelech & Phillip, 2011). Desalination plants can be powered by renewable energy sources (Kalogirou, 2005), which will become increasingly important in reducing the overall environmental impact of this process in the future. In an era of projected increased irrigation demands due to climate change impacts on agriculture (Lobell et al., 2008), making desalination sustainable is necessary in the FEW nexus.

12.2.11 Refrigeration

Refrigeration is important in urban food systems, touching all parts of the FEW nexus. Even though refrigeration requires both energy and water for cooling, it helps prevent food losses. Despite preventing some pre-consumer losses, refrigeration has been associated with overbuying behaviours and increased household food waste (Garnett, 2007). Accordingly, developing and developed regions waste approximately the same proportion of food, about 30–40% (Godfray et al., 2010).

Food waste represents a lost opportunity within the FEW nexus, with wasted calories and a further loss of resources in the form of energy and water embedded in this waste. Indeed, the full implications of refrigeration for food sustainability have not yet been fully investigated, with some key knowledge gaps remaining for a technology that is rapidly spreading globally (Heard & Miller, 2016).

12.2.12 Infrastructure

Finally, urbanisation is likely to increase the number of buildings in larger cities, but not necessarily in denser ones (Andrews, 2017). This is a problem for reducing building energy use, as more compact urban form has typically meant less overall energy use (Güneralp et al., 2017). For example, in residential buildings in the United States, while building operations are primarily responsible for emitting greenhouse gases, construction activities are disproportionately impactful in terms of air toxins and hazardous waste (Ochoa et al., 2002). Therefore, metrics that relate not only to direct building energy use but also to so-called "embodied" emissions from construction materials are relevant (Ramaswami et al., 2008). That is why urban infrastructure evolution should allow technological progress, such as the introduction of an electric vehicle recharging facility or inductive recharging of electrified trolleybuses (Bi et al., 2015; San Román et al., 2011). Finally, with the growing access to electric vehicles, the environmental impacts of their batteries will have to be carefully considered (Casals et al., 2017).

The next pressing issues that arise concern the use and management of groundwater for urban infrastructure planning centres. Groundwater is being utilised heavily for both domestic and industrial purposes (Foster, 2001), in most developing cities, raising serious problems, such as land subsidence due to excessive extraction for both urban and agricultural uses (Chaussard et al., 2013).

12.3 FUTURE DIRECTIONS

12.3.1 Future Directions of the FEW Nexus 4.0

Studying the future trends of the FEW nexus is vital for addressing increasing difficulties in feeding people, generating energy, and obtaining water in cities in a sustainable and resilient manner. Here are some key future directions regarding the FEW nexus summarised from the works of Rhouma et al. (2024), Caixeta et al. (2023), and Ghosh et al. (2024):

i. **Integrated Approaches**: FEW nexus continues to evolve through processes and innovative approaches that address the interaction among FEW systems. This includes the enhancement of models and frameworks that further explain the interactions and feedback loops within the FEW nexus.
ii. **Policy and Governance**: Policy improvements are needed for the management of food, energy, and water resources at the local, national, and global levels. This involves identifying ways to address conflicts and balance trade-offs between different sectors for trade and development, while also taking equity into consideration.
iii. **Technological Innovations**: Technological developments aim to enhance efficiency and optimise the use of resources related to the FEW nexus. This is found in renewable energy technologies, such as solar panels, wind turbines, high tech farming (precision agriculture), water recycling and reuse, and the development of smart cities.
iv. **Resilience and Adaptation**: To enhance coping and adaptive capacities in urban environments that support people's ability to recover from threats arising from climate change, population growth, and resource constraints. This may entail embracing various concepts such as nature-based solutions, green infrastructure, and decentralised systems of food and water management.
v. **Data and Monitoring**: Improving data collection, surveillance, and evaluation functions to enhance understanding and assessment of changes, their magnitude, and recommendations in the FEW complex. This includes

creating measures and benchmarks for sustainability that will strengthen the sustainability assessment.

vi. **Public Awareness and Education**: Raising public awareness and strengthening civil society's active involvement in matters concerning the FEW nexus. Therefore, this entails encouraging the practice of moderation, buying from within the region, and participating in campaigns for water use and energy conservation.

vii. **Cross-Sectoral Collaboration**: Promoting the concept of collaboration and integration between different sectors and stakeholders directly or indirectly involved in the management of FEW nexus. Strategies should encourage collaboration among academia, government, industry, and civil society to ensure integrated and approaches are developed to address the problem.

viii. **Social Justice and Equity Consideration**: It implies that policies and interventions in the FEW nexus take into consideration the issues of equity, social justice, and inclusivity by reducing disparities in access to food, water, and energy resources, especially in vulnerable and marginalised communities.

ix. **Economic Considerations**: Integrating the contribution of economic principles to managing water, energy, and food security, both independently and holistically. This includes using price mechanisms that capture the true costs and benefits, incentives for resource-efficient practices, and promoting circular economy approaches.

x. **International Cooperation**: Strengthening international cooperation and collaboration to address transboundary challenges associated with the FEW nexus, such as water-sharing agreements, sustainable trade practices, and collaborative research initiatives.

The forward path shown here emphasises that holistic, interdisciplinary managerial priorities are essential to address the problems of food, energy, and water in urban areas in order to actualise the goals of sustainability, resilience, and fair development.

12.3.2 Future Directions in Water Management and Treatment

The challenges faced in water management and treatment have led to the rise of innovations focused on resilience and equity to guarantee that all populations are granted access to safe and reliable water resources. One of these future trends and innovations is smart water management systems, which make use of advanced technologies like IoT, AI, big data analytics, and many more to improve the monitoring, control, and maximisation of water resources (Gohil et al., 2021; Li et al., 2023). It is also crucial to improve water treatment technologies in response to emerging contaminants and adjust policy and governance frameworks to ensure sustainability in the regulation of water resources (Gogoi et al., 2018). Strengthening policies and regulations regarding water reuse, conservation, and pollution, and standards as to what determines water quality are essential (Bahri, 1999). Applying stakeholder engagement to address different needs and perspectives and fostering international cooperation is also critical (Nonet et al., 2022). The innovations don't stop here. More advanced innovations will be introduced, with the aim of making water management and treatment more resilient, efficient, and sustainable.

12.3.3 Future Directions in Energy Efficiency and Optimisation

The advancement in energy efficiency and optimisation lies in the innovation and deployment of technologies such as quantum computing and advanced materials. Quantum computing shows potential for addressing complex optimisation challenges related to energy systems, enabling more efficient energy management and grid operations (Ajagekar & You, 2019). Advanced materials, including nanomaterials and metamaterials, offer novel properties for energy storage, generation, and conversion, paving the way for breakthroughs in sustainable energy technologies and energy-efficient devices (Surjadi et al., 2019; Tonkaev et al., 2022). Addressing energy efficiency challenges requires robust policy frameworks and global cooperation. Governments play a crucial role in establishing energy efficiency standards, providing incentives, and implementing regulations to promote investment in energy-efficient technologies and practices (Shen et al., 2016). International collaboration is essential for sharing best practices, collectively agreeing on standards, and scaling up successful energy efficiency initiatives across borders. These collaborative efforts are particularly important within the FEW Nexus 4.0 context, as they impact interconnected food, water, and energy systems (Borg & von Knorring, 2019). By addressing current challenges and adopting future trends and innovations, stakeholders can accelerate progress towards a more energy-efficient future, contributing to the overall sustainability and resilience of the FEW nexus.

12.4 GOVERNANCE, IMPACTS, AND FUTURE DIRECTIONS OF FEW NEXUS 4.0

The framework for understanding the FEW Nexus interdependencies includes its governance, impacts, and future directions. In our contemporary world, it is a common expectation that governance should provide strong policy and robust institutional frameworks that guide all interactions across food, energy, and water systems to effectively and efficiently manage resources across sectors housed by FEW (Liu et al., 2018). The impacts of the FEW Nexus 4.0 are grounded in a wide range of fields, ranging from those related to the environment to those that have to do with resource depletion and climate change. Social concerns derived from food security and the nagging issues of equitable access to resources, economic preoccupations about the proper management of such resources, and technological innovations for their sustainable use are also not left out (Zarei et al., 2021). Building the future, syntheses, and recommendations point to fruitful directions ahead, highlighting the challenges that remain in many areas, including the full integration of technological advances, policy fine-tuning, resilience building, and community engagement tasks in these next steps for sustainable development. The complexities of such nexuses can be overcome by addressing these components holistically to foster resilience, sustainability, and growth in the future. The IEA's 2012 World Energy Outlook includes a specific chapter on water and energy connections and emphasises water supply. Currently, demands, resources, and obstacles transcend national borders. These demands require improved technological capabilities (*World Energy Outlook 2012 – Analysis – IEA*, n.d.). The *Thirsty Energy* report was published by the World Bank in 2013, providing member countries with solutions for possible technology development and governance strategies to ensure water and energy security (*(PDF) Thirsty Energy*, n.d.). The United Nations published the "World Water Development Report" in 2014. This research highlights the importance of combining energy, water, and agriculture. It advocates constructing a hierarchical structure for public affairs management by investigating frameworks and generating reliable data and creative technology (*UN World Water Development Report 2014 | UN-Water*, n.d.).

12.5 SUMMARY

The FEW Nexus is constantly evolving, and there will be challenges along the way, such as rapid, urbanisation, urban metabolism, food, and land use. These challenges are covered in this chapter. Despite the challenges, there'll be more opportunities for growth and development, such as in urban agriculture. This chapter also highlights the interconnectedness of the FEW nexus challenges and opportunities, for example, the shift in dietary patterns, which is a challenge as it could increase emissions from food manufacturing. Yet, this waste can also be utilised for energy production. This is further explained in the chapter, and so are the future directions of the FEW nexus.

REFERENCES

Ajagekar, A., & You, F. (2019). Quantum computing for energy systems optimization: Challenges and opportunities. *Energy*, *179*, 76–89. https://doi.org/10.1016/J.ENERGY.2019.04.186

Alsalemi, A., Sardianos, C., Bensaali, F., Varlamis, I., Amira, A., & Dimitrakopoulos, G. (2019). The role of micro-moments: A survey of habitual behavior change and recommender systems for energy saving. *IEEE Systems Journal*, *13*(3), 3376–3387. https://doi.org/10.1109/JSYST.2019.2899832

Andrews, C. J. (2017). The changing socioeconomic context of buildings. *Journal of Solar Energy Engineering, Transactions of the ASME*, *139*(1). https://doi.org/10.1115/1.4034911

Aneke, M., & Wang, M. (2016). Energy storage technologies and real life applications – A state of the art review. *Applied Energy*, *179*, 350–377. https://doi.org/10.1016/J.APENERGY.2016.06.097

Angelsen, A. (1999). Agricultural expansion and deforestation: Modelling the impact of population, market forces and property rights. *Journal of Development Economics*, *58*(1), 185–218. https://doi.org/10.1016/S0304-3878(98)00108-4

Angotti, T. (2015). Urban agriculture: Long-term strategy or impossible dream?: Lessons from prospect farm in Brooklyn, New York. *Public Health*, *129*(4), 336–341. https://doi.org/10.1016/J.PUHE.2014.12.008

Apeh, O. O., & Nwulu, N. I. (2024). The water-energy-food-ecosystem nexus scenario in Africa: Perspective and policy implementations. *Energy Reports*, *11*, 5947–5962. https://doi.org/10.1016/J.EGYR.2024.05.060

Bahri, A. (1999). Agricultural reuse of wastewater and global water management. *Water Science and Technology*, *40*(4–5), 339–346. https://doi.org/10.2166/WST.1999.0608

Bajželj, B., Richards, K. S., Allwood, J. M., Smith, P., Dennis, J. S., Curmi, E., & Gilligan, C. A. (2014). Importance of food-demand management for climate mitigation. *Nature Climate Change*, *4*(10), 924–929. https://doi.org/10.1038/NCLIMATE2353

Bazilian, M., Rogner, H., Howells, M., Hermann, S., Arent, D., Gielen, D., Steduto, P., Mueller, A., Komor, P., Tol, R. S. J., & Yumkella, K. K. (2011). Considering the energy, water and food nexus: Towards an integrated modelling approach. *Energy Policy*, *39*(12), 7896–7906. https://doi.org/10.1016/J.ENPOL.2011.09.039

Berbel, J., & Esteban, E. (2019). Droughts as a catalyst for water policy change. Analysis of Spain, Australia (MDB), and California. *Global Environmental Change*, *58*, 101969. https://doi.org/10.1016/J.GLOENVCHA.2019.101969

Bi, Z., Song, L., De Kleine, R., Mi, C. C., & Keoleian, G. A. (2015). Plug-in vs. wireless charging: Life cycle energy and greenhouse gas emissions for an electric bus system. *Applied Energy*, *146*, 11–19. https://doi.org/10.1016/J.APENERGY.2015.02.031

Bibri, S. E., Krogstie, J., & Kärrholm, M. (2020). Compact city planning and development: Emerging practices and strategies for achieving the goals of sustainability. *Developments in the Built Environment*, *4*, 100021

Borg, J., & von Knorring, H. (2019). Inter-organizational collaboration for energy efficiency in the maritime sector: The case of a database project. *Energy Efficiency*, *12*(8), 2201–2213. https://doi.org/10.1007/S12053-019-09822-X

Caixeta, F., Saraiva, P., & Freire, F. (2023). Future research trends on the water-energy-food nexus approach. In *Handbook of Sustainability Science in the Future* (pp. 1–20). https://doi.org/10.1007/978-3-030-68074-9_125-1

Cañedo-Argüelles, M., Hawkins, C. P., Kefford, B. J., Schäfer, R. B., Dyack, B. J., Brucet, S., Buchwalter, D., Dunlop, J., Frör, O., Lazorchak, J., Coring, E., Fernandez, H. R., Goodfellow, W., Achem, A. L. G., Hatfield-Dodds, S., Karimov, B. K., Mensah, P., Olson, J. R., Piscart, C., … Timpano, A. J. (2016). Saving freshwater from salts : Ion-specific standards are needed to protect biodiversity. *Science*, *351*(6276), 914–916. https://doi.org/10.1126/SCIENCE.AAD3488

Casals, L.C., García, B.A., Aguesse, F., et al. (2017). Second life of electric vehicle batteries: Relation between materials degradation and environmental impact. *The International Journal of Life Cycle Assessment*, *22*, 82–93. https://doi.org/10.1007/s11367-015-0918-3

Chai, K. H., & Yeo, C. (2012). Overcoming energy efficiency barriers through systems approach—A conceptual framework. *Energy Policy*, *46*, 460–472. https://doi.org/10.1016/J.ENPOL.2012.04.012

Chaussard, E., Amelung, F., Abidin, H., & Hong, S. H. (2013). Sinking cities in Indonesia: ALOS PALSAR detects rapid subsidence due to groundwater and gas extraction. *Remote Sensing of Environment*, *128*, 150–161. https://doi.org/10.1016/J.RSE.2012.10.015

Daigger, G. T. (2009). Evolving urban water and residuals management paradigms: Water reclamation and reuse, decentralization, and resource recovery. *Water Environment Research : A Research Publication of the Water Environment Federation*, *81*(8), 809–823. https://doi.org/10.2175/106143009X425898

D'Amour, C. B., Reitsma, F., Baiocchi, G., Barthel, S., Güneralp, B., Erb, K. H., Haberl, H., Creutzig, F., & Seto, K. C. (2017). Future urban land expansion and implications for global croplands. *Proceedings of the National Academy of Sciences of the United States of America*, *114*(34), 8939–8944. https://doi.org/10.1073/PNAS.1606036114

Darrel Jenerette, G., & Larsen, L. (2006). A global perspective on changing sustainable urban water supplies. *Global and Planetary Change*, *50*(3–4), 202–211. https://doi.org/10.1016/J.GLOPLACHA.2006.01.004

Defries, R. S., Rudel, T., Uriarte, M., & Hansen, M. (2010). Deforestation driven by urban population growth and agricultural trade in the twenty-first century. *Nature Geoscience*, *3*(3), 178–181. https://doi.org/10.1038/NGEO756

Despommier, D. (2011). The vertical farm: Controlled environment agriculture carried out in tall buildings would create greater food safety and security for large urban populations. *Journal Fur Verbraucherschutz Und Lebensmittelsicherheit*, *6*(2), 233–236. https://doi.org/10.1007/S00003-010-0654-3

Dodman, D. (2009). Blaming cities for climate change? An analysis of urban greenhouse gas emissions inventories. *Environment and Urbanization*, *21*(1), 185–201. https://doi.org/10.1177/0956247809103016

du Plessis, A. (2018). Primary water quality challenges, contaminants and the world's dirtiest places. In *Springer Water* (pp. 79–114). Cham: Springer International Publishing. https://doi.org/10.1007/978-3-030-03186-2_5

du Plessis, A. (2019). Water as an inescapable risk current global water availability, quality and risks with a specific focus on South Africa. *Springer Water*, 1–299.

Elimelech, M., & Phillip, W. A. (2011). The future of seawater desalination: Energy, technology, and the environment. *Science*, *333*(6043), 712–717. https://doi.org/10.1126/SCIENCE.1200488

Feeley, T. J., Skone, T. J., Stiegel, G. J., McNemar, A., Nemeth, M., Schimmoller, B., Murphy, J. T., & Manfredo, L. (2008). Water: A critical resource in the thermoelectric power industry. *Energy*, *33*(1), 1–11. https://doi.org/10.1016/J.ENERGY.2007.08.007

Forkes, J. (2007). Nitrogen balance for the urban food metabolism of Toronto, Canada. *Resources, Conservation and Recycling*, *52*(1), 74–94. https://doi.org/10.1016/J.RESCONREC.2007.02.003

Foster, S. S. D. (2001). The interdependence of groundwater and urbanisation in rapidly developing cities. *Urban Water*, *3*(3), 185–192. https://doi.org/10.1016/S1462-0758(01)00043-7

Garnett, T. (2007). Food refrigeration: What is the contribution to greenhouse gas emissions and how might emissions be reduced. *Food Climate Research Network, University of Surrey.*

Ghosh, B., Gubareva, M., Ghosh, A., Paparas, D., & Vo, X. V. (2024). Food, energy, and water nexus: A study on interconnectedness and trade-offs. *Energy Economics*, *133*, 107521. https://doi.org/10.1016/J.ENECO.2024.107521

Giri, S. (2021). Water quality prospective in twenty first century: Status of water quality in major river basins, contemporary strategies and impediments: A review. *Environmental Pollution*, *271*, 116332. https://doi.org/10.1016/J.ENVPOL.2020.116332

Godfray, H. C. J., Beddington, J. R., Crute, I. R., Haddad, L., Lawrence, D., Muir, J. F., Pretty, J., Robinson, S., Thomas, S. M., & Toulmin, C. (2010). Food security: The challenge of feeding 9 billion people. *Science (New York)*, *327*(5967), 812–818. https://doi.org/10.1126/SCIENCE.1185383

Gogoi, A., Mazumder, P., Tyagi, V. K., Tushara Chaminda, G. G., An, A. K., & Kumar, M. (2018). Occurrence and fate of emerging contaminants in water environment: A review. *Groundwater for Sustainable Development*, *6*, 169–180. https://doi.org/10.1016/J.GSD.2017.12.009

Gohari, S., Silvia, S. C., Ashrafian, T., Konstantinou, T., Giancola, E., Prebreza, B., Aelenei, L., Murauskaite, L., & Liu, M. (2024). Unraveling the implementation processes of PEDs: Lesson learned from multiple urban contexts. *Sustainable Cities and Society*, *106*. https://doi.org/10.1016/j.scs.2024.105402

Gohil, J., Patel, J., Chopra, J., Chhaya, K., Taravia, J., & Shah, M. (2021). Advent of big data technology in environment and water management sector. *Environmental Science and Pollution Research*, *28*(45), 64084–64102. https://doi.org/10.1007/S11356-021-14017-Y

Güneralp, B., Zhou, Y., Ürge-Vorsatz, D., Gupta, M., Yu, S., Patel, P. L., Fragkias, M., Li, X., & Seto, K. C. (2017). Global scenarios of urban density and its impacts on building energy use through 2050. *Proceedings of the National Academy of Sciences of the United States of America*, *114*(34), 8945–8950. https://doi.org/10.1073/PNAS.1606035114

Heard, B. R., & Miller, S. A. (2016). Critical research needed to examine the environmental impacts of expanded refrigeration on the food system. *Environmental Science and Technology*, *50*(22), 12060–12071. https://doi.org/10.1021/ACS.EST.6B02740

Heard, B. R., Miller, S. A., Liang, S., & Xu, M. (2017). Emerging challenges and opportunities for the food–energy–water nexus in urban systems. *Current Opinion in Chemical Engineering*, *17*, 48–53. https://doi.org/10.1016/J.COCHE.2017.06.006

Hecht, A. D. (2004). International efforts to improve access to water and sanitation in the developing world: A good start, but more is needed. *Water Policy*, *6*(1), 67–85. https://doi.org/10.2166/WP.2004.0005

Henderson, A. D., Asselin-Balençon, A. C., Heller, M., Lessard, L., Vionnet, S., & Jolliet, O. (2017). Spatial variability and uncertainty of water use impacts from U.S. feed and milk production. *Environmental Science & Technology*, *51*(4), 2382–2391. https://doi.org/10.1021/ACS.EST.6B04713

Hou, Y., Ma, L., Gao, Z. L., Wang, F. H., Sims, J. T., Ma, W. Q., & Zhang, F. S. (2013). The driving forces for nitrogen and phosphorus flows in the food chain of China, 1980 to 2010. *Journal of Environmental Quality*, *42*(4), 962–971. https://doi.org/10.2134/JEQ2012.0489

Hughes, J., Cowper-Heays, K., Olesson, E., Bell, R., & Stroombergen, A. (2021). Impacts and implications of climate change on wastewater systems: A New Zealand perspective. *Climate Risk Management*, *31*, 100262. https://doi.org/10.1016/J.CRM.2020.100262

Imran Khan, M., Bicer, Y., Asif, M., Al-Ansari, T. A., Khan, M., Kurniawan, T. A., & Al-Ghamdi, S. G. (2024). The GCC's path to a sustainable future: Navigating the barriers to the adoption of energy efficiency measures in the built environment. *Energy Conversion and Management: X*, *23*. https://doi.org/10.1016/j.ecmx.2024.100636

Jarvie, H. P., Sharpley, A. N., Flaten, D., Kleinman, P. J. A., Jenkins, A., & Simmons, T. (2015). The pivotal role of phosphorus in a resilient water–energy–food security nexus. *Journal of Environmental Quality*, *44*(4), 1049–1062. https://doi.org/10.2134/JEQ2015.01.0030

Jönsson, H. (2002). Urine separating sewage systems – Environmental effects and resource usage. *Water Science and Technology*, *46*(6–7), 333–340. https://doi.org/10.2166/WST.2002.0697

Kalogirou, S. A. (2005). Seawater desalination using renewable energy sources. *Progress in Energy and Combustion Science*, *31*(3), 242–281. https://doi.org/10.1016/J.PECS.2005.03.001

Kennedy, C., & Hoornweg, D. (2012). Mainstreaming urban metabolism. *Journal of Industrial Ecology*, *16*(6), 780–782. https://doi.org/10.1111/J.1530-9290.2012.00548.X

Leeson, G. W. (2018). The growth, ageing and urbanisation of our world. *Population Ageing*, *11*, 107–115. https://doi.org/10.1007/s12062-018-9225-7

Li, J., Herdem, M. S., Nathwani, J., & Wen, J. Z. (2023). Methods and applications for artificial intelligence, big data, internet of things, and blockchain in smart energy management. *Energy and AI*, *11*, 100208. https://doi.org/10.1016/J.EGYAI.2022.100208

Li, J., Li, X., Liu, H., Gao, L., Wang, W., Wang, Z., Zhou, T., & Wang, Q. (2023). Climate change impacts on wastewater infrastructure: A systematic review and typological adaptation strategy. *Water Research*, *242*, 120282. https://doi.org/10.1016/J.WATRES.2023.120282

Liang, S., & Zhang, T. (2011). Data acquisition for applying physical input-output tables in Chinese cities: The case of Suzhou. *Journal of Industrial Ecology*, *15*(6), 825–835. https://doi.org/10.1111/J.1530-9290.2011.00372.X

Liu, G., Ji, D., & Markides, C. N. (2024). Progress and prospects of low-grade thermal energy utilization technologies. *Applied Thermal Engineering*, *254*. https://doi.org/10.1016/j.applthermaleng.2024.123859

Liu, J., Hull, V., Godfray, H. C. J., Tilman, D., Gleick, P., Hoff, H., Pahl-Wostl, C., Xu, Z., Chung, M. G., Sun, J., & Li, S. (2018). Nexus approaches to global sustainable development. *Nature Sustainability*, *1*(9), 466–476. https://doi.org/10.1038/S41893-018-0135-8

Lobell, D. B., Burke, M. B., Tebaldi, C., Mastrandrea, M. D., Falcon, W. P., & Naylor, R. L. (2008). Prioritizing climate change adaptation needs for food security in 2030. *Science, 319*(5863), 607–610. https://doi.org/10.1126/SCIENCE.1152339

Manda, S., Tallontire, A., & Dougill, A. J. (2019). Large-scale land acquisitions and institutions: Patterns, influence and barriers in Zambia. *The Geographical Journal, 185*(2), 194–208. https://doi.org/10.1111/GEOJ.12291

Medellín-Azuara, J., Escriva-Bou, A., Gaudin, A. C. M., Schwabe, K. A., & Sumner, D. A. (2024). Cultivating climate resilience in California agriculture: Adaptations to an increasingly volatile water future. *Proceedings of the National Academy of Sciences of the United States of America, 121*(32), e2310079121. https://doi.org/10.1073/PNAS.2310079121

Merrett, S., Allan, J. A., & Lant, C. (2003). Virtual water – The water, food, and trade nexus useful concept or misleading metaphor? *Water International, 28*(1), 106–113. https://doi.org/10.1080/02508060.2003.9724812

Miller, R. E., & Blair, P. D. (2009). Foundations and extensions. *Input-Output Analysis*, 10–68. https://books.google.com/books/about/Input_Output_Analysis.html?id=SmFUl-6X1FUC

Mishra, R. K. (2023). Fresh water availability and its global challenge. *British Journal of Multidisciplinary and Advanced Studies, 4*(3), 1–78. https://doi.org/10.37745/BJMAS.2022.0208

Miyan, M. A. (2015). Droughts in Asian least developed countries: Vulnerability and sustainability. *Weather and Climate Extremes, 7*, 8–23. https://doi.org/10.1016/J.WACE.2014.06.003

Nair, S., George, B., Malano, H. M., Arora, M., & Nawarathna, B. (2014). Water-energy-greenhouse gas nexus of urban water systems: Review of concepts, state-of-art and methods. *Resources, Conservation and Recycling, 89*, 1–10. https://doi.org/10.1016/J.RESCONREC.2014.05.007

Nonet, G. A. H., Gössling, T., Van Tulder, R., & Bryson, J. M. (2022). Multi-stakeholder engagement for the sustainable development goals: Introduction to the special issue. *Journal of Business Ethics, 180*(4), 945–957. https://doi.org/10.1007/S10551-022-05192-0

Ochoa, L., Hendrickson, C., & Matthews, H. S. (2002). Economic input-output life-cycle assessment of U.S. residential buildings. *Journal of Infrastructure Systems, 8*(4), 132–138. https://doi.org/10.1061/(ASCE)1076-0342(2002)8:4(132)

Ogbolumani, O. A., & Mabaso, B. (2023). An IoT-based hydroponic monitoring and control system for sustainable food production. *Journal of Digital Food, Energy & Water Systems, 4*(2), 106–140.

Ogbolumani, O. A., & Nwulu, N. (2020). Integrated appliance scheduling and optimal sizing of an autonomous hybrid renewable energy system for agricultural food production. In *Lecture Notes in Mechanical Engineering* (pp. 651–660). https://doi.org/10.1007/978-981-15-5753-8_60

Ogbolumani, O. A., & Nwulu, N. I. (2021). Multi-objective optimisation of constrained food-energy-water-nexus systems for sustainable resource allocation. *Sustainable Energy Technologies and Assessments, 44*, 100967. https://doi.org/10.1016/j.seta.2020.100967

Ogbolumani, O. A., & Nwulu, N. I. (2022). A food-energy-water nexus meta-model for food and energy security. *Sustainable Production and Consumption, 30*, 438–453. https://doi.org/10.1016/J.SPC.2021.12.019

Ogbolumani, O. A., & Nwulu, N. I. (2024). Environmental impact assessment for a meta-model-based food-energy-water-nexus system. *Energy Reports, 11*, 218–232. https://doi.org/10.1016/j.egyr.2023.11.033

Okun, D. A. (1996). Water reclamation and nonpotable reuse: An option for meeting urban water supply needs. *Desalination, 106*(1–3), 205–212. https://doi.org/10.1016/S0011-9164(96)00110-5

Opeyemi, L., Clinton, D., Nnamdi, A., Omoseni, N., & Adepoju, O. (2024). Sustainable synergy: A digital framework for the water-energy-food nexus project delivery in developing economies. *Springer Nature*. https://doi.org/10.1007/978-3-031-72833-4

Paul, P., Al Tenaiji, A. K., & Braimah, N. (2016). A review of the water and energy sectors and the use of a nexus approach in Abu Dhabi. *International Journal of Environmental Research and Public Health, 13*(4). https://doi.org/10.3390/IJERPH13040364

(PDF) Thirsty Energy. (n.d.). Retrieved August 24, 2024, from https://www.researchgate.net/publication/273122919_Thirsty_Energy

Ramaswami, A., Hillman, T., Janson, B., Reiner, M., & Thomas, G. (2008). A demand-centered, hybrid life-cycle methodology for city-scale greenhouse gas inventories. *Environmental Science and Technology, 42*(17), 6455–6461. https://doi.org/10.1021/ES702992Q

Ramaswami, A., Weible, C., Main, D., Heikkila, T., Siddiki, S., Duvall, A., Pattison, A., & Bernard, M. (2012). A social-ecological-infrastructural systems framework for interdisciplinary study of sustainable city systems. *Journal of Industrial Ecology, 16*(6), 801–813. https://doi.org/10.1111/J.1530-9290.2012.00566.X

Rhouma, A., El Jeitany, J., Mohtar, R., & Gil, J. M. (2024). Trends in the water–energy–food nexus research. *Sustainability, 16*(3), 1162. https://doi.org/10.3390/SU16031162

San Román, T. G., Momber, I., Abbad, M. R., & Sánchez Miralles, Á. (2011). Regulatory framework and business models for charging plug-in electric vehicles: Infrastructure, agents, and commercial relationships. *Energy Policy, 39*(10), 6360–6375. https://doi.org/10.1016/J.ENPOL.2011.07.037

Santopietro, L., Solimene, S., Lucchese, M., Di Carlo, F., & Scorza, F. (2024). An economic appraisal of the SE(C)AP public interventions towards the EU 2050 target: The case study of Basilicata region. *Cities, 149*. https://doi.org/10.1016/j.cities.2024.104957

Sarkar, A., & Singh, J. (2010). Financing energy efficiency in developing countries—lessons learned and remaining challenges. *Energy Policy, 38*(10), 5560–5571. https://doi.org/10.1016/J.ENPOL.2010.05.001

Seto, K. C., & Ramankutty, N. (2016). Hidden linkages between urbanization and food systems. *Science, 352*(6288), 943–945. https://doi.org/10.1126/SCIENCE.AAF7439

Shen, L., He, B., Jiao, L., Song, X., & Zhang, X. (2016). Research on the development of main policy instruments for improving building energy-efficiency. *Journal of Cleaner Production, 112*, 1789–1803. https://doi.org/10.1016/J.JCLEPRO.2015.06.108

Srinivasan, V., Gorelick, S. M., & Goulder, L. (2010). Sustainable urban water supply in south India: Desalination, efficiency improvement, or rainwater harvesting? *Water Resources Research, 46*(10), 10504. https://doi.org/10.1029/2009WR008698

Surjadi, J. U., Gao, L., Du, H., Li, X., Xiong, X., Fang, N. X., & Lu, Y. (2019). Mechanical metamaterials and their engineering applications. *Advanced Engineering Materials, 21*(3). https://doi.org/10.1002/ADEM.201800864

Tambo, E. (2016). China-Africa health development initiatives: Benefits and implications for shaping innovative and evidence-informed national health policies and programs in sub-saharan African Countries. *International Journal of MCH and AIDS (IJMA), 5*(2). https://doi.org/10.21106/IJMA.100

Tilman, D., & Clark, M. (2014). Global diets link environmental sustainability and human health. *Nature, 515*(7528), 518–522. https://doi.org/10.1038/nature13959

Tonkaev, P., Sinev, I. S., Rybin, M. V., Makarov, S. V., & Kivshar, Y. (2022). Multifunctional and transformative metaphotonics with emerging materials. *Chemical Reviews, 122*(19), 15414–15449. https://pubs.acs.org/doi/abs/10.1021/acs.chemrev.1c01029

Torcellini, P., Long, N., & Judkoff, R. (2003). *Consumptive Water Use for U.S. Power Production*. https://doi.org/10.2172/15005918

UN World Water Development Report 2014 | UN-Water. (n.d.). Retrieved August 25, 2024, from https://www.unwater.org/publications/un-world-water-development-report-2014

Villarroel Walker, R., Beck, M. B., Hall, J. W., Dawson, R. J., & Heidrich, O. (2014). The energy-water-food nexus: Strategic analysis of technologies for transforming the urban metabolism. *Journal of Environmental Management*, *141*, 104–115. https://doi.org/10.1016/J.JENVMAN.2014.01.054

Wang, L., Aziz, T. N., & De los Reyes, F. L. (2013). Determining the limits of anaerobic co-digestion of thickened waste activated sludge with grease interceptor waste. *Water Research*, *47*(11), 3835–3844. https://doi.org/10.1016/J.WATRES.2013.04.003

Watanabe, T., Cullmann, J., Pathak, C. S., Turunen, M., Emami, K., Ghinassi, G., & Siddiqi, Y. (2018). Management of climatic extremes with focus on floods and droughts in agriculture. *Irrigation and Drainage*, *67*(1), 29–42. https://doi.org/10.1002/IRD.2204

World Energy Outlook 2012 – Analysis – IEA. (n.d.). Retrieved August 24, 2024, from https://www.iea.org/reports/world-energy-outlook-2012

Yi, Y., Ma, S., Guan, W., & Li, K. (2017). An empirical study on the relationship between urban spatial form and CO2 in Chinese cities. *Sustainability (Switzerland)*, *9*(4). https://doi.org/10.3390/SU9040672

Zarei, S., Bozorg-Haddad, O., Kheirinejad, S., & Loáiciga, H. A. (2021). Environmental sustainability: A review of the water-energy-food nexus. *Aqua Water Infrastructure, Ecosystems and Society*, *70*(2), 138–154. https://doi.org/10.2166/AQUA.2020.058

Zvobgo, L., & Do, P. (2020). COVID-19 and the call for 'Safe Hands': Challenges facing the under-resourced municipalities that lack potable water access – A case study of Chitungwiza municipality, Zimbabwe. *Water Research X*, *9*, 100074. https://doi.org/10.1016/J.WROA.2020.100074

Index

4IR 13, 14, 18, 29, 44, 65, 91, 347

abiotic 114, 185
acoustics 170
actuator(s) 68, 159, 167, 225, 318
adaptive 60, 67, 109, 157, 181, 207, 257, 291, 311, 324, 331, 348, 358
adoption 32, 74, 83, 87, 91, 94, 111, 113, 114, 157, 167, 169, 173, 174, 182–184, 190, 232, 255, 256, 296–303, 305–309, 314, 317, 319, 320, 327, 328, 335, 336, 339, 346, 364
aerodynamics 220
agribusiness 128, 185, 322, 328, 350
agrivoltaics 88, 115, 117, 166–167, 309
agroecology 212
anaerobic digestion 139, 144, 260, 261, 263–265, 269, 355, 367
anthropocene 209, 212
aquaponics 5, 165
artificial intelligence 14, 25, 47, 88, 98, 101, 104, 225–226, 243–244, 297, 327

biodiversity 4, 8, 35, 235, 274, 289, 316, 326, 342, 353, 362
bioenergy 7, 8, 10, 20, 54, 116, 142, 144–146, 166, 168, 171, 172, 176, 191, 267, 269, 328
biofuels 6, 8, 10, 20, 88, 115–117, 134, 135, 143, 145, 146, 167, 172, 175, 178, 187, 220, 275, 327, 353
biogas 17, 139, 145, 146, 244, 246, 259, 261, 264, 265
blockchain 14, 18, 24, 25, 67, 71, 81, 86–88, 92, 105–109, 113, 165, 167, 168, 170, 171, 173–175, 177–181, 183–185, 187, 190–194, 297, 298, 302, 303, 312–314, 316, 317, 347, 364

carbon cycles 20, 54, 109, 116, 141, 145, 166, 170, 176, 214, 217, 221, 228, 230–232, 259, 263, 268, 289, 315, 316, 327, 338, 341, 343
catalyst 214, 362
cellulosic biofuels 117, 146, 173, 187
cyber-physical 25, 190
cybersecurity 22–25, 45, 71, 82

fourth industrial revolution 13–14, 23–24, 29, 65, 91, 104, 298, 306, 347

genetic modification (GM crops) 35, 37, 44, 85, 116, 165, 171, 305
geothermal energy 6, 145, 214
greenhouse gases (GHGs) 39, 50, 59, 116, 156, 214, 227, 230–232, 259, 262, 275, 286, 315–318, 327, 338, 341, 343
grid integration 206, 229, 284, 286, 297, 299, 305, 312, 314, 324, 325, 328, 338, 344, 345, 349, 352, 357, 358, 360, 363, 366

hydrogen economy 139, 144, 223, 227, 312, 358, 362
hydroponics 5, 80, 162, 165, 178, 184, 319, 340, 365

industrial symbiosis 206, 316, 329
Industry 4.0 13–14, 21–25, 27, 29–31, 38–39, 42, 44–45, 62, 65, 90, 107, 109, 161, 302, 305, 306
Information and Communication Technologies (ICTs) 67, 70, 73, 77, 81, 82, 107, 114, 120, 156, 158, 167, 187, 254, 272, 308, 313, 352
internet of things (IoT) 18, 23, 40, 41, 43, 45, 47, 65, 104, 161, 228, 301, 303, 305–307

life-cycle assessment (LCA) 11, 22, 53, 54, 56–58, 66, 67, 69, 72, 73, 76, 79, 80, 83, 85–87, 127, 132, 133, 160, 166, 169, 174, 176, 181, 183, 187, 211, 227–230, 232, 266, 268–270, 277–280, 287–293, 313, 318, 320, 327, 332, 337, 342, 346, 349, 358, 359, 362, 365
logistics (supply chains) 86, 106–108, 123, 126, 171, 173, 174, 180, 202, 203, 207, 208, 212, 311, 328, 332, 347, 353, 354

machine learning (ML) 22, 43, 52, 78, 98–102, 104, 123–127, 159, 162, 163, 188, 225, 246, 313, 315, 347
microgrids 227, 229, 231, 232, 240

nanotechnology 125, 188, 266
nexus (food-energy-water) 2, 3, 17, 22, 32–35, 40, 44, 48, 66, 69–72, 74, 75, 77, 79–82, 84, 85, 87, 90, 94, 97, 100, 107, 108, 114, 115, 122–130, 132, 136, 141, 144–147, 154, 164, 166–181, 183–193, 195, 199, 203, 205, 207, 210, 211, 213, 226, 227, 236, 237, 246, 248, 250, 251, 256, 258–260, 264–267, 272, 274, 277, 284, 287–290, 293, 296, 305, 309, 311–316, 318–320, 323–330, 333–337, 339, 340, 347–350, 352, 354–358, 361, 363, 365

photovoltaics (PV) 115, 117, 166, 167, 188, 191, 309, 318, 327, 348
precision agriculture 66, 79, 83, 85, 114, 115, 122–124, 126, 129, 132, 136, 137, 144, 146, 154, 159–161, 163, 176, 190, 191, 195, 198, 200, 205, 206, 208, 209, 212, 233, 247, 260, 271, 272, 274, 276, 285, 288, 297, 303, 307–310, 312, 314–316, 325, 327–329, 331–335, 338, 342–345, 351–356, 359, 360
pumped hydro storage 137, 268, 282, 287, 324, 343, 363

quantum computing 360, 361

renewable energy systems 6, 8, 10, 21, 54, 116, 142, 144–146, 166, 168, 171, 172, 176, 191, 214, 230, 267, 269, 282, 287, 313, 324, 328, 338, 344, 349, 352, 357, 360, 363, 366
resilience (systems) 44, 60–62, 91, 254, 291, 331

smart grids 13, 15, 23, 24, 39–41, 45, 80, 96, 97, 109, 112, 181, 184, 221–223, 225, 229, 297, 298, 301, 302, 312, 315, 319, 347, 350
solar energy 88, 115, 117, 166, 167, 188, 191, 309, 318, 327, 348
sustainability transitions 24, 41, 53, 67, 75, 81, 93, 97, 101, 110, 159, 162, 177, 189, 195, 200, 206, 212, 216, 229, 291, 302, 306, 311, 314, 318–320, 325, 326, 331, 340, 342, 346

transdisciplinarity 272, 286
transportation (low-carbon) 220, 261, 268, 279, 319, 349, 365

waste-to-energy (WtE) 134, 144, 160, 214, 230, 265
water-energy-food nexus 2, 3, 17, 22, 32–35, 40, 44, 48, 66, 69–72, 74, 75, 77, 79–82, 84, 85, 87, 90, 94, 97, 100, 107, 108, 114, 115, 122–130, 132, 136, 141, 144–147, 154, 164, 166–181, 183–193, 195, 199, 203, 205, 207, 210, 211, 213, 226, 227, 236, 237, 246, 248, 250, 251, 256, 258–260, 264–267, 272, 274, 277, 284, 287–290, 293, 296, 305, 309, 311–316, 318–320, 323–330, 333–337, 339, 340, 347–350, 352, 354–358, 361, 363, 365
wind power 5, 7, 18, 54, 103, 134, 146, 166, 188, 191, 227, 268, 279, 319, 325, 327, 328, 335, 349, 365